Pheromones and Animal Behavior

Chemical Signals and Signatures

SECOND EDITION

Pheromones and other kinds of chemical communication underlie the behavior of all animals. Building on the strengths of the first edition, widely recognized as the leading text in the subject, this is a comprehensive overview of how pheromones work.

Extensively revised and expanded to cover advances made over the last ten years, the book offers a thorough exploration of the evolutionary and behavioral contexts of chemical communication, along with a detailed introduction to the molecular and neural basis of chemosensory perception. At a time of ever increasing specialization, Wyatt offers a unique synthesis, integrating examples across the animal kingdom. A final chapter critically considers human pheromones and the importance of olfaction to human biology. Its breadth of coverage and readability make the book an unrivaled resource for students and researchers in a range of fields from chemistry, genetics, genomics, molecular biology, and neuroscience to ecology, evolution, and behavior.

A full list of the references from this book is available for download from www.cambridge. org/pheromones.

Tristram D. Wyatt is a researcher at Oxford University's Department of Zoology, and an Emeritus Fellow of Kellogg College, Oxford. He is interested in how pheromones evolve throughout the animal kingdom, at both molecular and behavioral levels. These broad interests give him a unique vantage point, enabling him to draw together developments across the subject.

Pheromones and Animal Behavior

Chemical Signals and Signatures

SECOND EDITION

TRISTRAM D. WYATT

Department of Zoology and Kellogg College,
University of Oxford

CAMBRIDGE
UNIVERSITY PRESS

University Printing House, Cambridge CB2 8BS, United Kingdom

Cambridge University Press is part of the University of Cambridge.

It furthers the University's mission by disseminating knowledge in the pursuit of education, learning and research at the highest international levels of excellence.

www.cambridge.org
Information on this title: www.cambridge.org/9780521130196

© T.D. Wyatt 2014

First edition © Cambridge University Press 2003

This publication is in copyright. Subject to statutory exception and to the provisions of relevant collective licensing agreements, no reproduction of any part may take place without the written permission of Cambridge University Press.

First published 2003
Second edition 2014

A catalogue record for this publication is available from the British Library

ISBN 978-0-521-11290-1 Hardback
ISBN 978-0-521-13019-6 Paperback

Additional resources for this publication at www.cambridge.org/pheromones

To Robert

CONTENTS

PREFACE TO THE SECOND EDITION

This book is designed to bring together people already working on chemical communication and to encourage others, especially chemists (who have a vital role in this research), to take up the challenge. My aim has been to make an evolutionary understanding of chemical communication, including pheromones, accessible to a broad scientific and lay audience.

Pheromone research brings together scientists with many different areas of expertise, from a rich diversity of chemists to biologists of many kinds. Each area of expertise has its own jargon and concepts – a behavioral ecologist speaks a different language from a neuroscientist. The book recognizes that every scientist is a novice outside their own subject, even science close to their own, so I try to explain ideas in terms understandable by non-specialists while at the same time aiming to be up to date and detailed enough for the specialist. I also wanted to write a book that could be enjoyed by the majority of the world's scientists whose first language is not English and thus also clearer for everyone.

Pheromones offer exceptional opportunities to study fundamental biological problems. The rapid progress of the last decades comes from the convergence of powerful techniques from different areas of science including chemistry and animal behavior, combined with new techniques in genomics and molecular biology. These allow us to investigate questions at every level: molecular, neurobiological, hormonal, behavioral, ecological, and evolutionary. The discoveries from molecular biologists have greatly expanded our knowledge of the evolutionary biology of olfactory communication. Equally, molecular biology only makes sense in the context of evolution.

I wrote the first edition to provide the overview of chemical communication we were then missing, covering the whole animal kingdom, integrating approaches from ecology to neurobiology, and all with an evolutionary perspective. I have kept the same overall structure for the book in the new edition. As before, the book is organized around themes such as sex, speciation, and social organization, rather than taxonomically. The book also covers the perception and processing of chemosensory information. In each topic I have aimed to integrate examples from across the animal kingdom. In the same paragraph you may read about nematodes, moths, snakes, and mice. I explore the often convergent ways evolved by different kinds of animals to solve the same communication needs.

All chapters have been comprehensively updated and most chapters have been completely rewritten. The changes are perhaps most significant, as you might expect, in those parts involving molecular biology, especially in the chapter on perception of pheromones. Recent results include the surprising discovery that insect chemoreceptors have evolved independently of vertebrate ones. However, there has also been much new to discuss in evolution and ecology, including results coming from the application of molecular techniques as well as detailed field work.

Different parts of the book emphasize examples from different taxa. As in the first edition, mammals feature more strongly than invertebrates in the sections on individual variation and hormonal effects of pheromones for example, but invertebrates dominate the sections covering mechanisms of searching behavior.

Chapter 1 defines pheromones and looks at evolution of pheromones as signals. I raise a pragmatic distinction between pheromones and the chemistry of individual or colony odors. I also look at the role pheromones play in speciation. The importance of both common ancestry and convergence in molding chemical signals is a key theme.

Chapter 2 is about the development of analytical tools and how these are changing the study of chemical communication, allowing us to identify types of molecules previously hard to work with. New genomics techniques can be used to identify genes involved in both production and perception of molecules and not just in model organisms such as *Drosophila*. On the behavioral side I emphasize the importance of proper randomization of treatments and "blinding" of experimenters wherever possible. Progress will depend on productive partnerships between chemists and biologists.

The following six chapters cover different aspects of pheromones in the ecology and behavior of animals. Chapter 3 is about the evolution of pheromones in sexual selection, drawing out the many parallels between animals in a wide range of taxa. Among the new material featured in the chapter is work on *Drosophila* and moths as well as developments in evolutionary theory.

Chapter 4 covers Allee effects and the roles pheromones have in spacing organisms, bringing them together, and keeping them apart.

Chapter 5 reviews territorial behavior, largely in terrestrial vertebrates. The discovery of the male mouse pheromone, darcin, offers fascinating insights into female mouse behavior (Roberts *et al.* 2010, 2012). Darcin prompts her to learn the individual odor of the territorial male and where the scent mark is.

The parallels between complex social behaviors mediated by chemical communication in social insects and social mammals are explored in Chapter 6. The queen pheromones of an increasing number of social insects are being identified. It seems, however, that mammals do not use pheromones to "suppress" reproduction by subordinate members of the group.

Recruitment in social insects for foraging and nest building and for defense are covered by Chapters 7 and 8 respectively. One major change in our understanding is a clearer distinction between alarm pheromones and cues. The molecules involved in fish alarm are likely to be cues rather than pheromones.

Chapter 9 explores how olfaction works and how the olfactory receptors themselves evolve, in enormous variety. Vertebrates and invertebrates are similar in the way they detect and process chemical cues, by combining inputs from neurons carrying different olfactory receptors, but they achieve this with quite different receptor families, which evolved independently.

The mechanisms that animals have evolved to find an odor source are discussed in Chapter 10. We understand more about the mechanisms that fish and birds use than when the first edition was written. There are some interesting uses of genetically manipulated *Drosophila* larvae to explore the ways they orientate in chemical gradients.

Broadcast signals can be eavesdropped. Chapter 11 covers a world of deception and spying, including new players and a clearer understanding of selection in some classic examples.

Chapter 12 discusses how an understanding of chemical communication can be used for agriculture and to control disease vectors. Whereas insects formed the main examples of pheromone control, pheromones are showing promise for the possible control of vertebrate pest species, notably the sea lamprey.

Chapter 13 covers the roles of chemical communication in human beings. I discuss the smells we produce and the ones we can perceive. One of the most surprising things that has emerged from genomics studies on humans is the enormous variation between us as individuals in what we can smell. Our olfactory receptor repertoires are individually quite different: it is likely that we each experience unique olfactory worlds. I conclude the chapter by exploring some of the limitations of current research on human pheromones and how we could take it forward.

Finally, the appendix explains the common chemical terminology you will come across.

While some of the molecules important in chemical communication are shown in the figures, there are too many mentioned in the text to illustrate them all. Instead, you can see them on sites such as www.chemspider.com, which allows you to search by common name and shows synonyms as well as the systematic names. Many pheromone molecules, together with some background, are included on Pherobase www.pherobase.com (El-Sayed 2013).

Choice of literature for the second edition

This book necessarily offers a selective distillation of an enormous literature. I have attempted to reflect our consensus understanding of each topic. For reviews, I have generally used the most recent I could (though I reference earlier reviews if they continue to be influential). The papers cited have been chosen to reflect both their contributions to the subject but also because they offer good entry points to the literature (do use Google Scholar™ or Web of Science™ to find papers citing these leads). Sometimes you will find a review and a particular experimental paper both referenced, for example (Cardé & Haynes 2004; Liénard *et al.* 2010), which will be obvious, I hope, when you look them up. The references for this edition are also available at www.cambridge.org/pheromones.

Wherever possible, I have chosen sources that you will be more likely to be able to find. Where I have had a choice between equally good papers I have gone for the one in an open access journal or one that the authors have made available on the web, for example on their own website. It may be worth searching on an article title to see if it is available. If an article is not available and you do not have institutional access to the journal, you might courteously write to the author to see if they have a PDF to send. Most people are pleased to be asked – I know I am.

Wikipedia^{TM}

Have you considered helping edit Wikipedia's entries in our subject? It might seem surprising for a textbook to recommend its readers to consider contributing their expertise to Wikipedia, the world's largest online encyclopedia, but this is where the greatest influence for our subject will be. As Wikipedia is where most people look first, Bateman and Logan (2010) encourage scientists to seize the opportunity to make sure that Wikipedia articles are understandable, scientifically accurate, well sourced, and up to date. Bond (2011) makes such a call to his fellow ornithologists and presents many advantages of getting involved. Pheromones and some aspects of chemical communication are briefly covered in Wikipedia but not to the depth and range of many other areas of science. You might be able to improve this. Logan *et al.* (2010) give tips for getting started and guidance on good practice.

If you would like Microsoft PowerPoint^{TM} slides of the illustrations in the book for teaching or talks, do email me, tristram.wyatt@zoo.ox.ac.uk, letting me know which chapters' figures you would like.

ACKNOWLEDGMENTS

I would particularly like to thank the following for generously reading the whole book in draft: Bruce Schulte, Jagan Srinivasan, Joan Wyatt, and Vivian Wyatt. I am also grateful to many other friends and colleagues for help with various chapters and recent writing projects, which helped me develop ideas explored in the book, including Olle Anderbrant, Richard Benton, Thomas Breithaupt, Patrizia d'Ettorre, Monica De Facci, Dick Doty, Heather Eisthen, Maud Ferrari, Jean-François Ferveur, Kevin Foster, Tom Getty, Stephen Goodwin, Alan Grafen, Christina Grozinger, Penny Hawken, Matthieu Keller, Jae Kwak, Jean-Marc Lassance, Darren Logan, Jocelyn Millar, Dan Rittschof, Benoist Schaal, Peter Sorensen, Számadó Szabolcs, Robert Taylor, Kevin Theis, Martin Thiel, Tobias Uller, Marc Weissburg, Tom Wenseelers, Danielle Whittaker, Brian Wisenden, and Ben Wyatt.

Any remaining errors are mine of course, and I would welcome comments and suggestions for corrections. You can contact me at tristram.wyatt@zoo.ox.ac.uk.

I would like to thank all the scientists in addition to those listed above who advised me on their areas of expertise and kindly sent reprints and pre-prints of their work. The book would not have been possible without their help and generosity. In keeping the range of animal groups represented as wide as possible, I have had to be selective. Inevitably I have not been able to include many examples that I would have liked to. I apologise to authors whose research I was not able to describe here despite its high quality.

Many colleagues generously helped me with high-resolution copies of their illustrations. I would like to give additional thanks to colleagues who produced new or especially adapted figures for me, including Christina Grozinger, Harland Patch, Troy Shirangi, Jagan Srinivasan, and John Terschak.

It is a pleasure to thank Martin Griffiths, Megan Waddington, Abigail Jones, Kath Pilgrem, Vania Cunha, and other colleagues at Cambridge University Press for their encouragement and assistance at all stages of producing the second edition.

I would like to thank the publishers and societies listed at the end of the book for permission to reproduce figures and tables, particularly those which did not charge fees.

SI PREFIXES

Factor	Name	Symbol
10^{-2}	centi	c
10^{-3}	milli	m
10^{-6}	micro	μ
10^{-9}	nano	n
10^{-12}	pico	p
10^{-15}	femto	f

ABBREVIATIONS

2MB2	2-methyl-but-2-enal
AOB	accessory olfactory bulb
AOS	accessory olfactory system
BNST	bed nucleus of the stria terminalis
cAMP	cyclic adenosine monophosphate
cGMP	cyclic guanosine monophosphate
CHC	cuticular hydrocarbon
CNG	cyclic nucleotide-gated channel
CNV	copy number variant
cVA	*cis*-vaccenyl acetate
ESP1	exocrine gland-secreting peptide 1
GC	gas chromatography
FPR	formyl peptide receptor
GABA	γ-aminobutyric acid
GPCR	G-protein-coupled receptor
GR	gustatory receptor (invertebrates)
GSN	gustatory sensory neuron (insects)
GUR	gustatory receptor (*Caenorhabditis elegans*)
HPLC	high-performance liquid chromatography
iGluR	ionotropic glutamate receptor
IR	ionotropic receptor
JH	juvenile hormone
MHC	major histocompatibility complex
MGC	macroglomerular complex
MOB	main olfactory bulb or OB
MOE	main olfactory epithelium
MOS	main olfactory system
MOT	medial olfactory tract
MTMT	(methylthio)methanethiol
MUP	major urinary protein
MPOA	medial pre-optic hypothalamus
OB	olfactory bulb (MOB)
OR	olfactory receptor
ORCO	olfactory receptor coreceptor (insects)
OSN	olfactory sensory neuron (also termed olfactory receptor neuron, ORN)
SEM	scanning electron micrograph
SEM	standard error of the mean
SNP	single nucleotide polymorphism
SPME	solid phase micro extraction
T1R	taste receptor type 1
T2R	taste receptor type 2
TAAR	trace amine-associated receptor
TRC	taste receptor cell
TRPC2	transient receptor potential channel 2 (= TRP2)
V1R	vomeronasal receptor type 1
V2R	vomeronasal receptor type 2
VNO	vomeronasal organ
VNS	vomeronasal system (= accessory olfactory system)

1 Animals in a chemical world

When two dogs meet and sniff, they gain a wealth of information from each other's smells. Each dog will discover the sex, maturity, and hormonal state of the other; some of these smells will be species-wide dog pheromone signals. Each dog also detects the individual smell of the other, which it learns as a "signature mixture" to remember in case they meet again.

When two ants meet and sweep antennae over each other, they have an olfactory exchange of information similar to that of the dogs, discovering age, sex, ovarian stage (reproductive or not), and caste (worker, soldier, queen), all signals from species-wide pheromones. They also detect the colony odor of the other ant, enabling them to decide by the "signature mixture" whether the other ant is a nestmate or not.

All animals produce a chemical profile, present on the body surface, released as volatile molecules, and from scent marks that they deposit (by dogs on lamp-posts for example) (Figures 1.1, 1.2, 13.2). As chemical senses are ancient and widespread, shared by all organisms including bacteria, animals are pre-adapted to detect chemical information in the environment (Box 1.1). Across the animal kingdom, animals of all kinds gain chemosensory information from other organisms. Chemical senses are used to locate potential food sources and detect predators. Chemical senses also mediate the social interactions that form the focus of this book, as illustrated by the dogs and ants above. We can probably say that more organisms use chemosensory communication than any other mode.

A chemical involved in the chemical interaction between organisms is called a semiochemical (Box 1.2). Some of the semiochemicals emitted by animals are pheromones, evolved as signals for communication. Other semiochemicals, such as the carbon dioxide in exhaled breath, did not evolve as a signal, but can be exploited as a cue by blood-sucking mosquitoes as a way of finding a host. Some of the other molecules emitted by animals, such as odors due to infections, may also be cues. The distinction between signals and cues is explored further in Section 1.3.

Pheromones and signature mixtures are semiochemicals used *within* a species. Semiochemicals acting between individuals from different species are called **allelochemicals** and are further divided depending on the costs and benefits to signaler and receiver (Box 1.2) (Chapter 11) (Nordlund & Lewis 1976; Wyatt 2011). Pheromone signals can be eavesdropped ("overheard") by unintended recipients: for example, specialist predatory beetles use the pheromones of their bark beetle prey to locate them. The predators are using the bark beetle pheromones as **kairomones**. Animals of one species can emit fake, counterfeit signals that benefit themselves at the cost of the receiving species. Chemical signals used in such deceit or propaganda are termed **allomones**: for example, bolas spiders synthesize particular moth pheromones to lure male moths of those species. Semiochemicals benefiting both signaler and receiver in mutualisms, such as those between sea anemones and anemone clownfish, are termed **synomones**. The multiplicity of terms is only useful as shorthand and the terms are clearly overlapping, not mutually exclusive (for example, a molecule used as a pheromone *within* a species can be used as a kairomone by its predator).

My aim in this book is to focus on patterns across the animal kingdom. I have tried to include examples from as many animal taxa as space allows, but for more detail see the suggestions in further reading and references in the text. This chapter introduces the ways in which animals use semiochemicals and many of the topics are explored at greater length in later chapters (see Preface for overview and rationale).

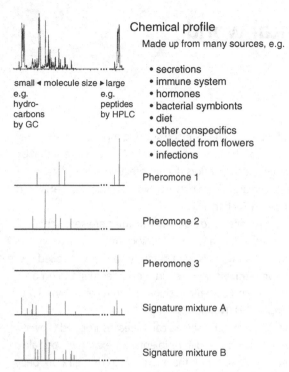

Chemical profile

Made up from many sources, e.g.

small ◄ molecule size ► large
e.g.   e.g.
hydro-   peptides
carbons   by HPLC
by GC

• secretions
• immune system
• hormones
• bacterial symbionts
• diet
• other conspecifics
• collected from flowers
• infections

Pheromone 1

Pheromone 2

Pheromone 3

Signature mixture A

Signature mixture B

Figure 1.1 Pheromones occur in a background of molecules that make up the chemical profile consisting of all the molecules extractable from an individual. The chemical profile (top) is an imaginary trace from an imaginary column capable of analyzing all the molecules (at one side is high-performance liquid chromatography (HPLC) with large proteins, at the other is gas chromatography (GC) with small volatile molecules). Each peak represents at least one molecule.

Much of the chemical profile is highly variable from individual to individual. The sources of the molecules in the chemical profile include the animal itself as well as its environment, food, bacteria, and other individuals etc. It is this complex background that makes identifying pheromones so challenging in many organisms.

The pheromones could include sex pheromones or ones related to life stage or caste. The pheromones would be the same in all individuals of the same type in a species (dominant male, worker ant, forager, etc.); that is, they are anonymous, common across the species. As examples, I have included some possible kinds of pheromones that are known from organisms (not necessarily in the same species): a specific combination of large and small molecules (Pheromone 1), a combination of small molecules

1.1 Intra-specific semiochemicals: pheromones and signature mixtures

Modern pheromone research could be said to date from 1959, when the chemist Adolf Butenandt and his team identified the first pheromone, the silk moth's sex pheromone bombykol, which prompted the coining of the word "pheromone," from the Greek *pherein*, to transfer; *hormōn*, to excite (Butenandt *et al.* 1959; Karlson & Lüscher 1959). Butenandt's discovery established that chemical signals between animals exist and can be identified (Chapter 2). From the start, Karlson and Lüscher (1959) anticipated pheromones would be used by every kind of animal, from insects and crustaceans to fish and mammals. Since then, pheromones have been found across the animal kingdom, in every habitat on land and underwater, carrying messages between courting lobsters, alarmed aphids, suckling rabbit pups, mound-building termites, and trail-following ants (Wyatt 2009). They are also used by algae, yeast, ciliates, and bacteria. It is likely that the majority of species across the animal kingdom use them for communication of various kinds. Much is known about the pheromones of

Figure 1.1 (cont.)
(Pheromone 2), or a particular large molecule by itself such as a peptide (Pheromone 3).

The signature mixtures (A and B) are subsets of variable molecules from the chemical profile that are learned as a template for distinguishing individuals or colonies. Different receivers might learn different signature mixtures of the same individual. For example, a male might learn a different signature mixture of his mate than the one her offspring might learn. Hypothetically it is conceivable that the male might learn different signature mixtures for the same female in different contexts, say immune-system associated molecules in one context and more diet influenced molecules in another. In other words, signature mixtures seem to be a "receiver-side" concept.

Adapted from Wyatt (2010). The layout is inspired by Figure 1 of Schaal (2009).

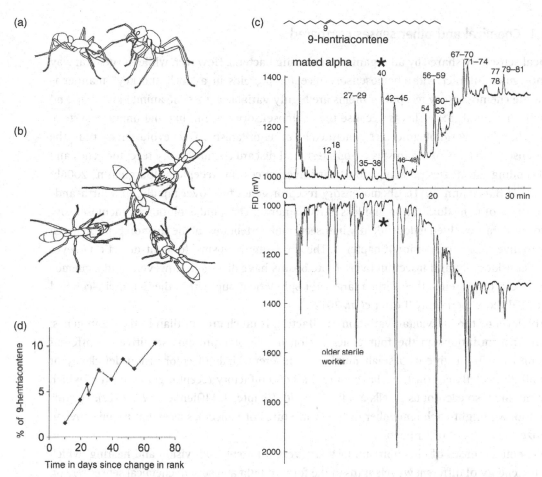

Figure 1.2 The "queenless" ant, *Dinoponera quadriceps*, lives in small groups headed by an alpha female, the only egg-laying individual in the colony. The hierarchy is maintained by physical aggression. This can include gaster rubbing (a) in which the alpha female rubs the antenna of the subordinate on the cuticular hydrocarbons, which include the alpha's pheromone "badge" of dominance, 9-hentriacontene (c, top). This molecule is characteristic of alpha females in all colonies of the species.

(b) If a subordinate female becomes reproductive and starts to produce the molecules characteristic of an alpha female, other ants in the colony detect this and immobilize her (an example of an honest signal maintained by punishment, Section 1.6).

(c) The colony profile of ants in the colony can be shown in a solid phase micro extraction (SPME) gas chromatographic analysis of their cuticular hydrocarbons (Monnin *et al.* 1998) (Chapter 2). As well as the many-peaked hydrocarbon chemical profile shared by the other ants in the colony, the alpha female also has the additional peak #40 (indicated by the asterisk) which is the pheromone 9-hentriacontene. Below, her fellow colony members have the same colony profile as her but lack this peak.

(d) Non-destructive SPME sampling allowed changes in the percentage of 9-hentriacontene in the cuticular hydrocarbons of an individual ant to be followed in the days after she became the alpha female. In a larger sample of ants undergoing the transition, the significant difference was between the quantities at 15 and 30 days.

(a) and (b) from Monnin and Peeters (1999), (c) chromatograph from Monnin *et al.* (1998), (d) from Peeters *et al.* (1999).

insects, fish, and mammals, but some other taxa have not been well studied. For example, crabs and other Crustacea make extensive use of pheromones but relatively few of these have been chemically identified (Breithaupt & Thiel 2011). Birds, too, have now been shown to have a rich olfactory life though we are only

Box 1.1 **Chemical and other senses compared**

Chemical senses are shared by all organisms including bacteria. However, while the general way that molecules interact with chemosensory receptor proteins in a "lock and key" manner is shared, the chemosensory receptor proteins are highly variable across the animal kingdom and even within animal taxa. This is because the chemosensory system, like the immune system, tracks a changing world of molecules generated by other organisms. Over evolutionary time, the chemosensory systems of organisms co-opt, test, and discard chemosensory receptor genes and neural coding strategies, leading to great divergences in receptors (Bargmann 2006b; Bendesky & Bargmann 2011). Chemosensory receptor genes turn over rapidly, in a birth-and-death process of gene duplication and loss (see Chapter 9). The rapid evolution of chemosensory receptor proteins, evolved independently in insects and vertebrates, made chemoreception much harder to investigate than vision (Chapter 9). The key proteins (opsins) for light-detection in eyes do vary considerably and insect and vertebrate opsins have diverged. However, unlike chemosensory receptor proteins, they form a large monophyletic group within the G-protein-coupled receptor (GPCR) superfamily (Porter *et al.* 2012).

At the level of the individual, variation in olfaction is much greater than in the opsin genes. For humans, mutations in the four genes for opsin receptor proteins sensitive to different wavelengths of light give us a small number of different kinds of color vision deficiency or "color blindness." By contrast, we have more than 400 olfactory receptor genes, each of which can be mutated, so each of us smells a unique world (Chapter 13) (Olender *et al.* 2012). For this reason too, we might each remember different mixtures of molecules as signature mixtures to recognize the odors of other people.

The chemical senses of olfaction and taste are very different from vision and hearing, which detect the energy of different wavelengths in the form of light and sound: chemical senses rely on the physical movement of molecules from the signaler to the sense organ of the receiving animal. This requires either diffusion, only likely to be important for small organisms at the scale of millimeters, or flow of currents (Chapter 10). Either way, the time taken for molecules to travel to the receiver means that chemical signals are rarely instantaneous in the way that visual and acoustic signals can be.

Challenges remain for studying chemical communication (Chapter 2). We can record and play back the sound signals of an animal easily enough, but we do not have devices to do the same for chemical signals. Each molecule needs to be correctly synthesized, in every detail (see Section 1.4.3 and Appendix), before it can be "played back" to the animal. This can be challenging for a team of biologists and makes chemist partners invaluable. For example, methyl-branched alkanes, important components of ant CHCs, are not commercially available and synthesizing these is a costly and time-consuming process (van Zweden & d'Ettorre 2010).

Yet, perhaps more than other modalities such as sound or vision, chemosensory systems are amenable to molecular manipulation: in model systems we can now study communication at the

Box 1.1 (cont.)

level of the genes involved in signal production (e.g., enzyme pathways) and signal reception (genetics of receptors, brain, and behavior) especially in model animals such as *Caenorhabditis elegans*, moths, *Drosophila*, and the mouse.

just beginning to discover what molecules their pheromones might be (Campagna *et al.* 2012; Caro & Balthazart 2010; Hagelin & Jones 2007; Zhang *et al.* 2010). Research on human semiochemicals is at a similarly early stage; I review our current state of knowledge in Chapter 13.

The idea of chemical communication was not new in 1959. The ancient Greeks knew that the secretions of a female dog attracted males. Charles Butler (1623) warned in *The Feminine Monarchie* that if a beekeeper accidentally crushes a honeybee, the bees "presently finding it by the ranke smell of the poisonous humor, will be so angry, that he shall have work enough to defend himself." In *The Descent of Man, and Selection in Relation to Sex* (1871), Charles Darwin included chemical signals alongside visual and auditory signals as outcomes of sexual selection, describing the strong smells of breeding males in moths, pythons, crocodiles, musk ducks, goats, and elephants. Jean-Henri Fabre (1911), also writing in the 1870s, described how male great peacock moths, *Saturnia pyri*, flocked around a female moth hidden behind wire-gauze, but ignored visible females sealed under glass. A female moth's smell could be collected on a cloth and males would flock to that too. Many other scientists in the nineteenth century and first half of the twentieth century, including Niko Tinbergen, had worked on phenomena we would recognize as being mediated by pheromones (some are mentioned in Karlson & Lüscher 1959). However, because the quantities emitted by an individual animal were so small, the chemistry of the day could not identify them, until the inspired idea of using domesticated silk moths, which could be reared in the hundreds of thousands necessary to collect enough material for analysis using the techniques available at that time (Chapter 2).

The enormous variety of organic molecules identified as pheromones since the first, bombykol, in 1959 is as diverse as the animal kingdom, and offers an ongoing challenge for chemists interested in the identification, synthesis, and exploration of natural functions of novel compounds (Cummins & Bowie 2012; El-Sayed 2013; Francke & Schulz 2010). The likely explanation for the diversity of pheromone chemistry is that these signals have evolved from chemical cues naturally released by organisms, facilitated by the broad tuning of olfactory receptors (Chapter 9) (Section 1.3).

Invertebrates and vertebrates, in a wide range of habitats, use chemical communication in similar ways. Animals as different as moths and elephants may share the same molecule(s) as part of their pheromones. However, there are more fundamental parallels in sensory processes, even if we are not always sure whether this has occurred by convergence or via shared ancestors. The parallels include the combinatorial way that the sense of smell is organized in the brain: olfactory sensory neurons with the same olfactory receptor all collect at the same spot (glomerulus) in the brain; the information from different glomeruli is combined to identify the molecule (the combinatorial mechanism) (Chapter 9).

1.1.1 Pheromones

Pheromones are molecules that have evolved as a signal between organisms of the same species. The signal elicits a specific reaction, for example, a stereotyped behavior (releaser effect) and/or a developmental process (primer effect) from a conspecific (member of the same species) (Box 1.2) (Section 1.9) (Wyatt 2010). Many, probably most, pheromones (including the sex pheromones of most moths and some mammal pheromones) are *not* single compounds, but rather a species-specific combination of molecules in a precise ratio. This combination *is* the pheromone (though sometimes called a multicomponent pheromone or pheromone blend). A pheromone can elicit a variety of effects, depending on the context and the receiver (Section 1.8). Responses to pheromones usually seem to be innate (though this is not a part of the definition). In the few instances where learning is first required for a pheromone to act, all animals normally learn the same molecule(s), which is what defines it as a pheromone (Section 1.2).

Box 1.2 Definitions of chemical mediators

Pheromones are signals. The other categories of semiochemicals in this box are cues that can be used for information but did not evolve for that function (Section 1.3). Adapted from Wyatt (2010, 2011) based on Nordlund and Lewis (1976).

See Wyatt (2011) for a discussion of the origins and usage of these terms. I discuss interspecific interactions mediated by allelochemicals in Chapter 11. "Infochemical" as an alternative to "semiochemical" was proposed by Dicke and Sabelis (1988) though its main change was to replace "produced or acquired by" with "pertinent to biology of" in each case for allelochemicals.

A. **Hormone:** a chemical agent, produced by tissue or endocrine glands, that controls various physiological processes within an organism. (Nordlund & Lewis 1976).

B. **Semiochemical:** a chemical involved in the chemical interaction between organisms. (Nordlund & Lewis 1976) (from the Greek: *semeion*, mark or signal).

1. **Pheromone:** molecules that are evolved signals, in defined ratios in the case of multiple component pheromones, which are emitted by an individual and received by a second individual of the same species, in which they cause a specific reaction, for example, a stereotyped behavior or a developmental process. (Wyatt 2010, modified after Karlson and Lüscher 1959). (From the Greek: *pherein*, to carry or transfer, and *hormōn*, to excite or stimulate).

2. **Signature mixture:** a variable chemical mixture (a subset of the molecules in an animal's chemical profile) learned by other conspecifics and used to recognize an animal as an individual (e.g., lobsters, mice) or as a member of a particular social group such as a family, clan, or colony (e.g., ants, bees, badgers). (Wyatt 2010; derived from Johnston's "mosaic signal" *sensu* 2003, 2005; Hölldobler and Carlin's, 1987 ideas; and Wyatt's, 2005 "signature odor").

3. **Allelochemical:** chemical significant to organisms of a species different from their source, for reasons other than food as such. (Nordlund & Lewis 1976).

Box 1.2 (cont.)

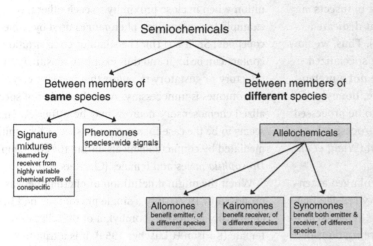

Diagram showing the relationships between different kinds of semiochemicals. Inspired by Box 7.1 in de Brito-Sanchez *et al.* (2008) and other sources.

Karlson and Lüscher (1959) predicted that most pheromones would act via the conventional senses of olfaction or taste, but that some pheromones might be ingested and act directly on the brain or other tissues. We would call these allohormone pheromones (Section 1.11). They speculated that royal jelly in honeybees might contain such a pheromone, and indeed an active molecule (royalactin) has been identified, which causes larvae receiving it to develop into queens rather than workers (Chapter 9) (Kamakura 2011).

Pheromones include the familiar sex attractant pheromones, and numerous others that serve a wide variety of functions. Some pheromones are specific to different life stages or castes. One key feature of pheromones is that they are "anonymous," that is, a given pheromone is the same in all individuals within a species of the same type (e.g., male or female) or physiological state, and it conveys a stereotyped message that is independent of the individual producing it (Hölldobler & Carlin 1987).

However, quantities of pheromone can differ between individuals or in the same individual over time. Some male mouse pheromones, the farnesenes, are produced only by dominant male territory holders, not subordinates (Hurst & Beynon 2004). In the ant *Dinoponera quadriceps*, when an ant becomes the top (alpha) female, she starts to produce the standard chemical badge of a "top female" in her species, 9-hentriacontene (Figure 1.2) (Peeters *et al.* 1999). However, in the male mouse and the top female ant of these examples, the pheromones are still anonymous (Hölldobler & Carlin 1987; Hölldobler & Wilson 2009, p. 270). They indicate the presence of, for example, a dominant male mouse or an alpha female ant, not a particular individual.

Some of our expectations of pheromones have been heavily influenced by the well studied response of male moths to the sex attractant pheromones of conspecific females. For example, the antennae of male moths have thousands of highly specialized receptors for the pheromone and specific areas of the brain dedicated to processing the pheromonal signal. However, other pheromone processing in insects may involve less specific receptors, without dedicated brain areas (glomeruli) (see Chapter 9). Thus, we now know that narrowly tuned and highly specialized receptors and dedicated glomeruli are not a prerequisite for pheromone use. For example, honeybee alarm pheromone components seem to be processed by receptors and glomeruli that also process other, non-pheromone molecules (Chapter 9) (Wang *et al.* 2008b).

Similarly, male moths' enormously enlarged antennae, covered with thousands of olfactory sensilla that are tuned specifically to the pheromone, reflect selection for extreme sensitivity to low concentrations of female pheromone, necessitated by the scramble competition to be the first to reach the female (Chapters 3, 9, and 10). Based on the great body of work on male moths, we tend to expect all receivers of pheromones to be very sensitive to them and to respond at great distances. However, other animals may not use attractant pheromones at all, although they may still use sex-specific contact pheromones for sex and species recognition when in close proximity to each other (for example the contact sex pheromones used by some copepods; Snell 2011b). The stimulus concentration on contact can be high and thus exquisite sensitivity in the olfactory or gustatory receptors that perceive contact pheromones is unnecessary. A small number of specialized chemosensory neurons may be sufficient. This seems to be the case for short range species recognition mediated by contact chemicals during the courtship of *Drosophila* males and females (Chapters 3 and 9).

When the original definition of pheromone was proposed in 1959, only a single pheromone had been chemically identified: bombykol of the silk moth female (Karlson & Lüscher 1959). It is a tribute to Karlson and Lüscher, and their wide consultation, that the definition has held up so well (Wyatt 2009). It is not surprising that the definition has needed to be updated slightly since then (Box 1.2) (Wyatt 2010). (See Box 1.3 and Box 1.4 for why words matter and how distinguishing the concepts can be helpful).

Box 1.3 Pheromones and signature mixtures: why words matter

Definitions matter because they can provide useful generalizations and predictions. My purpose in separating pheromones from signature mixtures is pragmatic and based on the heuristic (rule of thumb) value of separating these kinds of chemical information. When we say something is a pheromone, the reader can anticipate that it is a molecule (or a particular combination and ratio of molecules for a multicomponent pheromone) that will be found, for example, in all sexually mature females. Quantities of the pheromone may differ between individuals, and this may be important in mate choice (Chapter 3), but not in ways that allow an individual female to be recognized as an individual. In Hölldobler and Carlin's (1987) terms, the pheromone signal is "anonymous," it could be any female (see also Hölldobler & Wilson 2009, p. 270). (See also Box 1.4 Operational definition of pheromone.)

In contrast, if a phenomenon, such as a male distinguishing his mate from other females, relies on a learned signature mixture, it would be fruitless to search for a single combination of molecules eliciting individual mate recognition across the species: it is precisely the great

Box 1.3 (cont.)

differences *between* females' chemical profiles that makes learning signature mixtures by males possible.

In the first edition of this book, I included signature mixtures within the definition of "pheromones" (Wyatt 2003, pp. 2–4). I now think it is more helpful to explicitly separate signature mixtures as it is emerging that their characteristics are different, in particular the variability of signature mixtures and the need for learning (Tables 1.1 and 1.2) (Wyatt 2010). It seems to be a useful distinction, which has helped understand phenomena best explained by species-specific pheromone molecules appearing on a background of variable chemical profiles from which signature mixtures are learned, in situations as varied as the male effect in sheep (Hawken & Martin 2012) and trail pheromones in stingless bees (Reichle *et al.* 2013).

So, to be clear, not all molecules included in this book are pheromones. I will discuss many molecules that are not pheromones (Section 1.3), including the highly variable signature mixtures used to avoid mating with kin (Chapter 3) and learned by ants to distinguish nestmates from non-nestmates (Chapter 6), as well as chemical cues such as barnacle settlement cues (Chapter 4) and fish alarm cues (Chapter 8).

Box 1.4 Operational definition of pheromone

The formal definition of a pheromone includes both evolved emission and reception of the signal for that function (Section 1.3) (Table 1.1) (Maynard Smith & Harper 2003, p. 3). However, for many otherwise respectable pheromones, we do not know enough about the ways in which production and/or reception may have evolved. So, I propose we formalize an operational definition of pheromone, which most people already use in practice, as "fully identified molecule(s), the same across a species, in all lactating mature females for example, which when synthesized elicit the same characteristic response in the conspecific receiver as the natural stimulus."

To legitimately assert that a molecule or specific combination of molecules qualifies as a pheromone for a species (or in a genetically defined subpopulation within a species):

1. The synthesized molecule/combination of molecules (combination) should elicit the same response as the natural stimulus in the bioassay.

2. It should act in this way at realistic concentrations similar to the natural stimulus.

Box 1.4 (cont.)

3. For multicomponent pheromones, experiments should demonstrate that all compounds in the combination are necessary and sufficient.

4. Only this molecule or the proposed combination of molecules elicits the effect (and other similar molecules or combinations that the animal would encounter do not).

5. There should be a credible pathway for the pheromone signal to have evolved by direct or kin selection.

6. Quantities may vary between individuals (e.g., subordinate and dominant males).

The requirements follow those explored in Chapter 2. They are the equivalent of "Koch's postulates" for establishing causal relationships for pheromones: initial demonstration of an effect mediated by a pheromone, then identification and synthesis of the bioactive molecule(s), followed by bioassay confirmation of activity of the synthesized molecules. It can be equally important to show that other similar molecules do *not* have the effect of the proposed pheromone.

How the response develops (ontogeny) in an individual is a separate question (Section 1.2). Normally we do not know the details. Fish alarm substances are thought to be cues rather than pheromones (Chapter 8) as they fail to satisfy criterion #5.

Sadly, the experimental literature on humans, and other mammals, includes many unidentified extracts or molecules that have never been rigorously demonstrated to be biologically active by the full bioassay evidence and synthesis process. It is misleading to call them even "putative pheromones" (Chapter 13).

1.1.2 Signature mixtures

Returning to the dogs and ants that opened this chapter, the individually distinctive mixture of molecules that allows dogs to tell each other apart by smell and allows ants, at a colony level, to distinguish nestmate from non-nestmate, are *not* pheromones and were not included in the original definition.

We need a different term for the molecules that animals learn and use to distinguish other individuals or colonies. I have proposed "signature mixture" (Wyatt (2010) inspired by Johnston's (2003, 2005) "mosaic signal," Hölldobler and Carlin's (1987) ideas, and based on Wyatt's (2005) "signature odor"). I think

some of the early doubts about mammal pheromones (Box 1.5) came from treating signature mixtures as if they were pheromones. Be aware when reading the past and current literature that the term "pheromone" is still used ambiguously and may be used in contexts where "signature mixture" or "chemosensory cues" would be more accurate or helpful.

Signature mixtures are the subsets of variable molecules from the chemical profile of an individual (Figure 1.1) that are learned as templates by members of the same species (conspecifics) and used to recognize an organism as an individual or as a member of a particular social group such as a family, clan, or colony

Box 1.5 **Mammal pheromones**

There is good evidence that mammals have pheromones that fit well with the original definition (Brennan & Zufall 2006; Wyatt 2010), despite doubts from some authors (Beauchamp *et al.* 1976; Doty 2010). The many small-molecule mammal pheromones include the rabbit mammary pheromone 2-methylbut-2-enal (2MB2) (Schaal *et al.* 2003) (see below), male mouse pheromones such as (methylthio)methanethiol (MTMT) (Lin *et al.* 2005), trimethylamine (Li *et al.* 2013), dehydro-*exo*-brevicomin and 2-*sec*-butyl-4,5-dihydrothiazole (Novotny 2003; Novotny *et al.* 1985), and Asian elephant pheromones including frontalin (1,5-dimethyl-6,8-dioxabicyclo [3.2.1]octane) and (*Z*)-7-dodecen-1-yl acetate (Rasmussen *et al.* 2003). Many of these molecules (or ones similar) are also used as pheromones by insects (Section 1.4.1) (Table 1.2). As well as small-molecule pheromones, mammals also have large-molecule pheromones, such as, in mice, exocrine gland-secreting peptide 1 (ESP1) (Haga *et al.* 2010) and the protein pheromone darcin (Chapter 9) (Roberts *et al.* 2010, 2012). See Figure 9.7 for some of the large and small molecules used by mice.

Part of the reason for the earlier doubts was confusion between pheromones and the highly variable chemical profiles of mammals (for example, see Figure 13.2). In addition, over the years, expectations about pheromones in mammals have built up, perhaps based on misconceptions about insect pheromones (e.g., Doty 2010): contrary to these expectations, as shown in a variety of animals including insects, pheromones do not have to be unique species-specific molecules (Section 1.4.1), animals' responses to pheromones can vary (Section 1.8), and pheromones can involve elements of learning (Section 1.2).

A separate set of problems came from some scientists in the 1990s onwards, using mice as a model system to study mammal pheromones, who seemed to assume, despite earlier evidence to the contrary, that (a) pheromones would be exclusively detected by the VNO system, and that (b) all molecules detected by the VNO were pheromones (Baxi *et al.* 2006; Wyatt 2009). As explored in Chapter 9, it is now confirmed that (i) pheromones are detected by both the VNO and the main olfactory system, depending on species and pheromone, (ii) that the VNO also responds to other odorants, and (iii) that there is extensive integration of inputs from the two olfactory systems.

An example of a small-molecule pheromone perceived by the main olfactory system is the rabbit mammary pheromone, 2MB2, which stimulates rabbit pups to suckle (Charra *et al.* 2012; Schaal *et al.* 2003). The pups respond to pheromone from their mother's nipple region, which elicits stereotyped searching, usually successful in just six seconds. The pheromone also prompts the pups to learn their mother's signature mixture (Coureaud *et al.* 2010). Newly born humans use olfactory stimuli, possibly including a pheromone, to find their mother's nipple (Chapter 13) (Doucet *et al.* 2009, 2012).

Box 1.5 (cont.)

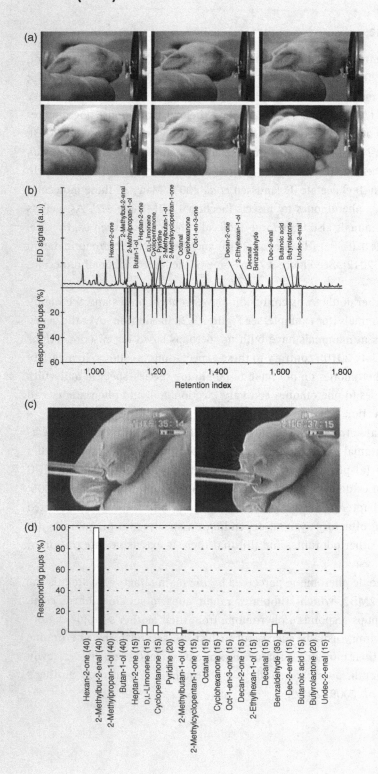

Box 1.5 (cont.)

(Figure facing). The discovery of the rabbit mammary pheromone, 2-methylbut-2-enal (2MB2), used a linked gas chromatograph-olfactory (GCO) assay, which allowed concurrent detection by neonatal rabbits and by a flame ionization detector (FID) (Chapter 2). (a) Photographs show the sequence (duration 5 s) of a two-day-old pup's searching–grasping response directed to the glass funnel of the GC sniff-port. (b) Typical chromatogram of rabbit milk effluvium (upper panel) and concurrent percentage of pups responding with searching–grasping responses (lower panel; inverted scale). The regions of the chromatogram eliciting more than 20% of responses (summed across 25 GCO runs) and the compounds eluting in these regions are shown. (c) Screening, by presenting a glass-rod, of milk volatiles presumed to have behavioral activity: two-day-old pup at rest (left) and exhibiting grasping (right) to a glass rod carrying 2MB2. (d) Frequency of searching (open bars) and seizing (solid bars) directed to the glass rod carrying one of the 21 compounds identified in milk. Numbers in parentheses indicate the numbers of pups tested. (Schaal et al. 2003).

The only significant mammal group for which chemical communication has not been demonstrated are whales and dolphins (cetaceans). However, it is possible and may be discovered in future, as baleen (mysticete) whales have a good olfactory system, which they may also use to detect upwind concentrations of plankton by smell (as albatross do, Chapter 10) (Thewissen et al. 2011). Some baleen whales have a Harderian gland, which in some rodents produces pheromones, though it may have other functions (Funasaka et al. 2010).

(Box 1.2). "Signature" is used as it denotes individuality.

The signature mixture is the mixture of molecules (and likely, their relative ratios) that are learned. The template is the neural representation of the signature mixture stored in the memory of the learner (after van Zweden & d'Ettorre 2010).

There are two distinguishing characteristics of signature mixtures: first, a requirement for learning and, second, the variability of the cues learned, allowing other individuals to be distinguished by their different chemical profiles (see Section 1.7 for a more detailed explanation). Other comparisons between pheromones and signature mixtures are explored in Table 1.1.

The signature mixture molecules in the chemical profile, learned by receivers as the template for recognition, can be produced by the organism itself, acquired from the diet, shared local environment, or other organisms (Section 1.7). In ants, the chemical profile may have species-characteristic types of molecules, but each colony produces different combinations and ratios of these (van Zweden & d'Ettorre 2010). For example, different colonies of the ant *Formica exsecta* have different colony-specific combinations of various (Z)-9-alkenes, under genetic influence (Martin & Drijfhout 2009b; Martin et al. 2008c). What makes this different from a pheromone is that each colony has a different set of ratios of these shared molecules – it is what allows the colonies to be distinguished. By contrast a multicomponent pheromone would be expected to have a uniform ratio across a population, the same in each colony.

Different receivers might learn different combinations of molecules from an individual's profile as the

Table 1.1 **Contrasting pheromones with signature mixtures.** Hölldobler and Carlin (1987) introduced the idea of anonymous signals (pheromones) contrasted with variable signature mixtures (though their terminology was different) (see also Hölldobler & Wilson 2009, p. 270). The anonymous pheromone signals are uniform throughout a category (e.g., species, male, female, and perhaps molt state and dominance status). In contrast, signature mixtures vary between individuals or colonies and can be used to recognize the organism as an individual or member of a particular social group such as a family or colony. From Wyatt (2010).

	Pheromone	Signature mixture
Stimulus	A species-wide molecule (or particular defined combination of molecules).	A combination of molecules, never a single molecule. Combination of molecules varies between individuals or colonies. Possible "receiver side effect": there may not be one signature mixture for each individual, as different conspecifics (receivers) may learn different subsets of molecules in the individual's chemical profile (Figure 1.2).
Type of information	Anonymous (independent of the source individual).	Variable (allows recognition of an individual or group such as a colony).
Molecule size	Any size or type, depending on habitat, medium, signal duration, and phylogeny.	Any size or type, depending on habitat, medium, signal duration, and phylogeny.
Source	Make self or acquire/modify. Usually genetically based.	Make self or acquire/modify. Use chemical mixtures, genetically based or from the environment or a combination.
Learning	Little requirement for learning of the signal molecule(s). Innate, stereotyped, or hardwired (with the caveat of developmental constraints).	Cues learned.
Response	Elicits a stereotyped behavior and/or physiological response. May be context dependent.	Learned and can be used to distinguish individuals or groups (can lead to stereotyped response e.g., aggression). May be context dependent.
Olfactory receptor proteins	Some (e.g., moth sex pheromones) have high specificity olfactory receptor proteins (and the "labeled lines" and "dedicated glomeruli" that result). Many other pheromones do not.	Low specificity, broadly tuned receptors.

Table 1.1 **(cont.)**

	Pheromone	Signature mixture
Processing	Mostly combinatorial across glomeruli.	Combinatorial across glomeruli.
Detection system: olfaction or taste or act directly.	Mostly by glomerularly organized olfactory system(s). A minority of pheromones by other chemosensory routes e.g., taste (gestation). Allohormone pheromones act directly on tissues or nervous system.	Glomerular olfactory system(s).
In vertebrates with a vomeronasal system (VNS).	Detection by the VNS or main olfactory system or both, depending on pheromone and species.	Detection by the VNS or main olfactory system or both, depending on species.

signature mixture to recognize that individual (see legend to Figure 1.1). In other words, signature mixtures seem to be a "receiver side" phenomenon, existing as a "template" in the nervous system of the receiver. Even if all receivers perceived the world in the same way, they could each still learn different subsets of molecules from the chemical profile as the signature mixture of an individual. A further complication comes from the way that each receiver smells a slightly different world, because one of the characteristics of olfaction is the variability of olfactory receptors between individuals – each of us smells a different world (Box 1.1) (Chapters 9 and 13). For this reason too, the learned signature mixtures could differ between receivers.

With perfect knowledge, one could know the whole chemical profile of an animal, which molecules from this profile are learned as the signature mixture by the receiver, and how the signature mixture is represented as a template in the nervous system of the receiver. An outline of what can already be achieved, treating the system as a "black box," is shown by experiments with the ant *Formica japonica*, which showed that the nestmate "label" could be reproduced with synthetic

hydrocarbons matching the colony cuticular hydrocarbons (CHCs) (Akino *et al.* 2004).

1.1.3 Which sensory systems, olfaction or gustation, are used to detect pheromones and signature mixtures?

In both invertebrates and vertebrates, gustatory (taste) receptors come from different families of receptors from olfactory receptors, and link to the brain in different, simpler, ways from the olfactory pathways (in mammals olfaction includes both the main olfactory system and vomeronasal organ-accessory olfactory system; see Chapter 9 for more explanation). Most pheromones seem to be processed by olfaction. However, a small but significant proportion of pheromones in invertebrates are processed by taste (gustation) (Chapter 9). These include the response of a male *Drosophila melanogaster*'s gustatory receptors on its front leg to CHC pheromones important in sex and species recognition (Chapters 3 and 9). Some allohormone pheromones in both vertebrates and invertebrates may act directly on the brain or other organs (see Section 1.11).

All signature mixtures are likely to be processed by the combinatorial processes of olfaction rather than taste (gustation), in part because discrimination learning is likely to be involved (see Box 9.1).

1.2 "Innateness" of pheromones

Generally speaking, pheromones do not require learning: they seem to be "innate," "hardwired," predisposed, or "work out of the box" (Table 1.2). However, being innate is not part of the original pheromone definition (Karlson & Lüscher 1959) or its updated version (Box 1.2) (see also Sections 1.8 and 9.6) (Wyatt 2010). Instead, the defining point for pheromones is that they are species-wide rather than that they are innate (for example, it would be possible for a pheromone to rely on early learning so long as typically all individuals imprint on the same molecule (s) in normal circumstances).

The idea of "innate" behavior is itself a long-debated question in animal behavior. Seemingly innate behaviors often have developmental and environmental requirements for full expression (Bateson & Mameli 2007; Mameli & Bateson 2011). Part of the problem is that the term "innate" covers many different phenomena (Mameli & Bateson 2011). In practice, trying to separate the contributions of nature (genes) and nurture (environment) to the development of a behavior is like asking whether the area of a rectangle is due more to its length or width (ascribed to psychologist Donald Hebb, in Meaney 2001). Gene–environment interactions on behavior are explored by Bendesky and Bargmann (2011).

Just as a mammal's visual cortex does not form correctly if the eyes do not receive visual stimuli during critical periods after birth (Hensch 2004), olfactory stimuli in amniotic fluid before birth can influence olfactory bulb development (e.g., Todrank et al. 2011). Normal responses to pheromones may not develop unless species-specific conditions are met. These usually occur as a matter of course in normal development. Experiments that "dissect" the

developmental process can expose the normally hidden mechanism(s) by which a response develops. For example, perhaps surprisingly, early imprinting on species-specific odors can be important in recognition of a mate of the correct species when adult (Chapters 3 and 9) (Doty 2010, p. 39 ff.; Owens et al. 1999). Normally, as the parents are of its own species, this leads to appropriate courtship choices, but cross-fostering experiments can demonstrate that these olfactory choices are learned in some species such as pygmy mice, house mice, sheep, and deer (Doty 2010, p. 39 ff.). Cross-fostered young are attracted to the species of their foster parents.

In the wild, this learning can be important in sexual selection and speciation in some species (Verzijden et al. 2012). Cross fostering showed that early olfactory imprinting by young fish (learning at a sensitive period; Chapter 9) normally contributes to sexual isolation in two stickleback species by influencing adult mate choices (Kozak et al. 2011). When adult, learning may also be involved: male mammals such as rats and mice may need sexual experience before they can distinguish estrous from diestrous female odors (Chapter 9) (Swaney & Keverne 2011).

In some cases, developmental effects have been shown to act at the periphery of the sensory system: for example, the behavioral response of young worker bees to queen mandibular pheromone depends on exposure to the pheromone soon after pupal emergence, via an effect on dopamine receptor gene expression in the olfactory sensory neurons (Vergoz et al. 2009).

Pheromones themselves can prompt learning. While the response of rabbit pups to the mammary pheromone 2-methylbut-2-enal seems hardwired (Box 1.5), the pheromone stimulates learning of other odors, which will then also stimulate suckling (Coureaud et al. 2010; Schaal et al. 2009). Contact with the male mouse protein pheromone, darcin, prompts a female mouse to learn both his volatile individual signature mixture and the location of the scent mark (Chapter 9) (Roberts et al. 2010, 2012).

Table 1.2 **Biochemical convergence of pheromones among ants, bees, moths and termites, and other animals including mammals. In some cases, the same or related compound is used for similar functions in different species. More commonly, the arbitrary nature of signals is revealed by different uses for same or similar compound. See other chapters for more details of the functions of these pheromones. After Blum (1982), with additional information from Kelly (1996), Novotny (2003), Mori (2007) and Breithaupt and Hardege (2012). See Appendix for notation.**

			Occurrence
Compound	Function	Occurrence Animal	Genus
Benzaldehyde	Trail pheromone	Bee, Apidae	*Trigona*
	Defense	Ant, Formicidae	*Veromessor*
	Male sex pheromone	Moth, Amphipyrinae	*Pseudaletia*
2-Tridecanone	Alarm pheromone	Ant, Formicidae	*Acanthomyops*
	Defense	Termite, Rhinotermitidae	*Schedorhinotermes*
(*R*)-(−)-5-Methyl-3-heptanone; (*S*)-(−)-5-Methyl-3-heptanone	Female sex pheromone, male sex pheromone	Nereid worm	*Platynereis*
2-heptanone	Male and female pheromones	Mammal	Mouse, *Mus*
Dehydro-*exo*-brevicomin	Male sex pheromone	Mammal	Mouse, *Mus*
Exo-brevicomin	Aggregation pheromone	Insect	Bark beetle, *Dendroctonus*
(*Z*)-7-Dodecen-1-yl acetate	Female sex pheromone	Mammal	Female Asian elephant *Elephas maximus* & 140 species
		Insect	of moth (as one component of a multicomponent pheromone)
(1*S*, 5*R*)-(−)-Frontalin	Aggregation pheromone	Insect	Bark beetles
	Sex pheromone	Mammal	Male Asian elephant *E. maximus* (both +/− enantiomers)

1.3 How pheromone signals evolve from chemical cues

The ubiquity and chemical diversity of pheromones can be explained by natural selection and are the evolutionary consequences of the powerful and flexible way the olfactory system is organized (Chapter 9); gustation (taste) does not have such flexibility. This may explain why most pheromones are detected by the olfactory system (in terrestrial vertebrates with both, it includes the main and accessory olfactory systems). The olfactory systems of most species have a large range of relatively non-specific "broadly tuned" olfactory receptors (Chapter 9). This means that almost any chemical cue in the rich chemical world of animals will stimulate some olfactory receptors and can potentially evolve into a pheromone signal.

Pheromones are evolved signals. Signals alter the behavior of other organisms, have evolved because of that effect, and work because the receiver's response has also evolved (Maynard Smith & Harper 2003, p. 3; Seeley 1995, p. 248).

If the signal alters the behavior of the receiver it must, on average, pay the receiver to respond in this way, otherwise receivers would evolve not to respond (Maynard Smith & Harper 2003, p. 3). Signaling is synonymous with communication as narrowly defined by Ruxton and Schaefer (2011, p. 2583) in a helpful discussion of recent debates about animal communication and information.

In contrast, a cue is any feature of the world, animate or inanimate, that can be used by an animal as a guide to future action, but has not evolved for this purpose (Maynard Smith & Harper 2003, p. 3). For cues, only the receiver's response is evolved. For example, the CO_2 released by an animal as it breathes can be used as a cue by a blood-sucking insect to find its host. The mosquito's response is certainly evolved (and indeed it has highly specialized receptors to detect CO_2), but the release of CO_2 by the host did not evolve to have the effect of attracting mosquitoes so it does not count as a signal. The molecules learned in signature mixtures of kin or familiar animals are probably best seen as cues rather than signals, as their production has usually not evolved for that purpose (see discussion in Section 1.7).

Scenarios for how chemical cues become pheromone signals fall into two main categories: *sender pre-adaptations*, with signals starting from chemical cues released by the sender, and *receiver pre-adaptations* selecting for molecules that match the already existing sensitivity of the receiver (Bradbury & Vehrencamp 2011, p. 377). If the receiver benefits, it may refine its tuning system. If the sender benefits from the response, the cue will undergo the evolutionary process of ritualization, becoming a pheromone signal (Bradbury & Vehrencamp 2011, p. 378; Tinbergen 1952). The changes in ritualization could include increasing its conspicuousness (e.g., making it more different from other similar molecules) and reducing its variability (stereotypy). Thus pheromones evolve from compounds originally having other uses or significance, for example from hormones, host plant odors, chemicals released on injury, or waste products (Steiger *et al.* 2011; Wyatt 2010). There will also be selection for functional signal features such as longevity and specificity (Section 1.4). The original functions of the molecules may or may not be eventually lost.

1.3.1 Pheromone signals derived from sender precursors

In the sender-precursor model of signal evolution, pheromones can evolve from any reliable chemical cue(s) to the sender's condition (Figure 1.3) (Bradbury & Vehrencamp 2011, p. 377 ff.). For example, if there are molecules leaking from a mature female about to lay eggs, then mutant males better able to detect them will find her first and gain more matings (Figure 1.3). Over generations this would result in selection for increasing sensitivity to the female's molecules (with multiple copies of such receptors) and changes in the receptors for greater specificity. The opportunistic co-option of chemosensory receptor proteins for detection of new molecules is discussed in Chapter 9. Although a given odorant may be unlikely to fit any one receptor

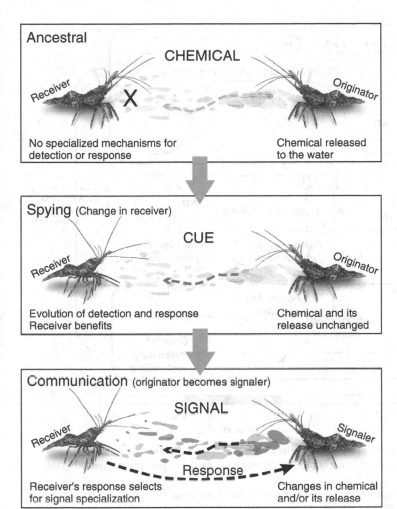

Ancestral

CHEMICAL

Receiver

X

Originator

No specialized mechanisms for detection or response

Chemical released to the water

Spying (Change in receiver)

CUE

Receiver

Originator

Evolution of detection and response
Receiver benefits

Chemical and its release unchanged

Communication (originator becomes signaler)

SIGNAL

Receiver

Signaler

Response

Receiver's response selects for signal specialization

Changes in chemical and/or its release

Figure 1.3 The sender-precursor model of signal evolution. Proposed stages in the evolution of a communication function for molecules released by an "originator" animal (the potential sender). The × in the upper panel indicates that the receiving individual has no special adaptations to receive the cue beyond detecting it. The process starts with an association between a cue and a condition of the originator. Receivers must be able to perceive or evolve receptors for the cue, and then incorporate the information into a decision rule and a response. In this "spying phase" only the receiver benefits. The transition to bilateral benefit to both sender and receiver could occur later if there is a selective advantage to the sender, leading to ritualization of the signal to maximize information transfer.

An original figure by Ivan Hinojosa (www.flickr.com/photos/ivan_hinojosa) in Wyatt (2011), inspired by, and with text adapted from, Stacey and Sorensen (2006) (with permission from the authors). Additional text adapted from Bradbury and Vehrencamp (2011, p. 377).

perfectly, it is likely to stimulate some. In turn, if the sender benefits, in the fish example the female would benefit by attracting males to fertilize her eggs, then production and release of the molecules can evolve into a signal (pheromone). Molecules become a pheromone only if there is positive selection on both signaler and receiver.

Such a scenario has been suggested to explain the use of body-fluid molecules as pheromones by marine polychaete worms, released with their gametes, which immediately prompt the other sex to release its gametes (Chapter 3) (Figure 3.2) (Breithaupt & Hardege 2012). Similarly, hormones have been co-opted as sex

pheromones in fish, excreted in urine or across permeable membranes such as gills (Figure 1.4) (Chung-Davidson *et al.* 2011; Stacey & Sorensen 2011). Species specificity of these multicomponent fish sex pheromones comes from other molecules (Section 1.4) (Levesque *et al.* 2011; Lim & Sorensen 2012). In terrestrial animals such as elephants and mice, many pheromones are excreted in the urine and though not necessarily hormones themselves, the quantities and qualities reflect hormonal levels related to body condition (Section 1.6.2).

The aggregation pheromones used by bark beetles may have evolved from the molecules produced by the

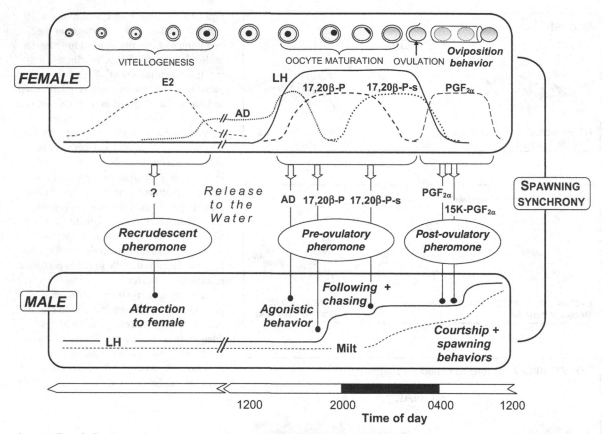

Figure 1.4 Female hormone pheromones co-ordinate reproduction in the goldfish, *Carassius auratus*, by primer and releaser effects on the male (Stacey & Sorensen 2009, 2011). It is likely that the hormones evolved into pheromones following the scenario presented in Figure 1.3.

In the female, the rise and fall of blood concentrations of a succession of hormones (top), from 17β-estradiol (E2) to luteinizing hormone (LH), steroids, and prostaglandin $F_{2\alpha}$ ($PGF_{2\alpha}$), stimulate release to the water of a succession of hormone pheromones that reflect her hormone levels. First, an unidentified recrudescent pheromone attracts males.

Second, a pre-ovulatory pheromone (the steroid androstenedione (AD), the maturation-inducing steroid 17,20β-P, and its sulfated metabolite, 17,20β-P-S) are released the night before ovulation. Androstenedione induces agonistic behaviors among males. As the 17,20β-P:AD ratio increases, males increase their own LH and begin to follow and chase conspecifics. Males exposed to the pre-ovulatory pheromone increase both the quantity and quality of sperm in the milt (semen), increasing the likelihood of reproductive success.

Third, post-ovulatory pheromone (prostaglandin $F_{2\alpha}$ ($PGF_{2\alpha}$)) and its major metabolite 15K-$PGF_{2\alpha}$ stimulate both male courtship and spawning behaviors and additional male LH increase.

All hormonal cues are released in quantities that range from 10 to >100 ng/h, are detected at concentrations in the picomolar range, and act in concert to synchronize male behavior and physiology with the female. Figure from and caption after Stacey and Sorensen (2009).

beetles' detoxification of the toxic monoterpenes used by the host trees as a defense against beetle attack (Chapter 4) (Blomquist *et al.* 2010). Some of the detoxification enzymes may have been co-opted as biosynthetic enzymes for synthesis of pheromones by the beetle.

Many alarm pheromones in social insects, which provoke fight or flight in receivers, appear to have evolved from defensive compounds released by fighting or injured conspecifics (Chapter 8). There will be a selective advantage to the potential receivers sensitive to these compounds and responding appropriately to protect the colony. Over evolutionary time, defensive compounds may gain a signal function: for example, many ant species use the same chemicals for defense and alarm, to repel enemies and to alert and recruit nestmates (Hölldobler & Wilson 1990, p. 260). One example is the use of volatile formic acid for both functions in *Formica* ant species (Blum 1996).

1.3.2 Pheromone signals derived from receiver sensory bias

Any secreted molecule from a sender that overlaps the receiver's pre-existing sensory sensitivities, such as for food odors, is likely to be selected over others and thus potentially become a signal (Figure 1.5) (Arnqvist 2006; Bradbury & Vehrencamp 2011, p. 391 ff.; Endler & Basolo 1998; Ryan 1998). For example, as female moths use plant odors to find host plants when egg laying, their olfactory system is already tuned to these odors: male moth pheromones appear to have evolved to exploit this female sensory bias (Chapter 3) (Figure 1.5) (Birch *et al.* 1990; Phelan 1997). The male sex pheromone of the European beewolf, *Philanthus triangulum*, (Z)-11-eicosen-1-ol, may exploit a pre-existing female sensory bias for this molecule as it is a characteristic volatile molecule given off by its honeybee prey (Chapter 3) (Kroiss *et al.* 2010; Steiger *et al.* 2011). In Iberian rock lizards *Iberolacerta cyreni* (formerly *Lacerta monticola*), a pre-existing sensory bias in females for a food chemical found in their insect prey, the lipid cholesta-5,7-dien-3-ol (provitamin D_3), may have driven selection of this molecule as a component of the pheromone secreted by males in their femoral glands (Martín & López 2008, 2010a) (though see Font *et al.* 2012). It is possible that this is also an honest signal (index) that only high-quality males can display (Section 1.6.2.1).

As well as the adaptive sensory biases above, there could also be *hidden preferences* ("receiver

psychology"), which are incidental side effects of how the sensory system is constructed (Arak & Enquist 1993; Arnqvist 2006; Bradbury & Vehrencamp 2011, p. 391 ff.; Guilford & Dawkins 1991, 1993). Such side effects include what a receiver finds easy to detect, easy to discriminate, and easy to learn. I wonder if an animal's range of olfactory receptors and olfactory "brain circuits" might lead to such effects.

Many sex pheromones that initially evolved by exploiting sensory biases may benefit the receiver, by speeding finding of a mate for example, and ritualization will refine the signal and tune its reception (3b in Figure 1.5) (Bradbury & Vehrencamp 2011). In other cases, where there are costs to the receiver in responding, there may be sexually antagonistic co-evolution in the subsequent evolutionary elaboration of sexual traits, as the receiver is selected to evade the "sensory trap" (Arnqvist 2006). In internally fertilizing species, one example may be the molecules that males pass to the female along with sperm (Chapters 3 and 9) (Arnqvist & Rowe 2005; Eberhard 2009; Poiani 2006). The sex peptides in the seminal fluid of *Drosophila* interact with the internal receptors the female uses to regulate reproductive rate and delay remating, at a potential cost to herself and a benefit in paternity to the male (Section 1.11). The great variety of sex peptides may reflect the continuing sexually antagonistic co-evolution between male and female *Drosophila*.

1.3.3 How do we know that a chemical signal has "evolved for that effect" in the sender?

The definition of a signal includes a requirement that the signal should have evolved in the sender for the effect it has on the receiver (above). In some cases we can identify evolved structures that produce the signal. For example, in moth females we can see the specialized pheromone glands evolved for secreting and releasing the female sex pheromones (Cardé & Haynes 2004; Liénard *et al.* 2010). The specific enzyme pathways for producing the pheromones are also well understood (Chapter 3). In other animals, we may not know the source of the pheromones other than a likely tissue, as

(a)

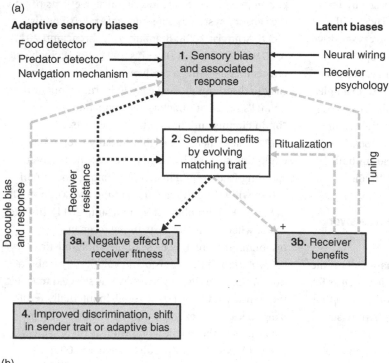

Adaptive sensory biases

Food detector
Predator detector
Navigation mechanism

1. Sensory bias and associated response

Latent biases

Neural wiring
Receiver psychology

Decouple bias and response

Receiver resistance

2. Sender benefits by evolving matching trait

Ritualization

Tuning

3a. Negative effect on receiver fitness

−

+

3b. Receiver benefits

4. Improved discrimination, shift in sender trait or adaptive bias

(b)

Figure 1.5 Signals that exploit the existing senses of the receiver will be selected for.

(a) Receiver–precursor model of signal evolution. Shaded boxes represent receiver steps, white boxes sender steps; dashed gray arrows (‑ ‑ ‑ ‑) indicate positive fitness effects, dotted black arrows (•••••) indicate negative effects. The first step (1) is the evolution of a sensory bias and coupled response in a non-communication context, such as food detection. Senders evolve a trait that matches or stimulates the sensory bias and exploit the associated response (Step 2). If the receiver benefits (Step 3b), then it may fine tune its sensory system, and the sender's trait may be ritualized to match the bias better. If the receiver experiences costs (Step 3a), it will attempt to resist exploitation by changing its sensory bias, which may be costly, and the sender may counter this move, initiating a cycle of antagonistic co-evolution. If the receiver can escape from the sensory trap (Step 4), the bias and the response will become decoupled. Figure and caption after Bradbury and Vehrencamp (2011).

(b) A male oriental fruit moth, *Grapholita molesta*, displays its hair pencils in courtship to a female. The male's hair pencils are loaded with plant-derived pheromones including ethyl *trans*-cinnamate (inset), a signal that may have evolved through sensory drive exploiting female sensitivity for odors present in their fruit food (Löfstedt *et al.* 1989). The females prefer males with the most cinnamate. Photograph by Tom Baker.

for example the male sex pheromone 3-keto petromyzonol sulphate of the sea lamprey, *Petromyzon marinus*, is produced in the liver and then released from specialized gland cells in the gills (Siefkes *et al.* 2003).

In other cases, the response to a cue is adaptive but it is not clear a signal has evolved. For example, male *Drosophila melanogaster* stop courting a female if they detect the *cis*-vaccenyl acetate left by an earlier mating male (Chapters 3 and 9). Similarly, red-sided garter snake males stop courtship if they detect volatiles from the ejaculate of a previous male (Chapter 3) (Shine & Mason 2012). In both species, this is an adaptive response from the other males as a female will not mate again for some days, given the effects of the sex peptides in the female *Drosophila* (Section 1.11) and with the copulatory plug in place in the case of the female snake. The male molecules prompting the response would certainly be cues. Whether or not you call these molecules pheromones will depend on whether the first male's molecules have evolved for this function.

All biological systems are evolving, so for a given species we may be at any point on the continuum from cue to signal. As when defining species in the process of diverging, we are "dichotomizing a continuum." In practice it may be difficult to establish that production and emission have both evolved, so I propose an operational definition for pheromones (see Box 1.4). However, for molecules to be treated as a signal under the "operational definition," there also needs to be a credible evolutionary pathway by direct or kin selection. For example, fish have evolved sensitive responses to molecules released when other fish are injured by predators. However, these molecules are probably alarm cues rather than an evolved signal as the responding fish are unlikely to be kin and may even be from a different species (Chapter 8) (Ferrari *et al.* 2010; Wisenden 2014).

1.3.4 Pheromone characteristics, transmission medium, and signal duration

Whether pheromones evolve from sender precursors or are derived from receiver sensory bias, which molecules become pheromones is also a product of the function of the message, as well as the medium the message will be carried in. For example, in air, ant alarm pheromones are volatile, with low molecular weights of between 100 and 200, diffusing rapidly and dropping below threshold quickly once the danger has passed (Chapters 8 and 10) (Hölldobler & Wilson 1990). In water, solubility of molecules is perhaps the functional equivalent of volatility in air. Aquatic pheromones range from small molecules such as the amino acid L-kynurenine used by masu salmon,

Oncorhynchus masou, as a female sex pheromone (Yambe *et al.* 2006) to polypeptides and proteins, which, despite their large size, can be highly soluble, such as those used as sex pheromones (attractin, enticin, seductin, and temptin) by the marine mollusc *Aplysia* (Cummins & Degnan 2010; Cummins *et al.* 2007).

Volatility or solubility is less important if pheromones are transferred directly from signaler to receiver: male *Danaus gilippus* butterflies drop crystals of the pheromone danaidone from their hair pencils directly onto the antennae of the female (Eisner & Meinwald 1995, 2003). The male of the terrestrial salamander, *Plethodon shermani*, directly transfers his high molecular weight glycopeptide pheromone from his chin gland to the nostrils of the female (Section 1.4.3.3) (Houck 2009; Woodley 2010).

Different durations of signal life can evolve. Whereas sound and visual signals only act at the time they are made, chemical messages can "shout" long after the signaler has left. Selection can act on the chemical characteristics of pheromones such as volatility and stability, giving signal durations from the seconds of ant alarm pheromones (above) (Chapter 8) to the months or years of some termite trail pheromones (Chapter 7) (Bordereau & Pasteels 2011). The molecules that add longevity to signals have been identified in some species. Dominant male rabbits, *Oryctolagus cuniculus*, secrete a molecule, 2-phenoxyethanol, in their chin secretion used to mark their territories (Hayes *et al.* 2003). The molecule does not seem to be part of the signal but it is used in the perfume industry as a fixative and seems to have the effect in tests of extending the life of volatile compounds in scent marks. In the male secretion of *Heliconius melpomene* butterflies a similar role is proposed for fatty acid esters that slow the evaporation of the volatile male anti-aphrodisiac pheromone, which includes (E)-β-ocimene (Chapter 3) (Schulz *et al.* 2008). Similarly, major urinary proteins in mouse urine slowly release bound volatiles, prolonging the life of signaling volatiles in scent marks from minutes to more than 24 hours (Hurst & Beynon 2004).

1.4 Pheromone diversity, specificity, and speciation

Pheromones are well known for their species specificity, with animals only responding to the pheromones of their own species. If there is such specificity, how do pheromones change as species diverge? This is explored in Section 1.4.3. First, I cover cases where pheromones are shared – in both related and unrelated species – and cases of multiple messages from one pheromone, before discussing the ways in which species can have unique pheromones. The broad-brush diversity of pheromone molecules comes from the processes of evolving from chemical cues, as molecules of all kinds are co-opted as signals (Section 1.3). A finer grain of diversity comes from the variations around a "chemical theme" as part of speciation.

1.4.1 When the same molecules are used as pheromones by different species or taxa

Species may share pheromone molecules if there is no evolutionary selection for species specificity. For example, there is usually little need for privacy in communication for alarm pheromones, and in ants these are often shared by related species (Chapter 8). Similarly, the alarm pheromone (E)-β-farnesene is shared by aphid species across more than 30 genera (Byers 2005) whereas aphid sex pheromones are species-specific multicomponent blends (Dewhirst *et al.* 2010). Oviposition pheromones in *Culex* mosquitoes, which lead other females to lay near previously laid egg masses, seem to show cross-species attraction, perhaps because the benefits of predator dilution are not species specific (Chapter 4) (Seenivasagan & Vijayaraghavan 2010). Larval lampreys of different species appear to release a common pheromone, petromyzonol sulfate and allocholic acid, to which adults of other lamprey species are attracted (Chapter 12) (Fine *et al.* 2004). Moth species that live in different places may share a pheromone blend as they will not meet. If moth species live in sympatry (in the same geographic area) they can have the same

pheromone blend so long as they use different calling times or host plants to avoid cross-attraction (Chapter 3) (Cardé & Haynes 2004).

A different phenomenon occurs with pheromones shared not because of near-relatedness but by convergence. Across the animal kingdom, species that are not closely related may share the use of a molecule as a signal, illustrating the independent evolution of particular molecules as signals (Table 1.2) (Kelly 1996; Novotny 2003); for example, variations of the terpene brevicomin are used by male house mice and some bark beetle species (Novotny, 2003). The Asian elephant female pheromone, (Z)-7-dodecen-1-yl acetate is a component of the female pheromone blend of some 140 species of moth, and the Asian male elephant's pheromone frontalin is also used by some bark beetles (Rasmussen *et al.* 1997, 2003). The use of the same molecules may reflect some constraints on the number of low molecular weight molecules that are volatile, stable, and relatively non-toxic. Such coincidences are also a consequence of the common origin of life: basic enzyme pathways are common to all multicellular organisms and most classes of molecule are found throughout the animal kingdom.

1.4.2 Different messages from the same pheromone molecules

Pheromonal parsimony, a species taking different meanings from the same molecule at different concentrations and/or different social contexts, is found in many animals and is common in social insects (Blum 1996; Bordereau & Pasteels 2011; Hölldobler & Wilson 2009, p. 179). Perhaps, once an animal has the receptors and neural circuitry for a specific pheromone, these can be co-opted by other communication needs. For example, in a number of termite species such as *Pseudacanthotermes spiniger*, the same molecule is used at low concentrations as a trail-following pheromone by foragers and by sexual males during tandem running (Bordereau & Pasteels 2011). When released at higher concentrations by female reproductives, it attracts males from long distances and elicits typical sexual excitement behaviors when males contact it.

The phenomenon is also well illustrated by the mandibular pheromone of queen honeybees, which is a sex pheromone for males (drones) and, with additional pheromone components, has a releaser effect in attracting workers as the retinue pheromone and also has primer effects suppressing worker reproduction (Chapters 6 and 9) (Section 1.9) (Grozinger 2013; Kocher & Grozinger 2011; Slessor *et al.* 2005). The primer and releaser effects may act via different receptors and nerve circuits (Kocher & Grozinger 2011). In the nematode *Caenorhabditis elegans*, overlapping sets of ascaroside molecules are the sex pheromone active at picomolar concentrations and, at about 10,000 times higher concentrations, the dauer pheromone, which induces a resistant resting stage in larvae (Figure 1.6) (Pungaliya *et al.* 2009; Srinivasan *et al.* 2008, 2012).

Pheromonal parsimony also occurs in mammals. The same molecules, such as male mouse pheromones, can have different effects on other males and on females (see Chapter 9). Rabbit mammary pheromone (Box 1.5) elicits suckling responses from pups and also stimulates learning of any co-occurring odorant such as the mother's odors (Coureaud *et al.* 2010).

1.4.3 Specificity and the evolution of pheromones

There are two main ways of gaining specificity in pheromone signals, making a unique signal. One, less common, way is to use a single unusual molecule (see below). The alternative, found in most species across the animal kingdom, is to use a **multicomponent pheromone**: a particular combination of molecules, which individually may not be unusual and may overlap with those used by related species. The combination makes the pheromone unique.

In most biological signaling systems, sexual selection (Chapter 3) leads to species-specific sex pheromones and responses, important for pre-mating isolation and speciation in both vertebrates and invertebrates (Smadja & Butlin 2009). Chemical communication achieves specificity in different ways from visual and acoustic communication, which are continuous spectra, varying in wavelength and temporal structure. By contrast, molecules can differ in many dimensions including stereochemistry. Stereoisomers are molecules that have the same atoms connected in the same order but differ in the arrangement of atoms in space, changing the shape of the molecule (Appendix). For both a chemosensory receptor detecting a molecule, and the enzymes synthesizing it, a molecule's shape is a key part of interacting with it (Chapter 9) (Reisert & Restrepo 2009). As a result, stereoisomers are usually treated by receptors as different molecules (so proper chemical identification of pheromone molecules must include stereochemistry). Some stereoisomers are enantiomers, mirror images of each other (said to be chiral, from the Greek meaning hand) (Mori 2007). Some pairs of species gain specificity by using different enantiomers of the same compound; for example, among sympatric scarab beetles in Japan, the Japanese beetle, *Popilla japonica*, uses (S)-japonilure as its female sex pheromone whereas the Osaka beetle, *Anomala osakana*, uses (R)-japonilure (see Appendix for notation) (Leal 1999).

1.4.3.1 Single unique molecule pheromones

A few species have a pheromone consisting of a single unusual molecule: for example, the female sex pheromone of the brown-banded cockroach, *Supella longipalpa*, is the single unique molecule supellapyrone (Gemeno *et al.* 2003). Most other cockroach species use multicomponent pheromones (Gemeno & Schal 2004) (see below). Animals using peptide pheromones can evolve peptides with amino acids in unique combinations and sequences. For example, the decapeptide pheromones of the related species of Japanese newt, *Cynops ensicauda* and *C. pyrrhogaster*, differ by just two amino acids (Chapter 3) (Toyoda *et al.* 2004). There are some single small-molecule mammal pheromones such as the rabbit mammary pheromone, 2-methylbut-2-enal (Box 1.5) (Schaal *et al.* 2003).

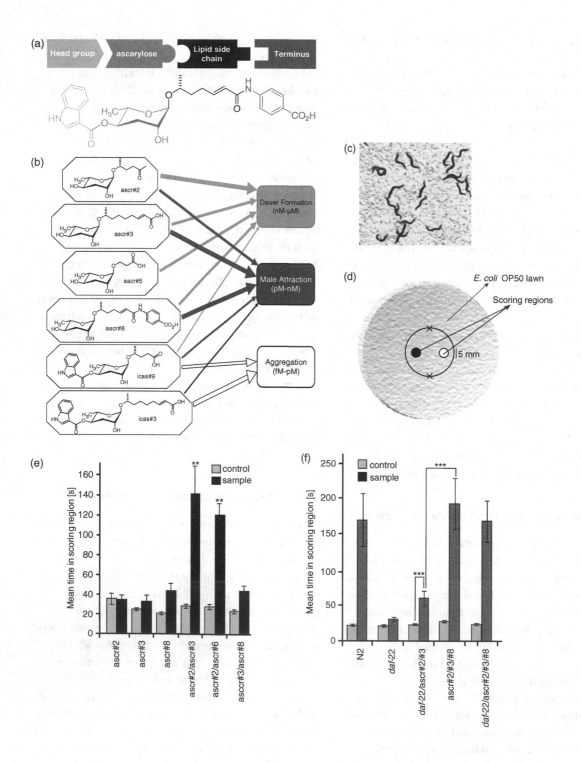

However, I think it is likely that when more mammalian sex pheromones are identified it may emerge that, if they are not peptides, many gain species specificity by being multicomponent (next section).

1.4.3.2 Multicomponent pheromones

Most pheromone specificity is achieved by using a combination of compounds as a multicomponent pheromone that only works as a whole (synergy, see below). (These are not the same as signature mixtures; Table 1.1.) The molecules of a multicomponent pheromone need not be unusual themselves. It is the particular combination that gives specificity. For example, female sex pheromones in moths usually consist of multicomponent blends of five to six hydrocarbons (10 to 18 carbons long) including unbranched fatty acids, alcohols, acetates, or aldehydes in particular combinations and ratios (Chapter 3) (Cardé & Haynes 2004; de Bruyne & Baker 2008). Because of the way that odor signals are carried in the wind, all the molecules travel together, so the whole multicomponent blend is perceived by a responding male, even far downwind (Chapters 3 and 10) (Linn & Roelofs 1989; Linn et al. 1987).

Vertebrate pheromones may be multicomponent too. Two compounds isolated from the urine of male mice, *Mus musculus*, provoke aggressive behavior in conspecific males: dehydro-*exo*-brevicomin and 2-*sec*-butyl-4,5-dihydrothiazole (Chapter 9) (Novotny 2003; Novotny et al. 1999b). For this effect, both compounds have to be present together and in addition they need to be presented in mouse urine. The closely related species of goldfish, *Carassius auratus*, and carp, *Cyprinus carpio*, share a set of five hormonal compounds that mediate pre-spawning hormonal surges and reproductive behavior (Figure 1.4) (Stacey & Sorensen 2009), but these form species-specific multicomponent pheromones with other molecules, as yet unidentified, so cross-attraction does not occur (Levesque et al. 2011; Lim & Sorensen 2012). Nematode worms from several different clades (branches) produce species-specific but partially overlapping mixtures of ascarosides, a family of molecules unique to nematodes (Figure 1.6) (Choe et al. 2012). These multicomponent pheromones mediate a variety of nematode behaviors including avoidance, sex, developmental diapause, and long-range attraction. Social insect pheromones of many kinds are multicomponent and the components for a pheromone

Figure 1.6 Multicomponent pheromones and synergy. Sex, aggregation, and dauer (resting stage) pheromones in the nematode *Caenorhabditis elegans* are made up of overlapping combinations of related molecules whose action also depends on concentration (a, b). Synergistic blends of non-indole ascarosides induce dauer at nanomolar to micromolar concentrations and function as a male attractant at picomolar to nanomolar concentrations, whereas indole ascarosides icas#3 and icas#9 act as hermaphrodite attractants and aggregation signals at femtomolar to picomolar concentrations (Srinivasan et al. 2012). (c) *C. elegans* on a gel in a bioassay (Srinivasan et al. 2012). Each worm is about 1 mm long.

Synergy: individual components of the sex pheromone are no more attractive to males than the control (e) when presented singly in the bioassay (d), crosses mark the starting positions of assayed animals (Pungaliya et al. 2009). However, when particular combinations of two ascarosides (ascr#2 and ascr#3) and (ascr#2 and ascr#8) are presented together, a strong attraction is observed, right columns in (e). This effect is synergy: the components together are the message. In (e) ascr#2 and ascr#8 were tested at 100 fmol and ascr#3 at 10 fmol. At these concentrations, ascr#3 and ascr#8 did not show a strong synergy.

(f) Wild-type (N2) *C. elegans* metabolite extract has strong male-attracting activity, whereas mutant daf-22 metabolite extract is inactive. A mixture of ascr#2 and ascr#3 in amounts corresponding to those present in the wild-type metabolite extract (20 fmol each) added to the inactive daf-22 metabolite extract partially restores activity but full male attraction is restored by adding ascr#2, ascr#3, and ascr#8 (20 fmol of each). Adding daf-22 metabolite extract does not further increase activity.

See Chapter 2 for the way ascr#2 and ascr#3 were identified by activity-guided fractionation, and how ascr#8 and the indole ascarosides were discovered by a different technique (Pungaliya et al. 2009; Srinivasan et al. 2008, 2012). Another naming convention uses "C1" etc., for these molecules, see Edison (2009) for a key.

can come from different glandular sources and different families of molecules, by different enzyme pathways (see Chapters 6, 7, 8, 9, and 10). The ant trail pheromone in one species, *Leptogenys peuqueti*, consists of a blend of as many as 14 compounds (Morgan 2009). Social insects may use components in slightly different combinations for co-ordination of colony dynamics (Chapters 6 and 9) (Section 1.4.2) (Le Conte & Hefetz 2008; Slessor *et al.* 2005).

The prevalence of multicomponent pheromones may reflect how pheromone signals diverge in speciation (Chapter 3). Within closely related taxonomic groups of moths (families or subfamilies), species often use combinations of the same or similar components, as a result of sharing biosynthetic pathways by ancestry (Cardé & Haynes 2004; de Bruyne & Baker 2008; Symonds & Elgar 2008). Even where unusual molecules are used as pheromones, closely related species tend to use variations on these, as if exploring chemical space from a new starting point. In cockroaches, each of the long-range sex pheromones identified to date from different cockroach genera belongs to a different chemical class, with species in each genus using different combinations of variations of the unusual molecule (Eliyahu *et al.* 2012; Gemeno & Schal 2004). For example, *Periplaneta* species use different combinations of molecules based on the unusual molecule periplanone. A recent identification follows the same pattern, with a previously unidentified natural product and a previously unknown pheromonal structure for cockroaches found for the main pheromone component of the cockroach *Parcoblatta lata*, a macrocyclic lactone, (4Z,11Z)-oxacyclotrideca-4,11-dien-2-one (Eliyahu *et al.* 2012). This molecule also forms a component of the pheromones of related species in the genus.

Synergy: a natural outcome of multicomponent pheromones

Synergy describes the phenomenon when any one component shows little or no activity by itself and only the complete synthetic mixture has an activity comparable to the pheromone (Figure 1.6). This is how multicomponent pheromones are detected. The discovery of nematode sex pheromones is a classic example. The multiple components of the *C. elegans* sex pheromone were revealed during activity-guided fractionation (Chapter 2), as none of the fractions showed activity when tested alone, activity came only when brought together in combination. This indicated that active components were split between the fractions (Figure 1.6) (Srinivasan *et al.* 2008).

Synergy is to be expected from multicomponent pheromones, which gain their specificity by the combination (above). I would suggest that synergy is a natural outcome of the way multicomponent pheromones are processed in the brain. It reflects the combinatorial organization of olfaction (see Chapter 9). For example, in the male moth, the message "fly upwind" in response to female pheromone is only sent to the higher brain if all the correct molecules stimulate their antennal olfactory sensory neurons and the glomeruli in the brain to which these lead (Chapter 9) (Haupt *et al.* 2010). The neural circuits can be thought of as acting like digital logic "AND" gates: if a component is missing or at the wrong ratio, the stimulus does not go higher in the brain. Conversely, the circuit gives a "STOP" if there is activation of olfactory sensory neurons sensitive to a pheromone component of the wrong species (e.g., Lelito *et al.* 2008). Nematode multicomponent pheromones are processed by simpler circuits, without glomeruli, but on these same principles.

1.4.3.3 How does evolutionary change in pheromones occur?

The details of speciation and pheromone evolution have been explored in moths, allowing us to dissect the genetics of both pheromone production and signal reception (the genes for chemosensory receptors and neural wiring). We know less about the evolution of vertebrate pheromones. However, detailed studies of North American salamanders show rapid and sometimes cyclical changes in their protein courtship pheromones. In mice the evolution of some of the sex-dependent changes in expression of enzymes that lead to production of a male chemosignal, trimethylamine, has been explored along with its specific receptor (trace amine-associated receptor 5, TAAR5) (Li *et al.* 2013). I will start with moths then turn to the salamanders (see Chapters 3 and 9 for more details about moth speciation and also the evolution of pheromones in *Drosophila* species).

Signal divergence with new multicomponent pheromone blends in insects (with new ratios or presence or absence of components) can occur either with changes in a small number of genes or, in polygenic systems, changes in many genes (Cardé & Haynes 2004; Symonds & Elgar 2008). The changes in pheromone production can involve *cis*-regulatory DNA sections controlling gene expression or changes within genes leading to changes in enzyme binding sites and thus substrate specificity. Both kinds of changes ultimately affect which pheromone molecules are produced and in what ratios and quantities.

A small number of genes affecting substrate specificities of the pheromone-producing enzymes separate two strains of the European corn borer moth, *Ostrinia nubilalis* (Chapter 3). Females of the two strains produce, and respective males respond to, different ratios of the same components of the pheromone: the Z-strain uses a 97:3 mix of (Z)-11-tetradecenyl acetate (11–14:0Ac) and (E)-11-tetradecenyl acetate whereas the E-strain uses a 1:99 mix of Z/E 11–14:0Ac. The change in blend can be traced to alleles that give different versions of an enzyme in the biosynthetic pathway (Chapter 3) (Lassance 2010; Lassance *et al.* 2010).

The Asian corn borer moth, *Ostrinia furnacalis*, diverged from the common ancestor shared with *O. nubilalis* about a million years ago. One suggestion is that there was a "resurrection" of a long dormant desaturase gene, for an enzyme that changed the position of a double bond in the pheromone above (Roelofs & Rooney 2003; Roelofs *et al.* 2002). An alternative possibility is that the desaturase gene was always active but may have changed from being expressed in males, as in *O. nubilalis*, to being

expressed in females in *O. furnacalis*, changing the female blend (Chapter 3) (Lassance & Löfstedt 2009).

However, in many moth species, the differences between multicomponent pheromone blends of females result from polygenic changes, such as in the related sympatric species *Heliothis virescens* and *H. subflexa* (Chapter 3) (Groot *et al.* 2009). In these *Heliothis* species, quantitative trait locus (QTL) studies showed that genes on at least nine of the 31 *Heliothis* chromosomes contribute to the pheromone differences between the species, which may also involve genes other than those encoding the enzymes themselves.

The change leading to the Asian corn borer, *O. nubilalis*, pheromone has been called a "saltational" shift, and it has been suggested that "sudden major switches in pheromone blend and male response appear more likely than accumulation of small changes" (Roelofs *et al.* 2002). However, the basis for suggesting a greater likelihood of major "saltational" shifts seems to rest on one simulation model (Butlin & Trickett 1997). While a comparison of aggregation pheromones in some bark beetles could fit this idea, the pattern of aggregation pheromones in *Drosophila* species appeared to support gradual shifts (see Symonds & Elgar 2008). My own feeling is that the general pattern is likely to be gradual change as evidenced by the radiation of related molecules as pheromones within genera discussed in moths and cockroaches (Section 1.4.3.2). The polygenic changes, such as those in *Heliothis* moths, above, also suggest that modifiers and gradual changes are often at work (Chapter 3). Dramatic "saltational" changes may be uncommon. When they occur they are simply part of a continuum of change, and a large change in blend can just reflect a genetic change in an enzyme high "upstream" in the biosynthetic pathway (see Figure 3.15).

When pheromone blends change, will any males respond to the new pheromone blend? The responses of males, in moths at least, seem wide enough to cover some changes (Chapter 3) (Martin *et al.* 2011a). A screening of European corn borer, *Ostrinia nubilalis*, males in the laboratory showed that some rare, broadly tuned individuals would fly upwind to the new blend of the Asian corn borer, *O. furnacalis*, as well as to the blend of their own females (Chapter 3) (Linn *et al.* 2003; Roelofs *et al.* 2002). Similar results were found with some male cabbage looper, *Trichoplusia ni*, moths responding to a novel pheromone blend produced by mutant females (Cardé & Haynes 2004; Domingue *et al.* 2009) (Chapter 3). Over the generations, *T. ni* males with a greater response to the mutant blend could be selected for in the laboratory. Changes can be at the level of olfactory receptor sensitivity but also in the wiring in the brain. In the European corn borer moth, *Ostrinia nubilalis*, the "wiring" of the pheromone circuits of the brain of males in the two strains with opposite ratios of Z/E 11–14:OAc is simply mirrored (Chapter 9) (Karpati *et al.* 2008, 2010).

Salamanders

The evolution of courtship behavior, morphology, and male pheromones in North American plethodontid salamanders shows change on long and short time scales (Figure 1.7) (Houck 2009; Woodley 2010). The male courtship pheromone, which includes three unrelated proteins, is produced by his chin ("mental") gland and increases the receptivity of the female, shortening courtship time. Males in most of the 300 or so species show the ancestral courtship behavior that emerged about 100 million years ago: they deliver the pheromone to the female by depositing the chin secretion on the female's back while simultaneously scratching her skin with enlarged pre-maxillary teeth. The pheromone passes through the skin into the capillary blood system. About 19 million years ago, one clade (branch) of salamanders, now represented by about 30 species of *Plethodon*, evolved a different delivery route, and the male instead taps his chin gland directly on the female's nostrils, delivering the pheromone to the vomeronasal olfactory system (VNO) (a "second nose" that many terrestrial vertebrates have) (Chapter 9) (Figure 1.7). Along with the change in behavior, the males of these species also lost their elongated teeth.

The relative stability of courtship behavior over millions of years on either side of the major changes in delivery contrasts with the repeated, periodic episodes of rapid molecular evolution and diversification of the pheromone in many species, driven by positive selection acting on one or more of the three proteins of the

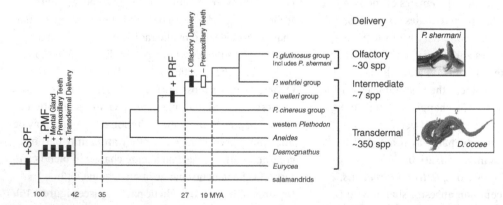

Figure 1.7 The evolution of courtship pheromone delivery in plethodontid salamanders in North America. Ancestrally, all the plethodontid salamanders had the sodefrin-like precursor factor (SPF) protein pheromone and the plethodontid modulating factor (PMF) protein produced in the mental (chin) gland, protruding pre-maxillary teeth and scratching (transdermal) delivery of courtship pheromones. Later, in some clades another protein, plethodontid receptivity factor (PRF), and, later still, olfactory delivery of courtship pheromones and loss of the pre-maxillary teeth evolved. Photographs to the right show olfactory pheromone delivery by the red-legged salamander, *Plethodon shermani*, and transdermal delivery by the Ocoee salamander, *Desmognathus ocoee*. Photographs Stevan J. Arnold. Figure adapted from Woodley (2010) and data from Kiemnec-Tyburczy *et al.* (2011). The cladogram shows the evolution of characters (for more details see Palmer *et al.* 2007a). The phylogeny at group level is still in flux, with some phylogenies making the *P. wehrlei* and *P. welleri* groups into sister groups.

pheromone (Figure 1.7) (Houck 2009; Palmer *et al.* 2010; Woodley 2010). Comparisons of DNA changes across 27 species for one of the proteins, plethodon receptivity factor (PRF), showed that some *Plethodon* lineages had neutral divergence and purifying selection with little change over time (Palmer *et al.* 2005, 2007b). Other lineages showed rapid, repeated, cyclical evolution driven by positive selection, probably resulting from sexual selection leading to co-evolution of the male pheromone variants with VNO receptors in the female (it is supposed). Several of the varying codons appear to be involved in a "molecular tango" in which the male signal and female receptors co-evolve on a "dance floor" constrained by the limited number of allowable amino acid substitutions that still allow the pheromone protein to interact with the receptor (Palmer *et al.* 2005). The same mutations seem to come and go cyclically, over time. The "molecular tango" is likely driven by sexual selection, which may be female preference or, perhaps, sexual conflict as in the fast evolving sex peptides in *Drosophila* (Chapter 3) (Sections 1.3.2, 1.11) (Arnqvist 2006). The other characteristics of the tango include gene duplication, hyperexpression in the mental (chin) gland, and abundant polymorphism within populations arising from the tendency to both retain and reinvent sequence variants (Palmer *et al.* 2010; Woodley 2010). A similar and more extreme pattern of change over time and polymorphism within individual males is shown for a second protein in the pheromone, plethodontid modulating factor (Palmer *et al.* 2010; Wilburn *et al.* 2012). What we don't have yet is the receptor side of the story for any of these pheromone proteins (unlike the moths).

1.5 Production of pheromones

Most pheromones are synthesized and secreted by the signaler, often from specialized glands (see Section 1.7 for signature mixtures). However, as long as they are a consistent signal across a species, pheromone molecules or precursors may be collected rather than synthesized from scratch (hence I have changed the verb in the definition in Box 1.2 to *emit* rather than secrete). For example, specialist moth and butterfly species (Lepidoptera) harvest pyrrolizidine alkaloids (PA) from plant species containing them (Boppré 1990; Conner 2009). In some specialist lepidopteran species only the larvae sequester the alkaloids; in others, such as the milkweed danaine butterflies (Nymphalidae), adults also feed on these PAs. Courtship in PA-sequestering species usually involves presentation to the female of derivatives of these alkaloids. Males without evidence of chemical gifts are rejected (see Chapter 3) (Section 1.6.2). Likewise, to display successfully and attract females, male euglossine orchid bees in the American tropics must fill specialized hind-leg pockets with fragrances such as limonene from orchid flowers and other sources (Ramírez *et al.* 2011; Zimmermann *et al.* 2009). The males get their species-specific pheromone mix by collecting from flowers of the correct orchid species. You could say orchid bees use takeaways rather than cooking for themselves.

So long as the molecules are consistent across the species, animals may use molecules produced by bacteria as pheromones. Among locust phase change pheromones are guaiacol (2-methoxyphenol) and phenol, produced by locust gut bacteria (Box 4.1) (Pener & Simpson 2009).

The independent and multiple evolution of pheromones is illustrated not only by the diversity of molecules used (Section 1.4) but also by the enormous variety of specialized secretory glands used to produce them. Among male mammals and male Lepidoptera (moths and butterflies) the variety is probably largely the result of sexual selection (Chapter 3) (Andersson 1994; Darwin 1871).

There is an enormous variety of glands and secretions across the social insects (Box 6.1) (Billen 2006). Genes associated with gland development are among the most rapidly evolving genes across eusocial bees and may be related to the convergent evolution of advanced systems of chemical communication used to organize eusocial colonies (Chapter 6) (Woodard *et al.* 2011). The diversity of hundreds of molecules produced by ants has been termed chemical sorcery for sociality

(Morgan 2008) and it is matched by the diversity of glands involved: more than 40 anatomically distinct exocrine glands have been found so far across the ants (Billen 2006; Hölldobler & Wilson 2009, p.180). The same gland may produce different molecules in different castes of the same species (Box 6.1) (Grozinger 2013; for example the queen and worker honeybees, Kocher & Grozinger 2011; Le Conte & Hefetz 2008; Slessor *et al.* 2005). The components of a social insect multicomponent pheromone can come from different glandular sources and different families of molecules, by different enzyme pathways.

1.6 Pheromones: signal honesty and costs

What is to stop a subordinate male mouse giving off the pheromones of a dominant male? This is a long-running question in animal communication: what keeps signals honest or reliable, so that the receiver can rely on the signal to reflect the real quality of the signaler? Generally, it seems that intra-specific signals are honest[1] (Bradbury & Vehrencamp 2011, p. 397; Greenfield 2006). What keeps them honest?

In many research papers you will read statements along the lines of "signals must be costly to make them honest." This is not true. It comes from widespread misinterpretations of the literature about animal communication (Maynard Smith & Harper 2003; Számadó 2011a, b). This is not to say that signaling cannot be costly – it can be, and I give some examples below. However, the "must be costly" statement refers instead to a theoretical idea, the handicap principle, an idea that is starting to be questioned again (see Box 1.6). Honest signals do not necessarily need to be costly. Showing that a signal has a cost does NOT demonstrate a handicap (Számadó 2011a,b).

[1] Reliable or "honest" signals reveal the relevant quality of the signaler to the receiver, with the intensity of signal reliably correlated with the quality (Maynard Smith & Harper 1995; Számadó 2011a,b).

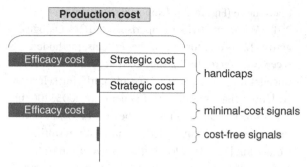

Figure 1.8 Signal types as a function of the cost of producing them. From the bottom up, signals with almost zero efficacy cost (the minimum cost needed to ensure the information can be reliably perceived) are called "cost-free signals." An example would be the individual body odor, used by other animals to recognize an individual. "Minimal-cost signals" have only efficacy cost. This could be the energetically expensive secretion of proteins into mouse urine for marking of territories (but no more expensive, though, than it needs to be for efficacy). Handicaps have wasteful "added cost" ("strategic cost") on top of whatever cost is needed for efficacy (getting the message across). We currently have no experimental way of separating efficacy and strategic costs of a given signal and thus no evidence of a strategic cost. Figure and caption after Számadó (2011b). Terminology from Guilford and Dawkins (1991); Maynard Smith and Harper (1995, 2003).

I need to explain briefly what costs we are talking about. The straightforward cost of signaling is called the "*efficacy cost*," the minimum cost needed to ensure the information can be reliably perceived by the receiver, for example a cricket song loud enough for a female to hear (Figure 1.8) (Guilford & Dawkins 1991; Maynard Smith & Harper 1995, 2003). Some signals are effectively free, with an efficacy cost of almost zero. The "*handicap cost*" (also called the "*strategic cost*") is the idea of a specifically wasteful cost on top of any straightforward efficacy cost of signaling (Box 1.6) (Figure 1.8).

1.6.1 Efficacy costs of pheromones compared with other modalities

How best to measure costs is itself a major question: a signal could take a lot of energy but have little fitness

Box 1.6 **The problems of signaling costs and the handicap principle**

In this box, I explain the more technical background to the conclusions that (1) honest signals do not necessarily need to be costly and that (2) showing a signal has a cost does not demonstrate a handicap (Számadó 2011a,b).

Zahavi's (1975) counter-intuitive "handicap" idea was that signalers, even honest ones, need to pay an extra wasteful cost, in addition to the efficacy cost (the simple cost of making a signal that can be perceived by the receiver), to ensure a signal is honest: "waste can make sense, because by wasting one proves conclusively that one has enough assets to waste and more. The investment – the waste itself – is just what makes the advertisement reliable" (sic) (Zahavi & Zahavi 1997, p. 229) (Figure 1.8).

Grafen's (1990a) models showed, but only for signals between animals that have conflicting interests, that Zahavi's ideas could work in a model of evolutionarily stable strategies (ESS): an additional (strategic) cost for honest signalers at the ESS equilibrium makes their signals reliable indicators of quality, and it costs a better male less to make the same signal (a differential cost). Despite the limited scope of Grafen's model support, restricted to signals between animals with a conflict of interest (Grafen 1990a, p. 530), Zahavi (2008, p. 2) has claimed "the handicap principle is an essential component in all signals" (and similarly in Zahavi & Zahavi 1997, pp. 40, 229–30).

However, there are now many alternative models, of different kinds or using different assumptions, which show that cost-free or efficacy-cost-only honest signals can evolve without the need for handicap costs, even for unrelated individuals with conflicting interests. Számadó (2011a,b) notes these alternative models have variously shown (1) that differential cost criteria are neither necessary nor a sufficient condition of honest signaling (Getty 2006); (2) that higher quality signalers need not waste more at the equilibrium than lower quality ones (Getty 2006); (3) that it is the weak signalers that will use the costlier signal, and not the strong signalers, if there are no alternatives (Hurd 1997); and that (4) honest equilibrium signals need not be handicaps (Bergstrom & Lachmann 1998; Bergstrom *et al.* 2002; Hurd 1995; Lachmann *et al.* 2001; Számadó 1999, 2003, 2008).

The assumption that all signals have to be wastefully costly to be honest has dominated discussion of animal communication and it has in turn skewed the investigation of costs in pheromones (Section 1.6). Alternative models (above) (reviewed by Számadó 2011a,b), which support non-handicap solutions tend to be ignored in standard texts on animal communication (see Further reading). This puts pheromone researchers at a disadvantage as they rely on these accounts to understand the theoretical basis of communication and how it might relate to pheromones. Since the challenges to the handicap idea are rarely mentioned, it is no surprise that individual researchers and reviews of pheromone signals tend to accept the handicap principle's ideas without reservations. This has led many researchers, despite the evidence, to conclude mistakenly that any pheromone costs they find must be "handicaps."

Box 1.6 (cont.)

Despite "strategic costs" being a crucial part of the handicap principle, there is currently no methodology for splitting the costs of a signal into its efficacy costs (just to get the message out) and strategic ones (the added wasteful costs for a handicap) (Számadó 2011a,b). Indeed the predictions of the handicap model and index models cannot be separated in many experimental systems.

The arguments above do not rule out the possibility that the handicap model could apply in some situations but I think the blanket "signals must be costly to be honest" statement is surely no longer useful.

cost in evolutionary terms such as survival or future mating opportunities (Clark 2012; Kotiaho 2001; Moreno-Rueda 2007). However, given the difficulty of measuring fitness costs, energy costs are most commonly measured.

The size of the efficacy cost, needed simply to convey the information (Figure 1.8) (Guilford & Dawkins 1991; Maynard Smith & Harper 1995, 2003), depends in large measure on the modality of the signal (whether it is using sound, light, or chemicals for example). Creating acoustic signals takes muscular activity and typically the cost is ~8 times higher than resting metabolic rate in ectotherms such as insects and amphibians, and ~2 times higher in birds (Ophir et al. 2010). Trilling male katydids (Orthoptera) have among the highest energy consumptions per unit mass of any acoustic signaler (Stoddard & Salazar 2011). These high costs are reflected, for example, in male crickets devoting up to half their daily respiratory budget to acoustic signaling (Prestwich & Walter 1981).

In contrast, the metabolic cost of most pheromone signaling is likely to be low compared with that of other signals, in part because the quantities of material needed are so small and because, generally, pheromones are released into the wind or current for passive transport, not actively pushed by muscle action to the receiver (Chapter 10). For example, just 40 nanograms of the peptide pheromone of the magnificent tree frog, *Litoria splendida*, released

into the water one meter from a female will attract her to the source in minutes (Wabnitz et al. 1999). The costs of production are probably similarly low for many invertebrate pheromones. The lifetime cost to a male boll weevil beetle, *Anthonomus grandis*, to produce its monoterpene sex pheromone is estimated at only 0.2% of its body weight (Hedin et al. 1974). Male Caribbean fruit flies, *Anastrepha suspensa*, can have their pheromone production doubled by application of synthetic hormone (methoprene) (Teal et al. 2000). In laboratory experiments this doubled their sexual success, without an increase in mortality (Pereira et al. 2010b). Adding protein to their sucrose diet similarly doubled pheromone production and these effects were additive, so combined methoprene and protein supplement produced males four times more sexually successful than untreated males (Pereira et al. 2010a) (incidentally this shows the positive effect of condition on pheromone production). Why then do males not already double their pheromone production? The cost of a small quantity of juvenile hormone (JH) does not seem a likely reason (Pereira et al. 2010b). Rather, outside the laboratory, perhaps exaggeration is prevented by the cost imposed by predators that are attracted by releasing more pheromone (Section 1.6.2.5).

Nonetheless, pheromone signals *can* have significant efficacy costs. For example, some mammals spend significant amounts of energy on carrier proteins for pheromone signals: in mice, the territorial male's urine marks contain 20 to 40 mg ml^{-1} of protein, largely major urinary proteins (MUPs) (Box 5.2) (Hurst & Beynon 2004). Marking with such large amounts of protein may have significant metabolic costs, reflected in lower rates of growth compared with animals marking less (Gosling et al. 2000). The MUPs bind the small molecule pheromones, thiazole and dehydro-exo-brevicomin, slowly releasing them and thus increasing the attractive volatile lifetime of the signal from minutes to perhaps 24 hours. One of the MUPs, darcin, which binds thiazole, is a pheromone in its own right (Box 1.5) (Chapter 9) (Roberts et al. 2010).

Efficacy costs could also include such things as the time for a territory owner to revisit and maintain its scent marks (Chapter 5). Similarly, in species that do not synthesize their pheromones themselves, efficacy costs could include the time and energy used to collect plant materials used as pheromones or pheromone precursors (Section 1.5).

1.6.2 Reliable signals without handicap

There are many ways for reliable signals to evolve, without the need for handicaps (Grafen 1990a; LaPorte 2002; Maynard Smith & Harper 1995, 2003; Számadó 2011b). These include index signals (plural: indices), individual recognition, shared interest, and punishment of cheaters. A given signal could involve more than one mechanism. None of these mechanisms require costs beyond the efficacy cost: no wasteful handicaps are needed.

1.6.2.1 Indices: unfakeable signals

An index signal is one that cannot be faked. Its reliability is maintained by a mechanistic link (physical connection) between signal intensity and a quality characteristic of the signaler (Maynard Smith & Harper 1995, 2003). It has an inherent honesty that makes it unfakeable. For example, male giant pandas,

Ailuropoda melanoleuca, do a handstand to get their urine marks as high as possible: only a genuinely large panda can get its mark high on a tree (Nie et al. 2012; White et al. 2002). Mouse territory scent markings are an honest index of territory ownership, as only the owner can exclusively cover the territory with his urine marks (Chapter 5) (Roberts 2007). A subordinate male, even if he produced the pheromones of a dominant male, could not hold and mark a territory.

The quantity of pheromone produced by animals may be an index reflecting quality, leading to "the success of the smelliest" (Chapter 3) (Wyatt 2009). For example, female tiger moths, *Utetheisa ornatrix*, choose a male with the most pheromone (Chapter 3). His pheromone is derived from a proportion of the alkaloid plant poison store that he will pass to the female at mating, and which she will use to protect the eggs (Section 1.5). His pheromone load is correlated with the alkaloid gift he will give (Chapter 3) (Kelly et al. 2012).

Among fish, male peacock blennies, *Salaria pavo*, offer parental care of eggs. The male blennies advertise with a pheromone produced by the same gland that produces protective protein secretions for the eggs (Chapter 3) (Barata et al. 2008a; Serrano et al. 2008). Bigger glands produce both more pheromone and more protein secretions (see also below).

Other indices may be related to body condition, reflecting environmental factors such as nutrition as well as genetic background (Chapter 3) (Cornwallis & Uller 2010; Pizzari & Bonduriansky 2010). Links between condition and quality can simply reflect efficacy costs: handicaps are NOT necessarily required (Getty 2006; Hill 2011; Maynard Smith & Harper 2003). The links could involve trade-offs of energy allocation (perhaps mediated by hormones – which need not be a handicap), or shared pathways for pheromone production and vital physiological processes (Hill 2011). There are many examples of pheromones related to condition that influence mate choice. For example, the attractiveness of a male *Nauphoeta cinerea* cockroach is increased with better body condition (influenced by greater carbohydrate intake), because he produces more pheromone (South *et al.* 2011). Meadow vole, *Microtus pennsylvanicus*, males on a higher protein diet produced more attractive chemosignals in their urine marks (Ferkin *et al.* 1997; Hobbs & Ferkin 2011). In rock lizards, *Iberolacerta cyreni*, the proportion of oleic acid, attractive to females, in a male's scent marks is dependent on his body condition (Martín & López 2010a,b) (though see Font *et al.* 2012). Chapter 3 gives more details of these and other examples.

1.6.2.2 Individual or colony identity

The signals and cues that allow individual or colony recognition are expected to be cheap and not related to condition (Tibbetts & Dale 2007). These can be cost free or minimal-cost signals (Figure 1.8) (Section 1.7). The molecules may even be cues rather than signals, using co-opted variation, as in molecules associated with the MHC. These are likely to be largely cost free. In some animals, the molecules may be evolved signals produced for recognition, as may be the case for part of the chemical profile of CHCs in social insects, which form the major part of the signature mixtures learned by nest mates (Section 1.7) but there is no need to assume a high cost for these (Tibbetts & Dale 2007).

1.6.2.3 Shared interest and relatedness

Signals between animals with a shared interest can be honest at minimal or no cost (Maynard Smith 1991; Maynard Smith & Harper 2003). For example, female moth sex pheromones for long-distance attraction may be such a signal, as both male and female moths have a common interest: both gain from meeting to mate and it does not benefit females to attract any males apart from ones of their own species. However, better fed female moths may release more pheromone (Foster & Johnson 2011), perhaps by an index effect via hemolymph blood sugar.

Minimal-cost signals are perhaps even more likely to evolve when the signaler and receiver share a

common interest through being related as kin. The young of subsocial insects such as burrower bugs, *Sehirus cinctus*, release a condition-dependent solicitation pheromone when begging for food from their parents (Kolliker *et al.* 2006; Mas & Kölliker 2008). Exaggerated begging by the signaler may be limited because the extra resources gained by the begging would be at a cost to its siblings (and hence to its inclusive fitness) (Moreno-Rueda 2007). Cuckoos are not restrained in this way as the costs of exaggerated begging are to the host offspring.

A strong shared interest through relatedness in social insects such as ants, wasps, bees, and termites probably makes much of their communication minimal cost (Maynard Smith & Harper 2003). These include alarm and trail pheromones but also the queen's fertility signal pheromone and egg-marking pheromones (Chapter 6). In the presence of the queen pheromone, workers' ovaries do not develop and workers do not lay eggs (Heinze & d'Ettorre 2009; Le Conte & Hefetz 2008; Peeters & Liebig 2009). This is likely to be an honest signal rather than "control," with the honesty maintained by a combination of relatedness and worker policing (see Chapter 6 for more details). Worker policing includes destroying worker-laid eggs, recognized because they are not tagged with queen egg-marking pheromone. Workers are selected to respond to queen egg-marking pheromone in this way as it allows them to rear more related brothers rather than the sons of sisters (see next section) (Chapter 6) (Ratnieks *et al.* 2006).

1.6.2.4 Punishment of cheaters (social cost)

Honesty can be maintained by punishment of cheating individuals. For example, in social insects, cheating workers who develop eggs and thus show the fertility signal CHCs of a fertile female are punished (see worker-policing, Chapter 6) (Figure 1.2) (Liebig 2010; Peeters & Liebig 2009). In bulldog ants, *Myrmecia gulosa*, non-laying workers immobilize workers starting to develop their ovaries, revealed by their CHC profiles (Dietemann *et al.* 2005). The queen of the ant

Aphaenogaster cockerelli herself also detects characteristic fertility signal CHCs on reproducing workers, and marks them with queen-specific secretion from her Dufour's gland, which leads other workers to attack them (Smith *et al.* 2012).

Subordinate mice are attacked if they produce the pheromones characteristic of dominant males and also if they "challenge" the urine marks of the dominant male with their own urine (Chapter 5) (Hurst 2009; Hurst & Beynon 2004).

1.6.2.5 Other costs

Animals may avoid exaggerating their signals not because of physiological costs but instead because making oneself more conspicuous brings greater costs from predation or parasitism (Chapter 11) (Zuk & Kolluru 1998). For example, the pheromone of Mediterranean fruit fly males, *Ceratitis capitata*, attracts yellowjacket wasps, *Vespula germanica*, which eat large numbers of signaling males (Hendrichs & Hendrichs 1998). This may be what normally limits pheromone production in males of another tephritid fruitfly, *Anastrepha suspensa*, able to greatly increase their pheromone production in response to hormone treatment in the laboratory (Section 1.6.1) (Pereira *et al.* 2010b).

Exaggerated advertisement of paternal gifts by male peacock blennies or tiger moths (above), at the expense of real care or poison-gift protection to offspring, may have its own costs as a dishonest male would presumably suffer greater egg losses.

1.7 Chemical profiles from which signature mixtures are learned for individual and colony recognition

This section emphasizes learning for kin or group recognition but the principles apply to other kinds of learning. How animals distinguish members of their group from non-members is a key behavior allowing them to favor offspring and other relatives (kin) or fellow group members (see reviews by Holmes 2004; Penn & Frommen 2010; Sherman *et al.* 1997; Wiley 2013). Kin recognition is also important for optimal outbreeding by avoiding close kin as mates (Chapter 3). Recognition of non-kin individuals, such as mates or neighbors is also important (Wiley 2013). Odor cue recognition can be used to recognize and reject previous mates (the Coolidge effect, Chapter 3). The mechanisms are the same and involve the learning of cues.

Chemical cues are widely used for recognition, perhaps because even the earliest organisms had the receptor mechanisms for receiving and processing the information and perhaps also because of the enormous variety of compounds available, which allows an effectively unlimited number of possible combinations.

Signature mixtures are the subsets of variable molecules from the chemical profile (Figure 1.1) that are learned as a template by other conspecifics and used to recognize an organism as an individual or as a member of a particular social group such as a family, clan, or colony (Chapter 6) (van Zweden & d'Ettorre 2010; Wyatt 2010). A key difference between pheromones and signature mixtures is that in all taxa so far investigated it seems that, with few exceptions, all recognition systems involve learning and all appear to use the olfactory rather than the gustatory system for detection (Chapters 6 and 9). Different receivers might learn different combinations of molecules as the signature mixture of the same individual (see legend to Figure 1.1) (Section 1.1.2).

The chemical signature mixtures learned by vertebrates and invertebrates may be seen best as cues rather than signals: although the response may be highly evolved, the signature mixture molecules may not be evolved specially for this function and may instead be "co-opted" for this use (Wyatt 2010). For example, the enormous variability of the major histocompatibility complex (MHC) is likely to be driven by its immune system function (Box 3.1) (Section 1.7.4) and so the best analogy might be with human fingerprints, not evolved for the purpose of individual recognition but potentially useful for human identification.

The signature cues learned may simply be mixtures sufficiently stable and individually different to enable an animal to recognize the same individual on another occasion as "familiar" (previously met) or, in some species, a particular individual (Johnston 2008; Thom & Hurst 2004; Wiley 2013). Individual recognition by smell is found in many organisms. Lobsters recognize each other by smell and avoid fighting a lobster they previously lost a fight to in the previous week (Atema & Steinbach 2007). Dominant male mice mark their territories (Chapter 5). If an experimenter adds a small urine mark from a resident subordinate, the dominant male soon attacks that individual (Hurst 1993). In some ant species, unrelated founding queens use chemical cues to recognize each other individually (d'Ettorre & Heinze 2005).

In some social insects that use largely hydrocarbon "labels" under genetic control, colony recognition blurs the signal–cue boundary (Section 1.3) as these evolved labels would count as signals (yet the highly variable labels are characterized by inter-colony variation and changes over time due to other molecules such as diet, so they do not match the "species-wide" requirement for a pheromone).

1.7.1 Learning and recognition

Perhaps surprisingly, recognition cues are usually learned through behavioral "rules," such as "learn the odor of your nestmates." There are three main potential mechanisms that animals use to recognize others as kin (Figure 1.9): first, by learning the characteristics of surrounding individuals (by direct familiarization with nestmates); second, by using this learning to allow phenotypic matching with unfamiliar kin; and third, by using self-inspection – the armpit effect (Dawkins 1982) – to allow phenotypic matching with unfamiliar kin. All three mechanisms rely on learning a memory template. The different mechanisms are not mutually exclusive and different ones may be used by

the same animal, for example, in different contexts or at different ages (Mateo 2004; Penn & Frommen 2010).

Self-matching may be favored in species where the young grow up alone (e.g., crickets, Chapter 3) and lack contact with kin for learning, or if the available relatives in the nest would give error-prone templates, as for example when nestmates include full- and half-siblings from multiple matings (Mateo 2010; Sherman *et al.* 1997). This second situation applies to the golden hamster, *Mesocricetus auratus*, which mates multiply and produces multiply sired litters. Recognition of kin seems to be by self-referent ("armpit") matching: hamsters that were reared only with non-kin since birth responded differently to the odors of unfamiliar relatives and non-relatives (Mateo & Johnston 2000, 2003). Post-natal association with kin was not necessary for this discrimination.

Olfactory learning of signature mixtures for familiarization and phenotypic matching often occurs at particular sensitive periods in life, a phenomenon termed imprinting (explored in Chapter 9) (Hudson 1993). In mammals this tends to occur as a young animal, say a young mouse pup in the nest learning the odors, including those related to the MHC, and other characteristics of its siblings in order to avoid them as mates when adult (reviewed by Brennan & Kendrick 2006; Penn & Frommen 2010). Such learning has been demonstrated by cross-fostering experiments with young pups (if reared with a foster family, the pups treat foster-family members as siblings) (see Figure 3.10). Cues need only be a reliable statistical indicator of kinship or group membership (Sherman *et al.* 1997). As an adult, learning occurs with the bonding with newly born offspring, as in the now classic system of mother sheep and lambs (Chapter 9) (Lévy & Keller 2009; Sanchez-Andrade & Kendrick 2009). It also occurs at mating in the female mouse, which remembers the signature odor of its mate, preventing pregnancy block (Chapter 9) (Brennan 2009). The neonatal imprinting and odor-based recognition of offspring occurs in humans too (Chapter 13) (Schaal *et al.* 2009).

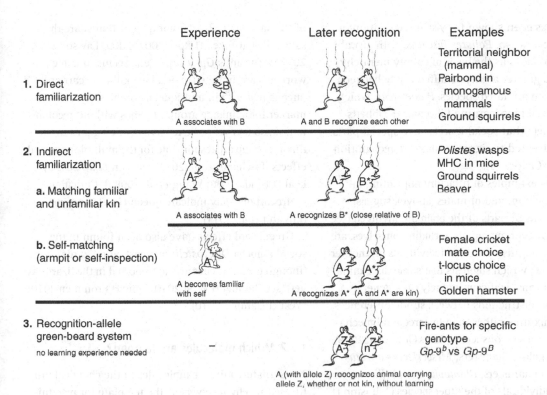

Figure 1.9 Kin recognition mechanisms in almost all animals, vertebrates and invertebrates, seem to involve learning a signature and then matching this template against the chemical profile of other animals. (The diagram is somewhat anthropomorphic as mice do not have smelly armpits – but humans do.) Three mechanisms are represented:

(1) Direct[‡] familiarization, by learning the characteristics of nestmates and recognizing these animals later.

(2) Indirect[‡] familiarization (phenotypic matching): (a) learning the characteristics of nestmates and using the template to allow phenotypic matching with unfamiliar kin; and (b) learning the odor of self to allow phenotypic matching of self with others (self-referent or "armpit" phenotypic matching).

(3) Recognition allele ("green beard"). A (with allele Z) recognizes an animal carrying allele Z, whether or not kin, without learning.

[‡]Note: "direct" and "indirect" are used as by Porter and Blaustein (1989). The same words are used in a very different way by other authors who use "indirect" for kin recognition rules using location e.g., "any baby in the nest is treated as kin," compared with "direct" for learning phenotypes, which would allow recognition away from the location, e.g., Pfennig and Sherman (1995) and Waldman *et al.* (1988). Figure after Porter and Blaustein (1989) with modifications and additions. See Penn and Frommen (2010) for more examples.

Olfactory imprinting occurs in social insects (Chapter 6). Ants, wasps, and bees learn their colony odor after emerging as callow adults from their pupae (Bos & d'Ettorre 2012; Breed 1998a; van Zweden & d'Ettorre 2010). In ants, just as in mammals, the learning can be demonstrated by cross-fostering a pupa or newly emerged adult: the transferred ant will learn the colony odor of its new hosts (Lenoir *et al.*

2001). Similarly, newly emerged *Polistes* wasps learn the odors of the nest rather than their own odor. However, individuals constantly need to reinforce and fine tune their template with nestmate odors over time (see below).

Some of these examples of individual or colony recognition suggest there may be some selection for receiver specialization for recognition (Tibbetts & Dale

2007). This has been shown for visual recognition: a species of social wasp, *Polistes fuscatus*, with visual recognition of faces as the basis of colony hierarchy has evolved a greater ability to differentiate between wasp faces than a related species, *P. metricus*, which lacks specialized face learning (Sheehan & Tibbetts 2011). It is likely that social insects have special parts of the brain devoted to signature mixture recognition and memory (Chapter 9).

Some of the examples of olfactory imprinting in mammals, of young, and of mates (above), suggest particular circuits or parts of the brain are involved (see Chapter 9). However, mammals, including ourselves, are also able to distinguish different individuals of another species by smell, which suggests that some discriminations between conspecifics might rely on a general ability to make distinctions between subtle differences in complex mixtures rather than to perceptual mechanisms specialized for conspecific odors (Johnston 2005). For example, golden hamsters, *Mesocricetus auratus*, and Djungarian hamsters, *Phodopus campbelli*, can distinguish individuals of the other species (Johnston & Robinson 1993). We do not know if they are learning the same molecules as the other species would use, of course (but then we do not know this for different hamster individuals smelling conspecifics).

1.7.1.1 An exception to learned recognition: greenbeards

The one theoretical exception that does not require learning for kin recognition is the "greenbeard effect," proposed by Hamilton (1964) and named by Dawkins (1976), with a hypothetical "supergene" or closely linked genes with three effects that code (1) for a conspicuous phenotype signal, (2) the genetic ability to recognize it in others, and (3) a genetically determined appropriate response. A greenbeard gene would simultaneously give the owner a green beard and prompt the greenbearded individual to look after others with green beards (or harm those without one).

The first example found may be in the fire ant, *Solenopsis invicta*: workers carrying one allele ($Gp-9^b$)

of the supergene $Gp-9$ favor queens that share the same allele (Gotzek & Ross 2007, 2009; Lawson *et al.* 2012a). The ant's $Gp-9$ supergene seems to make workers carrying the $Gp-9^b$ allele kill non-carrier queens ($Gp-9^{BB}$) in multiqueen colonies. $Gp-9$ is a marker for a linkage group of genes with no recombination, so it is yet to be resolved which genes in the linkage group are responsible for the multiple observed effects (Fischman *et al.* 2011; Lawson *et al.* 2012a; Leal & Ishida 2008; Wang *et al.* 2008a). Cuticular hydrocarbons may indicate queen $Gp-9$ genotype (Eliyahu *et al.* 2011).

Greenbeard effects have also been found in the social amoeba *Dictyostelium*, yeast, and lizards (though chemical cues are not reported in the lizards as yet) (see Gardner & West 2010; Penn & Frommen 2010; West & Gardner 2010).

1.7.2 Which molecules are learned?

The signature mixture molecules in the chemical profile, learned by receivers as the template for recognition, can be produced by the organism itself, acquired from the diet, shared local environment, or other organisms.

In mammals, genetically controlled cues produced by the individual include odor cues related to the MHC or lipocalin MUPs (Hurst 2009; Kwak *et al.* 2010). Family members of badgers, *Meles meles*, also mark each other with secretions during allomarking, when they back up to each other and smear from their anal and subcaudal glands (Buesching *et al.* 2003; Roper 2010, p. 198 ff.).

In many mammals, the fermenting of secretions by microbes may provide some of the individually varying odors (Archie & Theis 2011). For example, which molecules are produced in our armpits is affected by what we secrete and by which bacteria thrive in our armpits (both are influenced by, for example, the MHC, other genes, and factors such as diet) (Chapter 13) (Figure 13.2) (Grice & Segre 2011; Human Microbiome Project Consortium 2012). Males of the neotropical greater sac-winged bat,

Saccopteryx bilineata, appear to have individually different odors coming from their fermenting wing pouches and such differences may come from bacterial species combinations that differ markedly between individuals (Voigt *et al.* 2005).

The chemical profiles of mammal family groups change as their diet and bacterial flora change. Some of the exchanges of bacteria in family groups are encouraged by behaviors such as allomarking by badgers, *Meles meles* (above) (Buesching *et al.* 2003; Roper 2010, p. 198 ff.). Marking behavior in hyenas may have the same effect (Chapter 5) (Theis *et al.* 2012). A social insect colony's shared label is also constantly changing so the template has to be constantly updated (Chapter 6) (Bos & d'Ettorre 2012; van Zweden & d'Ettorre 2010). This constant change is another reason for regarding signature mixtures as different from pheromones.

In social insects, the colony chemical profiles are determined partly by the insect's own genes but also by sharing molecules with other colony members, the environment (e.g., nest, food, symbiotic fungi), or, in some species, molecules from the queen (for more detail see Chapter 6) (Breed & Buchwald 2009; Nash & Boomsma 2008; Sturgis & Gordon 2012; van Zweden & d'Ettorre 2010). Different taxa have characteristic surface molecule types that vary and appear to be involved in colony recognition: waxy molecules in bees and CHCs in ants. Ant CHCs are typically complex mixtures of alkanes, alkenes, and methyl branched alkanes, and the number, weight range, and chemical families of hydrocarbons differ between taxa, including between species (Chapter 6) (Martin & Drijfhout 2009a; van Wilgenburg *et al.* 2011). Within a species, different colonies present different chemical profiles based on the relative abundance of the same genus- or species-characteristic components (Hefetz 2007; van Zweden & d'Ettorre 2010).

However, not all the CHCs on the surface of an ant are involved in colony recognition. For example, the CHC profile of the ant *Formica exsecta* is composed of two independent parts: a colony-specific (Z)-9-alkene profile under genetic influence and an environmentally influenced task-related *n*-alkane profile (Martin & Drijfhout 2009b). It is the ratio of different (Z)-9-alkenes on an ant's surface that is "monitored" by other conspecifics to determine if it is a member of the colony. The *n*-alkanes, which increase if the ant has been foraging outside rather than working underground, are disregarded in nestmate recognition by the ants (Greene & Gordon 2003; Martin & Drijfhout 2009b. (See Appendix for chemical terminology and examples).

For nestmate recognition there may be selection over evolutionary time for particular types of branched hydrocarbons, which are easier to distinguish by shape and offer the scope for more variation than straight-chain hydrocarbons (Chapter 6) (this will in part be a co-evolution with the receptor sensitivities of the receivers, Chapter 9). Argentine ants, *Linepithema humile*, learned to distinguish different tri-methyl alkanes more easily than single-methyl or straight-chain alkanes (van Wilgenburg *et al.* 2012). The ants also found it easier to discriminate between hydrocarbons with different branching patterns and the same chain length, than between ones with the same branching patterns but different chain length.

1.7.3 Is there selection for greater diversity in the molecules offered in the chemical profile?

Most signature mixture-based recognition seems to rely on co-option of variability that exists for another reason (e.g., MHC) or has no selective advantage (e.g., diet). However, in the systems where the molecules in the chemical profile are directly or indirectly under genetic control (such as the ants, above) there might be selection for greater diversity of molecules in the "label" to allow greater distinctiveness either of individuals or of social insect colonies. An advantage for visual distinctiveness, by for example reducing fights because individuals are more easily distinguished, may explain why, in *Polistes* paper wasps, only species with complex social interactions have the variable facial markings used in individual recognition (Tibbetts 2004; Tibbetts & Dale 2007).

Do social species have more complex chemical profiles? In mammals, an investigation of chemical complexity of male and female glandular secretions in a clade of eight related species of *Eulemur* lemurs suggested greater complexity in species which live in multimale–multifemale groups rather than in pair-bonded species (delBarco-Trillo *et al.* 2012). The great diversity of MUPs in the house mouse, *Mus musculus domesticus*, may have been selected for in high-density breeding populations with a higher chance of encountering kin as potential mates (the polymorphic MUPs can be used as a cue to reject mates sharing MUP alleles with the chooser and thus likely to be kin; Sherborne *et al.* 2007). In contrast, other mouse species, such as *Mus macedonicus*, living at low densities have only one MUP isoform in their urine (see in Sherborne *et al.* 2007).

If the driver for selection for variety in the house mouse polymorphous MUPs is mate choice (whether the learned molecules are the MUPs themselves or smaller molecules associated with them), the MUPs could count as evolved signals rather than cues. However, I would suggest that the need to learn the variable MUPs (even for self-referent comparison) and their great variety would have them count as contributors to signature mixtures in chemical profiles.

An indication that social insect species might have more complex chemical profiles than solitary species comes from almost 1,000 different hydrocarbons found in just 78 ant species spread across 5 ant subfamilies (Martin & Drijfhout 2009a), which can be contrasted with the 20 to 50 different hydrocarbons typically found in non-social taxonomic families of insects of all kinds (Chapter 6) (S. Martin, unpublished data in Martin *et al.* 2008b). However, a stronger comparison would be *within* hymeopteran groups, such as halictid and allodapine bees, which show the full range of social patterns from solitary to eusocial among closely related species (Schwarz *et al.* 2007).

However, rather than *selection* for diversity, might some of the diversity in hydrocarbon blends (labels) between colonies come from a *relaxation* of the selection for conformity (i.e., permitting variation) in

contrast to the stabilizing selection for species recognition signals that reduces deviation from a norm (Chapter 3)?

1.7.4 How is diversity of chemical profile maintained?

If molecules related to genetic labels are used for recognition of nestmates, whether in ants responding to CHCs or family groups of mammals responding to MHC odors, how is this diversity of labels maintained?

Genetic variability (polymorphism) in labels is essential to allow distinction between nestmates and non-nestmates. However, frequency-dependent selection against rare labels may remove label diversity from recognition systems, leading to uniformity and making distinction impossible, a situation known as "Crozier's paradox" (Penn & Frommen 2010; Tsutsui 2004) after Crozier (1986, 1987). The selection against rare labels could take the form of being rejected (for example, being more likely to be seen as a non-nestmate and prevented from re-entering the ant colony or a mammal family's burrow). Individuals that are less different from the norm will have a selective advantage as they will be less likely to be rejected as possible outsiders.

Crozier (1986, 1987) suggested that genetic marker diversity used in recognition systems may be piggy-backing on variation maintained by other forms of selection such as parasites, pathogens, or mate choice. This suggestion is supported by models of various kinds (Gardner & West 2007; Penn & Frommen 2010). In ants, greater diversity of CHC labels appears to occur under greater parasite pressure (Chapter 11) (Martin *et al.* 2011b). For mammals, the MHC is the basis of the immune system and is under direct selection by parasites and disease (Chapter 3).

Mate choice to avoid inbreeding favors rarer genetic markers, maintaining diversity. In the ant *Leptothorax gredleri*, the cuticular hydrocarbons of both unmated queens and reproductive males are colony specific and this could in principle be used to avoid mating with siblings (Oppelt *et al.* 2008). Similarly, in vertebrates, mating choices for difference in the MHC may contribute to maintaining MHC diversity (Chapter 3) (Milinski 2006). Similar arguments apply to MUPs in mice (above).

1.8 Differences in response to pheromones

Different individuals may respond differently to the same pheromone stimulus. While responses to pheromones are characterized by being "innate" (Section 1.2), the responses can vary according to context, time of day, and many other factors including the receiver's genetics, age, sex, hormonal state, dominance status, and experience (Chapter 9). For example, honeybee responses to alarm pheromone may depend on how close to the nest they are (Chapter 9). Honeybee responses to the many other honeybee pheromones also change with age, as do the tasks undertaken (Chapters 6 and 9) (Le Conte & Hefetz 2008). Different, overlapping, subsets of the molecules give different messages depending on the receiver (for example young or older workers) and context (Box 6.3).

After mating, male *Agrotis ipsilon* moths stop responding to female pheromone for up to 24 hours, the time needed to replenish their accessory glands, though their antennae still detect the female's pheromone (Chapter 9) (Anton *et al.* 2007; Barrozo *et al.* 2010). Some changes in response to sex pheromone are mediated by responses to signature mixtures: animals do not respond to otherwise attractive sex pheromones if they remember they have mated with that individual, recognized by signature mixture (Coolidge effect) (Chapter 3).

1.9 Releaser and primer effects of pheromones

Wilson and Bossert (1963) introduced the terms releaser effects (immediate behavioral responses to pheromones) and primer effects (longer lasting physiological or developmental changes, sometimes mediated by hormones). They recognized that some pheromones had both effects. Later researchers tended

to refer to "releaser pheromones" and "primer pheromones." It is clear now that the effects form a continuum, so I think it is better to return to primer and releaser *effects* rather than primer and releaser *pheromones*. There are many examples of pheromones or their components having both kinds of effect at the same time or one effect depending on the context or receiver. Releaser effects may be accompanied by longer lasting primer effects: the principal component of honeybee alarm pheromone, isopentyl acetate, elicits a quick defensive response from honeybees (Chapter 8) and also induces gene expression in the antennal lobes, perhaps underlying the lasting changes in behavioral response to the pheromone (Chapter 9) (Alaux & Robinson 2007; Alaux *et al.* 2009b). Similarly, the suckling response to rabbit mammary pheromone by a rabbit pup (Box 1.5) is accompanied by learning of maternal odors, reflected in widespread immediate early gene activation in the rabbit pup brain (Charra *et al.* 2012; Courcaud *et al.* 2010).

The male pheromones of mice, dehydro-*exo*-brevicomin and 2-*sec*-butyl-4,5-dihydrothiazole appear to have the releaser effects of eliciting aggression from other males and attracting females, as well as the developmental (primer) effects of apparently inducing estrus in mature females and accelerating puberty in young females (Chapter 9) (Novotny 2003). The honeybee queen's mandibular pheromone attracts males during her nuptial flight, a releaser effect, but when she is queen of her own nest, the mandibular pheromone plus additional components have the releaser effect of attracting her retinue of workers around her (Chapters 6 and 9) (Grozinger 2013; Kocher & Grozinger 2011; Slessor *et al.* 2005). The queen mandibular pheromone also has a primer effect as a signal to the worker bees, her daughters, that she is present and laying eggs (with the physiological effect that the workers do not themselves lay eggs). The multiple use of a pheromone within a species for different functions is sometimes termed pheromone parsimony (Section 1.4.2).

The multiple effects of a pheromone may act by different receptors or nerve circuits. For example, the modes of action of various primer and releaser effects of different components of the honeybee queen mandibular pheromone on worker bees can be differentiated experimentally (Chapter 9) (Grozinger *et al.* 2007a). Primer effects can be mediated via chemosensory neurons such as olfactory sensory neurons or by acting directly on tissues (Chapter 9) (Section 1.11).

Though the physiologies of mammals and insects are very different, primer effects may work in similar ways (see Chapter 9). For example, in mammals, dominance hierarchies are reflected in blood gonadal hormone concentrations (e.g., Saltzman *et al.* 2009). In social insects juvenile hormone (JH) is often important in pheromone-mediated effects (Alaux *et al.* 2010; Le Conte & Hefetz 2008).

While primer effects may act over days or longer, some responses to endocrine-mediated pheromone signals can be rapid. For example, the odors of estrous female rats cause the release of hormones into the blood in sex-experienced male rats, which give them erections and elicit sexual behaviors within minutes (Sachs 1999).

1.10 Multimodal signals

Multimodal signals involve more than one sense (modality) and many include pheromones along with sound or visual signals (Bradbury & Vehrencamp 2011, pp. 296; Hebets & Papaj 2005; Partan & Marler 2005).

Signals may involve different modalities sequentially (though some authors might not count this as multimodal) (Partan & Marler 2005). Male butterflies use visual cues to find females at long range and then in many species, at short range they communicate with pheromones (Chapter 3) (Allen *et al.* 2011). Many carnivores, such as dogs, add their scent marks to visually conspicuous sites or landmarks. Ultraviolet (UV) absorbing molecules in the scent marks of the desert iguana, *Dipsosaurus dorsalis*, visually attract distant conspecifics and once at the scent mark, the lizards tongue-flick to pick up the non-volatile pheromone molecules (Alberts 1990).

Some multimodal signals feature "redundancy," in which the signal in some modes can be omitted without changing the message, as when we nod when speaking the word "yes." For example black-tailed deer alarm signals are transmitted not only as an odor, but also as sounds and visual signals (Chapter 8). Any one of these may be effective in alerting other deer in the group. This redundancy in signal can make dissecting the role of pheromones much more difficult (Chapter 2).

Some multimodal signals are non-redundant. Male *Drosophila melanogaster* fruit flies require a combination of chemical and visual stimuli from the female for successful courtship; pheromones are necessary but not sufficient alone (Chapters 3 and 9) (Dickson 2008). Some multimodal combinations can change the meaning of signals. In the snapping shrimp, *Alpheus heterochaelis*, male responses to visual threat signals are changed if they are accompanied by female pheromones (Hughes 1996). There may be modulation of the signal intensity by addition of other signals. For example, in desert ant *Aphaenogaster* species, recruitment of nestmates to a new food source is faster when the scouts release pheromone and stridulate (Chapter 7) (Hölldobler & Wilson 2009, p. 231).

Stimuli from different modalities are integrated in the higher parts of the brain in both invertebrates and vertebrates. For example, in the moth brain, neurons integrate olfactory and visual inputs to give the motor outputs for the flight response to a pheromone plume (Haupt *et al.* 2010) (see also Chapter 10). In Crustacea, hydrodynamic cues and chemical information are integrated (Chapter 10) as are inputs from sensilla in both the olfactory and distributed chemoreceptor systems (Mellon 2012; Schmidt & Mellon 2011).

Stimulation in one modality can affect response in another. Exposing male *Spodoptera littoralis* moths to the ultrasonic clicks of predatory bats increases the moths' behavioral response to female pheromone and the sensitivity of central olfactory neurons in the antennal lobe (Anton *et al.* 2011). Pup odors increase the sensitivity of neurons in the primary auditory cortex in mouse mothers to the ultrasonic distress calls of pups (Cohen *et al.* 2011).

The integrated inputs will be further modulated by the animal's internal state and experience, for example the hormonal state of male hamsters affects their responses to female odors (Section 1.8). An organism in a natural setting interweaves signals from its external and internal environment to yield an experience more complex than the sum of the individual inputs (Stein & Meredith 1993).

The final outcome of semiochemical stimulation comes from the integrated signals from the brain prompting immediate behaviors, or changes over the medium term such as increased alertness, or a complete developmental switch affecting the rest of the animal's life as in phase change in locusts (Chapter 4) (Section 1.9).

1.11 Allohormone pheromones bypassing olfaction and taste

Some allohormone pheromones are passed directly to another individual and have their effect directly on the recipient's tissues or sensory neurons but bypassing the usual external sensory systems of taste and olfaction (Chapter 9). This was anticipated by Karlson and Lüscher (1959) in their example of termite pheromones with primer effects on caste development, passed by mouth around the colony (see Box 6.3). In honeybees, nurse workers produce "royal jelly" containing an allohormone pheromone, royalactin, which is fed to larvae, switching them to develop as queens (Chapter 9) (Kamakura 2011).

Hormones or other molecules may be directly transferred by the male to the female when mating, causing the female to reject other males; these include prostaglandins in the semen of the red-sided garter snake, *Thamnophis sirtalis parietalis* (Mason 1993), and sex peptides in the seminal fluid of *Drosophila* (Avila *et al.* 2011). The *Drosophila* sex peptide changes the female's behavior so that after mating she rejects other males and starts to lay eggs (Chapter 3). The sex peptide activates a specific receptor protein on specific chemosensory neurons in her uterus and oviduct

(Chapter 9) (Häsemeyer *et al.* 2009; Rezával *et al.* 2012; Yang *et al.* 2009).

The term *allohormone* was proposed by Koene and ter Maat (2001, 2002) though Ruther and Steidle (2002) argued against. I think allohormones, if the term is seen to be useful, should be used as a subclass of pheromone. Such an approach would allow us to avoid classifying otherwise similar amphibian peptide pheromones differently depending on their route of transmission: peptide pheromones are wafted in currents by aquatic newt species, deposited on the openings to the VNO in some terrestrial salamanders mating on land (e.g., *Plethodon shermani*), and in most plethodontid species, such as *Desmognathus ocoee*, applied transdermally into the bloodstream of the female (via skin scratches made with enlarged premaxillary teeth) (Figure 1.7) (Section 1.4.3.3) (Houck 2009; Woodley 2010). The peptide delivered through the skin would thus be an allohormone pheromone.

1.12 Pheromones and signature mixtures in humans?

Sight and hearing are arguably our most important senses, which probably make us different from many other mammals. Nonetheless, olfactory signals and chemosensory cues may be more important to us than once supposed. For example, they may enable an important part of the bond between parents and babies and perhaps influence our choice of partner. However, despite what is claimed in the wild west of the Internet, no human pheromones have yet been properly chemically identified and validated. These topics are explored in Chapter 13.

1.13 Pollution disrupts chemical communication in aquatic organisms

Aquatic organisms seem to be particularly vulnerable to interference in chemical communication ("infodisruption") by human pollution (Lürling 2012; Olsén

2011; Zala & Penn 2004). Local effects include endocrine-disrupting chemicals such as 17β-estradiol, entering the environment via sewage outflows, which has negative effects on male goldfish responses to female pheromones. Invertebrates are also affected. For example, crustacean male responses to female odors are reduced by medetomidine, a molecule used in antifouling coatings, or naphthalene from motor boat fuel.

However, the most ubiquitous and global danger probably comes from the atmospheric pollutant CO_2 through its effect on ocean acidity (Doney et al. 2009, 2012). Largely a result of human fossil fuel combustion, CO_2 levels are rising at a rate about ten times faster than has occurred for millions of years. About a third of the CO_2 is absorbed by the oceans, reducing their pH. If atmospheric CO_2 concentrations reach an anticipated 800 ppmv by 2100 as predicted, the pH will drop from the current and historic pH of between 8.15 and 8.25 to about 7.8 or below (Doney et al. 2009, 2012). This acidification is likely to have significant effects on chemical communication by aquatic animals, which have evolved over 50 million years under relatively constant pH levels.

Lowering the pH affects both the semiochemical molecules themselves and their interaction with chemosensory receptor proteins. The way ligands (odor molecules) interact with chemosensory receptors changes with pH, as pH can affect the number, type, and alignment of intermolecular forces (e.g., hydrogen bonding, electrostatic potential, hydrophilic/hydrophobic regions) on both the chemosensory receptor and the ligand (Hardege et al. 2011a; Kaupp 2010; Reisert & Restrepo 2009). These include peptides, nucleosides, thiols, and organic acids in nereid polychaete worms; amino acids and peptides in *Aplysia* sea hares; bile acids in fish; and nucleotides in crustaceans such as shore crabs (Hardege et al. 2011a). For example, many aquatic sex pheromones have acid dissociation constant (pK_a) values in the range that is likely to be affected by the lower pH values (JD Hardege, pers. comm.). Experimental exposure to lower pH levels of between 7.6 and 7.8, forecast to occur by 2100, did indeed disrupt chemosensory responses of a diverse range of species, from North Sea polychaete worms to Caribbean shrimp species (JD Hardege, pers. comm.). The disrupted pheromones and cues related to sexual reproduction, feeding, sperm attraction, fertilization, social interactions, and larval settlement. Vertebrate chemical senses are also affected by lower pH: coral reef anemonefish larvae, for example, no longer respond appropriately to predator odors (Dixson et al. 2010). The fast rate of change of pH is likely to outstrip the speed that chemosensory systems can evolve.

Might the overall effects of lowering ocean pH be the chemosensory equivalent of the blinding of the world's human population by a meteor shower at the start of John Wyndham's classic (1951) science fiction novel *The Day of the Triffids*?

Summary

Across the animal kingdom, more interactions are mediated by pheromones and chemical cues than by any other kind of modality. Many different kinds of compounds are used as pheromones but there are many examples of the same compounds being used by different species for different functions. Signature mixtures, learned as a template, enable animals to distinguish each other as individuals or colony members. Pheromones tend to be innate.

The design of the olfactory system makes evolution of pheromones likely because there is selection for any odor cue that increases reproductive success or survival. Thus pheromones evolve from compounds originally having other uses or significance, for example from hormones, host plant odors, chemicals released on injury, or waste products. Other signals may evolve because they match previously existing sensitivities of the receiver. There is selection for functional signal features such as longevity and specificity. There is also evolution in the senses and response of the receiver. The original functions of the chemicals may or may not be eventually lost.

There is less difference between vertebrates and invertebrates, in both the pheromones produced and in the range of behaviors that pheromones influence, than was once thought. Given the ubiquity of chemical communication among animals, pheromones and chemical cues are likely to emerge as key criteria that animals use for mate choice.

The broad-brush diversity of pheromone molecules comes from the processes of evolving from chemical cues, as molecules of all kinds are co-opted as signals. A finer grain of diversity comes from the variations around a "chemical theme" as part of speciation. Multicomponent pheromones of related molecules are often a result. The changes in pheromones that occur in the process or speciation can involve large effects from a small number of genes or many genes working together, or a combination of these.

Synergy describes the phenomenon when any one component of a multicomponent pheromone shows little or no activity by itself and only the complete mixture has an activity comparable to the natural pheromone. It is to be expected from multicomponent pheromones and may be a natural outcome of the combinatorial way they are processed in the brain of both vertebrates and invertebrates.

Honest signals do not necessarily need to be costly and showing that a signal has a cost does not demonstrate a handicap. Currently there is no experimental way of separately measuring the two kinds of costs: efficacy (just to get the message out) and strategic (added costs for a handicap). Signals can be kept honest by a variety of non-handicap mechanisms including unfakeable indices, shared interest, and punishment of cheaters.

Acidification of the oceans due to rising carbon dioxide levels may seriously disrupt chemical communication in aquatic organisms of all kinds because pH affects the interaction between signal molecules and receptor proteins.

I hope that distinguishing between signature mixtures and pheromones (Table 1.1) could help guide research strategies and help clarify what we have discovered so far. Karlson and Lüscher (1959) ended their paper introducing "pheromones" by throwing the definition open for discussion, saying that they hoped it would prove itself in practice, which 50 years on, it certainly has. In a similar spirit, I welcome comments and suggestions for improving the ideas presented in this book.

Further reading

For pheromones in particular taxonomic groups see Müller-Schwarze (2006) and chapters in East and Dehnhard (2013) on vertebrates in general; Stacey and Sorensen (2011), Chung-Davidson *et al.* (2011), and Sorensen and Wisenden (2014) for pheromones in fish; Hölldobler and Wilson (1990, 2009) on ants and other social insects; Grozinger (2013) on honeybees; Allison and Cardé (2014) on moths; and Hardie and Minks (1999) for other insects. Gaskett (2007), Schulz (2004) and Trabalon and Bagnères (2010) cover various aspects of spider pheromones. Chapters in Breithaupt and Thiel (2011) cover chemical communication in Crustacea in detail. Brönmark and Hansson (2012) cover chemical communication in aquatic vertebrates and invertebrates.

Be aware when reading the past and current literature that the term "pheromone" is often used ambiguously and may be used in contexts where "signature mixture" or "olfactory cues" might be more accurate or helpful. Johnston (2003, 2005) gives a good overview of the ways mammals use smell, in particular the way that individuals are recognized (describing "mosaic signals," which inspired the term "signature mixtures").

For an excellent and comprehensive overview of communication see Bradbury and Vehrencamp (2011), and also books by Maynard Smith and Harper (2003) and Searcy and Nowicki (2005). However, for a fresh look at communication that challenges the "handicap" mechanism, see Számadó (2011a,b).

The Nobel lectures of Richard Axel and Linda Buck offer clear, freely available, overviews of how smell works (www.nobelprize.org) (Axel 2005; Buck 2005). For developments since then, see Chapter 9.

For an insight into the diversity and molecular structure of pheromones, you can spend an enjoyable and informative time browsing Pherobase www.pherobase.com developed and maintained by El-Sayed (2013). See also the Appendix for a short guide to the terminology (available for free download from the website associated with this book).

You can see the molecular structures of most molecules on sites such as www.chemspider.com, which allows you to search by common name and shows synonyms as well as the systematic names.

True to its title, this book focuses on animals. However, the social life of bacteria also involves chemical communication, including quorum sensing, and is explored in a number of good reviews including Keller and Surette (2006), Diggle *et al.* (2007), and Foster (2010).

2

Methods for identifying and studying semiochemicals

Progress in science depends on new techniques, new discoveries and new ideas, probably in that order.

Sydney Brenner (2002)

Since Butenandt's landmark identification in 1959 of the silk moth sex pheromone bombykol, there have been spectacular developments in our ability to identify semiochemicals (Meinwald 2009). In their work over some two decades, Butenandt's team needed more than 10 metric tonnes of female moths, providing 500,000 pheromone glands from which they extracted ~12 mg of the pheromone to identify (Billen 2006; Butenandt *et al.* 1959). The identification of the first ant trail pheromone in 1971 still required 3.7 kg of *Atta texana* ants (Tumlinson *et al.* 1971).

Today, it is possible to work with far, far less than a single moth's pheromone gland. The revolution has come from chromatographic techniques in particular and the direct coupling of these with mass spectrometers, and other detection devices including animals' own sensors. The ever-increasing power of nuclear magnetic resonance (NMR) spectroscopy has made complete structure determination (though perhaps not complete stereochemistry) possible on a microgram scale (Meinwald 2009). Insect pheromone identifications can be made from picogram to femtogram quantities, using gas chromatography–electroantennogram detector (GC-EAD, with the insect's antenna) to get retention indices, and microchemical reactions to determine presence/absence of functional groups (e.g., with a gall midge sex pheromone, Gries *et al.* 2002).

In the approach taken to identify bombykol, Butenandt demonstrated the "gold standard" for pheromone identification used to this day. It is the equivalent of "Koch's postulates" for establishing causal relationships for

pheromones: initial demonstration of an effect mediated by a semiochemical, then identification and synthesis of the bioactive molecule(s), followed by bioassay confirmation of activity (Wyatt 2009). Chemosensory receptors interact with the three-dimensional structure of molecules, including their stereochemistry, so it is critically important to both identify and synthesize the correct stereoisomer(s) in high purity for testing (Appendix) (Chapter 9).

Sensitive, reliable, and reproducible bioassays are important for pheromone identification, and for dissecting the details of the behaviors or physiological changes mediated by pheromones. Bioassays are also important for studying the odor cues used for recognition, which differ between individuals or social insect colonies (Chapter 1).

As well as the advances in chemistry, which are changing pheromone research, this chapter also introduces the increasingly important tools of metabolomics, "next generation sequencing" (NGS) of DNA and RNA, and gene manipulations.

2.1 Bioassays

The first step in the study of a new pheromone is the observation of a behavioral or physiological response that appears to be mediated by semiochemicals. A bioassay is then developed as a repeatable experiment for measuring this response. It could be time spent chin-rubbing (Figure 2.1), sniffing (Figure 2.2), settling by

(a)

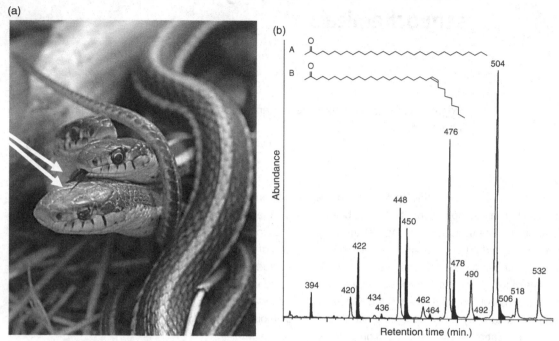

Figure 2.1 (a) Male red-sided garter snakes, *Thamnophis sirtalis parietalis*, tongue-flick (arrows) and chin-rub a female to detect the contact pheromones on her skin. In a bioassay, different fractions were presented on a paper towel allowing the pheromones to be identified. (b) A gas chromatogram of the contact sex pheromone blend from a female red-sided garter snake (see Chapter 3). The blend is composed of 17 unique methyl ketones (mw 394–532 Da). Most peaks in the chromatogram occur in pairs, with the first (unshaded) peak in each pair representing the unsaturated methyl ketone (B), and the second (shaded) being the saturated methyl ketone (A) of the same chain length (Mason and Parker 2010). Photograph courtesy Chris Friesen. (b) from Mason and Parker (2010).

marine planktonic larvae (Figure 2.3), or longer term physiological and behavioral changes such as becoming sexually mature. There are as many potential bioassays as there are animal species and responses to study (see Haynes & Millar 1998; Touhara 2013).

A key feature of any bioassay is that it should be a reliable measure of the behavior you ultimately want to assess. Simple bioassays such as wing fanning responses by male moths are easier to do than flight tests, but they can give misleading results because wing fanning is only the first in a series of steps that results in upwind flight and contact with a pheromone source. Thus, a more complex design may be needed to tease out the entire set of behaviors involved. In *Solenopsis* fire ants, different trail

pheromone components have different roles, some being involved in recruitment (alerting) and others in orientation (the tendency to follow a trail once alerted) (Chapter 7) (Morgan 2009; Vander Meer & Alonso 1998).

In an important paper, Wolff (2003) casts doubt on the ecological relevance of some effects observed in laboratory studies of rodent models, which did not appear in semi-natural tests (Chapter 9). However, the changes that occur to animals on domestication (or laboratory culture) can sometimes be exploited as natural loss-of-function or gain-of-function mutants of pheromone-mediated behaviors. These mutants can be useful tools in teasing out the molecular or genetic basis of pheromone systems, as in the case of *C. elegans* dauer-pheromone

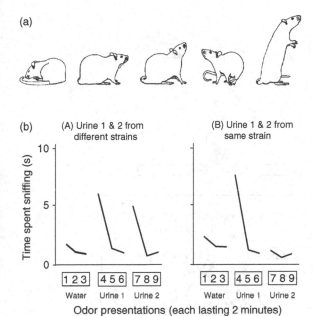

2.1.1 Bioassay design and variables can change results

Different bioassay methods can give different results, perhaps because they are testing different things. For example, studies of discrimination of odors by rats found different results depending on whether the study used a trained discrimination task (with a reward for choice of one odor over another) or used a habituation task (which measures what the animals notice naturally on their own) (Figure 2.2) (Schellinck *et al.* 1995). For laboratory rodents, the strain type and housing conditions before and during experiments can affect results in social recognition sniff tests, among many others (e.g., Macbeth *et al.* 2009).

Another problem arises when testing molecules on only one of an animal's many chemosensory subsystems, such as the mammalian main olfactory system or accessory olfactory system (AOS) (Chapter 9). Given the difficulties of studying even one subsystem, it is easy to understand why both are not studied. However, it is probably not fair to conclude that a given molecule only acts via one system when it has not been tried on both. This is why experiments such as those of Xu *et al.* (2005), which involved functional magnetic resonance imaging (fMRI) on both the main and olfactory bulbs of mice in the same experimental setup, are especially valuable (see Chapter 9).

An important consideration when running bioassays is to ensure that conditions are standardized, so that, for example, animals are tested at the same time in their day. Apart from external, environmental variations, organisms used in bioassays should have a standardized physiological status, i.e., the same age, sex, social status, and mating history. However, standardization can bring its own problems for external validity (generalizability): by reducing within-experiment variation, it may limit causal inference to the specific experimental conditions (Richter *et al.* 2009, 2011).

Experience may change behavior in subtle ways. Prior experience and learning can have effects on insect behaviors such as host choice (Dukas 2008; Huigens *et al.* 2009). Studies of vertebrates routinely

Figure 2.2 Bioassays can use habituation as a tool. (a) When a resting rat is presented with a new odor from above, he rears and sniffs. This response can be used to investigate which mixtures/individual odors the test animal treats as the same. The time spent sniffing when the animal is repeatedly presented with the same odor declines rapidly (habituation), but increases again if a new odor is offered (dishabituation). Figure from Sundberg *et al.* (1982). (b) Urine odors from rats from different major histocompatibility complex strains (Chapter 3) are treated as different at trial seven out of nine trials (A) whereas (B) urine from a second rat of the same strain does not cause dishabituation and is treated as if it were the same as the first. Figure redrawn from Brown *et al.* (1989).

insensitive laboratory populations (Chapter 9) (McGrath *et al.* 2011).

Bioassays for primer effects measure longer term physiological effects, such as an increase in milt (semen) volume in male goldfish, *Carassius auratus*, after overnight exposure to female odors (Chapter 1) (Stacey & Sorensen 2011). Primer effects offer particular challenges because they can be subtle and slow to develop. Metabolomic and genomic approaches may reveal the changes that occur on exposure to pheromone and narrow down the candidate molecules (Sections 2.2.1.3 and 2.3.3).

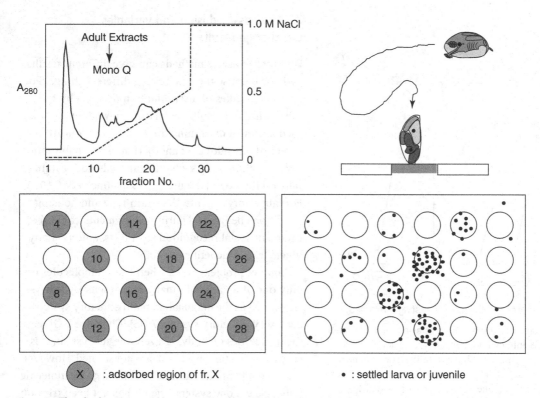

X : adsorbed region of fr. X   • : settled larva or juvenile

Figure 2.3 The planktonic cyprid larval stages of marine barnacles search for chemical evidence of other barnacles so that they can settle close to them. This bioassay separated the barnacle proteins by ion exchange fast protein liquid chromatography (FPLC) (top left) and then deposited the different fractions onto defined spots on a nitrocellulose membrane (lower left). The dosed plate was then exposed to searching cyprids in seawater for 48 hours to see which spots would cause them to settle and metamorphose (right). In this case, fractions 16 to 22 were judged to be active proteins. (See Chapter 12 for applied uses.) Figure from Matsumura *et al.* (1998).

require this factor to be taken into account. Olfactory learning could complicate studies of pheromones when successive exposures to test stimuli are required.

2.1.2 Studying signature mixtures

As soon as mammal secretions were analyzed with gas chromatography, the great variability between individuals was noted and this encouraged early doubts that mammal pheromones would be found (Box 1.5) (Chapter 1) (Beauchamp *et al.* 1976). The similarity between the mammal variability and the differences between the colony odors of social insects was noted

by Wilson (1970, p. 144) (Chapters 1 and 6) (see Figures 1.2 and 13.2).

In mammals, much of the work has been behavioral demonstration of the ability to distinguish different individuals. Aspects of bioassay design, underlying assumptions, and the distinction between the ability to discriminate and acting on it are discussed by Thom and Hurst (2004) and Johnston (2008) (see also Section 2.1.3). Which molecules are being used by animals, including ourselves, to distinguish individuals is not known, although in mice, major urinary proteins (MUPs), major histocompatibility complex (MHC) peptides, or small molecules associated or related to them, can all be used (Hurst 2009; Kwak *et al.* 2010).

In social insects, researchers have explored differences between colonies using techniques such as principal components analysis (Chapter 6) (van Zweden & d'Ettorre 2010). However, to see which molecules the insects are using as signals and cues, researchers are starting to do more direct experiments, using synthetic hydrocarbons, or hydrocarbons from other colonies, that are added to individuals or models such as glass beads (see van Zweden & d'Ettorre 2010). New non-lethal ways, using a water emulsion, to transfer cuticular hydrocarbons between ants could help (Roux et al. 2009). Aggression bioassays in arenas are commonly used to study ant colony interactions. A comparison of four bioassay designs with the Argentine ant, *Linepithema humile*, showed differences between them in replication consistency and in the ability to predict whole colony aggressive interactions (Roulston et al. 2003). Quantifying the aggressive responses of mandible opening by a harnessed ant to chemical cues touched to its antennae offers a more subtle and repeatable alternative to arena tests (Guerrieri & d'Ettorre 2008).

2.1.3 The importance of randomization and blind experiments

Unintended bias in experiments can lead to unreliable conclusions. To reduce potential bias, animals must be randomly allocated to treatments, and the scorers should not know which treatment they are observing (Burghardt et al. 2012; Evans et al. 2011). It is reassuring to see descriptions such as "In all experiments, the investigator was blind to the identity of the peptide stimulus as well as the genotype/surgical condition of the test animal" (Spehr et al. 2006a). For some experiments, automated behavioral analysis software can remove observer bias and allow subtle differences to be detected. Schaefer and Claridge-Chang (2012) describe recent successes using commercial and open source software, notably with mice, *C. elegans*, and *Drosophila*, as well as discussing the need for validation and potential pitfalls.

Randomized controlled trials and systematic reviews of the literature are rare in the kinds of chemosensory research described in this book. A systematic review of pheromone mediation of cat and dog behavior found many trials were flawed (Chapter 12) (Frank et al. 2010).

2.1.4 Replication

Apparently conflicting results may be obtained when a complex topic is studied over many years by different research teams using different bioassays. One example is female puberty acceleration by male pheromones in mice (Chapter 9). Flanagan et al. (2011) offer a valuable example of systematically repeating the bioassays of different laboratories in one set of experiments, using the same strain of mouse, which helped to resolve some of the inconsistencies. Such replications are rare.

The lack of true replication is one reason that systematic reviews (Evans et al. 2011) are not a common feature in chemosensory, ecological, and evolutionary research. The need for novelty to obtain funding or for maximum impact from publication discourages replication. Instead of true replication, we tend to use quasireplication, i.e., the "replication" of previous studies but using different species, systems, or conditions; many heterogeneous studies are then compared, either qualitatively or quantitatively, via meta-analysis (Palmer 2000). However, quasireplication provides less testing rigor than true replication, and it is less effective at resolving disputed hypotheses or results (Palmer 2000). To an evolutionary biologist like me, the repeated independent occurrence of a phenomenon across taxa is highly suggestive that the underlying traits have been subject to evolution. Nonetheless, we could do with more true replication.

2.1.5 Appropriate concentrations and delivery of odorants

Pheromone molecules should be active at the concentrations released under natural circumstances.

Otherwise, results may be spurious: almost any volatile compound extracted from ants will elicit alarm responses in ants if the concentrations are high enough (Hölldobler & Wilson 1990, p. 261).

The meanings of chemical signals are often affected by concentration. In the nematode *C. elegans*, the same molecules act as a sex pheromone in adults at picomolar concentrations but induce the dauer resting response in larvae at concentrations 10,000 times higher (Chapters 1 and 9) (Section 2.2.1.3) (Srinivasan *et al.* 2008). In many termite species, the trail-following molecule(s) used by workers is also used in quantities about 1,000 times greater as a sex-attractant pheromone by reproductive castes of the same species (Chapter 7) (Bordereau & Pasteels 2011).

Concentration is important for mammals too. The response of newborn rabbit pups to the mammary pheromone 2-methylbut-2-enal is limited to a range of concentrations (from 2.5×10^{-9} to 2.5×10^{-5} g/ml in milk) (Coureaud *et al.* 2004; Schaal *et al.* 2003). One of the weaknesses of some research on "putative" human pheromones is that a million times the natural amounts have been used (see Chapter 13).

Bioassays can depend on the method of delivery of chemical stimuli, which is more challenging than sound or light signals. Various designs of olfactometer seek to deliver chemical stimuli reliably (see e.g., Lundström *et al.* 2010). Special techniques may be needed to reach the right part of a sense organ such as the vomeronasal organ in mammals (Ben-Shaul *et al.* 2010).

2.1.6 Complications for discovery: signal redundancy and multimodal signals

The classic test of a possible role of a stimulus is to remove it and see if the response continues. Unfortunately, some signals may be redundant, meaning an animal can use one of a number of apparently synonymous signals to receive the message (Chapter 1) (Bradbury & Vehrencamp 2011, p. 298). Redundancy can occur with the multicomponent sex pheromones of moths: in the cabbage moth, *Trichoplusia ni*, each of the minor components could

be singly omitted from the blend without reducing male response (see Cardé & Haynes 2004). One reason this is rarely reported may be that, with subtractive bioassays in which only a single component is removed at a time (Section 2.1.7), redundant components would not be revealed (as they could be omitted without effect). A different kind of complication arises when the molecules can have multiple functions, and/or come from more than one source: in honeybees, pheromone molecules such as ethyl oleate are produced by both the brood and the queen (Chapters 6 and 9).

Conversely, many animal behaviors require combined inputs from many senses (multimodal or compound signals, Chapter 1). Multifactorial tests may show interactive effects of stimuli that would be missed had each been tested alone (Harris & Foster 1995). *Drosophila* males mate for longer, delivering more ejaculate, if they detect the presence of competitor males, revealed by a combination of any two of auditory, olfactory, or tactile cues (Chapter 3) (Bretman *et al.* 2011).

The challenge for the experimenter when dealing with possible multiple stimuli is to devise bioassays that can either separate out the functional roles of different stimuli, or that can integrate several different types of stimuli (e.g., chemical and acoustic signals) into one bioassay.

2.1.7 Multicomponent pheromones and subtractive approaches for analysis

Most pheromones are multicomponent, requiring a number of molecules in the correct ratio and concentration for activity (Chapters 1 and 9). Examples of multicomponent pheromones include those of nematodes, female moths, many ant trail pheromones, and some mouse and fish sex pheromones.

Multicomponent pheromones present particular problems for fractionation and bioassay because fractions of an extract may contain only one or some of the components of the signal, and so may show no behavioral activity unless they are recombined with other fractions. As the number of fractions rises, the complexity of bioassaying the potential combinations increases as a power function. The answer may be a subtractive approach in which individual fractions are dropped out one at a time. Fractions containing active components are revealed when removal of the fraction reduces the activity of the whole sample (Byers 1992) (for Byers' software for calculating an optimal strategy see www.chemical-ecology.net/papers/jce92.htm). It was fortunate that the first pheromone identified, bombykol, had a main component that was active by itself, otherwise Butenandt's bioassays might not have been successful in tracking the activity. Similarly the discovery of the rabbit mammary pheromone 2-methylbut-2-enal would have been harder if other components were needed for response (Chapter 1).

Metabolomic approaches, which do not require fractionation, offer some alternative solutions to the problem and have been used to identify multicomponent pheromones in nematodes (see Section 2.2.1.3) (Pungaliya *et al.* 2009; Srinivasan *et al.* 2012).

2.1.8 Using the animal's sense organs as detectors

The most direct methods to determine what the animal itself senses record the electrical impulses produced by biological receptors in the sense organs as they are stimulated by pheromones (see chapters in Touhara 2013). The techniques use the animal's own chemosensory system(s), which are tuned to the chemicals important to it. The first use of such techniques was in the 1950s, with antennae of male silk moths, by Dietrich Schneider who called it an electroantennogram (EAG) (Schneider 1999). Techniques similar to the EAG have been developed for vertebrates to give an electrovomerogram (EVG) or electro-olfactogram (EOG) (see chapters in Touhara 2013). The techniques are particularly powerful when they are coupled with separation techniques such as gas chromatography (GC) or high performance liquid chromatography (HPLC), so that as compounds elute from the chromatograph, they are passed over the sensory organ.

However, EAG or EVG/EOG sum the responses of many cells so they may give only weak signals to pheromone components for which there are few receptors. Single cell recording (SCR) may be needed to find cells sensitive to these components. In addition, electrophysiological techniques only indicate sensitivity to a molecule, not the behavioral response that may result. For example, recordings from males of the moth *Helicoverpa zea* show sensitivity to a compound characteristic of the female sex pheromone of a sympatric species but the behavioral response is to stop flying upwind (see Chapters 9 and 10) (Baker 2008).

Electrical recording from the sensory tissues of animals apart from insects is more complicated because the sensory organs are often less easily accessible. An alternative is to record from chemosensory neurons in tissue slices, for example in the VNO in mice (e.g., Leinders-Zufall *et al.* 2000) or to record from single mitral cells, also in mice (see Chapter 9) (e.g., Lin *et al.* 2005).

Instead of recording electrical impulses, optical imaging of living brain activity, visualized with calcium-sensitive dyes, can be used at every scale from microns, tissue slices, to whole brains (see also Section 2.3.1). Optical imaging in insects is discussed in Galizia and Vetter (2004) and Sandoz (2011). For rodents, optical imaging (e.g., Brechbühl *et al.* 2011) and other brain imaging of various kinds including fMRI (e.g., Xu *et al.* 2005) have been used (Chapter 9) (Frostig 2009; Kerr & Nimmerjahn 2012; Pain *et al.* 2011). Studies of human responses to odors have also used fMRI and positron emission tomography (PET) scans (Chapter 13) (Lundström & Olsson 2010).

2.2 Collection and analysis of semiochemicals

Having determined that a particular behavior is mediated by a semiochemical, the first step in the identification of the compound(s) is to collect them and confirm that the extract elicits the

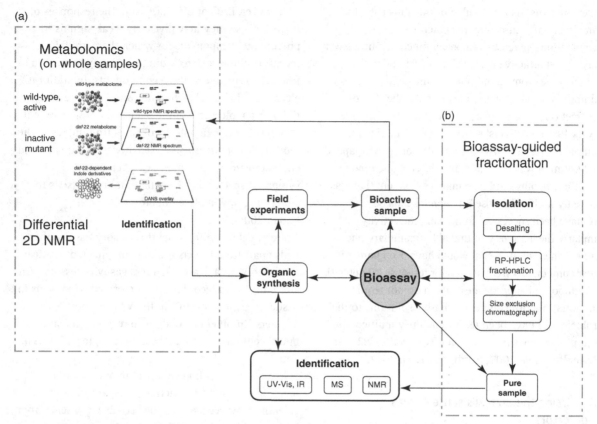

Figure 2.4 Pheromones can be found and identified by metabolomic approaches on whole samples (a) or by classical bioassay-guided fractionation (b). After identification by either route, the proposed molecules are synthesized and tested in laboratory and field bioassays (center).

(a) **Metabolomics.** Differential analysis via 2D-NMR spectroscopy (DANS) is one technique that can be used to compare samples from, for example, wild-type and non-signaling mutant animals (Section 2.2.1.3) (Forseth & Schroeder 2011; Robinette *et al.* 2011). Statistical comparison reveals the pheromone signals present in the wild-type but missing in the mutant. 2D-NMR may be sufficient to propose a structure for the molecule though other techniques may also be needed. For a larger version of this DANS analysis of nematode pheromones and explanation see Figure 2.5.

(b) **Fractionation.** Iterative bioassays to test the activity of the fractions as they are separated lead to relatively pure samples from which molecules are identified (Section 2.2). This flow chart is for isolation of molecules in seawater from marine organisms (Hardege & Terschak 2011). The usual route for analyzing insect pheromones might involve fewer steps because solvent extraction of gland tissues or SPME to collect headspace odors is often followed directly by GC–MS.

Center and part (b) adapted from Hardege (1999); (a) adapted from Srinivasan *et al.* (2012).

desired behavioral response. Traditionally the next step has been activity guided fractionation to locate the fraction(s) containing the active compounds(s) (Figure 2.4) (Millar & Haynes 1998). Bioassay-guided fractionation separates compounds in a crude extract into fractions based on a variety of properties, including solubility in solvents of different polarity, molecular mass, and molecular charge (Derby & Sorensen 2008). The details of the techniques used will depend on the physical and chemical properties of the bioactive compounds, and the medium from which they need to be separated and purified (see chapters in Touhara 2013).

With modern chromatographic and spectrometric techniques that require only nanogram to microgram quantities of a compound for full identification, large-scale fractionation methods such as solvent partitioning are often not required. Instead, a crude extract can be used either after one quick cleanup step to remove the worst of the co-eluting impurities, or even directly in the various chromatographic techniques that are coupled directly to a spectrometric technique, such as liquid chromatography–mass spectroscopy (LC–MS) and gas chromatography–MS (GC–MS), infrared, or NMR spectrometry, sometimes in combination with microchemical tests to determine the presence or absence of specific functional groups or other molecular features.

2.2.1 Collection and analysis without fractionation

Traditional fractionation processes, and even preliminary separations (chromatographic or otherwise), risk the loss of unstable, volatile, or irreversibly absorbed components (Meinwald 2003). Thus, techniques that do not require fractionation and cleanup steps can offer significant advantages.

2.2.1.1 Solid-phase microextraction

Solid-phase microextraction (SPME) traps molecules on an adsorbent polymeric coating, then desorbs them directly into an analytical instrument such as a GC or HPLC, combining sampling, analyte enrichment, and analysis into one step (Ouyang et al. 2011). In headspace sampling, the coated fiber or syringe is inserted into the headspace above the sample, for example collecting volatiles given off by male butterflies (Andersson et al. 2003) or mammalian urine or gland secretions (delBarco-Trillo et al. 2013).

With this non-destructive technique, the same organism can be sampled repeatedly, to track changes over time. For example, SPME wiped on the cuticle was used to follow cuticular hydrocarbon changes in *Drosophila* with age (Everaerts et al. 2010) and in an

ant after she became the dominant reproductive (Figure 1.2) (Peeters et al. 1999).

Stir-bar sorptive extraction (SBSE) is an analogous technique for larger samples, using a magnetic stir bar coated with an adsorbent (Ouyang et al. 2011; Soini et al. 2005). This technique has been used in studies of bird preen gland (Whittaker et al. 2010), human skin secretions (Penn et al. 2007), and oviposition cues in mosquitoes (Carson et al. 2010).

2.2.1.2 In situ analysis

Near-infrared spectroscopy (NIRS) generates absorption spectra for samples (Newey et al. 2008; Workman & Weyer 2012). Whole insects can be scanned rapidly with little or no preparation, generating a "chemical signature" of the insect cuticle and its hydrocarbons, though without identifying the molecules (Newey et al. 2008). Near-infrared spectroscopy has been used to distinguish different colonies of ants (Chapters 5 and 6) (e.g., Newey et al. 2010) and to distinguish cryptic ant species (e.g., Klarica et al. 2011).

On a different scale altogether, down to a resolution of 200 microns and 10 pmol sensitivities, the molecules on different parts of a single insect can be chemically identified using ultraviolet laser desorption/ionization orthogonal time-of-flight mass spectrometry (UV-LDI-TOF MS) (Yew et al. 2011). A UV laser is used to desorb and ionize compounds directly. The method has revealed new cuticular compounds on *Drosophila melanogaster* including oxygenated hydrocarbons; among them is a previously unknown male pheromone CH503 found on the cuticle around the genitalia (Chapter 9) (Yew et al. 2009, 2011).

2.2.1.3 Metabolomics

Metabolomics characterizes and quantifies all the compounds in the whole sample, such as the water the animal has been living in (for an introduction, see Nicholson & Lindon 2008). Candidate pheromone components, or biomarkers of a particular biological state, can then be pinpointed by comparing the compounds that are different in, for example, males vs. females or wild type vs. non-signaling mutants.

The use of such comparisons has a long history. For example, candidate mouse pheromones were identified by comparing, by eye, the GC profiles of urine volatiles from castrated and intact males, and mature and immature females (Schwende *et al.* 1986). (See Figure 1.2c for additional peak in the GC profile of an alpha ant.) However, metabolomics has been transformed by combining powerful statistical techniques with NMR spectrometry, which, unlike mass spectrometry, can be used with partially purified mixtures (Forseth & Schroeder, 2011; Robinette *et al.* 2011). Two-dimensional (2D) NMR spectroscopy enables a skilled spectroscopist to directly infer atom connectivity and spatial arrangements for novel structures (Forseth & Schroeder 2011; Robinette *et al.* 2011).

Metabolomics has a number of potential advantages over activity-guided fractionation: (1) when the molecules of interest are unstable and do not survive chemical isolation techniques such as chromatography; (2) when searching for diverse components of a multicomponent pheromone, which individually show little activity when separated by fractionation; (3) when the multiple bioassays needed for a fractionation approach are not practical because primer (developmental) effects of pheromones are being investigated and each bioassay takes a long time. Metabolomics may prove helpful for the identification of pheromones that have proven difficult so far, such as those of Crustacea (Derby & Sorensen 2008; Kamio & Derby 2011).

A study of *C. elegans* pheromones is a good example of using "differential analysis by 2D-NMR spectroscopy" (DANS) to find molecules of a synergistic multicomponent pheromone that were missed in earlier activity-guided fractionation studies (Pungaliya *et al.* 2009). Extracts of wild-type *C. elegans* cultures contain components that attract males and keep larvae at the dormant "dauer" stage (see Chapters 1 and 9). Activity-guided fractionation studies of extracts had previously allowed identification of three ascarosides (ascr#2, ascr#3, and ascr#4) that gave some but not all the activity of the *C. elegans* extracts, suggesting that important component(s) of the dauer and mating pheromone had been missed. DANS-based comparison of the metabolomes of wild-type *C. elegans* and the signaling-deficient *daf-22* mutant, known to lack the dauer and mating pheromones, revealed the missing molecules (Figure 2.5). As well as confirming the three previously known ascarosides, DANS revealed four previously undescribed ascarosides, perhaps missed because of their lack of activity when presented individually (Chapter 1) (Pungaliya *et al.* 2009). A synergistic combination of the three most active ascarosides (ascr#2, ascr#3, and the newly found ascr#8) resulted in the full mate-attracting activity of the wild-type pheromone extract in bioassays (Chapter 1). Each of the *daf-22*-dependent metabolites accounted for much less than 0.1% of the entire wild-type metabolite sample, demonstrating the sensitivity of 2D-NMR spectroscopy (Forseth & Schroeder 2011).

2.2.2 Horses for courses: techniques have different strengths

There is no one ideal approach. Which is best for your study will depend on many factors. New cuticular molecules, including some with pheromone activity, were found when *Drosophila melanogaster* was sampled with SPME or a laser technique (above). However, it is likely that previous work with solvents extracted the same, or possibly more, compounds than SPME, but these compounds were in a larger, more dilute sample, and so some of the trace components were missed during analysis. Among the disadvantages of SPME are that adsorption of different classes of compounds can be biased by the particular fiber used, and that it is essentially a sampling technique, i.e., once the sample is desorbed into an instrument, it is gone, so there is no material left for bioassays or for microchemical work. Similarly, the UV-laser techniques found new molecules but this technique itself

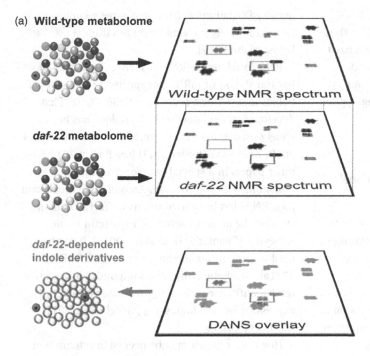

(a) **Wild-type metabolome**

Wild-type NMR spectrum

***daf-22* metabolome**

daf-22 NMR spectrum

***daf-22*-dependent indole derivatives**

DANS overlay

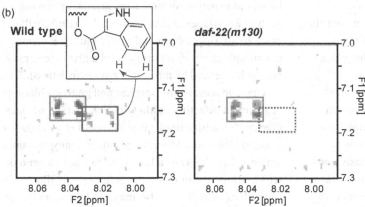

(b)

Wild type

daf-22(m130)

Figure 2.5 A metabolomics approach using differential analysis by 2D NMR spectroscopy (DANS) revealed missing pheromone components in the nematode *Caenorhabditis elegans* (Pungaliya *et al.* 2009; Srinivasan *et al.* 2012). Schematic representation of DANS. (a) *C. elegans* wild-type metabolite extract has dauer-inducing and male-attracting activity, whereas *daf-22* metabolite extract is inactive. A comparison of wild-type and *daf-22* metabolomes via DANS revealed candidate molecules for biological evaluation (squares). See Figure 1.6 for the *C. elegans* pheromone molecules. (b) Small section of the actual wild-type and *daf-22* NMR spectra used for DANS. Signals of indole carboxylic acid are present in both spectra (left small box), whereas another indole-derived signal (right small box) is only present in the wild-type, but not the *daf-22* spectrum. Figure from Srinivasan *et al.* (2012).

misses the saturated hydrocarbons that form >90% of the hydrocarbons.

Metabolomics has limitations, including a likely need for molecules to be present in larger amounts than needed for analysis using coupled GC–EAD for example, so metabolomics is unlikely to replace current methods for identifying sex pheromones in insects even though it has demonstrated its usefulness for nematode pheromone work.

2.3 Using genetic and other techniques from molecular biology

Genetic and molecular biology techniques, as well as traditional genetic crosses and mutants, have long been used to study pheromone production and perception. The methods range from switching genes on/off, labeling neurons to map circuits, through to genomics approaches to see when and where genes are active (Bendesky & Bargmann 2011; Lois & Groves 2012; Stapley *et al.* 2010; Touhara 2013). These techniques tend to be centered on well characterized model systems such as *Drosophila*, the nematode *Caenorhabditis elegans*, mice, and zebra fish, but as costs drop, many of these techniques are being applied to other organisms.

2.3.1 Manipulating genes for pheromone production, perception, and other pheromonal effects

Genetic engineering methods offer powerful tools to dissect signaling systems by selectively eliminating or modifying pheromone production, chemosensory perception, or related gene functions (Hardy *et al.* 2010; Touhara 2013). For example, in *Drosophila* an ingenious set of transgenic manipulations was used to eliminate CHC pheromones by ablating the oenocytes, specialized cells required for their production (Chapters 3 and 9) (Billeter *et al.* 2009). In mice, VNO perception was largely eliminated by knocking out the gene for TRP2, an ion channel expressed in most vomeronasal sensory neurons but not by chemosensory neurons in the main olfactory epithelium (Chapter 9) (Leypold *et al.* 2002; Stowers *et al.* 2002).

Temperature- or drug-sensitive promoter regions can be used to activate constructs, for example at a particular developmental stage (e.g., Billeter *et al.* 2009). Optogenetic methods use light-responsive proteins, introduced by genetic manipulation, to achieve gain- or loss-of-function of well defined events within specific cells of living tissue (Knöpfel & Boyden 2012; Miesenböck 2009; Vaziri & Emiliani 2012) (see Kasparov 2011 and other papers in a special issue of *Experimental Physiology* 96(1)). Light-sensitive ion channels, such as channelrhodopsin, can be expressed in olfactory sensory neurons, for example producing *Drosophila* larvae that "smell" blue light (Störtkuhl & Fiala 2011) or allowing light pulses to reveal mitral cell coding of olfactory circuits in mice (Chapter 9) (e.g. Dhawale *et al.* 2010). Optical imaging of light emitted by fluorescent proteins can be used to report neural

activity (Yizhar *et al.* 2011) and some techniques can be used in freely moving mammals (Kerr & Nimmerjahn 2012).

RNA interference (RNAi) "silences" genes by blocking translation of mRNA for specific proteins in particular tissues (for a review see Bellés 2010). First developed in nematodes, the technique has been widely used in insects and vertebrate systems (Bellés 2010; Zhuang & Hunter 2011) (see Poulin 2011 and other papers in a special issue of *Briefings in Functional Genomics* 10(4)). Among numerous examples, RNAi has been used to great effect in demonstrating the mode of action of royalactin in the honeybee (Chapter 9) (Kamakura 2011). It has been used to explore sex pheromones in rotifers (Snell 2011a). Synthetic morpholino oligonucleotides are a related antisense technology for modifying gene expression in vertebrates (e.g., Bedell *et al.* 2011; Hardy *et al.* 2010).

However, caution may be needed in interpreting knockout experiments and other genetic interventions such as RNAi (see e.g., Bonthuis *et al.* 2010; Schellinck *et al.* 2010). Genes commonly have multiple, diverse phenotypic effects (pleiotropy) and a given gene may also be influenced by many other genes (epistasis). This may be one reason why surgical and genetic ablations of the main olfactory epithelium and VNO sometimes give contradictory results (see Tirindelli *et al.* 2009 for a useful table comparing the results of many mammal studies). The effects of a knockout can also depend on which inbred strain or species is being used (Bonthuis *et al.* 2010). Knockouts also may not be complete: for example, some VNO sensory detection continues even in animals without a functioning TRP2 ion channel (Kelliher *et al.* 2006). Similar cautions surely apply to invertebrate knockouts.

2.3.2 Visualizing neural circuits (the "connectome")

Traditionally, chemosensory neural circuits have been traced by the Golgi method of cell staining or by filling neurons with dyes during single-cell recording (see in

a) Sequencing Whole genome, transcriptome or targeted regions can be sequenced.

Contig

Reads

NGS data consists of many short (<400bp) partially overlapping sequences (reads) that are assembled into longer continuous sequences (contigs) or aligned against a reference genome sequence. The overlapping nature of NGS data provides new opportunities to analyse genetic variation as outlined below.

b) Identify loci These methods often identify large genomic regions containing many genes but NGS is making it easier to annotate these regions and identify potential candidate genes.

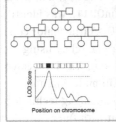

QTL mapping uses phenotype and genotype data from a pedigree or cross to identify chromosomal regions (loci) that explain variation in traits.

Population genomics uses genotypic data from individuals from divergent populations to identify outlier loci (grey). Adaptation loci can be identified with or without detailed phenotypic data.

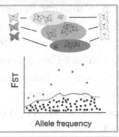

Genome-wide association studies find associations between traits and markers in large population samples. This method requires an existing linkage map or genome sequence to order markers.

Depth of coverage analysis quantifies the number of reads (read depth) between loci and populations to identify genetic variation (i.e structural variation). When transcriptome sequence is analysed, read depth can be used to identify divergent expression patterns and splice variants.

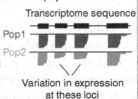
Variation in expression at these loci

c) Candidate gene NGS targeted sequencing of hundreds of genes or large genomic regions provides a useful method to identify adaptation genes.

NGS sequencing of targeted regions (i.e. genome enrichment) can be used to identify adaptation loci. Targeted genes can be selected *a priori* based on knowledge of gene function, or genomic regions that were identified via the methods outlined above (B) can be targeted.

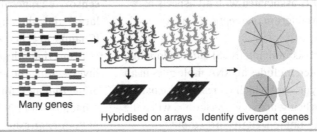

Figure 2.6 Next-generation sequencing (NGS) makes finding loci and genes underpinning adaptation in non-model species much more feasible. These techniques also help the identifications of candidate genes and variations between the sexes, life stages, and states (e.g., virgin, mated). (a) It is possible to obtain large amounts of sequence data quickly and cheaply; this can be used to create dense single nucleotide polymorphism (SNP) marker panels. (b) These are ideal for methods commonly used to identify loci involved in adaptation. (c) NGS can also be used for targeted sequencing of large candidate regions to identify particular genes. Perhaps most important, several complementary approaches can be combined in a single study, e.g., quantitative trait loci (QTL) mapping and population genomics (See Chapter 3). From Stapley *et al.* (2010).

Jefferis & Livet 2012). Genetic methods have used labeling with chromogenic enzymes (e.g., β-galactosidase) or green fluorescent protein (GFP) (e.g., Jefferis et al. 2007). Fluorescent proteins are now available in many colors and offer the advantages of genetically encoded labels that are directly detectable, and that can be restricted to specific cell types or compartments by appropriate promoters or targeting sequences (Jefferis & Livet 2012). A combinatorial color method called "Brainbow" marks individual neurons in one of >100 colors by combinatorial and stochastic expression of multiple fluorescent proteins from a single transgene, allowing neurons to be tracked like different colored wires (Jefferis & Livet 2012; Lichtman et al. 2008). First developed for mice, Brainbow is also available for *Drosophila* (see in Jefferis & Livet 2012). Cell labeling can be coupled with genetic manipulations that reveal or perturb cell function (Section 2.3.1).

2.3.3 Genomics

Genomics using DNA and RNA sequence data contributed to the search for the first vertebrate olfactory receptors, exploiting their anticipated sequence similarity to other G protein-coupled receptors (Chapter 9) (Axel 2005; Buck 2005). With the continuing development of faster and cheaper sequencing methods, genomic techniques will allow us to identify the genes and genetic variants associated with all aspects of chemical signal production, perception, and processing.

Next generation sequencing (NGS) can explore the transcriptome, which includes all RNA molecules in an organism or cell type, and which represents the part of the genome that is being actively transcribed when the study organism is sampled. High-throughput sequencing technologies have revolutionized transcriptomics by allowing RNA sequencing at massive scale (RNA-seq), after RNA conversion to cDNA (Martin & Wang 2011; Ozsolak & Milos 2010). Modern platforms sequence up to billions of DNA strands in parallel, substantially increasing throughput and minimizing the need for the fragment-cloning methods often used in the older Sanger sequencing techniques (Ozsolak & Milos 2010). The huge drop in costs allows transcriptome or whole genome sequencing of non-model species and even of individuals (Hawkins et al. 2010; Stapley et al. 2010). The approaches used to identify genes from sequence data are summarized in Figure 2.6.

Differential RNA expression was used to help identify candidate genes, before the current fast sequencing methods were available. For example, the honeybee olfactory receptor, AmOR11, was identified as the male receptor for 9-ODA (the major component in honeybee queen pheromone) by using microarrays (Slonim & Yanai 2009), which revealed greater expression of this gene in male antennae (Wanner et al. 2007). Modern sequencing techniques enable this type of investigation with finer resolution and unlike microarrays, RNA-sequencing does not require you to know the gene of interest in advance.

A typical RNA-sequencing experiment offers a near-complete snapshot of a transcriptome, including the rare transcripts that have regulatory roles. This means that it can be used to explore the transcriptional and epigenetic differences between individuals of different types. For example, to investigate the molecular mechanisms and chemosensory changes during barnacle larval settlement, larval and adult transcriptomes were compared (Chapter 4) (Chen et al. 2011).

Transcriptomics could be particularly useful for investigating the complex and delayed primer effects of pheromones, for example in social insects (Chapters 6 and 9) (Zayed & Robinson 2012). Changes in RNA expression could be used to monitor the initiation (and development) of physiological changes, which are often hard to measure in bioassays. For example, microarray studies of honeybee brain gene expression showed that queen bee mandibular pheromone, brood pheromone, and alarm pheromone alter worker-bee brain gene expression patterns in ways consistent with their effects on behavioral maturation (see Chapter 9) (Kocher & Grozinger 2011; Zayed & Robinson 2012).

Broader taxonomic comparisons can be made using whole-genome comparisons, candidate gene approaches, and a genome-scale comparative analysis of protein-coding sequences to explore, for example, the evolution of chemical signaling and sociality across the social insects (Fischman *et al.* 2011). The use of genomics approaches in understanding the evolution of queen pheromones in social insects is reviewed by Kocher and Grozinger (2011) (Chapters 6 and 9).

Summary

The study of chemical signals and signature mixtures provides many examples supporting Brenner's (2002) observation on the importance of new techniques for progress in science. You may invent some of the techniques that change our understanding in the future.

However, whatever the latest technology, success in working with semiochemicals comes down to good bioassays and a sound understanding of the biology and behavior of your study species.

The special challenges presented by understanding both the chemistry and biology of pheromonal signaling and chemosensory cues have been addressed most successfully by research teams that have combined the expertise of scientists in the relevant disciplines. One such fruitful collaboration is described by the chemist Jerrold Meinwald (2009) and his biologist colleague, the late Thomas Eisner (2003), but there have been numerous others over the past several decades, across all taxa.

Further reading

For comprehensive information on well established techniques in chemical ecology see Millar and Haynes (1998), Haynes and Millar (1998), and Touhara (2013). Chapters in Touhara (2013) also cover modern techniques in molecular biology applied to pheromone research. Ouyang *et al.* (2011) give a comprehensive and practical review of SPME techniques for biological samples of various kinds. Techniques for collecting and analyzing complex chemical mixtures from vertebrates are reviewed in chapters in East and Dehnhard (2013). For a discussion of techniques for analysis of crustacean pheromones, see chapters in Breithaupt and Thiele (2011). For analysis of cuticular hydrocarbons see Blomquist (2010). Francke and Schulz (2010) review methods for analysis of insect pheromones in general. Johnston (2003) and Drea *et al.* (2013) have good advice on exploring mammal semiochemicals and behavior, applicable to many vertebrate systems.

It is also worth looking at recent articles in journals such as the *Journal of Chemical Ecology* or *Chemoecology*. For advice on good practice, see the current versions of journal advice to authors (e.g., Anon 2008, 2009, 2010). See the International Society for Chemical Ecology www.chemecol.org for information about courses.

For a stimulating and readable introduction to the importance of using well designed experiments with "fair tests" including "blinding" and randomization, and systematic review, see Evans *et al*. (2011) (www.testingtreatments.org). A number of books cover behavioral methods: Lehner (1996), Dawkins (2007), Touhara (2013), and Martin and Bateson (2007).

For molecular structures see sites such as www.chemspider.com, which allows you to search by common name and shows synonyms as well as the systematic names. See also the Appendix for a short guide to the terminology (available for free download from the website associated with this book). Many pheromones are illustrated on Pherobase www.pherobase.com (El-Sayed, 2013).

Pheromones, chemical cues, and sexual selection

For species across the animal kingdom, pheromones and chemical cues play essential roles in successful mate location and mate choice. Given the ubiquity of chemical communication among animals, chemical cues are likely to emerge among the key criteria animals use for mate choice (Johansson & Jones 2007; Smadja & Butlin 2009; Thomas 2011; Wyatt 2010). These include a wide variety of species-specific sex pheromones but also non-pheromone chemical cues such as those associated with genetic compatibility, with the immune system, and with health or disease, which are also important in mate choice.

Darwin proposed sexual selection to explain why the males of so many species have conspicuous bright colors or long tail feathers even though such extravagant features or behavior might reduce survival by natural selection (Darwin 1871). He included chemical signals alongside visual and sound signals as outcomes of sexual selection, describing the distinctive odors of breeding male insects, reptiles, birds, and many mammals. Darwin concluded that the development of elaborate odor glands in male mammals is "intelligible through sexual selection, if the most odoriferous males are the most successful in winning the females, and in leaving offspring to inherit their gradually perfected glands and odors" (Darwin 1871, Vol. 2, p. 281). I have called this "success of the smelliest" (Wyatt 2009).

Darwin suggested that sexually selected signals in animals would have many features in common, including elaboration or expression of the signals in only one sex, development only in adults, often only in the breeding season, and use primarily or exclusively in mating (Darwin 1871, Vol. 2, p. 279). Chemical signals, and the scent glands that create them, commonly show all these characteristics (in humans too, Chapter 13).

Sexual selection is shorthand for selection that arises due to differences in reproductive success caused by competition over mates (Andersson 1994; Andersson & Simmons 2006; Jones & Ratterman 2009; Shuster 2009). Sexual selection is one part of natural selection (Figure 3.1). The other parts are viability selection, for survival, and kin selection, for the ability to affect the reproductive success and survival of related individuals (Cornwallis & Uller). The three parts of natural selection operate in fundamentally the same way, by acting on hereditary phenotypic variation among individuals, which translates into different numbers of offspring directly or indirectly (in kin selection).

Sexual selection can take many forms (Figure 3.1) (Table 3.1) (Andersson 1994; Jones & Ratterman 2009). In intra-sexual selection, members of the same sex compete with each other, in *scrambles* to be first to arrive at a potential mate (Section 3.3) or in *same-sex "contests"* in fights to be the top animal (Section 3.5). Inter-sexual selection involves *mate choice*, in which members of one sex bias their mate choice with respect to some phenotypic character of the other sex (Sections 3.6, 3.7, and 3.8). Scrambles and mate choice continue after copulation, in aspects of sexual selection missed by Darwin: *post-copulatory* sperm competition and cryptic female choice of sperm (Section 3.11) (Eberhard 2009). Any characteristic influencing success in competitions for fertilizations, before or after mating, can be sexually selected.

Sexual selection also includes basic factors affecting reproductive success including choosing a mate of the right species, sex, and reproductive state (Andersson 1994; Andersson & Simmons 2006; Smadja & Butlin 2009; Thomas 2011). Sex pheromones and speciation are discussed in Section 3.12.

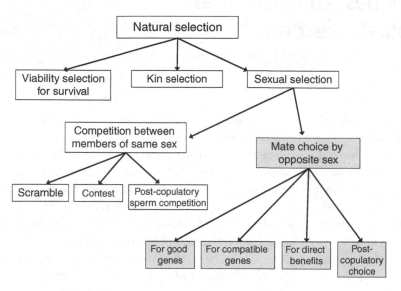

Figure 3.1 Classification of modes of natural selection (Andersson 1994; Cornwallis & Uller 2010; Jones & Ratterman 2009). Intra-sexual competition and inter-sexual selection are different kinds of sexual selection. Mate choice for the opposite sex includes species recognition and choice of direct benefits, which could be health, territory, or other resources. Competition and choice continue with post-copulatory competition between sperm from different males and female choice of sperm. For examples of each of the kinds of sexual selection, see text.

The enormous variety of specialized secretory glands among male mammals and male Lepidoptera (moths and butterflies) is probably largely the result of sexual selection. In mammals there are secretory glands in species-specific positions such as the flanks, around the eye, around the genitals, and anus. The equivalents in male Lepidoptera are a profusion of brushes, inflatable balloons (coremata), and other structures on the wings, legs, and abdomen for exposing the pheromones produced in associated glands (Allen *et al.* 2011; Birch *et al.* 1990). There is convergence in the special lattice-like hairs (osmetrichia) to soak up and emit pheromone in some moths and mammals (Müller-Schwarze 2006, pp. 56–7). Feathers in birds may have the same effect: birds preen them with an oily secretion from the uropygial gland near the tail (as well as waterproofing, secretions differ between the sexes) (Campagna *et al.* 2012; Caro & Balthazart 2010; Zhang *et al.* 2010).

Sexual selection and mate choice is not just about success of the smelliest male with the most pheromone, it is also about genetic compatibility (Section 3.8) (Mays & Hill 2004). The most compatible males will be different for each female. Chemical cues that vary between individuals are commonly used to identify genetically compatible mates. These are not species-wide pheromones but instead are cues (Chapter 1).

3.1 Which sex should advertise?

Which sex should signal to attract mates? In many animals, such as crickets, frogs, and birds, males spend large amounts of energy singing to attract females (Chapter 1) (Kokko & Rankin 2006; Kokko & Wong 2007). Among vertebrates as a whole, there are not many examples of pheromones used to attract mates from a distance (Müller-Schwarze 2006) but there are many examples among fish: male lampreys release sex pheromone to attract females and both visual and pheromone signals are used by territory-holding male gobies to attract females (Figure 12.3) (Chung-Davidson *et al.* 2011). Some frogs have male pheromones (Chapter 1) (Poth *et al.* 2012; Wabnitz *et al.* 1999). There are female signals too: female hamsters leave trails of vaginal secretions (Petrulis 2009) and female dogs, as dog owners will know, produce long-range attractant pheromones (as yet unidentified).

Table 3.1 **Examples of mechanisms involved in sexual selection involving chemosensory cues and pheromones. See text for additional examples.**

Evolutionary mechanisms (not mutually exclusive)	Chooser prefers...; Benefit (B) to choosy females	Chemosensory example
Direct benefits		
Healthy mate selection	Avoid disease cues; B = avoid disease	Female house mice choose urine of uninfected males (Kavaliers *et al.* 2005)
Parental care	Gain good parent; B = parental care for young	Mediterranean peacock blenny, *Salaria pavo*, females choose males secreting pheromone (Barata *et al.* 2008a; Serrano *et al.* 2008)
Resources	Signal of greater resources; B = resources	Female tiger moths, *Utetheisa ornatrix*, choose males with most alkaloids, receive most protective poison (Dussourd *et al.* 1991)
Indirect benefits		
Fisherian sexy sons (Runaway)	Sexually attractive; B = sexy sons, daughters inherit majority mate preference	Female tiger moths, *U. ornatrix*, choose males with most alkaloids, which are larger, and whose offspring are large and attractive (Iyengar & Eisner 1999a)
		Female mice prefer odors of dominant mice, and the sons of dominant mice (could be good genes?) (Drickamer 1992)
Indicator, "good genes"	Indicative of mate's viability; B = offspring inherit viability advantages	Female tiger moths, *U. ornatrix*, choose the pheromones of larger males and the daughters inherit this larger size and thus lay more eggs (Iyengar & Eisner 1999a)
Genetic compatibility	Non-kin, complementary genes; B = healthier offspring	Cockroach, *Blattella germanica*, use CHC signature mixture cues to avoid mating with their sisters (Lihoreau & Rivault 2009) In three-spined stickleback fish, *Gasterosteus aculeatus*, selection for mates with particular MHC alleles to enhance resistance to local parasites (Eizaguirre *et al.* 2011)
Other mechanisms		
Sensory bias	Exploitative of pre-existing sensory bias; B = possibly no benefit (Chapter 1)	Nuptial feeding in cockroaches on male abdominal pheromones/secretions (Gemeno & Schal 2004; Vahed 2007)

Table 3.1 **(cont.)**

Evolutionary mechanisms (not mutually exclusive)	Chooser prefers... ; Benefit (B) to choosy females	Chemosensory example
Sexual conflict		Sex peptides in seminal fluids change female reproductive behavior and physiology e.g., *Drosophila* (Avila *et al.* 2011; Chapman 2008; Wolfner 2009)
Post-copulatory		(Eberhard 2009)
Sperm competition		A male meadow vole, *Microtus pennsylvanicus*, doubles the number of sperm if rivals' scent marks are detected (delBarco-Trillo & Ferkin 2004; Ferkin 2011)
Cryptic female choice		Female flour beetles, *Tribolium castaneum*, preferentially use sperm of males with more pheromone (Fedina & Lewis 2008; Lewis & Austad 1994)
		Female red junglefowl, *Gallus gallus*, retain fewer sperm following insemination by brothers (Pizzari & Snook 2004)

After tables and ideas in Alcock (2009); Andersson and Iwasa (1996); Andersson and Simmons (2006); Jones and Ratterman (2009); Solomon and Keane (2007) and other authors

Only a minority of insect species have male pheromones as long-range attractants and the majority of these are beetles. Apart from species in which males display together in leks (Section 3.7.4), male insects producing long-range pheromone signals tend to be associated with patchy resources needed by the female (Landolt 1997). Females can get both mating and resources at the same place. For example, males of the scorpion fly, *Panorpa*, pheromone-call when they have an insect food item gift (Kock *et al.* 2007; Thornhill 1979) and male papaya fruit flies call from fruits (Landolt *et al.* 1992). The pheromones of plant-feeding and storage product insects are often synergized by plant or food volatiles (Bartlet 2010; Dewhirst *et al.* 2010; Reddy & Guerrero 2004). *Drosophila melanogaster* males have a sex-specific receptor that is stimulated by plant odor molecules and that feeds into the sex pheromone brain circuits (Chapter 9) (Grosjean *et al.* 2011). Male *Nicrophorus* burying beetles typically pheromone-call for females when they have found a dead mammal (see below). The *N. vespilloides* pheromone, ethyl 4-methyl heptanoate, also attracts other eavesdropping males (Haberer *et al.* 2008). Attraction of both sexes to male pheromones occurs in many other species of insects and the pheromones concerned are often called aggregation pheromones even if their original function appears to be sex attraction (Chapter 4).

In moths, pheromone calling is famously by females and it may even have contributed to their diversification and success as a group (Section 3.12.3). Female moths signal, rather than males, probably because of the low energy cost of advertising with pheromones compared with sound (Chapter 1) and possibly the relatively small risk of predation to the caller (compared with the larger number of predators responding to sound or visual signals) (Chapter 11) (Zuk & Kolluru 1998). The major costs are borne by the moth males in expensive flight and the risk of bat predation on the wing (Cardé & Haynes 2004; Greenfield 2002, p. 61 ff.). Males compete to be first (scramble competition, Section 3.3).

Though female calling with pheromones is ancestral for moths (Cardé & Haynes 2004; Greenfield 2002, p. 61 ff.), not all species follow this pattern. For example, in the cabbage looper moth, *Trichoplusia ni*, both males and females produce pheromones (Landolt *et al.* 1996). The male pheromone, which is released from the hair pencils around dusk, attracts males and females. The female pheromone, released in the latter half of the night, attracts males.

The conflict between the sexes can be significant at the point of signaling for mates when the attraction of additional mates is in the interests of only one party (Hosken *et al.* 2009). A male burying beetle, *Nicrophorus*, calls with pheromone to attract females (Eggert & Müller 1997). If the carcass is big enough to support more offspring than a single female can produce, the male could benefit by pheromone-calling to attract another female, but at a potential cost to the first female's larvae. In *N. defodiens* the first female tries to prevent the male pheromone-calling again, enforcing monogamy (Eggert & Sakaluk 1995).

A converse example is the conflict between the female spider, *Linyphia litogiosa*, which might benefit from additional males attracted to her pheromone, and the first male arriving, who would benefit from reducing the competition from other males. The first spider male neatly bundles up her web to prevent further release of pheromone from the silk (Watson 1986).

3.2 External fertilization and chemical duets

Animals that rely on external fertilization by releasing eggs, or sperm, or both into the water, such as marine worms, sea urchins, and most fish, must precisely synchronize gamete release and many use pheromones for this (Breithaupt & Hardege 2012). Sperm do not live long and are quickly diluted by the huge volume of the sea, so the sexes need to be close together at gamete release (which is one reason many marine organisms form aggregations, see Chapter 4) (Gascoigne *et al.* 2009; Levitan & McGovern 2005; Serrão & Havenhand 2009). Responses to cues such as tides, lunar cycles, temperature, and day length enable all members of a species to reach sexual maturity together for an annual synchronized spawning event. The exact moment of spawning is co-ordinated by a chemical duet of pheromones released into the water by males and females. The exchange of chemical signals is well explored in polychaete worms (Figure 3.2) (Breithaupt & Hardege 2012; Lindsay 2009). Mass spawnings occur on coral reefs, for example, involving animals from many phyla. Animals may respond to the cues from another species to co-ordinate their own spawning but may use species-specific pheromone cues for the precise courtship and gamete release (Breithaupt & Hardege 2012). Species specificity despite shared pheromone molecules can also occur if species spawn at different times of day (Mercier & Hamel 2010).

3.3 Scramble competition

Sexual selection can act on the ability to scramble, to get to a mate ahead of rivals. In the great majority of moths, males make long-distance flights to find females emitting minute quantities of pheromones, in pico- or nano- grams per hour (Cardé & Haynes 2004). A low pheromone release rate may select for strong, effective searchers: a test of quality in both Fisherian and good-genes senses (Greenfield 2002, p. 61 ff.). Females mating with the first arrival(s) would produce winning sons in the next generation. The intense selection to be first has led to selection for great sensitivity to detect pheromones, reflected in

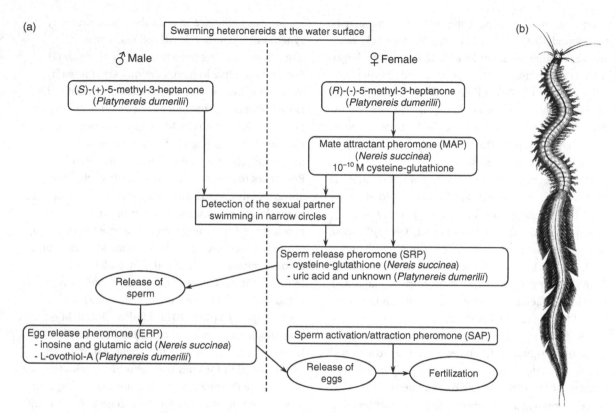

(a)

Swarming heteronereids at the water surface

♂ Male

♀ Female

(S)-(+)-5-methyl-3-heptanone
(*Platynereis dumerilii*)

(R)-(-)-5-methyl-3-heptanone
(*Platynereis dumerilii*)

Mate attractant pheromone (MAP)
(*Nereis succinea*)
10^{-10} M cysteine-glutathione

Detection of the sexual partner
swimming in narrow circles

Sperm release pheromone (SRP)
- cysteine-glutathione (*Nereis succinea*)
- uric acid and unknown (*Platynereis dumerilii*)

Release of
sperm

Egg release pheromone (ERP)
- inosine and glutamic acid (*Nereis succinea*)
- L-ovothiol-A (*Platynereis dumerilii*)

Sperm activation/attraction pheromone (SAP)

Release of
eggs

Fertilization

(b)

Figure 3.2 Animals that release their eggs and sperm into the sea ensure co-ordination by a chemical duet of pheromones. (a) The marine polychaete worms *Nereis succinea* and *Platynereis dumerilii* spawn in "nuptial dances" timed by the phases of the moon, initiated by sundown, and choreographed by their species-specific pheromones (Breithaupt & Hardege 2012; Hardege *et al.* 2004). The mature heteronereids leave their burrows and swarm near the surface of the water. The females swim slowly in circles, releasing a mate attractant pheromone (MAP) and sperm-release-pheromone (SRP), which attracts many males who start to swim fast around her. As they swim, the males release small quantities of sperm and coelomic (body cavity) fluids that contain the egg release pheromone (ERP). On sensing the egg release pheromone, the female swims fast in small circles (the nuptial dance), and spawns, discharging a massive cloud of eggs and coelomic fluid, which contains MAP. Males respond immediately to MAP by swimming through the egg cloud and releasing large clouds of sperm. (b) Heteronereis reproductive stage of a nereid worm. The rear part of the body is full of gametes. Worm from Harter (1979). Figure courtesy of John A. Terschak. Caption adapted from Breithaupt & Hardege (2012).

the huge feathery antennae of males (Box 9.2) (Symonds *et al.* 2012), and in neurosensory adaptations enabling faster flight and/or greater ability to track a complex pheromone plume (Chapters 9 and 10). It seems that in many moth species, the female mates with the first male to arrive (Cardé & Haynes 2004). However, in some species there may be female choice between the chemical displays of males (Section 3.12.3.3).

The pheromone-controlled mating system in the goldfish, *Carassius auratus* (Chapter 1), provides another example (Appelt & Sorensen 2007). When she is ready to spawn, the female goldfish releases pulses of urine containing pheromone (a mixture of prostaglandin $F_{2\alpha}$ ($PGF_{2\alpha}$) and its metabolite, 15-keto prostaglandin $F2_{a\alpha}$ ($15K-PGF2_{a\alpha}$)) (Figure 1.4). Her pheromone attracts males. As she releases her eggs, males compete to be closest to her by trying to push other males away (a behavior stimulated by a male-released pheromone, androstenedione, Sorensen *et al.* 2005).

3.4 Pre-copulatory mate guarding

In many species, females can only be fertilized at particular times and males compete to be there at the right time. Males attempt to monopolize access by pre-copulatory mate guarding (see Estrada *et al.* 2010 and references therein). Recognizing when a female is about to be fertilizable is often done by chemical cues (Thomas 2011).

In many insects, males emerge before females (possibly as a result of sexual selection among males to be first), and males then attempt to find the pupae of females about to emerge, which they guard, mating with the female as she emerges (Thornhill & Alcock 1983). In the nymphalid butterfly *Heliconius charithonia* males can distinguish the sex of pupae around a day before emergence, using the volatile monoterpenes secreted by the pupae (linalool by male and linalool oxide (furanoid) by female pupae Estrada *et al.* 2010). Males compete to guard the female pupae, and mate with the female which emerges. Estrada *et al.* (2010) suggest there may be some benefits to the female so her linalool oxide could be a pheromone signal, not simply a cue exploited by males.

Male crustaceans and many mammals share the problem that their females can only be fertilized at the molt or estrus, respectively (events controlled by hormones). In both types of animal, urine cues advertise hormonal state. The phenomenon has been well studied in Crustacea (Breithaupt & Hardege 2012). In shore crabs, *Carcinus maenas*, uridine diphosphate is a major component of a urine-borne pheromone mixture which elicits cradle-carrying of the pre-molt female by the male (Hardege & Terschak 2011; Hardege *et al.* 2011b). Males of the lobster *Homarus americanus* create urine odor currents from their shelters, "singing" a chemical song (Figure 3.3) (Aggio & Derby 2011; Atema & Steinbach 2007).

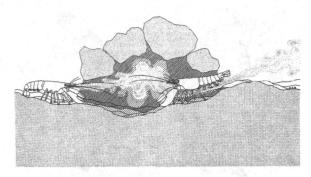

Figure 3.3 A female lobster (left), attracted by a male's chemical "song," jets her own urine toward a male in his shelter. He responds by retreating to the opposite entrance and fanning his pleopods. The female will join the male until her molt and copulation. Figure from Atema (1986).

Females choose the locally dominant male by his urine signals. Males accept pre-molt females of all sizes, recognized by chemical cues that stimulate a high-on-legs response in the male. However, the male only accepts one female at a time. After some days with the male, the female molts and mates. Apart from any indirect benefits, the guarded female profits by protection from predation and cannibalism during her vulnerable soft-shell post-molt condition.

Decapod shrimps show a variety of mate guarding patterns, from temporary guarding prompted to contact with a pre-molt reproductive female, long term monogamous mate guarding, through to "neighborhood of dominance" attraction of females to dominant males, as in lobsters (Bauer 2011). Pheromones appear to be involved in most systems studied so far.

Many mammals show estrus testing and pre-copulatory mate guarding by dominant males, sometimes called "consorting." For example, male kangaroos sniff the pouch and cloacal opening of females and follow them if they are near estrus (Figure 3.4) (Miller *et al.* 2010a). Male Asian elephants, *Elephas maximus*, can detect pheromones in the urine of females coming into estrus and about to ovulate (Chapter 1) (Rasmussen *et al.* 2003; Rasmussen *et al.* 2005). Male African elephants, *Loxodonta africana*, are able to do this too (but the molecules are likely to be

Figure 3.4 Male mammals use smell to identify females approaching estrus and then stay close to them, guarding them from rivals. Here, an adult male kangaroo smells the pouch and cloaca of a female. Figure from Russell (1985). Drawing by Priscilla Barrett.

different from the Asian elephant female's signal) (Bagley *et al.* 2006). Dominant African elephants only monopolize females during the estrus phase, when it matters. Females appear to choose to consort with the dominant males in musth (Section 3.7.1).

3.5 Contests

Sexual selection by contest has males fighting each other (for access to females or for ownership of resources needed by females). Pheromones are not weapons but may be reliable signals of health, fighting ability, or territory ownership (Chapter 1).

The pheromones associated with dominant males and their role as "badges of status" have been investigated in the cockroach, *Nauphoeta cinerea* (d'Ettorre & Moore 2008; Kou *et al.* 2009). Sexual selection occurs contemporaneously through male–male competition and female–male choice. Both processes are mediated by a male sex pheromone consisting of three molecules: 3-hydroxy-2-butanone, 2-methylthiazolidine, 4-ethyl-2-methoxyphenol (Figure 3.5) (Sirugue *et al.* 1992; Sreng 1990). Males fight for status and, while fighting, the dominant male releases thousands of nanograms of 3-hydroxy-2-butanone (Kou *et al.* 2009). Thereafter, threats from an aggressive posture and

further release of 3-hydroxy-2-butanone serve to keep subordinate males submissive. Subordinates have large quantities of 3-hydroxy-2-butanone in their sternal gland but they do not release it (until, after some weeks, they fight the aging dominant to take over). Females prefer males with higher levels of all three molecules (for condition effects, see Section 3.7.1) (Moore & Moore 1999).

Males of the lobster, *Homarus americanus*, fight to establish dominance and thus ownership of preferred large shelters under rocks, attractive to females (Section 3.4) (Aggio & Derby 2011; Atema & Steinbach 2007). During the fights, males release pulses of urine. Losers stop urine release as soon as they have lost. For the next week, almost no fights take place as the former loser tends to avoid the winner, recognized by individual signature mixture odor cues in urine (Aggio & Derby 2011; Atema & Steinbach 2007).

Urine signals are also important in interactions between freshwater tilapia fish, *Oreochromis mossambicus*, on leks (Section 3.7.4). Dominant males release urine, containing a pheromone, probably a sulphated amino-sterol, during fights with other males (Barata *et al.* 2007, 2008b).

Male squid, *Loligo pealeii*, fight other males for access to females at spawning grounds. The male squids are first attracted by visual stimulus of the egg capsules and contact with the 10 kDa protein pheromone on the surface of eggs leads to the aggressive response (Cummins *et al.* 2011). Receptive females are nearby and the largest and the most dominant males appear to gain the greatest number of copulations.

Male mice fight for exclusive territories, which the winner marks frequently with his urine signal to give an honest signal of ownership (Chapter 5) (Hurst 2009). Females sniff his urine marks, and prompted by a protein pheromone in male urine (darcin) learn his signature mixture odor and the location of the scent mark (Chapter 9) (Roberts *et al.* 2010, 2012). Females choose to mate with the territory owner, recognized by his individual odor.

While many of these examples show benefits for females that choose dominant males, this may not be

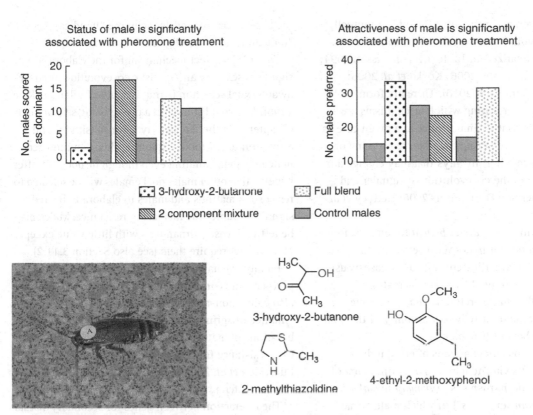

Figure 3.5 The multicomponent pheromone blend of the cockroach, *Nauphoeta cinerea*, has different effects under two different mechanisms of sexual selection. Males with higher levels of 2-methylthiazolidine and 4-ethyl-2-methoxyphenol, which are genetically correlated, and lower levels of 3-hydroxybutanone (possibly resulting from release in antagonistic interactions, see text), which is genetically independent, are more likely to dominate other males. In contrast, females prefer males with higher levels of 3-hydroxybutanone. The pheromone is thus under balancing sexual selection. In the figures, each of the comparisons is pairwise with control males (gray columns). Figures adapted from Moore *et al.* (1997) and Moore and Moore (1999) by P. J. Moore. Figure and adapted caption from d'Ettorre and Moore (2008).

the case in all species (Kokko 2005; Qvarnström & Forsgren 1998).

3.6 Mate choice: overview

Typically, males compete for matings with females because males in most species invest less in their off-spring. As a result of investing less, males tend to have greater potential reproductive rates than females and so compete for the limiting, choosier sex (females) (Clutton-Brock 2007; Shuster 2009). Thus there are

usually more males seeking females than vice versa, so the ratio of males to reproductively available females (the operational sex ratio, OSR) at a given time and place is usually greater than 1:1 (Kokko *et al.* 2006). Mating systems are also influenced by the ecology of feeding and resource distribution, among other factors (Clutton-Brock 2007; Shuster 2009).

Mate choice seems to matter for reproductive success. Animals able to make their own choice of mates produce more and/or healthier offspring than if assigned mates by an experimenter (for example, Consuegra & Garcia de Leaniz 2008; Gowaty *et al.*

2003). However, a number of different but mutually compatible evolutionary mechanisms could be involved (summarized in Table 3.1) (Andersson 1994; Andersson & Simmons 2006; Kokko *et al.* 2006; Pizzari & Bonduriansky 2010). There are formidable challenges in determining which mechanisms are at work and their relative importance in a given case as they are likely to be acting together, reinforcing or countering each other, in ways that may change during the course of the co-evolution of signaler and receiver (Andersson & Simmons 2006; Hettyey *et al.* 2010).

One mechanism is a *direct benefit* to females, for example by choosing males with better nuptial gifts (Section 3.7.2) or avoiding contagious disease by using odor cues to choose healthy males as mates (Section 3.7.3). Female preference for an ornament can evolve if the ornament reflects the ability of the male to deliver the benefit (Chapter 1).

Next are various mechanisms offering indirect (genetic) benefits via the female's offspring. First of these indirect mechanisms is "*runaway sexual selection*" ("*Fisherian sexy sons*") in which male attractiveness is itself heritable (so choosy females get the indirect benefit of producing attractive sons). The original female preference can be for indicators of genuine quality or an exploitation of sensory bias (Chapter 1) (Arnqvist 2006). In the later "runaway" phases, "sexy sons" are not necessarily better at surviving but have a high reproductive value simply because they are attractive to females in that population. Runaway elaboration of odor might be selection for stronger, longer lasting odors, by analogy with visual and acoustic signals, which females prefer stronger and more rapidly repeated (Ryan *et al.* 2010).

The second indirect mechanism is the *indicator* mechanism. If males display a trait, for example quantity of pheromone (Section 3.6), in proportion to their health and viability, and the viability is heritable, then females choosing males with stronger odors will gain genes for high viability to pass to their offspring (an "additive genetic benefit") (Andersson 1994, p. 53; Andersson & Simmons 2006; Grether 2010). The idea is

the basis of many models including "handicap," "indicator," and "good genes."

The third indirect mechanism for the elaboration of signals is sexually antagonistic co-evolution or chase-away sexual selection (Arnqvist 2006; Holland & Rice 1998). This could start with a pre-existing sensory bias (Chapter 1) in the female, say color sensitivity and attraction to red food. Exploiting this, a mutant red male may gain a mating even though he does not offer benefits that other males do. Females will be selected to resist such matings and males to elaborate this red signal to counter the increasing resistance. Males may be left with costly ornaments with little value except that females require them (see also Section 3.11.2).

Finally, we have sexual selection for choice of partners with compatible genes (Section 3.8). These are also called non-additive genetic effects, which enhance offspring fitness by creating favorable combinations of maternal and paternal genomes, for disease resistance for example (Hettyey *et al.* 2010; Puurtinen *et al.* 2009). Each female's ideal choice may thus be different.

The interaction between genes for secondary sexual traits and environmental factors may be more complex than the indirect genetic models, above, often assume, with variable effects in different seasons, different population densities, and on different spatial scales for example (Cornwallis & Uller 2010). For example, while models of sexual selection tend to assume a central role of "genetically determined" mate preference, there is growing evidence that in some species mate preferences are partly learned through imprinting in early experience (Chapter 1) (Owens *et al.* 1999; Verzijden *et al.* 2012; Wyatt 2010). This is particularly relevant to olfactory imprinting, for example to variable odors associated with the major MHC (Section 3.8) and colony odors, but also for species-specific odors that are pheromones (individuals normally learn the same odors) (Chapter 1). Imprinting could lead to choosing partners like their parents (as with species-specific odors), for example, daughter imprinting on father's odor and color in diverging stickleback species (Kozak *et al.* 2011) and in swordtail fish (Verzijden &

Rosenthal 2011). Sexual imprinting could induce preferences for exaggerated parental traits (e.g., ten Cate *et al.* 2006).

3.7 Mate choice for good genes, mate quality, and direct benefits

In many species, the most important roles for chemical communication occur in courtship, once the male and female are close to each other. Short-range phero-mones may be important in species and sex recognition. Once species identity is confirmed, females may then choose a male using chemosensory cues to his quality, which reflect factors, in mammals, such as his social status, the quality of nutrients he has consumed, his reproductive state, and his health (Johansson & Jones 2007; Thomas 2011). She may also assess the direct benefits such as gifts of protective poison or parental care he can offer her or her offspring (Section 3.7.2). Cues for genetic compatibility are discussed in Section 3.8.

3.7.1 Signals and cues to quality

The pheromone signals given by animals reflect their condition, itself a product of genetics and factors such as richness of diet and other environmental conditions (Cornwallis & Uller 2010; Hill 2011; Pizzari & Bonduriansky 2010).

Females of the tobacco moth, *Ephestia elutella*, are more likely to mate with larger males, which produce more than twice as much wing-gland pheromone as smaller ones (Phelan & Baker 1986). Matings with these larger males produced a greater number of offspring that survived to the pupal stage and these pupae were larger. Both Fisherian "sexy sons" and "good genes" effects are suggested by these data. Similarly, the size of male scent glands and frequency of marking in bank voles, *Clethrionomys glareolus*, and in mice are heritable (Horne & Ylönen 1998; Roberts & Gosling 2003). Female mice show a preference for the odors of dominant males and the odors of their sons (Drickamer

1992). In the crested auklet, *Aethia cristatella*, a socially monogamous seabird with a citrusy scent (notably octanal, (Z)-4-decenal, decanal among other molecules) produced by both sexes in the breeding season, concentration appears to correlate in males with social rank (in captivity), though the story is not yet clear (Hagelin 2007). The display may also involve the birds' facial crest of feathers in a multimodal signal (Chapter 1).

Male mating success in the oriental fruit fly, *Bactrocera dorsalis*, requires both sufficient protein in the adult diet and the ingestion of a plant molecule, methyl eugenol, which acts as a pheromone precursor (Shelly *et al.* 2007). The attractiveness of a male *Nauphoeta cinerea* cockroach is increased with better body condition (influenced by greater carbohydrate intake), because he produces more pheromone (all three molecules) though this does not increase his dominance (Section 3.5) (South *et al.* 2011). Meadow vole males, *Microtus pennsylvanicus*, on a higher protein diet produced more attractive chemosignals in their urine marks, probably via effects on their physiological condition (Ferkin *et al.* 1997; Hobbs & Ferkin 2011). In rock lizards, *Iberolacerta cyreni*, the proportion of oleic acid, attractive to females, in a male's scent marks is dependent on his body condition (Martín & López 2010a,b) (though the importance of female choice by chemical cues in this lizard family has been questioned, Font *et al.* 2012).

In vertebrates, an important internal measure integrating many of these effects, including dominance and feeding status, is the level of androgen hormones such as testosterone in the blood (Hill 2011). Production of pheromone is directly related to hormone levels and so scent marks will tend to be honest in reflecting their owner's condition (Chapters 1 and 5) (Hurst 2009; Roberts 2007). Testosterone stimulates many of the glands important in producing pheromones or other signal odors such as the leg (femoral) gland in the male lizard, *Iguana iguana*, the temporal glands in elephants (below) and many rodents. Testosterone implants in castrated male meadow voles, *Microtus pennsylvanicus*, show that the male scent glands respond in a graded way to testosterone levels (Ferkin *et al.* 1994). In the receiver, hormones also have an impact on how chemosensory cues are perceived and responded to (Chapter 9).

Only dominant male African, *Loxodonta africana*, and Asian, *Elephas maximus*, elephants in good physiological condition enter "musth," an annual period of heightened aggressiveness and highly elevated testosterone concentrations (Rasmussen *et al.* 2003, 2008). During musth, mature males secrete volatiles, in Asian elephants including frontalin, in fluid streaming copiously from the temporal gland on the head and dribble strong smelling urine (Rasmussen *et al.* 2003). Females prefer to mate with mature males rather than with non-musth males or young males (see also Section 3.4). Similarly, the male European corn borer moth, *Ostrinia nubilalis*, has a four-component pheromone blend on his hair pencils, used in courtship of females (Section 3.12.3.3). The proportions of the four components change as he ages and females find the blend produced by four-day-old males most attractive (Lassance & Löfstedt 2009).

While choice by females is emphasized in this chapter, males can be choosy too (Edward & Chapman 2011). In many species, males gain direct benefits by choosing larger, more fecund females (Edward & Chapman 2011). For example, male garter snakes select bigger females, in better condition, by the females' pheromones alone (Mason & Parker 2010). The males respond to the females' skin pheromone blend, which becomes dominated by the longest chain, unsaturated methyl ketones with increasing body length and better body condition (likely the result of higher estrogen levels, Parker & Mason 2012). Such females will produce more offspring per litter.

3.7.2 Paternal investment, pheromones, and female choice

The male's contribution of parental care or other resources can be important for a female's reproductive success. In insects and fish there are examples of a close relationship between paternal investment by the male and his pheromone signal. There are direct benefits to the female but in some insects inherited, genetic benefits have been demonstrated.

The males of some beetle, butterfly, and moth species offer protective anti-predator compounds to the female and advertise them with a pheromone (Conner 2009; Eisner & Meinwald 2003). All have convergent courtship behaviors that allow females to assess the value of potential gifts from each male. The value of his gift is indicated by an honest signal with a pheromone derived from the precursor of the defensive molecule(s).

The North American tiger moth, *Utetheisa ornatrix*, uses plant alkaloids for defense and the male produces an honest signal based on his store of alkaloids; this has a decisive role in sexual selection and a selective advantage for the survival of offspring (Eisner & Meinwald 1995, 2003). As larvae, both sexes sequester plant alkaloids (PAs) as they feed on their natural diet of *Crotalaria* plants. When adult, the male transfers the poisonous PAs to the female in his spermatophore when he mates. The gift of poisonous PAs supplements her protection from predators such as spiders and she uses them to protect the eggs from predators such as coccinellid beetles and ants.

The mating system is conventional for a moth: females pheromone-call at dusk. When the male arrives, he hovers around her, exposing his scent brushes (coremata) containing the male pheromone, hydroxydanaidal (HD) derived from PAs in the diet (Figure 3.6). The female can use the amount of pheromone on the coremata to assess the potential gift from a male since his HD pheromone content correlates with his PA levels and the quantity of alkaloid he would transfer (Figure 3.7) (Dussourd *et al.* 1991; Kelly *et al.* 2012). In nature, male PA content is proportional to male size – larger males have more PAs (Lamunyon & Eisner 1993). So by selecting males with high PAs, the female also selects big males, which transfer larger spermatophores, providing more nutrients for her eggs as well as more PA (Iyengar & Eisner 1999a). Body mass is heritable in both sexes, so by choosing larger males, females obtain genetic benefits for their offspring as well as direct

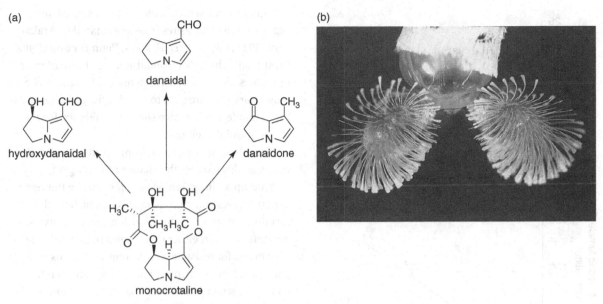

Figure 3.6 (a) Many Lepidoptera sequester monocrotaline, a pyrrolizidine alkaloid, from their host plants and most species convert it into another molecule as their pheromone (Conner & Weller 2004; Schulz 2009). Males of the queen butterfly, *Danaus gilippus bernice*, convert it to danaidone. (b) Male tiger moths, *Utetheisa ornatrix*, convert it to hydroxydanaidal and present it to females from their everted coremata during courtship (the "hairs" on the brushes are modified scales). Females choose males with more pheromone (Conner & Weller 2004, Eisner & Meinwald 1995, 2003). Chemical diagrams adapted from Conner and Weller (2004). Photograph courtesy of the late Tom Eisner in Eisner and Meinwald (1995).

benefits as larger sons and daughters are, respectively, more successful in courtship and more fecund (Iyengar & Eisner 1999b). The preferred male trait (size) and the female preference for that trait are correlated, as females with larger fathers have a stronger preference for larger males, fitting the predictions of both "good genes" and Fisherian models (Iyengar *et al.* 2002). The preference gene or genes lie mostly or exclusively on the Z sex chromosome, which is strictly paternally inherited by daughters (males are the homogametic ZZ sex in Lepidoptera).

Male fish offering parental care of eggs may also advertise paternal benefits. Parental males of the Mediterranean peacock blenny, *Salaria pavo*, defend a nest in a shell or rock crevice and use pheromone(s), including pheromone secretions from their anal glands, to attract females (Barata *et al.* 2008a; Serrano *et al.* 2008). Females choose to lay eggs with parental males that have well developed anal glands (Barata *et al.*

2008a). The anal glands are also the source of proteinaceous antimicrobial secretions, which the male regularly smears on the eggs as part of his parental care (the secretion appears to be effective at protecting the eggs from microbial growth, Pizzolon *et al.* 2010). Development of the anal glands is induced by increased levels of a sex steroid, which is also correlated with male nesting behavior and paternal care (Saraiva *et al.* 2010). The male's anal gland pheromones may be a sexually selected advertisement of his paternal behavior and the amount of anal gland secretion he can offer. Females can also use the perceived size of the anal glands as a cue (Pizzolon *et al.* 2010).

3.7.3 Mate choice and infection with pathogens and parasites

In many species, uninfected animals are preferentially selected as mates. This has been shown in studies of

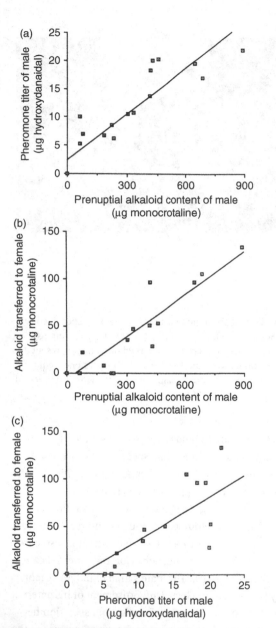

Figure 3.7 The male tiger moth pheromone provides an honest signal of the value of his alkaloid store, which he will transfer to the female during mating. Graphs (a), (b), and (c) show the various relationships between the prenuptial adult male, *Utetheisa ornatrix*, alkaloid titer (monocrotaline), the adult male pheromone hydroxydanaidal (HD) titer, and quantity of alkaloid transferred to the female at mating. Figures from Dussourd *et al.* (1991).

rodents infected with a variety of viruses, micro-organisms, and parasites such as nematodes (Arakawa *et al.* 2011; Kavaliers *et al.* 2005; Penn & Potts 1998a). Most studies have been about female choice of males but males also avoid diseased mates. Typically, chemosensory cues are used to avoid infected individuals; for example, female mice show a preference for the odors of uninfected males.

Direct benefits can come from avoiding infected mates as this reduces the chance of contagion, of picking up an infection. There may also be indirect benefits as resistance to parasites can be heritable (Section 3.8) and thus by choosing resistant mates, animals can gain either good genes in general or specific genes for resistance to current pathogens and parasites. More controversially, a link between the quality of secondary sexual characteristics and immunocompetence has also been proposed (see below).

Avoiding infected individuals as mates could be analogous to avoiding people with coughs and runny noses: these would be best seen as cues rather than evolved signals. I wonder if avoidance might also work by "expectancy violation," if infected animals give off cues that do not match the ones that have been learned to be expected from healthy animals, in the way that rats learn the likely flavor of a familiar food and attend to violations of it (Stevenson 2010). It is possible that animals learn the individual volatile signature mixtures of infected animals, later recognizing these individuals and avoiding contact. This social recognition learning appears to be influenced by the neurotransmitter oxytocin (Chapter 9) (Arakawa *et al.* 2011; Choleris *et al.* 2009).

Nonetheless rodents may also have particular receptors in their VNO (Chapter 9) that respond to the generalized molecular cues of infection or inflammation such as formylated peptides released from bacteria or mitochondria at sites of infection or tissue damage (Liberles *et al.* 2009; Rivière *et al.* 2009). These receptors are a new class of VNO receptor, formyl peptide receptors, which appear to have evolved from receptors that function in the immune system.

Choice of uninfected mates could also be due to positive attraction to the high quality of their secondary sexual characteristics including the quantity of pheromones in androgen-dependent glands. For example, only well fed and healthy males have testosterone levels sufficient for pheromone gland development (see Section 3.7.1, Chapter 9). Zala *et al.* (2004) found that male mice infected with the bacterium *Salmonella enterica* had reduced marking rates and their scent marks were less attractive to females. This may be an honest indicator of health, as an index (Chapter 1). It may be physiologically difficult for an individual to fake a healthy profile because of release of chemosensory cues such as formyl peptides (above) produced by pathogens, or parasites, or immune responses.

The immunocompetence handicap hypothesis proposes that the hormonal maintenance of sexual signals such as pheromones is at the expense of the simultaneous suppression of immune response to infection (Folstad & Karter 1992). The immunocompetence handicap hypothesis shares the problems of its parent handicap hypothesis (Box 1.6) (Chapter 1) (Hill 2011; Számadó 2011a,b). The immunocompetence theory suggests only high quality males (and/or those with best genes for resistance) could afford to signal without suffering large parasite loads. Though it was first proposed for vertebrates, the hypothesis has also been explored for pheromone signals in invertebrates (Lawniczak *et al.* 2007). The link between testosterone and pheromone production in rodents is well established (Section 3.7.1) but the effects of testosterone on the immune system and links to resistance to disease and parasites are in some doubt (Drury 2010; Roberts *et al.* 2004). Part of the problem for the work on both vertebrates and invertebrates may be the reliance on simple measures of immunocompetence. Though single causal factors are attractive hypotheses, they may not reflect biological reality.

3.7.4 Leks

Leks are arenas where displaying males apparently offer only sperm to attending females. These places thus offer an opportunity to test the fitness benefits of female choice, through direct benefits, "sexy sons" or "good genes" mechanisms, without the complication of paternal investment (Högland & Alatalo 1995). Mate choice is particularly strong on leks and typically a few males make most of the matings.

Males of the tephritid Mediterranean fruit fly (medfly, *Ceratitis capitata*) each defend a single leaf in leks (Shelly & Kennelly 2007). Once the male has a territory, he releases pheromone as a long-range attractant to females (one male cost is that predators are also attracted, Chapter 11). In this species male–male competition is not important and territories are only weakly defended (Whittier *et al.* 1994). In other species, lekking spots may be fiercely contested (see Shelly & Whittier 1997). Reproductive success, condition, and the cost of pheromone production have been explored in another tephritid fruit fly, *Anastrepha suspensa* (see costs, Chapter 1) (Pereira *et al.* 2010b).

Insects have been used to test the theoretical benefits of female choice on leks as it is possible to create leks artificially in some species. In the South American sandfly, *Lutzomyia longipalpis*, male pheromones attract females to leks on or near the host, a bird or mammal (Johansson & Jones 2007). A laboratory study of lekking in this species demonstrated that the most important benefit to females of choosing attractive males was the Fisherian one of "sexy sons": rearing the next generation showed that the sons of attractive fathers tended to be attractive when they in turn formed leks (Jones *et al.* 1998). Attractive males wing-fanned more and had more pheromone in their tergal gland.

Females may come close to a number of potential mates before choosing one (Högland & Alatalo 1995). For example, females of a mouth-brooding fish, the cichlid Mozambique tilapia, *Oreochromis mossambicus*, visit leks of males defending small nest territories (Barata *et al.* 2008b). When the males detect the pheromones in the urine of pre-ovulatory females, they respond with courtship and release pulses of urine containing a male pheromone, likely to be a sulphated amino-sterol. Dominant males release more urine, which contains a higher concentration of the putative pheromone, a signal that females may use to choose

(a)

(b)

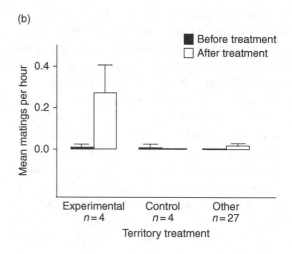

Figure 3.8 Uganda kob, *Kobus kob*, females use smell to identify the territories most often visited by other females (the urine deposited by the previous females impregnates the soil). (a) Males compete to hold these frequently visited territories. (b) Moving the soil from successful territories to previously unpopular ones ("experimental") greatly increases visits by females and mating rates in contrast to "controls" (with the earth only disturbed). "Other" indicates unmanipulated previously unsuccessful territories, which remain unsuccessful. Mean visits per territory per hour ± SE.

Figure redrawn from Deutsch & Nefdt (1992). Photograph of male kob, Frank Dickert (www.fdickert.de), Creative Commons license.

dominant males. Females spawn in the territories, but collect and brood the fertilized eggs away from the lek.

In two African antelope species, the Uganda kob, *Kobus kob*, and Kafue lechwe, *Kobus leche*, the territories preferred by estrous females were the ones most often visited previously by other females (Clutton-Brock & McAuliffe 2009; Deutsch & Nefdt 1992). Successful territories remained popular for months even though the male owners changed almost daily. Females sniff the ground to identify the territories most often urine marked by earlier females while soliciting the owner (Figure 3.8). Males compete for these heavily marked territories. Females choosing these territories would find the most vigorous and dominant males as mating partners (Clutton-Brock & McAuliffe 2009).

Among lekking insects, pheromones are commonly used by males to attract females (Johansson & Jones 2007). Shelly and Whittier (1997) found lekking species in some 52 genera in a broad range of insect taxa including Diptera, Hymenoptera, Lepidoptera, Homoptera, Coleoptera, and Orthoptera. Within these taxa, some groups were characterized by male pheromones: notably tephritid fruit flies, bumblebees, anthophorid bees, ants, sphecid wasps, and vespid wasps. For example, males of the European beewolf digger wasp, *Philanthus triangulum*, defend small territories that they scent-mark with a cephalic-gland pheromone (whose major component, (Z)-11-eicosen-1-ol, possibly evolved from a pre-existing sensory bias, Chapter 1) (Kroiss *et al.* 2010).

However, the benefits of female choice in leks are not always clear. In the medfly, *Ceratitis capitata*, despite non-random mating, no benefits were associated with female choice (either in increased fecundity or as "sexy sons") (Whittier & Kaneshiro 1995). In the lekking Hawaiian *Drosophila grimshawi*, females mating with the most vigorously courting males (which invested most in pheromone production) had lower fecundity (Droney 2003).

3.7.5 Fluctuating asymmetry

The idea that females could use body symmetry as a sensitive indicator of male genetic quality (a "good genes" benefit) seems increasingly doubtful. More symmetrical males are said to have low fluctuating asymmetry (FA), which is claimed to reflect the ability of the individual to develop perfect bilateral symmetry despite developmental instability caused by environmental perturbations such as food availability (review Møller & Thornhill 1998) (for other examples see Johansson & Jones 2007).

However, there is little evidence that FA in any secondary sexual trait is significantly heritable, so FA is unlikely to reveal genetic quality that can be transmitted to offspring (Maynard Smith & Harper 2003, p. 63 ff.; Polak 2008). Similarly, there is little evidence that FA is condition dependent or negatively correlated with the size of sexual traits and their genotypes (Polak 2008). It is possible that the measured symmetry might be correlated with a cue such as male size that is being used in female mate choice even though the females are not responding to FA (Maynard Smith & Harper 2003, p. 63 ff.).

3.8 Mate choice for genetic compatibility revealed by chemical cues

Females and males in many species use the chemical cues of potential partners to select mates for optimal outbreeding, choosing a mate that is genetically different from, or compatible with, themselves (Hettyey

et al. 2010; Mays & Hill 2004). It is thus different from other kinds of sexual selection (in which there is general agreement between females on which is the "sexiest" male), as each female will have her own optimal choice for compatibility.

A key selective benefit of choice for difference from self in mates, and thus choosing non-kin, is the avoidance of inbreeding depression, largely caused by the presence of recessive deleterious mutations in populations (Charlesworth & Willis 2009; Kempenaers 2007). Greater individual heterozygosity (the proportion of gene loci in an individual that are heterozygous) is often correlated with enhanced disease resistance, survival, and reproductive success (see below). The benefits of heterozygosity come generally from dominance (masking the deleterious alleles) (Charlesworth & Willis 2009; Penn 2002).

Across the animal kingdom, animals use whatever cues allow them to distinguish and avoid relatives as mates. Some species will use visual or sound cues for recognition of kin but the use of chemosensory cues seems to be most common, perhaps because all animals have chemosensory receptors. Whatever the cue used, these have to be learned (Chapters 1 and 9). For chemosensory cues these are learned signature mixtures, not pheromones (Chapter 1) though you will find these confused in the literature on mate choice. The signature mixture cues range from cuticular hydrocarbons (CHCs) in many insects, lipids in snakes, to the molecules associated with the MHC in many other vertebrates (see below, Box 3.1) or, in mice, potentially also the MUPs.

Males of the cockroach, *Blattella germanica*, use CHC signature mixture cues to avoid mating with their sisters by instead preferentially courting non-sibling females (Lihoreau & Rivault 2009). Males seem to use the rule: "choose different from self" (Chapter 1). In turn, females choose to mate with the most enthusiastically courting males (Lihoreau *et al.* 2008). The cricket, *Teleogryllus oceanicus*, also uses CHC profiles to avoid genetically similar potential mates, as CHC similarities correlate with genetic similarity (Thomas & Simmons 2011). Birds can distinguish between kin and non-kin by olfactory cues (Bonadonna & Sanz-Aguilar

Box 3.1 What odors are associated with the major histocompatibility complex?

Behavioral choices including mate choice may be based on chemical profiles (Chapter 1) (Figures 1.1 and 5.3), the sum total of molecules given off by an individual (see figure below). In vertebrates, along with the effects of other genes, diet, hormonal state, and parasites, some of the chemical profile is associated with the MHC (Kwak *et al.* 2010; Restrepo *et al.* 2006). The highly polymorphic MHC genes play a central role in controlling immunological self/non-self recognition (Penn & Ilmonen 2005). The MHC genes encode cell surface proteins (class I and II MHC molecules) that present small peptide antigens to T cells. These MHC genes are the most polymorphic loci known in vertebrates.

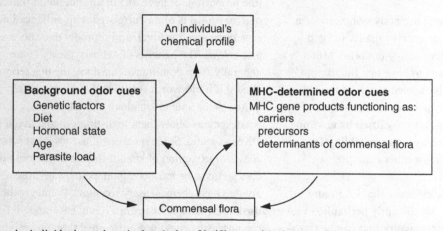

An individual vertebrate's chemical profile (Chapter 1) is the sum of the interplay of MHC-determined products, odors produced by the genetic background of the individual, diet, and its commensal microbial flora among other factors. Figure adapted from Zavazava and Eggert (1997).

How the MHC proteins influence odor profiles is still not clear (Penn 2002; Restrepo *et al.* 2006). The proteins themselves are too large to be volatile, but they may affect the bacterial flora and thus bacterial odors (Archie & Theis 2011). The MHC proteins may also act as specific carriers for smaller molecules perhaps derived from bacteria or the host animal (Chapters 1 and 13). In mice, the differences in MHC type are reflected in the relative concentrations of volatile carboxylic acids in the urine (Singer *et al.* 1997), though in humans no association of MHC with different armpit carboxylic acids was found (Natsch *et al.* 2010). In mice, non-volatile MHC peptides may also give identity (Leinders-Zufall *et al.* 2004) as can major MUPs (Hurst 2009; Hurst & Beynon 2013). However, identity can also come from peptide-free volatiles related to the MHC (Kwak *et al.* 2009). See Section 3.8.1 for further discussion.

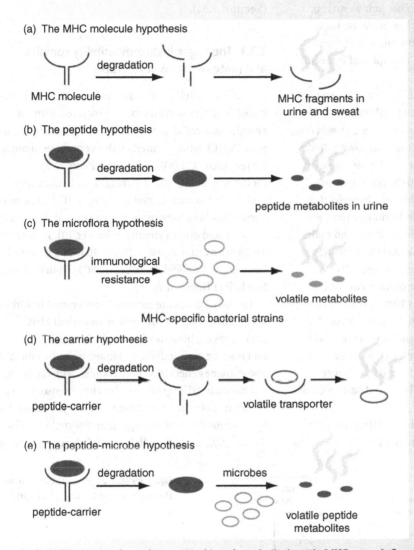

(a) The MHC molecule hypothesis

degradation

MHC molecule → MHC fragments in urine and sweat

(b) The peptide hypothesis

degradation

peptide metabolites in urine

(c) The microflora hypothesis

immunological resistance

MHC-specific bacterial strains → volatile metabolites

(d) The carrier hypothesis

degradation

peptide-carrier → volatile transporter

(e) The peptide-microbe hypothesis

degradation   microbes

peptide-carrier → volatile peptide metabolites

As the MHC proteins themselves are too big to be volatile, how do MHC genes influence individual odor? Among the possible mechanisms are: (a) fragments of MHC molecules in urine and sweat provide the odorants (the MHC molecule hypothesis); (b) MHC molecules may alter the pool of peptides in urine whose metabolites provide the odorants (the peptide hypothesis); (c) MHC genes may alter odor by shaping allele-specific populations of commensal microbes (the microflora hypothesis); (d) a 27 kDa soluble MHC fragment could carry volatiles (the carrier hypothesis); and (e) MHC molecules may alter odor by changing the peptides that are available to commensal microbes (the peptide-microflora hypothesis). Figure from Penn & Potts (1998a).

2012; Krause *et al.* 2012). On the basis of breeding records (which show that kin do not pair as mates), storm petrels, *Hydrobates pelagicus*, may use these olfactory cues in mate choice (Bonadonna & Sanz-Aguilar 2012). Other vertebrate examples of inbreeding avoidance are explored below.

Both inbreeding and outbreeding have costs, so many species appear to aim for optimum outbreeding (Kempenaers 2007). Female red mason bees, *Osmia rufa*, preferentially mate with males from their own population rather than more distantly related males, possibly using CHC cues (Conrad *et al.* 2010). Both male and female garter snakes, *Thamnophis sirtalis parietalis*, choose to court and mate with individuals from their own dens versus those from different dens, and males can make this choice based on detection of female chemical cues, including the presence and relative abundance of the methyl ketone components of the garter snake pheromone blend (which varies between populations) (Figure 2.2) (Mason & Parker 2010). An investigation of some 250 years of human population data from Iceland showed greatest reproductive success for third- to fourth-cousin marriages (Helgason *et al.* 2008), perhaps resulting from MHC-based partner choice (see below) (Chapter 13).

Female choice does not end at mating: cryptic female choice after copulation can lead to selection of sperm from compatible mates (Section 3.11).

3.8.1 The major histocompatibility complex and mate choice in vertebrates

Across the vertebrates, mate choice is heavily influenced by chemosensory cues associated with the enormously variable major histocompatibility complex (MHC), which underlies the vertebrate immune system (Box 3.1) (Milinski 2006; Penn 2002; Piertney & Oliver 2006; Yamazaki & Beauchamp 2007). First demonstrated in mice, MHC-linked mate choice has been shown in other mammals including humans and other primates (Chapter 13), fish, birds, amphibians, and reptiles (Drury 2010; Eizaguirre & Lenz 2010; Havlíček & Roberts 2009; Houck 2009; Setchell & Huchard 2010).

The MHC genes are extremely polymorphic with many alleles (50–70 for some mammalian MHC genes), giving billions of individual combinations and making each individual almost unique at the MHC. The extreme genetic diversity coupled with the ability to discriminate MHC genotypes by chemosensory cues makes it useful in mate choice. Diversity is maintained by a combination of ecological and sexual selection (Figure 3.9). What the MHC-related chemosensory cues

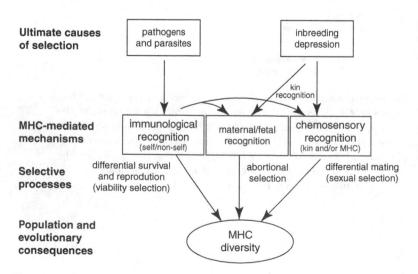

Figure 3.9 Many kinds of selection shape the evolution of MHC genes. Figure after Apanius *et al.* (1997).

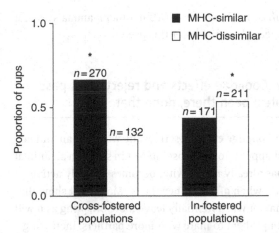

Figure 3.10 House mice, *Mus musculus domesticus*, use odor cues to avoid mating with individuals that are genetically similar at the major histocompatibility complex (MHC).When young they learn the odors of their close kin (familial imprinting) and when adult they compare this with the MHC identity of potential mates. Wild-derived female mouse pups cross-fostered into MHC-dissimilar families reversed their mating preferences compared with in-fostered controls. Genetic typing of their progeny indicated that females avoided mating with males carrying the MHC genes of their foster family (left), thus supporting the familial imprinting hypothesis. Stars indicate significant differences. Figure redrawn from Penn & Potts (1998b). Mice can also use variation in major urinary proteins (see text).

are is not clear, though some volatiles have been proposed in mammals (Box 3.1) (Kwak *et al.* 2010). Major histocompatibility complex-associated peptide ligands seem to be important in fish (Milinski *et al.* 2005) and may also have a role in mice (Chapter 9) (Leinders-Zufall *et al.* 2004; Spehr *et al.* 2006b).

The mouse learns the MHC-associated odors of its parents and siblings and compares potential mates with this template (Chapter 1). The MHC preferences of mice can be experimentally reversed by cross-fostering when newly born (Figure 3.10) (Penn & Potts 1998b). By contrast, other species such as stickleback fish, *Gasterosteus aculeatus*, use self-reference to compare potential mates for suitability (Aeschlimann *et al.* 2003).

Generally, animals choose mates that are dissimilar to themselves at the MHC. There could be a variety of

non-exclusive benefits from this disassortative mating for MHC genotype. First, it may be used as an indicator of genetic difference to avoid inbreeding with kin. Second, it might allow choice of a mate for greatest heterozygosity across the whole genome. Third, as the MHC underlies the immune system, the benefit may be increased resistance of progeny to infectious diseases, by increased diversity of MHC alleles in offspring (e.g., in mice, Thoss *et al.* 2011) or by providing a "moving target" against rapidly evolving parasites. Fourth, there might be selection for particular MHC alleles to enhance resistance to local parasites (e.g., in fish, Eizaguirre *et al.* 2011). In mammals there are higher implantation rates if the fetus and mother differ in MHC (possibly shown in human populations, Chapter 13). It is possible that different combinations of these benefits are important in different species and circumstances. Mate choice based on the MHC may also involve cryptic sperm selection after mating, for example in birds (see Section 3.11.2) (Gillingham *et al.* 2009).

The MHC choice of partners may depend on the MHC diversity of the chooser: females with low MHC diversity choose males that are more diverse in the MHC, a mating-up preference by allele counting to enhance offspring disease resistance shown, for example, in sticklebacks (Figure 3.11) (Aeschlimann *et al.* 2003; Milinski *et al.* 2005) and in house sparrows (Griggio *et al.* 2011). Optimal rather than maximal diversity of MHC can maximize lifetime reproductive success in sticklebacks (Kalbe *et al.* 2009).

Major histocompatibility complex-based mate choice is not always found in wild populations of other species. One reason may be the mating system of the species. For example, Soay sheep do not show disassortative mating by MHC type, perhaps because their harem mating system, based on male–male fighting, gives little opportunity for female mate choice (Paterson & Pemberton 1997). In salmon, *Oncorhynchus tshawytscha*, harassment by males may prevent female choice for MHC dissimilarity (Garner *et al.* 2010). Another reason, revealed by experiments on wild-caught individuals, is that extremely MHC-dissimilar mates may be

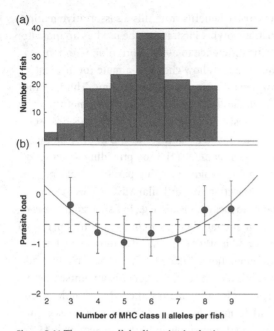

Figure 3.11 The mean allele diversity in the immune system of the three-spined stickleback fish, *Gasterosteus aculeatus*, is reflected in the optimum number of alleles to reduce parasite load. Females appear to choose mates to give their offspring the allele diversity to come close to this (see text). (a) Frequency distribution of the number of major histocompatibility complex (MHC) class II alleles detectable in 144 three-spined sticklebacks from a lake system. The mean number of MHC alleles was 5.8 (Reusch *et al.* 2001). (b) Relationship between the number of expressed MHC class II molecules and mean parasite load, expressed as summed residuals from general linear model (GLM) analysis. The function matches a quadratic polynomial with a minimum of 5.82 alleles (Wegner *et al.* 2003). Figure from Milinski (2006).

avoided as intermediate MHC-dissimilarity or particular alleles may give the greatest parasite resistance for example (for sticklebacks, see Eizaguirre *et al.* 2011). In tiger salamanders, *Ambystoma tigrinum*, indicators of male condition (tail length) were more important than MHC to female choice (Bos *et al.* 2009).

In mice, another protein system (the major urinary proteins, MUPs) is highly variable and may be used as well as or instead of the MHC for kin recognition to avoid inbreeding and to choose heterozygosity (Chapter 9) (Hurst 2009; Hurst & Beynon 2013; Thom

et al. 2008). However, MUPs in other mammals are not sufficiently variable for this use.

3.9 Coolidge effects and rejection of past mates: been there, done that

The Coolidge effect describes the way that an animal that appears to have lost interest in sex with an animal it has already mated with, becomes sexually active again when offered a new mate. Males that show this behavior will potentially leave more offspring as it will prompt them to mate with more partners, increasing the likelihood of compatibility, among other benefits. Females mating with more than one male can also gain in a number of ways including the material benefits from spermatophores, ensuring that at least one male is fertile, and possibly genetic benefits (Slatyer *et al.* 2012). For example, female snakes (adders, *Vipera berus*) mate up to eight times per season and avoid mating with the same male twice (Madsen *et al.* 1992).

The proximate cue in many kinds of animals, including adders, is likely to be recognition based on chemical cues, though the recognition mechanisms may differ (see Chapter 1). Coolidge effects mediated by odors have been demonstrated in many species of male rodents such as hamsters (see Section 9.8.3). Female crickets, *Gryllodes sigillatus*, leave their own individual CHCs on males when they mate and do not mate again with males that they have labeled in this way (Ivy *et al.* 2005). Male burying beetles, *Nicrophorus vespilloides*, instead seem to remember the CHC profile of the previous mate, and will not mate with females whose CHC profile they remember and instead chose novel females (Steiger *et al.* 2008a).

3.10 Alternative mating strategies

Staying "silent" to avoid predators drawn to mating calls, or pretending to be female, are both well known alternative strategies used by some males in a great variety of species to obtain matings without the risks of advertisement or direct confrontation with other males (see also Chapter 11) (Pizzari & Bonduriansky 2010; Shuster 2010; Taborsky *et al.* 2008).

In mouse social groups, the challenge from dominants leads to both physiological suppression of male pheromones in urine and suppression of marking behavior by subordinate males (Chapter 5) (Hurst 2009; Hurst & Beynon 2004, 2013). Males of the cockroach, *Nauphoeta cinerea*, which has male pheromones attractive to the female (Section 3.5), are able to control release of pheromone by either exposing the gland or not (Kou *et al.* 2009). Subordinate male cockroaches do not release pheromone but instead stay near calling males, as silent satellites, sometimes gaining sneaky matings (Moore *et al.* 1995).

Males in many species steal matings by mimicking female morphology, behavior, or both (Taborsky *et al.* 2008). In fish that have external fertilization, small males mimicking females can sneak between the dominant male and the female as she releases her eggs. Older, larger male gobies, *Gobius niger*, release a sex pheromone (primarily 5β-reduced androgens) that attracts females to their nests (Locatello *et al.* 2002). Sneakier male gobies are visually and chemically inconspicuous: they are colored like females and their ejaculate, which does not contain pheromone, goes unnoticed when they attempt to parasitize a spawning. The ejaculate of dominant males, however, does contain the pheromone and provokes an aggressive response from other dominant males.

Males in some species of snakes may camouflage themselves with female sex pheromones. During the spring breeding season and upon first emergence, almost all male red-sided garter snakes, *Thamnophis sirtalis parietalis*, in the Manitoba populations produce the female sex pheromone and are courted as if they were female (Mason & Parker 2010). Male garter snakes that mimic females ("she-males") may benefit simply because large mating balls of warmer, courting males form around them, transferring heat to the "she-males" and protecting them by reducing their exposure to predators. Over time the she-males start to produce the characteristic male lipids and are no longer courted.

In the ant, *Cardiocondyla obscurior*, there are two kinds of males, wingless males with sharp mandibles and peaceful winged males that chemically mimic females (Cremer *et al.* 2008). The wingless males mate inside their maternal colony and fight and kill other wingless males for access to females. The winged males mate mostly outside the colony and escape fighting by chemically mimicking females, perfectly for the first five days then just enough to protect themselves from attack for the last few days before they leave (Cremer *et al.* 2008).

3.11 Post-copulatory sexual selection

Sexual competition between males continues after copulation (Eberhard 2009). Males may attempt to reduce the effects of sperm competition by changing their ejaculate to match the risk, stopping or delaying the female mating again by applying anti-aphrodisiac pheromones (Section 3.11.1), or using seminal fluids to manipulate the female's behavior and physiology (Section 3.11.2). Female choice continues after copulation in the form of cryptic choice of sperm by females (Section 3.11.3) (Eberhard 2009).

3.11.1 Anti-aphrodisiacs

In response to "anti-aphrodisiac" molecules deposited by a previous mate, males in many species will not attempt to court already-mated females (Thomas 2011; Wedell 2005). For the previous mating male, this offers a form of chemical "mate-guarding." There may be an advantage to the female as she avoids further harassment by other males, though there is the possibility of sexual conflict if it were in the interest of the female to re-mate (Wedell 2005).

To be in the interest of the later males to respond to it, the presence of the anti-aphrodisiac needs to honestly reflect the low likelihood of mating success.

This seems to be the case in the staphylinid beetle, *Aleochara curtula*, in which the life of the male-applied anti-aphrodisiac molecule matches the time the female keeps the spermatophore before being ready to re-mate (Schlechter-Helas *et al.* 2011).

Male garter snakes, *Thamnophis sirtalis parietalis*, stop courting a female as soon as they detect the volatile molecule(s) from the ejaculate of a male that has started mating with her (Shine & Mason 2012). This is an adaptive response from the other males, as the mating male will deposit a gelatinous plug in the female's cloaca, which prevents her mating for some days (Mason & Parker 2010).

In *Drosophila melanogaster*, the male ejaculate includes a male-specific lipid, *cis*-vaccenyl acetate (cVA) (Chapter 9). This inhibits courtship from other males. There may be sexual conflict between the female, who would benefit from further matings, and the male's interest in preventing other matings. However, it is possible that the female can overcome this inhibition by increasing the secretion of her own sex-specific cuticular hydrocarbon, (7*Z*,11*Z*)-heptacosadiene (Billeter *et al.* 2009).

Pierid and heliconid male butterflies transfer species-specific volatile anti-aphrodisiac molecules to the female during mating. Both unmated and mated female show a "mate refusal" display to courting males, but mated females also release the anti-aphrodisiac during the display. In *Pieris napi* the male anti-aphrodisiac is methyl salicylate, in *P. rapae*, methyl salicylate and indole, and in *P. brassicae*, benzyl cyanide (Andersson *et al.* 2003). Across ten species of heliconid butterflies a variety of molecules are used as anti-aphrodisiacs (Estrada *et al.* 2011). In pierids and heliconids, the female cannot control the release of the anti-aphrodisiac so if males are selected to deposit more, lasting longer than the time it is in the interest of the female, there could be sexual conflict as she may gain nuptial gifts by re-mating sooner (Andersson *et al.* 2004; Estrada *et al.* 2011). (See Chapter 11 for chemical espionage by parasitoid wasps using the anti-aphrodisiac pheromone to locate host eggs.)

3.11.2 Seminal fluids

In species with internal fertilization, males can use molecules passed to the female in the seminal fluid, along with the sperm, to influence female behavior and physiology to reduce or bias sperm competition (Arnqvist & Rowe 2005; Eberhard 2009; Poiani 2006). The seminal molecules can cause her to refuse other courting males, transport sperm, or start laying eggs without further matings, all to the reproductive advantage of the first male. In turn, females would be expected to evolve traits that reduce the impact of the male. Both natural selection and sensory exploitation by the male followed by sexually antagonistic co-evolution could be involved, though distinguishing these mechanisms is difficult (Arnqvist 2006; Eberhard 2009). The story is complicated by the observation that in insects, the seminal fluids also contribute nutrients; more ejaculate leads to increased egg production even if some male peptides cause a shorter female lifespan (South & Lewis 2011).

The seminal molecules include mimics of the molecules females use to regulate their own reproduction. Many could be classified as allohormone pheromones, bypassing the sense organs to act directly on the receiver tissues (Chapters 1 and 9). Seminal fluids, which contain many peptides and hormones, are most explored in insects (Avila *et al.* 2011; Gillott 2003). For example, *Drosophila* semen contains a cocktail of more than 130 proteins and peptides, most of them from the male's accessory glands (accessory gland proteins, Acps) (Avila *et al.* 2011; Chapman 2008; Wolfner 2009). Accessory gland proteins are diverse in form, ranging from short peptides to prohormones through to large glycoproteins. The various Acps induce female *Drosophila* to reject further courting males, to eat more, to increase egg production/laying, and to store and release sperm. The first Acp to be identified and the most intensively studied is the 36-amino acid peptide Acp70A, known as the sex peptide. It affects many aspects of female biology including egg production, female receptivity, increased female feeding, and antimicrobial peptide production. Some of these effects may be mediated by raised juvenile hormone (JH) synthesis. A receptor for the sex peptide has been identified (Chapter 9) (Häsemeyer *et al.* 2009; Yang *et al.* 2009). Another accessory peptide, ovulin, stimulates octopamine production by the female, promoting ovulation.

In moths, as well as peptides and sperm, the male's seminal fluid also includes large quantities of JH and ecdysteriods (such as the "molting hormone" 20-hydroxyecdysone), molecules used by the female in controlling her reproduction (Wedell 2005). Cessation of calling in female moths appears in general to be triggered by these peptides and JH together with neural inputs. For example, female sex pheromone production and calling behavior in the moth, *Helicoverpa armigera*, stimulated by female pheromone biosynthesis-activating neuropeptide (PBAN), is largely shut down by a male sex peptide similar to the *Drosophila* sex peptide (Hanin *et al.* 2012). In the moth, *Heliothis virescens*, JH in the male's semen stimulates female egg maturation.

In vertebrates, as well as many peptides, hormones including cortisol, β-endorphin, and female pituitary gland hormones (such as luteinizing hormone and follicle-stimulating hormone) are reported in semen (Arnqvist & Rowe 2005; Poiani 2006) but the effects of seminal fluid on female mating behavior do not seem to have been investigated as thoroughly as in insects. Nonetheless there is rapid evolution of seminal fluid proteins and these are characterized, as are insect Acps, by rapid divergence and great variety suggesting similar evolutionary processes may be involved (e.g., Ramm *et al.* 2009). Sperm competition was implicated in generating such divergence for at least one major seminal fluid protein, responsible for copulatory plug formation in rodents.

3.11.3 Sperm competition

Sperm competition between the ejaculates of different males can be intense and is influenced by the number and types of sperm along with the volume and other molecules in the ejaculate (Section 3.11.2) (Kelly &

Jennions 2011; Parker & Pizzari 2010). Chemosensory cues may be used by males to match their ejaculate allocation to likely competition risks. For example, a male meadow vole, *Microtus pennsylvanicus*, doubles the number of sperm he gives the female if he detects the scent marks of competitor male(s) near her (delBarco-Trillo & Ferkin 2004; Ferkin 2011). Male *Teleogryllus oceanicus* crickets adjust their ejaculate according to the number of different males' CHC traces on the female, indicating the number of males she has mated with (Thomas & Simmons 2009). *Drosophila* males mate for longer, delivering more ejaculate, if they detect the presence of competitor males, revealed by a combination of auditory, olfactory, or tactile cues (Bretman *et al.* 2011).

3.11.4 Cryptic choice

However, a male's sperm are not just competing directly with other sperm. Females can use "cryptic female choice" to manipulate which male's sperm are most likely to fertilize her eggs (Eberhard 2009). Sexual selection by the female, using a wide range of mechanisms, may continue long after copulation.

Some of the cryptic choice can be for genetic compatibility or to avoid kin mating (Section 3.8). Female red junglefowl, *Gallus gallus*, retained fewer sperm following insemination by brothers (Pizzari & Snook 2004). Similarly, female guppies, *Poecilia reticulata*, a fish with internal fertilization, appear to select against sperm from brothers, an effect mediated by molecules in their ovarian fluid (Gasparini & Pilastro 2011). There can also be cryptic male choice: male red junglefowl allocated more sperm when mating with MHC-dissimilar females, possibly detected by olfactory cues (Gillingham *et al.* 2009).

Female flour beetles, *Tribolium castaneum*, are more attracted to the smell of some males, presumably those with more of the male pheromone, isomers of 4,8-dimethyldecanal (Fedina & Lewis 2008; Lewis & Austad 1994). Females also skew the likelihood of fertilization of their eggs toward the sperm of these attractive males.

Female insects may gain from multiple matings but may be selective about fertilization. Each additional mating gives the female arctiid moth, *Utetheisa ornatrix*, nutrients and protective alkaloids from each male's spermatophore (Section 3.7.2), increasing her egg output by 15% each time. However, after the matings she cryptically controls which sperm fertilize the eggs and is able to choose the sperm of the largest male, whatever the mating order (Lamunyon & Eisner 1994).

3.12 Sex pheromones and speciation

New species develop when previously interbreeding populations no longer interbreed successfully. There are two main alternative explanations for speciation (allopatric and sympatric mechanisms), which form a continuum (Coyne & Orr 2004; Futuyama 2009). In allopatric speciation, populations become geographically separated by a physical barrier such as topography or unfavorable habitat. While apart, many characters diverge by drift or by different selective conditions including ecological and sexual selection, so that the two populations do not interbreed if they come back into contact. This is likely to be the most common form of speciation (Coyne & Orr 2004). Sympatric speciation is the evolution of reproductive barriers within a single, initially randomly mating, population.

Sympatric divergence of mate recognition systems, either in populations that come back into contact or in populations that have always lived in sympatry, could occur if genetic divergence has reached a stage where hybrid offspring are less fit than within-population offspring. This could either be because hybrids are less viable, or by sexual selection, if hybrid offspring are less successful in getting matings than pure-bred individuals (see Andersson 1994, p. 214; Ritchie 2007). In either case, selection against hybridization (*reinforcement*) favors an increase in assortative mating (non-random mating resulting from a preference for similar partners) (Smadja & Butlin 2009). Assortative

mating will increase reproductive isolation, further reducing the exchange of genes between the populations. Sexual selection probably contributes most effectively alongside ecological selection or selection for species recognition than as a solitary process driving speciation (Bolnick & Fitzpatrick 2007; Ritchie 2007). The limitations of sexual selection on male CHC cues in the face of natural selection have been explored in laboratory experiments on *Drosophila serrata* and these support the suggestion that sexual selection is unlikely to lead on its own to speciation (Hine *et al.* 2011). Divergent selection on quantitative trait loci associated with ecological isolation traits in diverging populations may be one route to sympatric speciation (Via 2009).

A separate process from reinforcement can lead to divergence in mating recognition signals in species, perhaps from completely different taxa, which share the same communication channel (Smadja & Butlin 2009). This "reproductive character displacement" is comparable with ecological character displacements such as changes in beak shape in different bird species sharing the same niche. Reproductive character displacement may explain the greater divergence in odors of subspecies of *Mus musculus* where they come into contact (Smadja & Ganem 2008). *Mus musculus musculus* and *M. m. domesticus* diverged during approximately 0.5 million years in allopatry and re-established contact in Europe approximately 5,000 years ago. These odor cues can be used for assortative mating choices by the mouse subspecies (see Smadja & Ganem 2008).

Reproductive character displacement could lead to moths in different families that share some pheromone components diverging in reception or signal to reduce interference in communication (Cardé & Haynes 2004). The potential selection pressures were explored in experiments on the sympatric moths *Heliothis subflexa* and *H. virescens*. When females of *H. virescens* were genetically manipulated to produce less of their normal acetate pheromone components, which distinguish them from *H. subflexa* females, ten times more *H. subflexa* males were attracted (Groot *et al.* 2006). This could exert a strong directional selection pressure as hybrids are infertile, perhaps sufficient to cause divergence in moth pheromone blends despite the balancing selection against change from conspecific males.

Species-specific pheromone signals are important for pre-mating isolation and speciation in both vertebrates and invertebrates (Smadja & Butlin 2009). It is likely that most animals have chemosensory barriers at some level. Even animals better known for courtship songs or visual displays may also be using chemosensory cues for pre-mating isolation, for example grasshoppers (e.g., Buckley *et al.* 2003) and brightly colored cichlid fish (e.g., Plenderleith *et al.* 2005).

Behavioral changes may often be the initial triggers for population divergence, and altered recognition and response to chemical cues are expected to be involved in many behavioral changes (Smadja & Butlin 2009). These can lead to behavioral isolation, a lack of cross-attraction between members of different species, reducing attraction, or preventing courtship or copulation (Coyne & Orr 2004).

In contrast to insects, relatively few vertebrate species have full identification of the pheromones that differ between species. One exception is the newt genus *Cynops*, in which related species have species-specific male decapeptide sex pheromones that differ in just two amino acids (Chapter 1) (Houck 2009; Woodley 2010). However, the ability to use chemosensory cues to distinguish conspecifics from members of other species or populations of conspecifics, needed for chemosensory-based behavioral isolation, has been demonstrated in a wide variety of vertebrates including snakes and lizards (e.g., Gabirot *et al.* 2011; review Mason & Parker 2010), cichlid fish and goldfish (e.g., Levesque *et al.* 2011; Plenderleith *et al.* 2005), bats (e.g., Caspers *et al.* 2009), and rodents such as mice (e.g., Li et al. 2013; Smadja & Ganem 2008). Differences in chemical profile between related bird species have been shown (e.g., Mardon *et al.* 2010; Zhang *et al.* 2009); though likely, whether these are used by birds for species recognition has not been established. Similarly, different strepsirrhine primate

species have characteristic differences in the molecules in their urine (delBarco Trillo *et al.* 2011). See Smadja and Butlin (2009) for more examples.

Study of chemosensory speciation cues is especially powerful because in some species, we can investigate both the genetics of changes in signal production (pheromones), down to the level of enzyme pathways, and also the genetics of signal reception (the genes for chemosensory receptors and neural wiring) (see below). Genomic techniques, in particular the new RNA sequencing techniques (Chapter 2), allow gene families involved with each of these steps to be examined for loci that might contribute to divergence among populations or species (Smadja & Butlin 2009; Stapley *et al.* 2010). Among the two best understood systems are pheromones in moths and in *Drosophila* and I will use them to explore details of speciation processes by chemosensory mechanisms (Sections 3.12.3 and 3.12.4).

The genetic changes involved in changing the signal along the route to speciation can include changes to the gene product coding region or changes to the *cis*-regulatory sections controlling gene expression. For example, the coding region of a gene differs between species of the newt *Cynops*, thus producing peptide pheromones that differ in their amino acid sequence (Woodley 2010). Also, the changes to the binding site of an enzyme changes its specificity giving, in moth pheromone, blend changes in the *E* and *Z* strains of *Ostrinia nubilalis*. Changes to the *cis*-regulatory sections controlling gene expression in species and/or sex-specific ways are shown in the reawakening of "dormant genes" in the new species *Ostrinia furnacalis* and in the pattern of activation and inactivation of the *desatF* gene across the *Drosophila* genus in speciation events (Section 3.12.4). The loci and genes above, which seem to have made a large contribution to the evolution of reproductive isolation between diverging species have been referred to as "speciation genes" (see Nosil & Schluter 2011). In other situations many genes may be involved (Section 3.12.3.2).

Sexual selection may have played a major part in the explosive speciation of drosophilid fruit flies on the

Hawaiian islands, leading to more than 800 endemic species (many represented on only one island each), characterized by complex courtship behavior and elaborate secondary sexual features such as lengthened antennae or legs (Kaneshiro 2006; Kaneshiro & Boake 1987). Rapid speciation may be encouraged by the relaxation in sexual selection during the founding phase of new populations: with few males around, less choosy females will have a selective advantage as they can be more sure of finding a mate (Kaneshiro 1989, 2006) (see Coyne & Orr 2004 for an alternative view). Less choosy females will allow male sexual signals to diversify as the signals will be under weak selection. By the time that populations increase or other species invade, returning an advantage to choosier females, the male signals will have changed. There is limited evidence that newer species have females that are less choosy than their "ancestral" species. Many of the Hawaiian drosophilid species lek, with males displaying near each other (see Section 3.7.4). Males produce species-specific hydrocarbon pheromones in an anal droplet used in advertisement and courtship, and cuticular pheromones on the females probably enable species recognition by males, as for example in three species of the *Idiomyia* (formerly *Drosophila) adiastola* subgroup (Alves *et al.* 2010; Tompkins *et al.* 1993).

3.12.1 Individual and within-population variation, ecological races, and geographical variation

While much of the narrative about pheromones emphasizes the specificity of pheromones both in production and perception, ultimately no two animals are genetically identical. Pheromone biosynthesis pathways may be controlled by many genes on a complex genetic background so we can expect some differences between individuals, despite stabilizing selection on both receiver and sender. An indication of the range of pheromone blends produced by female moths is given in the histograms of female signal in Figures 3.12 and 3.13. In some species, such as the potato tuber moth, *Phthorimaea operculella*, the ratio between components varies greatly according to

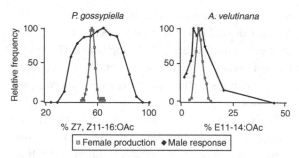

Figure 3.12 Male moths have much wider pheromone response windows than the range of variation in female pheromone production. This is shown in two species of moths: *Pectinophora gossypiella* and *Argyortaenia velutinana*. Frequencies within each class are calculated relative to the most common type of female production or male response respectively (taken as 100 by definition). Figure redrawn from Löfstedt (1990).

rearing temperature (or between females at the same temperature) (see Cardé & Haynes 2004). Males in that species have a correspondingly wide range of response. The amount of variation in the female signal may depend on the risk of heterospecific attraction in that habitat.

Ecological races and almost continuous geographical variation occur commonly in pheromone traits when explored in detail. At the population level such variation underlies signal divergence in incipient speciation. There may be more fluid boundaries between ecological races and species and between species than often thought (see discussion in Mallet 2008). For example, red sided garter snakes, *Thamnophis sirtalis parietalis*, have female chemical cues that vary between populations (Section 3.8) (Mason & Parker 2010). In one part of Japan, the newt *Cynops pyrrhogaster* has a sex pheromone peptide that differs by one amino acid from the main population (Nakada *et al.* 2007). Geographical subspecies of *Mus musculus* have different male urine signals (Smadja & Butlin 2009) (previous section). Cuticular hydrocarbon blends vary between regional races of *Drosophila melanogaster* (Section 3.12.4). Across Eurasia and Africa, turnip moth, *Agrotis segetum*, females show wide differences within and between populations in the ratios

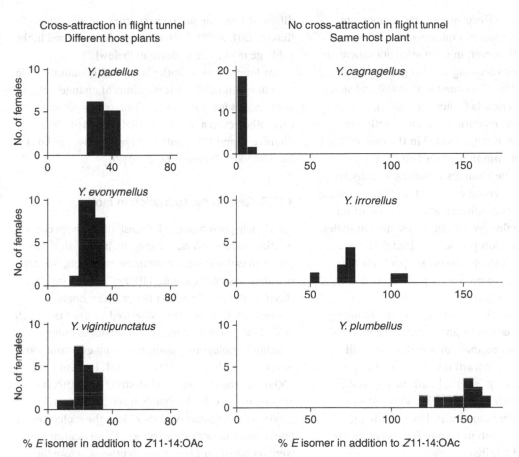

Figure 3.13 Species isolation and mating site. (Right) Three of six sympatric species of small ermine moths, *Yponomeuta*, share the same host plant and so have evolved different pheromone blends: males are only attracted to the blend of their own species of female (there is no cross-attraction in a flight tunnel).

(Left) The other three species have different host plants from each other and have similar pheromone blends, which show significant cross-attraction in the flight tunnel but *not* in the field. The data show the frequencies of (*E*)- to (*Z*)-11-tetradecenyl acetate blends produced by individual females (incidentally, showing the *intra-specific* range of blends). Figure redrawn from Löfstedt *et al.* (1991).

of the three main components forming their pheromone blend (Cardé & Haynes 2004; Löfstedt 1993). The proportions of male antennal receptors tuned to each component co-vary geographically with female blend production (Löfstedt 1993).

3.12.2 How receivers follow signal shifts

One of the key questions in the evolution of sexual communication is how signal and response are co-ordinated for species specificity, with stabilizing selection against mutational changes in either, but yet signals and response change during speciation. It is particularly challenging because the characters for signal production and response are only rarely genetically linked (Boake 1991; Butlin & Ritchie 1989; Smadja & Butlin 2009). For example, in most insects so far studied, pheromone production in females and response in males are controlled independently on different chromosomes (see below). There is a rare exception in *Drosophila*:

the *desat1* gene has effects on both pheromone production and male responses (pleiotropy) (Section 3.12.4). However, in other modalities there are examples of linked visual signals and preference in *Heliconius* butterflies (Kronforst *et al.* 2006) and similarly for songs in crickets (Wiley *et al.* 2011).

If the signal and reception are not genetically linked, how might change occur? In the case of the long-distance pheromones of the female moth (Section 3.12.3), the changes in mating signals are often treated as an equal co-evolution between male and female. However, Phelan and others point out that because females are a limiting resource to males, there is more selection pressure on individual males to have a slightly broader response spectrum than there is on females to produce a particular pheromone blend (Figure 3.12). In isolated populations, there may be chance changes in the female blend but the variation in pheromone receptors between males means that some males may still respond and evolve toward this new blend (the "asymmetric tracking" hypothesis) (Baker 2008; Löfstedt 1993; Phelan 1992, 1997). Variation in males is nonetheless constrained by stabilizing selection as males with unresponsive receptors are strongly selected against.

If the change in blend is large then there may be disruptive selection toward the extremes of male response leading to polymorphism, assortative mating, and possibly speciation. Asymmetric tracking has been modeled (Bengtsson & Löfstedt 2007; Butlin & Trickett 1997) but perhaps more important are the moth examples where some rare males have responded as predicted in asymmetric tracking despite a significant change in pheromone blend, for example in the hypothesized generation of a new species, the Asian corn borer moth, *Ostrinia furnacalis* (Section 3.12.3.1). While most discussions of asymmetric tracking have emphasized major blend alterations (see below), asymmetric tracking would work for males to follow more subtle pheromone shifts, such as a change in the emitted ratios of existing molecules used by related species, caused by a new mutant enzyme that has

different substrate preferences from the wild type (Baker 2002, 2008). Such a change has occurred in the cabbage moth, *Trichoplusia ni* (below).

The long-distance moth pheromone communication system might facilitate these kinds of changes, as pheromones make it possible for rare morphs to locate each other, even at a low population density (see Chapters 4 and 10) (Cardé & Haynes 2004; Kokko & Rankin 2006; Symonds *et al.* 2012).

3.12.3 Case study: speciation in moths

The female pheromones of almost all moth species are multicomponent blends of long hydrocarbon chain (10 to 18 carbons long) unbranched alcohols, acetates, or aldehydes (see Appendix) (Baker 2011; Cardé & Haynes 2004; de Bruyne & Baker 2008; Roelofs & Rooney 2003). They are synthesized *de novo* from C16 or C18 fatty acid precursors and produced from roughly homologous glands, reflecting evolution early in the lineage (Cardé & Haynes 2004; Liénard *et al.* 2008). Among species-specific structural differences are one to three double bonds inserted at specific locations and geometry (*E* or *Z*) into the molecules by one or more desaturase enzymes. These desaturases were recruited for pheromone synthesis before the expansion of the Lepidoptera into the second largest taxon in insects, with about 180,000 species (Liénard *et al.* 2008; Roelofs & Rooney 2003).

Within closely related taxonomic groups (families or subfamilies), species often use the same or similar components, as a result of sharing biosynthetic pathways by ancestry (Chapter 1) (Cardé & Haynes 2004; de Bruyne & Baker 2008; Symonds & Elgar 2008). Closely related sympatric species can gain blend specificity by using unique ratios or combinations of these components and in some closely related moth species we can see how the different pheromone blends have arisen by particular gene changes that affect enzymes in pheromone biosynthesis (see below). The multiple pheromone components are reflected in each component having specialized olfactory sensory neurons leading to their own glomerulus in the antennal lobe of the male (see Chapter 9).

Preventing cross-attraction between different species can come from blend differences, from calling at different times of the day (or season), from calling on host plants, or from a combination of these (Cardé & Haynes 2004; Greenfield 2002). A study of nine sympatric species of European small ermine moths, *Yponomeuta*, shows the use of all of these dimensions (Löfstedt *et al.* 1991). Six of the species use a mix of (*E*)-11- and (*Z*)-11-tetradecenyl acetate in different ratios as primary components of their pheromone. Three species that have different host plants appear to share very similar pheromone blends and were attracted to each other's pheromone in wind-tunnel tests, but not in their natural habitats (Figure 3.13). However, three species that shared a host plant, and thus would be likely to meet in the field, showed no overlap in pheromone blend and did not cross-attract in the wind tunnel. The present pattern of blends has occurred by loss and gain of minor components and shifts in ratio, which can be largely explained by a combination of changes in a particular pheromone-gland expressed enzyme, a fatty-acyl CoA reductase that processes pheromone precursors, along with changes in desaturase enzymes (Liénard *et al.* 2010). In addition, two species with similar blends call at different times in the night. The pheromone components of some species act as behavioral antagonists (see below) to males of other species.

The use of pheromone blends by female moths affects selection on male signal reception (Cardé & Haynes 2004). On theoretical grounds, perception of the whole blend would dramatically enhance the signal-to-noise ratio against background chemical noise and thus would enable a male to detect a conspecific female reliably from further away (Cardé & Baker 1984; Nehring *et al.* 2014). Important support for this came from the discovery that male moths are most sensitive, at the greatest distance, to the whole blend produced by the female, and not to the major component alone as had once been thought (Figure 3.14) (Linn & Roelofs 1989; Linn *et al.* 1987). The minor components might be less than 10% of the concentration of the major component, but they give

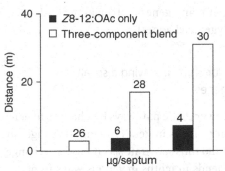

Figure 3.14 Far downwind, male oriental fruit moths, *Grapholita molesta*, are more sensitive to the whole female pheromone blend than to the major component alone. Wing-fanning activation of male moths was tested downwind from pheromone sources: major female pheromone component alone (*Z*8–12:OAc) or the three-component blend (*Z*8–12: OAc + 6% *E*8–12:OAc + 3% *Z*8–12:OH) at three doses. The experiment was done in the field and soap bubbles were used to track the path of pheromone plumes in the wind. Values above bars are the number of males of the 30 tested that activated. Figure redrawn from Linn and Roelofs (1989); data from Linn *et al.* (1987).

the signal its specificity. Because of the way pheromone plumes travel downwind as filaments of pheromone-rich air (Chapter 10), all the components are present in the ratio as released, even far downwind.

A flying male would benefit from early rejection of a female not from his species. He would then save the energy cost of flight, predation risks, and the time opportunity cost, as he is not likely to be accepted by the female when he reaches her, or if she accepts, the mating would be infertile. Males have specialized olfactory sensilla for these "behavioral antagonists" – key components that distinguish the blends of other species similar to their own – and they turn back if they detect antagonists (Figure 9.4) (Cardé & Haynes 2004; de Bruyne & Baker 2008). They also turn back if they encounter too much or too little of components of their own blend (as might be expected given that ratios can be the basis of specificity, as for example *Ostrinia nubilalis* strains) (Chapters 9 and 10) (Baker 2008).

Changes in pheromone blend can occur either by major shifts involving a small number of genes or a combination of many genes each making small changes (polygenic changes).

3.12.3.1 Major shifts involving a small number of genes

Changes in biosynthetic pathways by changes in gene expression, or changes in genes leading to changes in binding sites and thus substrate specificity, can change pheromone blends in moths in various ways from changing the ratios between components to the addition of new molecules to the blend (Baker 2008; Symonds & Elgar 2008). Changes in small numbers of genes also affect *Drosophila* pheromones (Section 3.12.4). Changes also occur in the receiver and these are discussed below.

Cabbage moth mutants, *Trichoplusia ni*

One example of how major shifts in pheromone blends and male response can occur comes from a chance observation of a change in pheromone blend produced by mutant females in a laboratory population of cabbage looper moths, *Trichoplusia ni* (Cardé & Haynes 2004; Domingue *et al.* 2009). The mutant females produced a blend with 50 times more of one of the minor components and a lower quantity of the major component. This resulted in a 40-fold reduction in the number of conspecific males attracted in the field, but nonetheless some wild males did arrive. In the laboratory, crossing experiments on the mutant population showed that genetic control of the female signal and male response were independent. The change in the female was due to a single recessive autosomal mutation in a gene for a chain-shortening enzyme in the pheromone biosynthesis pathway (Cardé & Haynes 2004; Jurenka *et al.* 1994). The males from the mutant colony had the same proportion of antennal olfactory sensory neurons sensitive to each of the components as did normal males. In flight tests to pheromone lures, the mutant males, like normal males, were initially much more responsive to the normal blend but after 50 generations the male response spectrum broadened to include the new blend, due to olfactory sensory neuron changes (Domingue *et al.* 2009). One feature of the changes was a desensitization of the olfactory sensory neurons responding to the minor component produced in excess by the mutant females. The way male response "followed" the change in female blend is in line with the "asymmetric tracking" hypothesis (Baker 2008; Löfstedt 1993; Phelan 1992) (above).

European corn borer moth, *Ostrinia nubilalis*

The European corn borer moth, *Ostrinia nubilalis*, has become a model species for the study of pheromone genetics and speciation in action (for a review see Lassance 2010). In North America the species has two sympatric populations (strains) with such limited gene flow between them that they could be considered semi-species. The females produce, and males respond to, different ratios of the same components of the pheromone: the more prevalent Z-strain uses a 97:3 mix of (Z)-11-tetradecenyl acetate and (E)-11-tetradecenyl acetate whereas the E-strain uses a 1:99 mix of Z/E 11-tetradecenyl acetate (11–14:OAc). Although other factors such as circadian rhythms and flying dates differ, pheromone preference by males is the strongest reproductive barrier between the strains (Dopman *et al.* 2010). There is strong assortative (same with same) mating in the field. However, some hybrid individuals are found in the field, and in the laboratory the two pheromone strains will hybridize, which has made it possible to investigate the genetics of pheromone signal and response.

The *O. nubilalis* pheromone signaling-perception polymorphism in females and males is controlled by unlinked genes on different chromosomes: the female pheromone production locus is autosomal, the male behavioral response is a sex-linked quantitative trait locus (QTL), and a third locus (autosomal but not linked to the female production locus) controls the male pheromone-sensitive sensilla on his antennae (Lassance 2010). The difference in pheromone production between the *E* and *Z* females is controlled by two alleles of a fatty acyl reductase gene *pgFAR*, giving versions of the enzyme with strain-specific substrate specificities

(Lassance *et al.* 2010, 2013). This gene may represent a speciation gene critical in the evolution of reproductive isolation between the diverging strains in *O. nubilalis* and potentially other lepidopteran groups (Lassance *et al.* 2010, 2013; Liénard *et al.* 2010).

Male response behavior and variation in antennal response, probably reflecting olfactory receptor differences, appear to be controlled by a combination of sex-linked and autosomal loci respectively (see Lassance *et al.* 2011; Olsson *et al.* 2010). The olfactory sensory neurons (and their olfactory receptors) may influence the antennal lobe morphology. The males of the two strains have an identical antennal lobe morphology but with the functional "wiring" for *E* and *Z* components simply reversed: in each strain, the olfactory sensory neurons responding to the major component of the pheromone go to the same large macroglomerulus and those responding to the minor component go to a particular small macroglomerulus (see Chapter 9 for an explanation of these terms) (Karpati *et al.* 2008, 2010).

Different male pheromones may also help in reinforcing the behavioral isolation between the two *O. nubilalis* strains (see below) (Lassance & Löfstedt 2009).

Asian corn borer: a new *Ostrinia* species, *O. furnacalis*

Ostrinia corn borer moths provide another example of pheromone blend changes leading to speciation, in this case the sibling species of the Asian corn borer moth, *Ostrinia furnacalis*, which diverged from the common ancestor shared with *O. nubilalis* about 1 million years ago (Roelofs & Rooney 2003; Roelofs *et al.* 2002). Instead of using the common lepidopteran Δ11-desaturase used by *O. nubilalis* to produce the (*Z*)- and (*E*)-11-tetradecenyl acetate blend (see above), the gene involved in *O. furnacalis* pheromone biosynthesis is a Δ14-desaturase (Figure 3.15) (Roelofs *et al.* 2002; Zhao *et al.* 1990). One suggestion is that the pheromone change and speciation event came with the "resurrection" of a long-dormant desaturase gene, for an enzyme that changed the position of a double bond in the pheromone above in females

expressing the gene (Roelofs & Rooney 2003; Roelofs *et al.* 2002). The change has been called "saltational" (see Chapter 1 for a discussion). An alternative possibility is that the desaturase gene was always active (in the context of male pheromone synthesis) but may have changed from being expressed in males, as in *O. nubilalis*, to being expressed in females in *O. furnacalis*, changing the female blend (Lassance & Löfstedt 2009). Biochemically, the Δ14-desaturase acts on a 16-carbon fatty acid and, after a subsequent round of chain shortening, produces a pheromone blend comprising (*Z*)- and (*E*)-12-tetradecenyl acetate (with the double bond moved to carbon 12 rather than the carbon 11 of the *O. nubilalis* pheromone). The gene for the Δ11-desaturase used by *O. nubilalis* is present in the *O. furnacalis* genome but not expressed, so it appears to have been switched off (Roelofs *et al.* 2002; Sakai *et al.* 2009). Desaturases form a multigene family that has undergone birth-and-death evolution with duplication, coding region change, and sometimes pseudogenization and resurrection (see Chapter 9 for discussion of the birth and death of genes in the context of olfactory receptor protein genes) (Fujii *et al.* 2011; Xue *et al.* 2007).

Would there be any males to respond to the new pheromone blend? A screening of *O. nubilalis* males in the laboratory showed that some rare, broadly tuned individuals would fly upwind to the new blend of *O. furnacalis* as well as to the blend of their own females, as predicted by asymmetric tracking (above) (Linn *et al.* 2003; Roelofs *et al.* 2002). A change in one amino acid in one pheromone receptor OR contributes to the new response specificity in *O. furnicalis* (Leary *et al.* 2012).

3.12.3.2 Polygenic changes involving many genes

Whereas the shifts in pheromone blend in *Ostrinia*, *Trichoplusia ni*, and *Yponomeuta* seem to be due to a small number of genes, the differences between multicomponent pheromone blends of females of many moth species, such as the related sympatric species *Heliothis virescens* and *H. subflexa*, cannot be explained by one enzyme. This is because

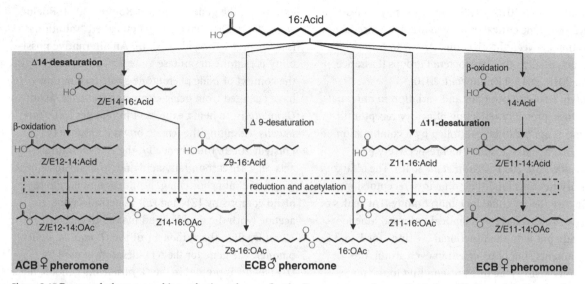

Figure 3.15 Proposed pheromone biosynthetic pathways for the European corn borer, *Ostrinia nubilalis*, and Asian corn borer, *O. furnacalis*. Unusually, the male pheromone in the European corn borer (ECB) moth is similar to the female pheromone. The male and female pheromone components in the ECB and the Asian corn borer (ACB) are biosynthesized *de novo* from the ubiquitous palmitic acid (16:Acid). The routes leading to the 14-carbon acetate components produced by females and to the 16-carbon acetate components used by males employ common key desaturation enzymes. The Δ14-desaturase appears to be common to female ACB (it is the resurrected gene, Section 3.12.3.1) and male ECB pathways, as well as the Δ11-desaturase that is shared in male and female ECB pathways. Note that the Δ9-desaturation leading to palmitoleic acid (Z9–16:Acid) also occurs in females of both species, but the reduction and acetylation of this substrate occur in males only. Figure from Lassance and Löfstedt (2009).

compounds with three different functional groups (alcohols, aldehydes, and acetate esters) and two different chain lengths (C14 and C16) are involved (Groot *et al.* 2009). In these *Heliothis* species, QTL studies with back-crosses demonstrated that genes on at least 9 of the 31 *Heliothis* chromosomes contribute to the differences between the species in the volatile compounds produced by the pheromone gland and indicate that there may be epistatic interactions among the QTL, which may also involve genes other than those encoding the enzymes themselves (Groot *et al.* 2009). Female response to male pheromones in *Heliothis* courtship is also important in species isolation (next section) (Hillier & Vickers 2011).

The differences in male response to the three female-produced pheromone components (Z9–14:Ald, Z9–16: Ald, Z11–16:OAc) that maintain sexual isolation of

these species are all controlled by a single QTL containing at least four tightly linked olfactory receptor (OR) genes (see Chapter 9) (Gould *et al.* 2010). The QTL for the male ORs is on a different chromosome from the QTLs for the female pheromone components. Some of these OR genes appear to be the product of duplication events in the birth-and-death pattern of OR evolution (Chapter 9).

3.12.3.3 Sexual selection of male lepidopteran pheromones and speciation

In many Lepidoptera, chemical signals from males are crucial for later courtship and mating success, even in moth species with long-range female pheromones and in day-flying butterflies, which use visual cues for early mate recognition (Allen *et al.* 2011; Cardé & Haynes 2004; Phelan 1997). These male pheromones, often called scents or "aphrodisiacs," are released from special structures such as modified wing scales (androconia), hair pencils, brush organs, and coremata (Allen *et al.* 2011; Birch *et al.* 1990). Females will reject males without the pheromones. In some species the male advertises not only his species but his quality. For example, in *Utetheisa ornatrix* the male's pheromone quantity predicts the gift of protective poison he will give the female (Section 3.7.2).

One strong selective force for the evolution of male pheromones and courtship displays is selection by females to avoid hybrid matings (Phelan 1997; Wedell 2005). Males of the sympatric moths *Heliothis virescens* and *H. subflexa* have abdominal hair pencils, which they expose during courtship to release species-specific volatile pheromones that seem to be important in preventing inter-specific matings (Hillier & Vickers 2011). In *Ostrinia nubilalis* the male pheromone blend, which differs between the strains in the amount of Z11–16:OAc, may provide a crucial step in mating isolation for the *E* and *Z* strains, as well as allowing females to distinguish *O. nubilalis* males from *O. furnacalis* males, which lack Z14–16: OAc (Figure 3.15) (Section 3.12.3.1) (Lassance & Löfstedt 2009). The male pheromones of the nymphalid butterfly, *Bicyclus anynana*, are important in mating success (Nieberding *et al.* 2008) and within the genus and wider subtribe, the number, morphology, and position of the pheromone-producing androconia are discriminating traits in the phylogeny suggesting that pheromones may have been important in the spectacular radiation of some 250 species (Allen *et al.* 2011).

Moth species sharing the same host plant, and thus more likely to meet, are more likely to have male pheromones. In each of five different moth families, all having some species with male scent-emitting organs, significantly more pairs of species sharing a host plant had male scent-emitting structures than not (Phelan & Baker 1987). When the data were pooled, 53% of 396 species sharing a host plant had male pheromone structures compared with only 28% of 419 species without a host plant in common.

Male lepidopteran pheromones are chemically diverse and the scent organs show enormous variety (like those of male rodents), reflecting multiple independent evolution across the taxon (Cardé & Haynes 2004). This is in contrast to the pheromones of female moths, which are largely similar compounds produced from roughly homologous glands, reflecting evolution early in the lineage (Cardé & Haynes 2004; Liénard *et al.* 2008). Many male pheromones are identical or related to their species' food-plant molecules so may have evolved by sensory exploitation (Chapter 1). Supporting this, Conner and Weller (2004) note that the female receptors for the male pheromone in *U. ornatrix* are housed in sensilla basiconica on the antennae, commonly reserved for receptors of plant odors (the same receptor is also good for finding host plants with the precursor molecules). An alternative model reverses the history and suggests a sexually selected female preference for male diet-related pheromones was instead exploited by maternal-effect genes that preferentially place offspring on a specific host plant (Quental *et al.* 2007).

3.12.4 Case study: speciation in *Drosophila*

Rapid divergence of pheromones appears to underlie rapid speciation in the *Drosophila melanogaster* subgroup, with behavioral isolation enabled by changes in a small number of genes coding for pheromone synthesis enzymes, as in moths. Both male and female sex pheromones are involved. Like other dipteran pheromones, these are mostly cuticular hydrocarbons (CHCs), relatively involatile molecules only detected at short range or on contact during courtship (see reviews by Ferveur 2010; Ferveur & Cobb 2010; Wicker-Thomas 2007).

Different species tend to have different characteristic combinations of CHCs (Figure 3.16) (Ferveur 2005). For both sexes, courtship follows a series of required steps, which ensure the potential mate is of the right species and sex, receptive and unmated. Courtship includes chemosensory signals from volatile and non-volatile pheromones along with visual, auditory, mechanosensory inputs (for more details see

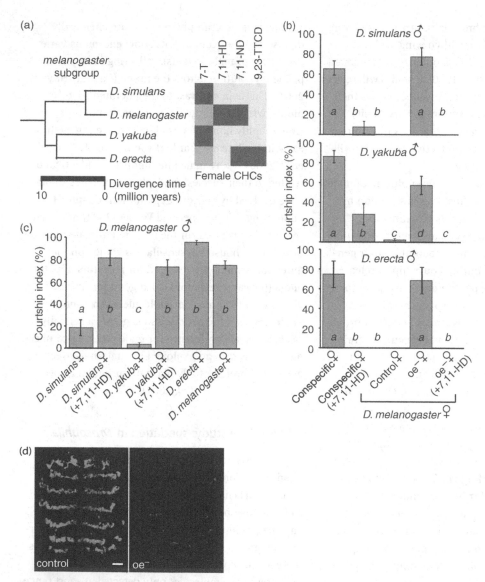

Figure 3.16 Cuticular hydrocarbon (CHC) pheromones contribute to the species barriers between different *Drosophila* species. The diene CHC 7,11 heptacosadiene (7,11-HD), is characteristic of *D. melanogaster* females and stimulates courtship from conspecific males.

(a) Phylogenetic relationships of the four *Drosophila* species used in this study, along with a time scale for evolution in this group (Russo *et al.* 1995). Partitioning of the four major female hydrocarbon compounds (7-T; 7,11-HD; 7,11-ND; 9,23-TTCD (9Z, 23Z)-tritriacontadiene) in these different species: light gray indicates absence, gray indicates minor amounts, and black indicates a major fraction of the total hydrocarbon (Jallon & David 1987).

(b) Wild-type males courtship behavior in intra- and inter-specific pairings with females of the indicated species and genotype. The *D. melanogaster* females marked "oe⁻" had no CHCs of their own as their CHC-producing oenocytes (oe) had been knocked out (d). N = 10–19. Males of all three species would court the *D. melanogaster* females lacking oe and thus CHC.

(c) Wild-type *D. melanogaster* male courtship toward females from the indicated species. N = 10–12. Females labeled (+7,11-HD) were treated with approximately 500 ng of synthesized 7,11-HD. *D. melanogaster* males do not court *D. simulans* or *D. yakuba* females unless the *D. melanogaster* female "badge" 7,11-HD was added.

Chapter 9) (Figure 9.10) (Dickson 2008; Ferveur & Cobb 2010). *Drosophila melanogaster* females and males can be genetically engineered by a variety of techniques to produce almost no CHCs on their surface, allowing the "blank canvas" of these flies to be "perfumed" with synthetic CHCs directly or by contact with donor flies of various kinds (Billeter *et al.* 2009; Savarit *et al.* 1999). A male *Drosophila* stops courtship if he detects CHC pheromones that indicate the potential mate is a female of the wrong species or is a male of his own (Figure 3.16) (Billeter *et al.* 2009; Savarit *et al.* 1999). For example, the characteristic CHC of female *D. melanogaster* in most of the world, 7,11-heptacosadiene (7,11-HD), which acts as an aphrodisiac to males of *D. melanogaster*, is a key compound causing males of other closely related species in the genus to reject her (Figure 3.16).

Much of the variation of CHC blends between different species is the result of rapid evolution of desaturases, as in some of the moth examples above (Fang *et al.* 2009; Shirangi *et al.* 2009). Unusually for a signal production-reception system, in *D. melanogaster* one of the genes, *desaturase1* (*desat1*), appears to control not only the production of sex pheromones in male and female flies but also the perception of sex pheromones by males (see in Bousquet *et al.* 2012). Regulatory regions of the *desat1* gene independently drive the gene's expression in pheromone-producing cuticle areas in both sexes and in the male the expression in chemosensory neurons potentially involved in pheromone reception, which project to the sexually dimorphic glomerulus DA1 in the male's brain (see Chapter 9) (Bousquet *et al.* 2012).

Two other fast evolving *desat* genes (*desat2* and *desatF*) originating from the ancestral *desat1* gene may also have created pheromonal diversity (Fang *et al.* 2009; Shirangi *et al.* 2009). Variation in *desat2* largely explains the regional differences in the CHC pheromones of females of the widely distributed species *D. melanogaster* (Ferveur 2005; Ferveur *et al.* 1996). Different alleles of *desat2* give rise to sympatric populations in west and east Africa, one with "African-type" females producing high quantities of 5,9-HD and the other with females producing the usual 7,11-HD. Given that only Zimbabwe strains produced a high level of CHCs desaturated on carbon 5 both in males (5-tricosene) and females (5,9-HD), sexual isolation between the populations seems to be driven by Zimbabwe-type females rejecting "normal" males carrying 7-tricosene (Coyne & Elwyn 2006; Grillet *et al.* 2012).

Changes in expression of the desaturase *desatF* are correlated with the presence or absence of long-chain diene CHC pheromones (with two double bonds) in species- and sex-specific patterns across the many species of the genus *Drosophila* (Shirangi *et al.* 2009). The evolutionary transitions include six independent gene inactivations of *desatF*, three losses of expression without gene loss, and two transitions in sex specificity (Figure 3.17). In species that express *desatF* only in females, *desatF* expression is activated by binding (to a *desatF*-regulatory site) of the female-specific isoform of the protein encoded by the sex differentiation gene, *doublesex* (*dsx*) (see Chapter 9). The rapid evolution of the *doublesex*-binding *desatF*-regulatory site, with successive activations or inactivations, explains which sex in a species (males, females, both, or neither) expresses *desatF*. Shirangi *et al.* (2009) wonder if *desatF* might play a sufficiently important role in pre-mating isolation to count as a "speciation gene," contributing to reproductive isolation (Nosil & Schluter 2011).

Figure 3.16 (cont.)

(d) Expression of green fluorescent protein (GFP) in the hydrocarbon-producing cells (the oenocytes) in the dorsal abdomen of female *D. melanogaster* shows their presence in the control (left) and absence in the oenocyte-ablated (oe$^-$) females. Green fluorescent protein was under the control of a promoter from the *desaturase1* gene expressed specifically in the oenocytes. Scale bar, 200 μm.

Means ± SEM. Bars labeled by the same letter (*a*, *b*, *c*, or *d*) in the histograms are not statistically different (ANOVA, $P < 0.01$). Caption and figure after Billeter *et al.* (2009).

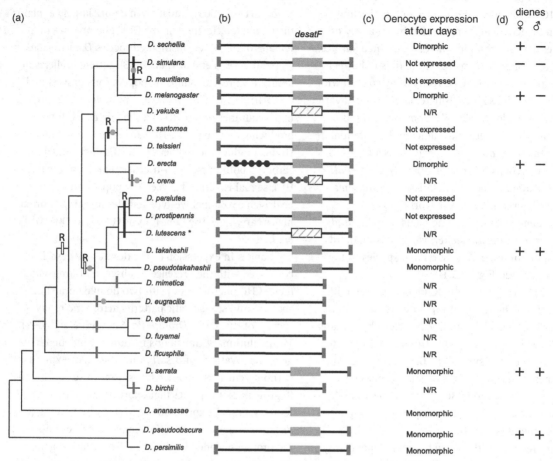

Figure 3.17 The *desatF* locus in *Drosophila* species is rapidly evolving. The locus codes for a key desaturase enzyme in the synthesis of cuticular hydrocarbon (CHC) sex pheromones. Adapted from Shirangi *et al.* (2009).

(a) Phylogenetic relationships of the 24 *Drosophila* species surveyed (adapted from Jeong *et al.* 2006). (b) Schematic of the *desatF* locus in these species. The rectangles at the end of the black horizontal bars indicate the landmarks used in cloning. In *D. ananassae*, *desatF* was found in the genome, but not on the same chromosome, which is indicated by the absence of the rectangles. The grey boxes indicate the coding region. A striped box indicates a mutation in the coding region leading to a loss of function of the protein (frameshift or nonsense mutation). Black full circles represent regions with repetitive DNA. The six independent gene losses are indicated by grey vertical bars on left-hand cladogram (a). Regulatory losses of expression without gene inactivation are marked by a black bar. Modifications in the sex-specificity of *desatF* expression are represented by a hollow bar. The "R" refers to regulatory transitions.

Altogether, 11 independent evolutionary changes in *desatF* expression occurred in the approximate 40 million years during which these species evolved. * Note that the gene inactivations in *D. yakuba* and *D. lutescens* are not counted as such in the tally. In *D. yakuba*, the regulatory loss of *desatF* expression appears to have preceded the pseudogenization event. In *D. lutescens*, the ambiguous phylogenetic relationships in the clade prevent the accurate inference of transitions. Gray full circles indicate independent losses of dimorphism. (c) Status of *desatF* expression in oenocytes (the CHC producing cells) in four-day-old adults. N/R: not relevant (because the gene was not functional, based on sequence information). (d) Diene production in males and females where known.

In addition to the single-gene changes noted above, polygenic changes, with multiple genes on the same or a variety of chromosomes, are also involved in CHC signal divergence between some species pairs in the *Drosophila melanogaster* subgroup (Ferveur 2005). For example, QTL affecting female CHCs in one such species pair, *Drosophila simulans* and *D. sechellia*, have been identified on the X and third chromosome, and a few candidate genes were potentially implicated, including some that coded for elongases (enzymes that lengthen CHC chains) in addition to desaturases (Gleason *et al.* 2009). Complex (epistatic) interactions between some of these genes may be also involved in mating isolation.

Summary

Chemical signals are sexually selected in the courtship behavior of many species of invertebrates and vertebrates. There are examples of pheromone-mediated mating systems that illustrate Fisherian ("sexy sons"), "indicator" ("good genes"), and direct benefit mechanisms for the evolution of female choice. Male moths, goldfish, and garter snakes respond to pheromones in scramble-mating systems.

Alternative mating strategies include "silent" satellite males and deception with cuticular pheromones in beetles and garter snakes. Assessment of male value by honest signals is shown in the cockroach, *Nauphoeta cinerea*, and the tiger moth, *Utethesia ornatrix*. Assessment of mates by olfactory cues is taken to the height of sophistication by mammals, ranging from the species, sex, and reproductive state to the genotype and health of a potential mate.

Pheromone signals are used by males in many insect species that lek. Odors are important in female choice of territories in some ungulate leks. Allohormone pheromones act in cryptic mate choice and may offer "chemical mate guarding."

Speciation mechanisms have been investigated in moths and the fruit fly, *Drosophila*. Pheromones in male moths have been selected for by female choice to avoid mating mistakes in sympatric species. Insect pheromone systems in particular offer the advantages of tractable genetics and laboratory populations for the study of speciation and sexual selection.

Further reading

Chapters in Székely *et al.* (2010) review the mechanisms involved in sexual selection, speciation, and the role of mate choice. Andersson (1994) gives a readable and comprehensive account of sexual selection. Futuyama (2009) provides a similarly readable introduction to these topics and evolutionary processes in general. Coyne and Orr (2004) provide the leading in-depth review of speciation. Chapters in Allison and Cardé (2014) cover speciation and the evolution of moth pheromones and include many case studies of key species.

Many chapters in Hurst *et al.* (2008) and East and Dehnhard (2013) cover pheromones and reproduction in vertebrates. Chapters in Breithaupt and Thiel (2011) discuss chemical communication, including sex pheromones, in Crustacea.

Johansson and Jones (2007) and Thomas (2011) are recommended as reviews (but be aware that Johansson and Jones, 2007, use "pheromone" for any kind of molecule including many that we would now classify as signature mixtures or chemosensory cues).

Articles in *Trends in Ecology & Evolution* regularly cover advances in sexual selection.

For molecular structures see sites such as www.chemspider.com, which allows you to search by common name and shows synonyms as well as the systematic names. See also the Appendix for a short guide to the terminology (available for free download from the website associated with this book). Many pheromones are illustrated on Pherobase www.pherobase.com (El-Sayed, 2013).

4 Coming together and keeping apart: aggregation pheromones and host-marking pheromones

Aggregation pheromones lead to the formation of animal groups near the pheromone source, either by attracting animals from a distance or stopping ("arresting") passing conspecifics (Chapter 10) (Wertheim et al. 2005). In contrast to sex pheromones (which attract only the opposite sex), aggregation pheromones by definition attract both sexes (and/or, possibly, larvae). However, the benefits of aggregation to individuals may be complex and two rather different mechanisms may apply. First, individuals may be aggregating for the benefits of living in a group. Second, what we call aggregation pheromones may be the response of eavesdropping conspecifics, for example males responding to sex pheromones released by other males to attract females, in order to mate with those females (Chapters 3 and 11).

By contrast, host-marking pheromones deposited on hosts by insects parasitizing fruits or caterpillars, for example, lead to dispersion as host-marking pheromones deter other females from laying eggs in that host. How host-marking pheromones evolved is discussed in Section 4.2.

4.1 Aggregation pheromones and Allee effects

While the negative effects of overcrowding are often obvious, it is also possible to be "undercrowded," to have a population density that is too low (Berec et al. 2007; Gascoigne et al. 2009). For example, a lone animal may be more vulnerable to predators or be unable to find a mate (Chapter 3). The benefits of conspecifics are broadly referred to as Allee effects (named after the pioneering work of W. C. Allee (1931)). Allee effects are often revealed by lower reproductive rates per individual at low numbers or densities of conspecifics. The classic beneficial effects of group-living include influences on one or more of the following: predation, reproduction, feeding, and local environment (Berec et al. 2007). Allee effects may explain the evolution of many strong aggregation responses to pheromones.

Pheromones play an important part in the life of desert locusts, one of the most famous aggregating animals (Box 4.1). The nematode *Caenorhabditis elegans* shows a strong response to its aggregation pheromone (Figure 1.6) (Srinivasan et al. 2012), though the benefits of aggregation are unclear as we are only starting to understand its natural ecology (Felix & Duveau 2012).

4.1.1 Defense against predation: safety in numbers

California spiny lobsters, *Panulirus interruptus*, form groups of up to 30 animals in dens during the day. The aggregations are caused by attraction to chemical cues in the urine from conspecifics of either sex (Figure 4.1) (Aggio & Derby 2011; Horner et al. 2006). Although vulnerable as individuals, as a group the lobsters wave their robust spiny antennae from the den opening to deter predatory fishes.

Two common benefits of safety in numbers come from the dilution of risk (any individual animal is less likely to be eaten) and the swamping of predators (when there are locally too many prey for predators to eat). The mating and egg-laying aggregations of the seaslug *Aplysia*, mediated by protein pheromones,

Box 4.1 Aggregation and other pheromones in the life of the desert locust

Pheromones are important in the biology of the desert locust, *Schistocerca gregaria*, though many details remain to be resolved (Hassanali *et al.* 2005; Pener & Simpson 2009; Simpson *et al.* 2011). *Schistocerca gregaria* can famously change between solitarious and gregarious phases (forms) in response to population density. In its gregarious phase, the desert locust forms swarms of thousands of millions of insects that can affect 20% of the Earth's land surface. Animals on the extremes of the phase continuum are strikingly different in appearance, physiology, and behavior (and were once thought to be separate species). The change from solitarious to gregarious starts subtly and is caused by crowding. The aggregation pheromones that bring the locusts together can thus have an important effect on gregarization by starting a powerful behavioral positive-feedback loop. Adults and nymphs seem to have separate aggregation pheromone systems. One feature of the gregarious phase is maturation synchrony, mediated by primer effects of volatile adult male pheromones (and possibly amplified by chemotactile signals), leading to swarming. Nymphal aggregation pheromones slow down the development of adults of both sexes. The combined primer effects of these pheromone(s) are to synchronize the development of the entire population. Cuticular compounds produced by gregarious nymphs may gregarize solitarious individuals.

Gregarization can be passed to the next generation in three ways involving pheromones (Hassanali *et al.* 2005; Pener & Simpson 2009; Simpson *et al.* 2011). First, adult females whose antennae contact cuticular compounds of adults, produce large eggs likely to produce hatchlings with gregarized features (Maeno & Tanaka 2012). Second, by the aggregation responses of laying females to oviposition pheromones in the froth of eggs already laid in the soil, the young will emerge to a concentrated population. Third, a chemically mediated maternal epigenetic effect: crowd-reared, gregarious female locusts secrete a hydrophilic "gregarizing factor" in the foam on the eggs, which promotes gregarization of their offspring hoppers.

may have this function (Cummins *et al.* 2007). Female blood-sucking insects such as mosquitoes and blackfly (Simulidae) may gain such benefits for their offspring by laying their eggs close to others. Many species have oviposition pheromones (Seenivasagan & Vijayaraghavan 2010). For example, *Culex* mosquito females are attracted to pheromones produced by egg rafts laid by congeneric females (Laurence & Pickett 1985). As the pheromone, (5*R*,6*S*)-6-acetoxy-5-hexadecanolide, found in a droplet at the top of the egg, is not produced for the first 24 hours, only surviving eggs will produce it. While the response to oviposition pheromone may have evolved as eavesdropping by the later females (Chapter 11), the "signaler" also benefits from the same effects so this could also be an evolved signal. Oviposition pheromones may offer ways to control these important insect vectors of disease (Chapter 12).

Figure 4.1 California spiny lobsters, *Panulirus interruptus*, in a shelter, using their second antennae to collectively protect the entrance from predators (Aggio & Derby 2011; Horner *et al.* 2006). Molecules in urine attract lobsters to the aggregation. Photograph courtesy Crow White.

Many species of insect with strong chemical defenses and bright warning colors (aposematic colors) form conspicuous aggregations mediated by pheromones. The aggregation pheromone of the seven-spot ladybird beetle, *Coccinella septempunctata*, is an alkylmethoxypyrazine (Al Abassi *et al.* 1998). This class of compounds acts as an olfactory warning signal in this and many other aposematic species – giving the olfactory equivalent of aposematic coloration (Guilford *et al.* 1987; Moore *et al.* 1990). The ladybird alkylmethoxypyrazines are persistent and may provide the chemical cues leading to the traditional over-wintering sites used year after year by large aggregations (Aldrich 1999). It seems likely that defense generally precedes the evolution of both aggregation and conspicuous warning signals, though debate continues (Ruxton & Sherratt 2006).

4.1.2 Ecophysiological benefits

Some aggregations may give ecophysiological benefits, related to the interaction between the animals' physiology and environment. For example, pheromone-mediated aggregation of conspecifics reduces water loss in ladybirds, cockroaches, house mites, ticks, and barnacles exposed at low tide (e.g., Bertness *et al.* 1999; Yoder & Grojean 1997). The pheromonal basis of these aggregations has been explored in some cases. Cockroach aggregation in response to conspecific odors is well known (Bell *et al.* 2007; Lihoreau *et al.* 2012). An arrestment (= assembly) pheromone stops wandering ticks, forming tight clusters of inactive ticks in sheltered locations where they sit out adverse conditions (Sonenshine 2004, 2006). In most tick species, purines are the major component of the arrestment pheromone. The purines, such as guanine, are excreted waste products of blood meals so places where ticks have successfully fed before, and to which hosts will return, will be marked most (Allan 2010). The pheromones are not species specific because it benefits any tick to aggregate. The response is only induced when the relative humidity is very low, thus supporting a role as behavior to reduce water loss (Hassanali *et al.* 1989).

4.1.3 Beetle aggregation pheromones: Allee benefits and same-sex eavesdropping?

While most bark beetles live in already dead trees, a minority of species can colonize and kill living trees, creating a moving front of killed trees turning from green to brown (Kausrud *et al.* 2011; Raffa 2001). Rising global temperatures, which weaken trees by water stress and lead to greater over-winter survival of bark beetles, have contributed to a continuing epidemic of tree-killing bark beetles over millions of hectares in North America (Bentz *et al.* 2010).

Colonizing a living tree is only made possible by mass attack of thousands of beetles attracted by aggregation pheromones (Blomquist *et al.* 2010; Raffa 2001). Only attack by large numbers can overcome the living tree's formidable defenses, which include toxic chemicals such as terpenes and phenols as well as physical defenses such as gums, latexes, and resins (Byers & Zhang 2011; Seybold *et al.* 2006). For the females of the bark beetles *Dendroctonus rufipennis* and *D. ponderosae*, these Allee effects explain why reproductive success on living trees

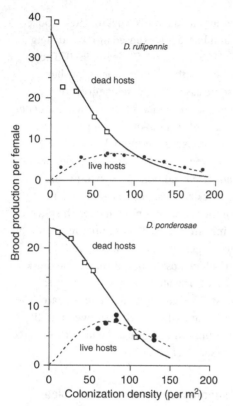

Figure 4.2 Allee effects shown in the reproductive success of bark beetles *Dendroctonus rufipennis* and *D. ponderosae* colonizing living host trees. Female beetles attacking a living tree (dashed line) have a higher individual reproductive success with increasing female density until competition starts to reduce success at greater densities. The effectiveness of the living tree's defenses is revealed by the contrasting high reproductive success of females in dead logs even at low population density (solid line). The region to the right of both lines indicates the effect of intra-specific competition, which reduces beetle reproductive success across all densities, but its net effect only becomes negative in previously live trees once their defenses are overcome. Figure and caption adapted from Raffa (2001). See Raffa (2001) for details of data sources and data fitting.

rises with increased density until intra-specific competition for limited larval food space overtakes the benefits (Figure 4.2) (Raffa 2001).

Mass attack in most bark beetles starts with the arrival of the first beetles on a suitable tree. The pioneers start to bore into the bark and release long-range aggregation pheromones that attract conspecifics of both sexes (see also Chapter 11). In most monogamous

species, this is the female (e.g., *Dendroctonus* spp.); among polygamous species, the male plays this role (e.g. *Ips* spp.) (Kirkendall *et al.* 1997).

By definition, aggregation pheromones attract both sexes. There has been much debate about the origin of these pheromones in bark beetles and other animals. Are they pheromones that invite fellow colonists of both sexes, for Allee benefits to the signaler? Or instead, are these sex pheromones, usually released by the male to attract females, but eavesdropped by males hoping to intercept some of the females attracted (Chapter 11)? The answer may depend on the lifestyle of the beetle species (Kirkendall *et al.* 1997; Raffa 2001). In tree-killing *Dendroctonus* spp., same sex (female–female) responses dominate during colonization. Only later during the attack do more males than females arrive. For less aggressive bark beetle species (the majority), which only attack dead trees, there may be little advantage in attracting fellow colonists of the same sex, and much to lose in competition. The shutting down of pheromone calling when the opposite sex arrives suggests that in these species eavesdropping is the reason for same-sex arrivals. For example, mating stops calling in the less aggressive *D. pseudotsugae* and there is little difference in the attractiveness of unmated and mated females, but pheromone calling continues after mating in the tree-killing southern pine beetle, *D. frontalis* (Raffa *et al.* 1993).

There are a few bark beetle species in which a single female beetle attacks a live tree by herself (e.g., *Dendroctonus micans* and *D. valens*). Her larvae cope with the resin defenses of the tree by aggregating, in response to larval pheromones, to form a feeding front of up to 50 larvae feeding side by side. Larvae reared in groups have higher survivorship and gain mass faster than isolated larvae (Storer *et al.* 1997).

In stored-product insects, beetle species with long-lived adults tend to have male-produced aggregation pheromones and mating occurs on the food source (for example, *Prostephanus truncatus* and *Sitophilus zeamais*) (Burkholder 1982; Phillips & Throne 2010). In *Prostephanus truncatus*, the males produce a sex pheromone probably directed at females, but which is eavesdropped by conspecific males in search of mates.

This is suggested by the way that the males shut off pheromone release when females arrive, like the bark beetles described above (Smith *et al.* 1996). This, incidentally, prevented the isolation of sufficient pheromone to analyze for many years until scientists tried collecting from males only, rather than from mixed-sex cultures. There may, however, be some advantages of aggregation: Allee's original work (1931) cited studies of the flour beetle, *Tribolium castaneum*, which showed optimum individual reproductive success at intermediate population densities – perhaps because of the warmth or high humidity created at these densities, offset by cannibalism at high densities.

By contrast, stored-product beetle species with short-lived adults (<1 month) generally have female-produced sex pheromones and usually do not feed as adults (for example *Trogoderma glabrum* and *Lasioderma serricorne*) (Levinson & Levinson 1995; Phillips & Throne 2010). However, Levinson and Levinson (1995) also provide examples of stored-product beetle species that break these predictions, sometimes in the same genera as species that conform to them.

4.1.4 Settlement of marine invertebrates

Marine invertebrates living on the seabed or seashore show some of the most conspicuous aggregation behavior known. Most of these organisms have a planktonic larval dispersal phase and it is crucial for their later survival and successful reproduction that the planktonic larvae settle in the right habitat, close to conspecifics.

Chemical cues produced by adult conspecifics or recently metamorphozed larvae are among the key stimuli (Clare 2011; Hay 2009). For example, when searching for settlement sites, oyster larvae sink into the bottom boundary layer where they detect soluble cues from the mantle cavity of adult oysters. A larva has to decide quickly whether to settle as the current sweeps it past potential sites. Barnacle cypris larvae respond to chemical cues from conspecific adults including a glycoprotein, settlement inducing protein complex (SIPC), which appears to have evolved from

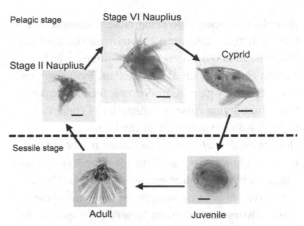

Figure 4.3 Developmental stages of the barnacle, *Balanus amphitrite*. The cyprid larvae settle from the plankton and search rocks for chemical cues from conspecific adults including a glycoprotein, settlement inducing protein complex (SIPC). The cyprid larvae also deposit SIPC as they move over the surface, which arrests other larvae (Dreanno *et al.* 2006a). The diameter of the adult barnacle is 2 cm. Scale bars = 100µm. Figure from Chen *et al.* (2011).

α_2-macroglobulin-like proteins in the adult barnacle cuticle (Figures 2.3. and 4.3)(Clare 2011; Dreanno *et al.* 2006b). A larva settling near conspecific adults has evidence that the site sustains that species – for example, a site with adequate water flow rates and food supply, and a reasonable exposure and temperature range if intertidal.

Many other advantages may come from settling close to conspecifics, including ones for reproduction. At maturity, adults need to be close enough to find or reach mates. This is especially important for internally fertilizing animals such as barnacles that are permanently glued to the rock. Barnacles have the longest penis in relation to body size of any animal: for example that of *Elminius modestus* is up to 5 cm long, impressive for an animal only 0.5 cm across.

For externally fertilizing animals such as sea urchins, which release their gametes (eggs and sperm) into the water, population density is still important. The ocean is big and quantities of gametes are small. Fertilization rates are greatly reduced if adults releasing gametes are far apart so there are significant reproductive benefits to being close to conspecifics (Crimaldi 2012; Levitan & McGovern 2005; Serrão & Havenhand 2009). Filter-feeding animals, which exploit currents, such as sand dollars, byrozoans, and barnacles, may gain hydrodynamic feeding advantages from living close to conspecifics (Vogel 1994, p. 246). Adult aggregation of the Pacific sand dollar, *Dendraster excentricus*, is facilitated by the specific metamorphosis response of the planktonic larval stage triggered by adult odors on the sand (Burke 1984).

If larvae only settle in response to chemical cues from existing aggregations, how do such species colonize new, empty sites? Marine invertebrates with non-feeding larvae become less discriminating with age, so that if ideal sites are not found, larvae will "settle for less," responding to suboptimal sites rather than dying unsettled (Jenkins *et al.* 2009). A different, bet-hedging solution, evolved by the colonial tube-dwelling polychaete worm, *Hydroides*, is to have two types of larvae: one type responds to chemical cues from adult conspecifics, the other type consists of "pioneer larvae," which colonize bare new substrate (Toonen & Pawlik 2001).

4.1.5 Aggregation in time: synchronization

Predators may be swamped if members of a prey species synchronize the release of their young. This seems to be a common adaptation in brachyuran crabs, which brood their eggs before releasing the hatching larvae into the sea for a planktonic stage (Christy 2011). In the mud crab, *Rhitropanopeus harrisii*, which lives in tidal creeks on the east coast of North America, all the larvae in the population are released on the same few nights (Forward 2009). The female carries the fertilized eggs glued to her abdomen until they are ready to hatch. Egg hatching is timed to occur at night, to reduce predation risk, and at the spring high tide, to avoid low-salinity stress to the larvae. The trigger for release comes from the eggs themselves: when ready to hatch, the eggs swell and the egg membranes break, releasing the "pumping pheromone." The pheromone stimulates a stereotyped "pumping" behavior in the

adult female, which vigorously flexes her abdomen, releasing all the larvae. The pheromone may be a short peptide (it can be mimicked by amino acids and short synthetic polypeptides), with activity at concentrations as low as 10^{-21} M (Pettis *et al.* 1993; Rittschof & Cohen 2004).

The egg-hatching pheromone in the barnacle *Semibalanus balanoides* comes from the adult. The barnacle broods its eggs in its mantle cavity over the winter. The eggs are ready to hatch for some months, just waiting for the signal – the egg-hatching pheromone released by the adult after feeding on the spring phytoplankton bloom (Oyarzun & Strathmann 2011). The system ensures that the barnacle nauplii are released when the phytoplankton they need for food have arrived. The hatching pheromone may include eicosanoids (called prostaglandins in vertebrates), which are C_{20} polyunsaturated fatty acids (PUFAs) (Rowley *et al.* 2005). Other eicosanoids are important aquatic pheromones for spawning in the abalone, a marine mollusc, and fish (see Chapters 1 and 3).

Externally fertilizing marine animals need to coordinate the timing of their gamete release, and pheromones are commonly used to achieve this (Chapter 3) (Breithaupt & Hardege 2012).

4.2 Host-marking pheromones

After laying an egg inside a host, such as a small fruit or a caterpillar, many species of phytophagus or parasitic insects leave a pheromone mark on or in the host (host-marking pheromone, HMP, previously often called oviposition-marking pheromone OMP) (Figure 4.4) (Liu *et al.* 2011; Roitberg *et al.* 2009). Other female conspecifics show "host discrimination": they avoid superparasitizing a host, that is, laying an egg in a host in which an egg has already been laid. Host-marking pheromones have evolved in several orders of insects (Diptera, Coleoptera, Hymenoptera, Lepidoptera, and Neuroptera), including more than 200 species of parasitoid wasps (wasps that lay their eggs in other insects) and more than 20 families of

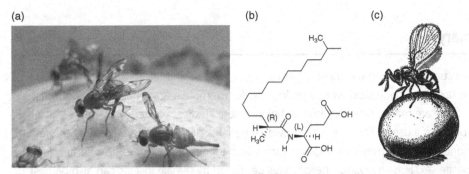

Figure 4.4 (a) The tephritid Mexican fruit fly *Anastrepha ludens* marks its host fruit with a pheromone (b) after egg laying (Aluja *et al.* 2009b). Photo by Jack Dykinga, USDA ARS. (c) An adult female parasitoid wasp, *Trichogramma pretiosum*, in the process of parasitizing an egg of the brown-tail moth. After laying her egg inside, she will mark the host egg with pheromone. Figure from Howard & Fiske (1911).

herbivorous insects (Liu *et al.* 2011; Nufio & Papaj 2001). (See Chapter 12 for the potential use of HMP in pest control.)

Host-marking species lay their eggs in hosts of limited size, which can only successfully support the development of a limited number of larvae; any further larvae will reduce the chance of the first surviving. Thus the primary reason HMP evolved is to enable a female to avoid laying a second egg in a host she has already laid in, as it would waste her second egg and risk the life of her first (Hoffmeister & Roitberg 2002; Roitberg *et al.* 2009). Conspecific females gain from taking notice of other females' marks because they do not waste an egg in an already parasitized host (the original female's offspring would have a head start). However, it may be adaptive for females to ignore the marks when there are few unparasitized hosts available (Hoffmeister & Roitberg 2002; Roitberg *et al.* 2009). Indeed, female wasps adjust their rate of superparasitism according to the proportion of unparasitized hosts they encounter (Nufio & Papaj 2001).

Some parasitoid species use only internal marks while others only mark externally and some use both mechanisms (Godfray 1994; Nufio & Papaj 2001: note their definition of marking pheromone is wider than HMP). In many species the Dufour's gland is the source of the mark and the variation in the chemical profile of gland secretion may allow distinction of self- vs. conspecific marks. Females of six out of nine parasitoid species could distinguish between hosts parasitized by themselves and those parasitized by conspecifics, perhaps by individual recognition of marks (van Djiken *et al.* 1992). Would these marks thus include molecules learnt as signature mixtures by the wasps (Chapter 1) as well as a species-specific mark? I am not aware of animals apart from insects that have HMP, but the situations in which to look for it would be ones where the developing larva needs the whole of a finite resource.

As well as marking hosts, some parasitoid wasps also pheromone-mark patches they have visited and spend more time searching unmarked patches (Hoffmeister & Roitberg 2002). Predatory insects similarly avoid trails left by other predators, which show an area already has many competitors present. For example, ladybirds lay fewer eggs if they detect chemical traces of conspecific or heterospecific ladybird larvae (Magro *et al.* 2010; Pasteels 2007). Nectar-feeding insects such as bumblebees use the chemical footprints left by previous bee visitors to avoid returning to a flower too soon, though these might be considered cues rather than evolved pheromone signals (Goulson 2009; Saleh *et al.* 2007; Wilms & Eltz 2008).

Summary

Chemical cues are among the most important cues used by animals to aggregate. While different taxa have been studied independently, for example marine barnacles and forest beetles, there are many common threads linking these studies. The renewed interest in Allee effects provides a useful focus on the individual benefits of aggregation. Many aggregation pheromones in insects may be explained by eavesdropping of sex pheromones intended for the opposite sex. Host-marking pheromones deposited by female insects can also be best explained by individual advantage, to both the marking and responding females.

Further reading

Allee effects are well reviewed in a modern context by Gascoigne *et al.* (2009) and at book length in Courchamp *et al.* (2008). Chapters in Hardie and Minks (1999) discuss aggregation pheromones in bark beetles, stored-product beetles, and other insects. Wertheim *et al.* (2005) review aggregation in arthropods in general.

For molecular structures see sites such as www.chemspider.com, which allows you to search by common name and shows synonyms as well as the systematic names. See also the Appendix for a short guide to the terminology (available for free download from the website associated with this book). Many pheromones are illustrated on Pherobase www.pherobase.com (El-Sayed 2013).

5 | Territorial behavior and semiochemicals

Marking a territory with scent marks is a conspicuous behavior of many mammals and other terrestrial vertebrates including lizards (Font *et al.* 2012; Martín & López 2010a; Müller-Schwarze 2006; Roberts 2007). The molecules in scent marks include species-wide pheromones as well as highly individual odors, which are cues to identity (Chapter 1; Wyatt 2010).

Mammals have an enormous variety of specialized scent glands but a common pattern of scent marking: glandular secretions, and often feces and urine, are placed at conspicuous places in their home ranges or territories, often in lines along paths or boundaries (Figure 5.1) (Gosling & Roberts 2001). Males tend to mark more than females, and dominant males or territory holders mark most, especially during breeding (Roberts 2007). Scent marking may form a central part of ritualized contests between territorial males or between competing groups, for example "stink fights" between neighboring groups of ring-tailed lemurs (Figure 5.2).

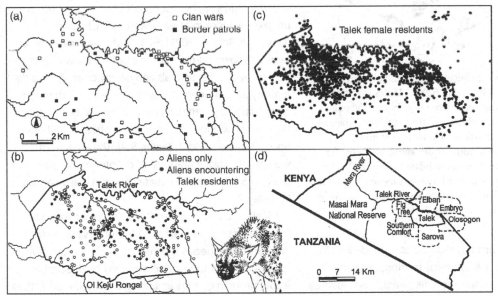

Figure 5.1 Members of the "Talek" clan of spotted hyenas, *Crocuta crocuta*, on border patrols (a) mark their territories with latrines and anal pouch scent marks (see Figure 5.5) (Boydston *et al.* 2001). The map shows (a) the northeastern portion of the Masai Mara National Reserve with watercourses indicated as well as locations of all clan wars and border patrols; (b) Talek territorial boundaries inferred from the data in (a), containing locations of 133 encounters (excluding clan wars) between Talek residents and alien hyenas observed within the Talek territory, and 160 sightings of aliens undetected by Talek residents; (c) 3,130 locations at which 13 Talek females were radio-tracked between 1996 and 1998; (d) the entire reserve showing the location of the Talek territory in relation to the approximate boundaries of territories defended by neighboring clans (dashed lines). Main figure, Boydston *et al.* (2001). Hyena inset, drawing by Priscilla Barrett (see Figure 5.5).

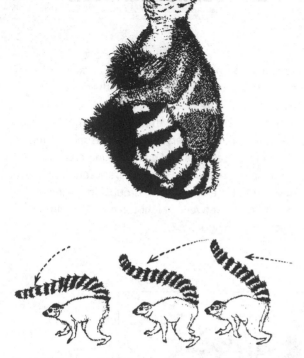

Figure 5.2 Groups of the Madagascan ringtailed lemur, *Lemur catta*, defend territories by "stink fights." Males load their tails with secretions from their wrist and shoulder glands (top) and then flick their tails toward their opponents (below). The scent from these glands, which are testosterone dependent, identifies individual males and is also used for intra-group dominance threats (Scordato & Drea 2007). The olfactory threat is complemented by the conspicuous visual display of striking tail and face markings. Figure from and by Dixson (1998) after Evans and Goy (1968).

Territories can be broadly defined as defended areas (Stamps 1994). They are often for feeding, but can include other resources such as a den site that is valuable for the owner or for attracting mates. Depending on the species, the territory might be owned by one male, such as a small antelope, or by a group, as in badgers. Many social insects are also fiercely territorial but they generally use scent marks in different ways from terrestrial vertebrates (Box 5.1). Aquatic animals do not seem to mark territories.

One benefit of scent marks for territorial behavior is that, unlike other signals, scent marks "shout" even when the animal is not there. However, while it may take less energy than sound signals (Chapter 1), scent marking in some species can have significant metabolic costs – in mice the territorial male's urine marks contain 20 to 40 mg/ml of protein, largely MUPs (Box 5.2) (Hurst & Beynon 2004, 2013). Other costs are in time and risk. For example, males of a territorial African antelope, the oribi, *Ourebia ourebi*, spend up to 35% of their time on marking or associated activities (Brashares & Arcese 1999a). Animals will go to some risk to investigate and over-mark intruders' scent marks. Beavers will leave the safety of water to investigate foreign scent marks on the shore (Sun & Müller-Schwarze 1998). Scent marks can be "eavesdropped" by predators and parasites (Chapter 11).

Scent marks are honest signals because they act as an unfakeable index in at least two ways (Chapter 1) (Gosling & Roberts 2001). First, fresh scent marks on a territory show that the marker is both successful in competition with other animals and has successfully held the territory long enough to mark it. Second, the chemical composition of scent marks reflects the biological state of the marker, including its social status, health, and nutrition (Chapters 3 and 9) (Figure 5.3). The content of scent marks and/or presence of counter-marks (Section 5.7) are used for mate choice in many species (Chapter 3).

Much of this chapter explores scent marking in the context of territorial behavior but it is important in non-territorial species too (Section 5.8). In group-living species many of the scent marks may be directed at fellow group members as scent marking also advertises dominance and reproductive status *within* social hierarchies (Roberts 2007). These may underlie the "chinning" marking behavior of male and female rabbits (Melo & González-Mariscal 2010). The scent marks may also advertise outside the group. For example, in high-density populations of the yellow mongoose, *Cynictis penicillata*, in southern Africa, which lives in social groups of ten or so animals, most marking is done by subordinate adults of both sexes rather than by the dominants (which usually mark in social species) (Le Roux *et al.* 2008; Wenhold & Rasa 1994). The boundary marks may advertise the sexual maturity and identity of the subordinates, important because in high-density populations these animals go outside the territory to mate with neighbors.

Scent-marking behavior can differ between related species and also varies between populations of the same species in ways that seem to be linked to habitat and the way habitat affects social structures.

5.1 Why scent mark on territories?

Among the three main theories to explain the evolution of scent marking on territories are that scent marks (1) deter potential intruders (the scent-fence hypothesis); (2) allow the intruder to recognize the owner, should it be met, by the match between its odors and those of the scent marks around the territory (the scent-matching hypothesis); or (3) establish boundaries with major competitors (the border-maintenance hypothesis) (Gosling & Roberts 2001; Roberts 2007). Scent marks could also provide information to intruding animals on the status of the resident through the intrinsic characteristics of the marks (for example, the pheromone molecules in the urine of dominant mice (Chapter 9) (Figure 5.3). Marks within territories might also attract or stimulate mates (Chapters 3, 6, and 9).

The predictions for the different theories (see below) are not mutually exclusive and indeed it is likely that scent marks have multiple functions. The diversity of marking patterns among closely related species such as antelopes, or populations of the same species in different habitats, for example yellow mongoose (Le Roux *et al.* 2008) or spotted hyenas (Section 5.5), shows the importance of ecological factors and suggests that it is probably a mistake to expect the selective forces for scent marking to be the same for all terrestrial vertebrates.

5.2 Scent-fence hypothesis

In theory, scent marks could work as a chemical "keep out sign" or "fence," which would repel all individuals of a given category, for example, adult males. However, even though marks are often placed around the edge of the territory, field observations in almost all species studied so far show that territory intruders do not avoid marked areas (Gosling 1990) (for example in hyenas, see the numerous intruders, "aliens", in Figure 5.1b). One exception is the European mole, *Talpa europaea*: males avoid the scent marks of other moles (whether familiar or unfamiliar), perhaps because escape is difficult in a confined burrow and the risk of injury in underground combat is high (Gorman & Stone 1990). There is some evidence that at least at low densities, beavers, *Castor canadensis*, will avoid marked sites (Welsh & Müller-Schwarze 1989). However, in some species scent marks paradoxically may make an area more attractive as they indicate that the habitat will support the species (Stamps 1994).

5.3 Scent-matching hypothesis

This hypothesis proposes that intruders learn the odor of the territory owner from the scent marks on the territory so that when they encounter another animal that matches the scent marks they recognize that this is the owner and can avoid fights they are likely to lose (Gosling & Roberts 2001; Roberts 2007). Scent matching helps to account for otherwise unexplained behavior such as the way territory owners and other resource holders often mark their own bodies with the substances used for scent marking and make themselves available for inspection by sniffing at the start of many encounters (Figure 5.4) (Gosling & Roberts 2001). Subordinates or intruders would be expected to search out scent marks, and this is the case in mice (see Section 5.3.1).

Owners are usually animals with high competitive ability and if they have more to gain by retaining their territory than an intruder has by taking it over (for example, because familiar territory is more valuable than unfamiliar territory), it will pay the owner to escalate the fight (Gosling & Roberts 2001; Roberts 2007). At stake are costs in time, energy, and risk of injury or death. Game theory predicts that if the identity of the territory holder is unambiguous, most intruders will retreat rather than become

Box 5.1 **Social insect territories**

The economics of territorial defense lead some ant species to defend trail trunk routes leading to valuable food resources rather than a wide area around the nest (Hölldobler & Wilson 1990). Some other ant species fiercely defend foraging territories (d'Ettorre & Lenoir 2010; Traniello & Robson 1995). Recruitment of major workers by alarm pheromones (Chapter 8) during territorial disputes is important in most species, but while many ant species mark the area immediately around the nest, few ant species have exclusive territories covered with scent marks in the way that many vertebrates do (Hölldobler & Wilson 1990, 2009). African weaver ants, *Oecophylla longinoda*, are an exception as they mark their territory with colony-specific fecal marks (see figure below) (Hölldobler & Wilson 1978). These marks likely include the colony odor (Chapter 1) (Grasso *et al.* 2005; Wyatt 2010).When new "territory" was offered to nests of *O. longinoda* in the laboratory, workers rapidly moved in and defecated systematically over the area. When contests between two ant colonies were staged, ants on an arena marked with their own colony's defecations were bolder, won more contests, and took over the whole arena. If neither ant colony had marked the arena then contests were even (Hölldobler & Wilson 1978). Similar results have been shown for the desert ant, *Cataglyphis niger* (Wenseleers *et al.* 2002).

Social insect colonies have the advantage over mammals as they can be in many places at once when defending a territory (Hölldobler & Wilson 1990, 2009). Large colonies of *O. longinoda*, consisting of a single queen and up to 500,000 workers, can control exclusive three-dimensional territories (Crozier *et al.* 2010; Hölldobler & Wilson 1978). One Kenyan colony included the canopies of 17 trees (with a leaf surface area equivalent to at least 100 km^2 if scaled to human dimensions). In the related weaver ant, *O. smaragdina*, older workers dominate in barrack nests around the periphery of the territory, where territorial invasion is most likely. This led Hölldobler and Wilson (1994) to say that, while human societies send their young men to war, weaver ants send their old ladies. These ants are particularly aggressive toward members of other weaver ant colonies and the territories of different colonies are often separated by "no ants' land," narrow zones into which few ants venture.

Box 5.1 (cont.)

(a)

(b)

(a) Territorial marking by weaver ants, *Oecophylla longinoda*, shown by anal spots on a paper surface in a laboratory foraging arena. (b) Two workers from different colonies in combat, conducted by rearing up on their legs and mandibles in a threat display (above), dodging each other, and seizing one another with their mandibles (below). The ant fighting on an area marked by her own colony enjoys an "owner advantage" typical in such cases. Figures from Hölldobler and Wilson (1978). Painting (b) by Turid Hölldobler.

involved in an escalated contest that can result in serious injury. If scent marking and allowing inspection behavior allows the territory owner to be identified unambiguously, then scent marks could reduce the cost of territorial defense for owners, by avoiding costly (and risky) escalations. Intruders also benefit from identifying whether an animal is the territory owner before deciding to escalate, which sometimes they may do after assessing the owner.

One major experimental difficulty in testing the scent-matching hypothesis is that the intruder's response may be determined by information about the competitive ability of the owner received when they meet, for example from the resident's size and odor, rather than by matching scent to marks perceived earlier in the territory. Rather than being part of a scent matching process, an alternative explanation for the resident allowing inspection could be that the resident is advertising its dominance status by odors characteristic of dominance.

5.3.1 Mice

Dominant male house mice, *Mus musculus domesticus*, mark their territories extensively (Box 5.2) (Hurst & Beynon 2004). If males do assess potential opponents by comparing the opponent's odor with local scent marks, they should be less likely to fight if these odors are the same, indicating that the opponent is the resource holder (of territory or females)

(Roberts 2007). Gosling and McKay (1990) tested this by observing the behavior of two unfamiliar males: an "intruding" male faced with a "resident" male on the other side of a mesh. An air flow meant that the "resident" could not detect the odors of the intruder. In some experiments the side of the intruding male was previously marked with urine from the "resident," in others not. As predicted, intruding male mice were less likely to fight another male if the urine on its side matched the odor of the "resident" male on the other side of the mesh. However, responses by subordinate mice to scent marks may also depend on their competitive ability: smaller mice tend to avoid marked areas, whereas relatively larger subordinates, with a greater chance of winning a contest with the owner, were more likely to investigate marked areas (Gosling et al. 1996). In experiments with larger, more natural arenas (Hurst 1993), intruders were likely to flee if they encountered the dominant resident (recognized at a distance by sniffing) (Box 5.2).

Box 5.2 Marking behavior of house mice

House mice, *Mus musculus domesticus*, are useful models for investigating the role of odor cues in structuring social groups (Hurst & Beynon 2004, 2008, 2013). At high densities mice live in large groups and dominant males tolerate subordinate males on their territory. Urine marks are made all over the territory and all new objects are quickly marked. The urine of dominant males is rich in major urinary proteins (MUPs), which add to signal longevity and identity, and small signal molecules (see Chapters 1 and 9). Dominant males mark at signal sites, which become "urine posts" up to 25 mm high. Urine posts develop near resources such as food or nesting sites and along pathways. Dominant resident males and breeding females mark at much higher rates, up to 100 times an hour, than subordinates (Drickamer 1995). Males counter-mark other males' scent marks and female mice may use over-marking patterns to choose between males (see Section 5.7).

  Within the group, low-level marking by subordinates away from the urine posts and diffusely across the territory helps to ensure continued tolerance by the resident dominant male. If an experimenter adds a small (mouse-sized) urine mark to a signal site, all males investigate these scent marks (Hurst 1993). If the urine was from a resident subordinate, the dominant male soon attacked that individual without warning. If the urine was from an unfamiliar subordinate, the resident dominant male investigated all subordinate males in his territory. An artificial mark from an unfamiliar dominant male prompted greater investigation of the mark by the resident dominant male, who counter-marked it more and, if challenged, was more likely to flee. A dominant neighbor's urine would almost be ignored (the dear-enemy phenomenon perhaps) (Section 5.6). The responses depend on recognition of the individual odor of other mice in the group as well as the dominance status of the animal. Sniffing darcin, the male MUP, prompts a male to also remember the location of a strange male's scent mark (Roberts et al. 2012). A female sniffing the darcin remembers the location of the scent mark and the identity of the male that left it (Section 5.7) (Roberts et al. 2010, 2012).

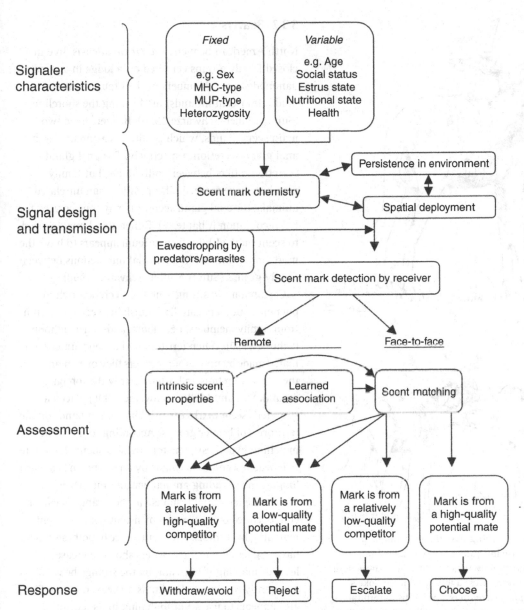

Figure 5.3 Summary of scent marking processes. The assessment and response phases are simplified for illustrative purposes, but in reality responses are complicated by many factors including the value to same-sex receivers of the marked resource, costs of fighting over it, and the probability that the receiver will meet the signaler. If receivers are potential mates, responses may vary according to, for example, sexual receptivity, genetic dissimilarity, and the suitability of the signaler in comparison to others already encountered. Figure and text from Roberts (2007).

(a)

(b)

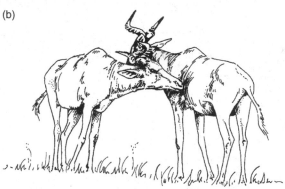

Figure 5.4 The scent-matching hypothesis would help explain two otherwise intriguing behaviors. First (a), self-marking by territory owners with the substances used to scent mark their territories: a territorial male Coke's hartebeest, *Alcelaphus buselapus cokei*, rubs antorbital gland secretion onto its side. Second (b), presentation for inspection: in encounters with intruders, owners make the odors used to mark the territory available for olfactory inspection. This behavior may allow intruders to identify the owner by matching the odor they detect with that of scent marks in the territory. A territorial male hartebeest stands immobile (right) while an intruding male smells and nibbles the scent impregnated fur of its neck and shoulder. Drawings by L. M. Gosling from Gosling (1982).

5.3.2 Beavers

North American beavers, *Castor canadensis*, live in closed family groups centered on a lodge in a pond or dammed stream. Members of the group scent mark mud piles (scent mounds) made along the shoreline (Sun & Müller-Schwarze 1999). Beavers have two main scent glands, which produce "castoreum" and anal gland secretion respectively. The anal gland secretion differs between individuals, but family members have more similar profiles than unrelated animals; the anal gland secretion may thus be used for kin recognition (Chapter 6). Either gland may be used to scent mark although castoreum appears to have the more important role in territorial interactions between family groups (Sun & Müller-Schwarze 1998).

Castoreum is a strong-smelling secretion rich in phenolics. Beavers can distinguish between castoreum from family members, neighbors, and non-neighbors (Schulte 1998). When territory owners encounter scent marks made by non-group beavers they over-mark with their own secretion and often destroy the foreign scent mound. Sun and Müller-Schwarze (1998) placed a stranger's scent marks on artificial scent mounds on the territories of beaver groups. According to the scent-matching hypothesis, territory owners should habituate to repeated scent marks made by an unknown beaver so long as the intruding animal was not met. Over six nights, the territorial response to the stranger's castoreum declined or was level, consistent with the scent-matching hypothesis. The scent-fence hypothesis would have suggested territory owners should increase the level of marking if intrusion by the strange beaver apparently continued. The Eurasian beaver, *Castor fiber*, also appears to use scent matching in its territorial marking (see Campbell-Palmer & Rosell 2010).

5.4 Border-maintenance hypothesis

In some species, males mark preferentially along borders adjacent to the most threatening rivals (see also Section 5.6). Scent marks placed along territory borders may serve to form a "property line" between neighbors, thereby preventing frequent and costly disputes between territory owners (Brashares & Arcese 1999b; Gosling & Roberts 2001). The East African oribi antelope, *Ourebia ourebi*, which has some territories containing a dominant male and subordinate males, scent marks with both its preorbital gland secretion and feces. Territorial males marked grass and shrub stems with their preorbital gland up to 45 times per hour (Brashares & Arcese 1999b). Almost all the marks were made along borders shared with other territorial males and males marked more often at borders shared with multimale groups than at borders shared with a single male. This suggests that males perceived neighboring male groups as a greater threat to territory ownership than neighboring males that defended their territories without the aid of adult subordinates; almost half of takeovers of territories are by neighbors (Arcese 1999). The subordinate males contribute up to half the scent marks and help to defend the territory. In klipspringer antelope, males placed more marks on contested boundaries and on tree branch tips facing likely intruders (Roberts & Lowen 1997).

5.5 Economics of scent-marking patterns in territories

Marking strategy is an outcome of a trade-off between costs of producing and positioning scent marks and the benefits to the owner from the behavior (Gosling & Roberts 2001). If a territory owner marked to ensure that any intruder would always encounter marks, it would have to mark the whole perimeter. Instead, animals tend to produce a limited number of marks, at heights and places likely to be found, such as along paths. There are many examples that suggest marking is limited –

either by material (feces, urine, gland marks) or by time; such economic limitations become more critical as the area of the territory increases. For example, even feces, which may be "free," are limited by the amount the animal can produce. The optimum position for a ring of scent marks may be some distance within the territory rather than at the actual boundary. This is found in klipspringer antelope and hyena latrines. Modeling suggested that this would maximize the probability of intruders encountering the marks and minimize the cost of intrusion (Roberts & Lowen 1997).

The distribution of scent marks in mammalian territories varies between species (Müller-Schwarze 2006, p. 159). Most antelope marks are "boundary" or "perimeter" marks, toward the edge of the territory. Many carnivores also show this pattern. For example, like North American grey wolves, *Canis lupus*, Ethiopian wolves, *Canis simensis*, make the most urine marks along or near territory boundaries (Sillero-Zubiri & Macdonald 1998). Some species instead mark throughout the territory or toward its center (hinterland marking) (see hyenas below) (Gosling & Roberts 2001). Among carnivores, which are noted for both inter-specific and intra-specific variation in behavioral ecology, there may be great intra-specific variation in marking patterns depending on habitat (Holekamp & Dloniak 2010; Macdonald 1985b). For example, spotted hyenas, *Crocuta crocuta*, living in large clans with small territories paste their marks along the perimeter (for example, in the Ngorongoro Crater, Tanzania and in the Masai Mara) (Figures 5.1 and 5.5). In contrast, in the Kalahari desert (southern Africa), where spotted hyenas live in small clans and occupy very large territories, marks are clustered toward the center of the territory (Gorman & Mills 1984; Holekamp & Dloniak 2010). Large territories with few defenders may make regular visits to the boundary uneconomical.

In high-density populations, such as those in southern England, European badgers, *Meles meles*,

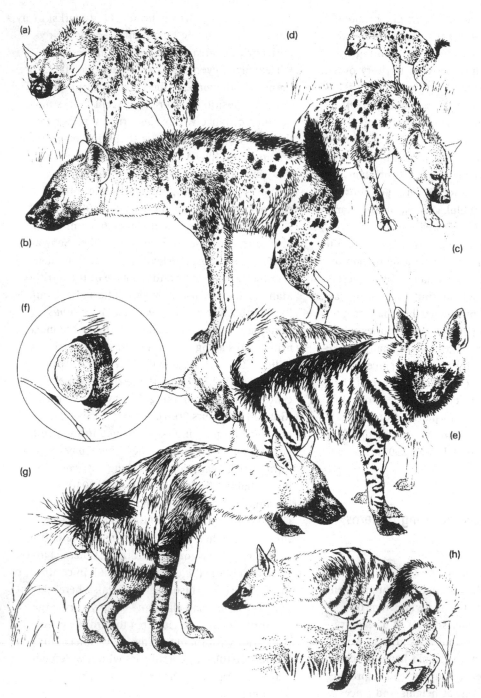

Figure 5.5 Anal scent marking among the Hyenidae: members of a clan of spotted hyenas, *Crocuta crocuta*, (a)–(d) in the Ngorongoro Crater visit pasting sites where they inspect (a) their secretions on long grass stalks and add more pastings from their anal scent pouches (b) (which may include cues for individual and group identity (Archie & Theis 2011; Burgener *et al.*

have both perimeter and hinterland latrines (Roper 2010, p. 201 ff.; Roper *et al.* 1993). Within the social group the two sexes have different marking behavior. Perimeter latrines are visited mostly by males, hinterland latrines by both sexes, but particularly by females. The early spring peak of perimeter marking by males during the main time for mating, suggests that male marking might be mate guarding or territory guarding of females rather than for food. Female marking on the other hand may be for protecting the den as much as for food. Marking and social behaviors in badgers, as in other mammals, are likely to be complex (Roper 2010). Understanding the territorial behavior of badgers is important for guiding management strategies in relation to bovine tuberculosis (Chapter 12) (McDonald *et al.* 2008).

Long-lasting marks with a low volatility may be missed by animals attempting to find them by olfaction from a distance. Across the animal kingdom, many animals have ways of making marks more conspicuous (Alberts 1992). For example, the marks made by the desert iguana, *Dipsosaurus dorsalis*, include molecules that strongly absorb ultraviolet light, which lizards see as dark patches (Chapter 1) (Alberts 1990). Antelope and hippopotamus dung heaps stand out visually. Many animals, including dogs, choose conspicuous objects such as lamp posts or trees to mark (Macdonald 1985b).

5.6 Dear enemies or nasty neighbors

Many territorial animals respond more aggressively to intrusions by non-neighbors (strangers) than territory-holding neighbors, which are treated as "dear enemies" (Temeles 1994). If animals learn to recognize neighbors and respect each other's territorial boundaries they can reduce the energy expended in territorial defense and instead focus defense against strangers in search of a territory.

In group-living species, however, neighbors may be treated more aggressively than strangers, with "nasty neighbor" effects being found for example in some social carnivores such as the banded mongoose, *Mungos mungo* (Müller & Manser 2007), and *Oecophylla* weaver ants (Crozier *et al.* 2010; Newey *et al.* 2010). With intense between-group competition, a neighboring group may be a significant threat.

5.7 Counter-marking and over-marking

Dogs sniffing the marks on lamp posts left by previous dogs and adding their own urine mark are perhaps the most familiar example of the common mammal behavior of scent counter-marking, creating a chemical bulletin board. Finding a mark from a stranger

---

Figure 5.5 (cont.)

2009; Theis *et al.* 2012). Paste is composed of lipid-rich sebum secretions that are bacterially fermented. Pasting by successive individuals may be the way that anal pouch bacteria are shared (Theis *et al.* 2012). Pawing at the ground (c) occurs frequently at pasting sites and at latrines (d), which also occur at the territorial borders, but are generally separate from pasting sites. Marking behavior and the pattern of pastings and feces differ between populations of spotted hyenas (Holekamp & Dloniak 2010; Kruuk 1972).

Despite inter-specific differences in their social organization, sniffing of the anal pouch is an important component of social greeting for all hyenids, as shown here for two striped hyenas, *Hyaena hyaena* (e). When everted, the anal pouch of the brown hyena, *Hyaena brunnea* (f)–(g), has a central crevice through which a grass stalk is drawn and coated with a white secretion. As the sac inverts a further smear of black secretion from its walls is left higher up on the grass stalk.

The smallest hyenid, the insectivorous aardwolf, *Proteles cristata* (h), pastes throughout its territory, but at a higher rate at the borders. Figure and caption adapted from Macdonald (1985b). Drawings by Priscilla Barrett.

often prompts the counter-marking behavior, particularly if it is on the owner's territory.

Over-marks are a special case of counter-marking in which the new mark is placed directly over the earlier one (Ferkin & Pierce 2007; Johnston 2008). Johnston, Ferkin and others used neat experiments to reveal the highly discerning responses of receivers. Both hamsters and voles have a preferential memory for the top scent whether or not the bottom scent is completely covered by another individual's scent (Johnston 2008). They appear to preferentially value or remember the top scent regardless of how much of the bottom scent is covered (though it is likely the odors of both individuals are remembered). This was true even if the top mark was applied only 30 seconds after the bottom mark. Mark freshness was not a factor in itself. Females of both species apparently use the strategy of determining "who is on top?" as a means to evaluate the relative vigor and quality of males (Johnston 2008). Females have a preference for the top-scent male, regardless of whether the bottom-scent males had a greater number of scent marks or covered more area with scent marks.

In mice, mark freshness matters: females are strongly attracted to territory owners that most recently counter-marked the scent-mark challenges of competitors (Hurst 2009; Hurst & Beynon 2013). Sniffing darcin, the MUP that is a male sex pheromone, in the scent marks prompts the female to learn the location of the scent mark and the marking male's more volatile signature mixture (Roberts *et al.* 2010, 2012). She will seek out his individual odors and will remain close to his scent marks, increasing the chance they will meet and mate.

Among primates, female pygmy loris, *Nycticebus pygmaeus*, prefer a male whose marks cover those of a second male (Fisher *et al.* 2003). This effect can be exploited to manipulate female choice for captive breeding of the species for conservation (Chapter 12).

In banded mongooses, *Mungos mungo*, Jordan *et al.* (2011) found no evidence that females chose to mate with males with high over-marking scores. Instead, the male over-marking may be more important in establishing male–male hierarchies.

5.8 Scent marking in non-territorial mammals

Scent marking is also a significant behavior of mammals that do not defend territories, including those with dominance and harem-defense mating systems such as bison and white-tailed deer (Gosling & Roberts 2001). Males may mark frequently as they move through their range and males in some species directly scent mark females, for example some species of antelope and rodents such as the South American mara, *Dolichotis patagonum* (Figure 5.6). Males in other species, such as goats, may pass on urine odors sprayed onto their under-parts to females when they mount. Direct marking of females may advertise the costs to other males of any attempt to compete for the female. Could this be a form of chemical mate guarding (Chapter 3)? As dominant males mark most frequently even in non-territorial species, scent matching could potentially be used by females to choose males.

Figure 5.6 Males in species with polygynous mating systems may mark females. For example, a male mara, *Dolichotis patagonum* (a South American rodent) marks a female with urine. Figure from Macdonald (1985a). Drawing by Priscilla Barrett.

Summary

Scent marking is one of the most conspicuous behaviors of many mammals and other terrestrial vertebrates. Scent marks appear to be involved in defending territories. The three main theories are that scent marks (1) deter potential intruders (scent-fence hypothesis or chemical "keep out" notice); (2) allow the intruder to recognize the owner, should he be met, by the match between his odors and those of the scent marks around the territory (scent-matching hypothesis); and (3) establish boundaries with major competitors (border-maintenance hypothesis). Scent matching seems to be the most supported idea currently. There are great opportunities for well designed laboratory or, in particular, field experiments to explore scent marking behavior further.

Further reading

Scent marking in mammals is reviewed by Gosling and Roberts (2001), Roberts (2007, 2012), and Müller-Schwarze (2006). Over-marking is discussed by Johnston (2008), Ferkin (2011), and Ferkin and Pierce (2007). Hölldobler and Wilson (1990, 2009) cover the behavior of ants. Crozier et al. (2010) describe the colony life of weaver ants. Roper (2010) offers a very readable account of badger behavior and its complexity.

For molecular structures see sites such as www.chemspider.com, which allows you to search by common name and shows synonyms as well as the systematic names. See also the Appendix for a short guide to the terminology (available for free download from the website associated with this book). Many pheromones are illustrated on Pherobase www.pherobase.com (El-Sayed 2013).

6 | Semiochemicals and social organization

The most complex animal societies described so far are found among the mammals and social insects. The individuals in these societies interact via a complex web of semiochemicals. In mammals and social insects, each individual's chemical profile consists of molecules produced by the animal itself, together with molecules acquired from the environment and from other group members. The resulting chemical profiles are complex and variable mixtures, giving a forest of peaks on a gas chromatograph trace, in contrast to the small number of defined peaks for the sex pheromones of moths and other insects (Chapter 1) (Figures 1.1, 1.2, and 13.2) (Wyatt 2010).

The chemical profile includes the colony or group odor that gives each member access to the group. In addition, within the overall chemical profile there are the species-wide pheromones. For example, on top of its individual odors, the saddle-back tamarin, *Saguinus fuscicollis*, a South American primate, produces pheromones that identify species, subspecies, sex, and social status within the group (Epple *et al*. 1993). As well as their colony odors, social insects carry pheromones on their cuticle that include information about their species, caste, age, and sex. In social insects, pheromones reflecting reproductive status, in particular fertility, may regulate reproduction of workers who do not themselves reproduce but instead help the queen (Section 6.2). Pheromones do not appear to have this role in social mammals.

There are two main themes in this chapter: first, how odor cues and learning of signature mixtures are used in colony and kin recognition (crucial for complex societies but also important for less social species); and, second, how pheromones act to affect reproduction in social groups: are they co-operative signals from the queen or irresistible control by the queen? Eusocial species show co-operative care of the brood,

overlap of adult generations (with offspring helping parents) together with reproductive division of labor, with some individuals specialized for reproduction (Wilson 1971).

6.1 Colony, kin, family, and individual recognition

Recognition of kin or fellow group members is central to social behavior and altruistic behavior toward kin (Hamilton 1964; Holmes 2004), whether the group is a colony of millions or a small family. Throughout the animal kingdom almost all recognition relies on learning (Chapter 1) and chemical cues predominate. The signature mixture molecules in the chemical profile, learned by receivers as the template for recognition, can be produced by the organism itself, acquired from the diet, shared local environment, or other organisms.

6.1.1 Colony and kin recognition in social insects

The ability to recognize nestmates has been found in almost all social insects that have been studied. Typically, colony members will accept nestmates but exclude and possibly kill non-nestmates (Bos & d'Ettorre 2012; Breed 1998b; Sturgis & Gordon 2012; van Zweden & d'Ettorre 2010). This can be seen at the entrance to a honeybee hive. The guard bees use their antennae to smell all bees attempting to enter. Strangers are identified by their cuticular odor signature and attacked. Guarding is the first line of colony defense in species of termites, ants, social wasps, and social bees. Guarding prevents conspecifics (members of the same species) from other colonies robbing food or brood and also prevents the entry of social parasite species (Chapter 11).

All mechanisms for nestmate recognition in social insects involve learning (Chapter 1) (Smith & Breed 1995). The signature mixture of nestmates is learned as a "template" (the neural representation of the colony odor stored in the memory): nestmates and non-nestmates can be distinguished by how well they match the "template" (Bos & d'Ettorre 2012; Breed & Buchwald 2009; Sturgis & Gordon 2012; van Zweden & d'Ettorre 2010). The "gestalt" colony odor is gained by the sharing of these surface chemical cues so that all members of the colony share the same chemical profile (see Lenoir et al. 2001). See Chapter 1 for more discussion of the terms "chemical profile," "signature mixture," and "template."

In ants, wasps, and termites, the main molecules used in recognition seem to be the hydrocarbons on the cuticle, though in honeybees other types of molecules may also be involved (see below) (Howard & Blomquist 2005; van Zweden & d'Ettorre 2010). Not all of the hydrocarbons are involved in colony recognition; some appear to have largely waterproofing functions and others may be species-wide pheromones. Until recently, most studies used correlation of hydrocarbon profiles and aggression between colonies. Now, the importance of different kinds of hydrocarbon for recognition has been tested by seeing the effect of directly adding synthetic hydrocarbons, or hydrocarbons from other colonies, to individuals or decoys such as glass beads (Chapter 2). The conclusion from experiments on different species of ants, bees, and wasps is that adding simple straight-chain hydrocarbons to an insect's chemical profile does not usually cause it to be treated as a non-nestmate but that this often results if a methyl-branched hydrocarbon or one with double bonds (an alkene) is added (van Zweden & d'Ettorre 2010) (see also Chapter 1).

The reason branched hydrocarbons may be important in colony recognition could be that simple straight-chain alkane hydrocarbons, differing only in chain length, offer far fewer variations. By contrast, more complicated hydrocarbons such as methyl-branched alkanes and unsaturated hydrocarbons (alkenes) can offer enormous possibilities of variation

with the addition of 1, 2, 3, or 4 methyl groups to the chain or the introduction of 1, 2, or 3 double bonds, and any of these can be in a variety of positions along the molecule (Martin & Drijfhout 2009a; van Wilgenburg et al. 2011). This potential variety was reflected in the almost 1,000 different hydrocarbons found in just 78 ant species spread across five ant subfamilies (Martin & Drijfhout 2009a). The linear n-alkanes were found in all 78 species, the more complex were found only in a smaller number of species. Dimethylalkanes were by far the most numerous group of compounds, with over 600 currently described. They are species specific, and can even be colony specific (Martin et al. 2008a). These large numbers are in contrast to the 20 to 50 different hydrocarbons typically found in non-social taxonomic families of insects (S. Martin, unpublished data in Martin et al. 2008b).

The differences used for colony recognition within a species tend to be quantitative rather than qualitative, i.e., different colonies present different chemical profiles based on the relative abundance of the same components (Hefetz 2007; van Zweden & d'Ettorre 2010). For example, this was demonstrated with synthetic hydrocarbons in the ant *Formica exsecta*, which appears to use different ratios of a homologous series of C21:1 to C33:1 Z9-alkenes for nestmate discrimination (Martin et al. 2008c). The hydrocarbons that are being "monitored" for nestmate discrimination are superimposed on all the other hydrocarbons, which act as waterproofing, and others, which are species-wide pheromones, the same in every colony, giving species specificity, sex, caste, and fertility status (worn like a crown by all queens in a species) (Chapter 1) (Figure 1.2) (see Section 6.2.1).

6.1.1.1 Honeybees

Honeybee guards respond to the gestalt colony odor on the surface of an incoming honeybee. The colony odor comes from cuticular molecules secreted by the bee itself, together with compounds picked up from the comb wax in the nest (Couvillon et al. 2007; Grozinger 2013). The compounds in the comb wax include molecules produced by other workers and floral scents brought back to the nest during foraging. Some of the cuticular molecules are under genetic control. In laboratory experiments with small numbers of bees reared in isolation, away from the nest, bees can distinguish between full sisters, half-sisters, or non-sisters, all reared in isolation (Getz 1991). However, laboratory-reared bees such as these, emerging isolated from the nest, are rejected by their sisters when introduced to their parental hive at five days' old so the bee's own cuticular molecules are not enough for acceptance. The key cues are the compounds that the bees pick up from the comb wax. As little as five minutes' contact with comb wax from a foreign nest is enough to contaminate a bee and cause attack by its sister guards (Breed *et al.* 1995). Cues are both picked up and learned from the comb: a bee exposed to comb from her own or another colony is rejected by unexposed bees reared in the laboratory, but will be accepted by laboratory-reared bees, whether nestmate or not, which have been exposed to the same comb.

Comb wax is an extraordinarily complex mixture of compounds from at least two glands of the bee – the wax glands plus mandibular gland secretions, which add hydrocarbons and fatty acids (Breed 1998b). These secretions are under genetic control so workers from different patrilines (with different fathers) will be contributing their potentially different secretions to comb building, with the overall odor reflecting the mixture of genotypes in the nest. Different nests will therefore have different profiles. The fatty acids and hydrocarbons are molded into the wax comb, then transferred to workers as they contact the comb. The result is a colony-level odor mixture that is also passed on from insect to insect and differs little among workers in the colony. The fatty acids may be key elements in honeybee nestmate recognition (Breed *et al.* 2004b), as well as alkene hydrocarbons (though less so, alkanes) (see Section 6.1.1) (Dani *et al.* 2005). In a cross-fostering experiment to partition genetic and environmental influences on guard acceptance, Breed *et al.* (1988) found that the environmental effects of hive exposure were so strong that no genetic

effect was observed. It thus seems that in nestmate recognition cues from comb and colony environment have priority over cuticular molecules produced by the bees although these can be used if other cues are absent. The guard bees may be taking their template from the wax comb rather than fellow colony members (Couvillon *et al.* 2007).

As the queen may mate with ten or more males on her nuptial flight, a honeybee colony consists of workers that are full sisters and half-sisters in different patrilines (Ratnieks & Wenseleers 2008). The recognition priority given by bees to the overall colony odor may be one proximate mechanism for the low levels of conflict within the nest despite these many patrilines (Ratnieks *et al.* 2006). Workers from different patrilines could in theory show favoritism (nepotism) toward rearing queen larvae sharing their own father but the evidence for honeybees (and other hymenopterans) suggests that queen-rearing nepotism may occur, but only weakly. One reason for the low levels of conflict is the potentially high cost of nepotistic queen rearing. Another could be the small benefits that result (possibly because workers are unable to discriminate full-sisters and half-sisters accurately (Ratnieks *et al.* 2006; Seeley 1995, p. 226). Cohesion of the colony would suit the queen, in order to avoid having conflict between daughter workers fathered by different males. Alternatively, kin discrimination of different patrilines could be a non-adaptive by-product of the inter-colonial recognition adaptive for defense against the invasion of parasites and non-colony members (Grafen 1990b). In some ants a similar blending of heritable recognition cues as a distinct colony *gestalt* odor appears to prevent within-colony nepotism or aggression despite different patrilines or even multiple queens (Helantera *et al.* 2011; van Zweden *et al.* 2010).

6.1.1.2 Sweatbees

The guarding behavior of a primitively eusocial sweatbee, *Lasioglossum zephyrum*, provided some of the first tests of kin recognition in Hymenoptera (see review by Smith & Breed 1995). This bee has small colonies consisting of a queen plus 2 to 20 daughters. Greenberg (1979) showed that the willingness of a guard bee to let in another bee is directly correlated with the relatedness between them. The recognition cues were learned by frequent contact with nestmates, but closely related bees were allowed to enter even if they had not been met before, which suggests a genetic basis to some of the cues.

The identity of the recognition cues is not known, but candidate compounds include lactones and hydrocarbons from the Dufour's gland. These have been shown to be correlated with species, familial, and individual differences (Hefetz *et al.* 1986). Males use Dufour's gland odors to choose mates for outbreeding (Chapter 3) (Smith & Breed 1995).

6.1.1.3 *Polistes* and other wasps

The primitively eusocial paper wasp, *Polistes fuscatus*, forms an excellent model system for investigating the development of kin recognition based on semiochemicals (reviewed by Gamboa 2004). After overwintering, queens tend to found nests with close relatives, which they recognize by odor cues. There is nonetheless reproductive conflict as one of the queens lays most of the eggs (Section 6.2). The colony will accept orphan kin (for example, from nearby nests destroyed by birds) but they need to keep away non-kin, which may come to rob the nest. Aunts, nieces, and a minority of cousins are accepted.

Though they have evolved eusociality independently, paper wasps and honeybees are strikingly similar in their use of nesting materials for transmitting cuticular molecules, used for recognition among colony members. Adult wasps (in particular the queen) deposit secretions, largely from the sternal glands, on to the nest surface. The secretions are hydrocarbons that can be extracted from the nest in hexane (wasp hydrocarbons are reviewed in Dani 2006; Espelie *et al.* 1994). The nest will thus have the queen's chemical signature as well as compounds from environmental sources including the nest material.

The learning of recognition cues in *Polistes* resembles imprinting (Chapter 9). When it emerges from its pupa, the young paper wasp rests on the natal comb and learns the odor template from the secretions laid down on the paper comb of the nest, not from its own cuticular chemistry. The young wasp recognizes nestmates via the odors she learned from the nest surface (without this learning experience the young wasp is not able to distinguish nestmates from nonnestmates). The hydrocarbons are colony specific and heritable. Close non-nestmate kin (aunts and nieces but not first cousins) may have CHC profiles sufficiently similar to the resident's nest profile to allow them to be accepted. Gamboa *et al.* (1996) were able to use hydrocarbon profiles to predict if a given wasp would be accepted or rejected.

As in honeybees, the use of the nest like an artist's mixing palette for the chemical secretions of the paper wasp nest members allows the genetic profiles of all members of the colony to be accommodated. In the tropical paper wasp, *Ropalidia marginata*, as in *P. fuscatus*, females discriminate nestmate from nonnestmate only after prior exposure to nest and nestmates (Venkataraman *et al.* 1988) (see Gadagkar 2009 for background). The use of the nest odors as the colony odor rather than genetic markers may have evolved because many of the wasps helping the colony are not genetically related (in this species young strangers may join the colony and are accepted because the colony benefits) (Arathi *et al.* 1997).

6.1.1.4 Ants

Cuticular hydrocarbons (CHCs) are the main chemical cues for nestmate recognition in ants (d'Ettorre & Lenoir 2010; Hefetz 2007; Lenoir *et al.* 2001; Sturgis & Gordon 2012; van Zweden & d'Ettorre 2010). Recognition cues are mixed throughout the colony by trophallaxis (transfer of liquid food between individuals) and mutual grooming, in combination with secretions from the postpharyngeal gland, thereby establishing and continuously updating the colony odor (Bos & d'Ettorre 2012; van Zweden & d'Ettorre

2010). Transfer of secretions deposited on nest materials seems less important than in bees and wasps (above). Recognition cues produced by the queen and spread to individuals in the colony seem to be important in some genera, such as *Camponotus* (Carlin & Hölldobler 1987), but not in others, such as *Solenopsis* (Obin & Vander Meer 1989). Genetically based variation in some CHCs between patrilines in *Acromyrmex* leaf-cutting ants can be detected but it is not known if the ants respond to these differences (Nehring *et al.* 2011). The great diversity of social organization in ants is likely to be matched in the variety of sources of recognition cues, with the relative importance of genetic, environmental, and queen-derived odors differing between species depending on the general ecology of the species, colony kin-structure, and (social) parasite pressure (d'Ettorre & Lenoir 2010; Sturgis & Gordon 2012; van Zweden & d'Ettorre 2010). Hydrocarbon mixtures also change with role, for example foragers have larger amounts of alkanes but this does not affect nestmate recognition (see discussion of colony odors and signature mixture molecules in Chapter 1) (Greene & Gordon 2003; Martin & Drijfhout 2009b).

As in bees and wasps, ant workers emerge from the pupa with almost no hydrocarbon cues on their surface. They pick up the colony odor from interactions with older workers and also start to synthesize their own molecules. Until they have gained the colony odor they avoid attack by moving little and avoiding aggressive behaviors (Vander Meer & Morel 1998).

6.1.1.5 Termites

Much less is known about colony recognition in termites but it seems to be based, as in Hymenoptera, on blends of CHCs; a similar role for genetically determined and acquired components in the colony odor is likely (Clément & Bagnères 1998; Dronnet *et al.* 2006; Kaib *et al.* 2004). Termites are capable of kin distinctions as subtle as any hymenopteran: DNA fingerprinting suggests that in polygynous and polyandrous termite colonies, workers departing from the nest to their foraging areas tend to form working parties with their kin (Figure 7.2) (Kaib *et al.* 1996).

6.1.2 Group, kin, family, and individual recognition in mammals

Mammal societies in species such as dwarf mongooses, ground squirrels, or hyenas form long-lasting, stable groups. Recognition of group members, which will often be kin, is important. They can then be favored in co-operative behavior, for example in alloparental care given by brown hyenas to young cousins, or sharing of limited food resources by spiny mouse siblings, or alliances by ground squirrel sisters to chase off territorial intruders (Porter & Blaustein 1989). The closest kin are parents and offspring: recognition of your own offspring is important for most species.

6.1.2.1 Mother–infant recognition

A key role for olfaction in mother–infant recognition has been shown in many mammal species, including humans, and has been intensively studied in sheep. Many lambs are born to the flock in a short period, so each mother (ewe) needs to recognize her offspring. She learns the odor of her lamb within the first few hours after birth (Chapter 9) (see reviews in Lévy & Keller 2008, 2009). In this early period, but not later, an orphan lamb will be accepted and adopted, particularly if coated with amniotic fluid to encourage licking by the ewe (a method traditionally used by farmers, see Chapter 12). The mother uses phenotype matching (Chapter 1) to recognize her lamb, not odors produced by "her genes" in the lamb or her secretions on it. This was shown by a neat study in which a surrogate mother sheep gave birth to identical twin lambs, from implanted monozygotic twin embryos (Romeyer *et al.* 1993). If allowed to learn the odor of one twin, she could recognize its identical twin, even though she was not related to either of them. This shows: first, that she does not use a comparison of the lamb's odor with her own for recognition; and, second, some of the lamb's odor is genetic because, while she could recognize an identical twin, she could not recognize non-identical twins. Nonetheless, environmental odor cues acquired by the lamb are also important and some of these may come from the mother's saliva and milk, for example. Although these "maternal labels" are not required for recognition, they are incorporated into the lamb's recognizable odor phenotype and learned by the mother (Lévy & Keller 2009; Lévy *et al.* 1996). Once the ewe has learned the odor of her lamb, adding a foreign scent (whether vanilla or that of another lamb) does not confuse her. Therefore it is her lamb's own complete odor that is important, not the absence of foreign scents.

Parents recognize their offspring, but offspring also learn the odors of their parents. For example, human babies will turn their head toward their mother's odor, learned soon after birth (Chapter 13) (Schaal *et al.* 2009).

6.1.2.2 Kin recognition

Many mammals use the rule of thumb "learn the odor of your nestmates," as in the wild these are likely to be kin (Chapter 1). The mechanisms of learning signature mixture odor templates can be studied in cross-fostering experiments. For example, young spiny mice, *Acomys cahirinus*, prefer to huddle with nestmates, which would normally be siblings, but if the experimenter creates mixed litters of non-siblings, they later treat the familiar but unrelated individuals as if they were siblings (Porter *et al.* 1981). Diet odors are an important source of cues used by juveniles to recognize animals as familiar or not, and for offspring and parents to recognize each other as kin; non-kin fed on the same distinctive diet may be treated as kin (Porter *et al.* 1989). Nonetheless, genetic components of the odor profile are important.

Learning a combination of environmental and genetic odor cues is also implicated in kin recognition by Belding's ground squirrels, *Spermophilus beldingi* (Holmes 2004). Mothers, daughters, and sisters favor kin (behave nepotistically), warning each other about predators and protecting their own and each other's pups against infanticide by establishing and defending a territory together. Nestmates learn each other's odors just before they wean (when litters normally begin to mix) and later treat each other as siblings regardless of actual relatedness. This works because unrelated pups are rarely in the same burrow until weaning (and until then mothers will accept any offspring in the burrow – after weaning they discriminate by odor). Cross-fostering and other odor manipulation experiments show how odors of nestmates or parents are learned during the sensitive period, so that as in spiny mice, ground squirrels are less aggressive toward familiar unrelated nestmates than unfamiliar unrelated ground squirrels (Mateo 2009, and references therein). However, despite having shared the same nest, littermate full-sisters and maternal half-sisters, resulting from multiple mating by the mother, seem to treat each other differently. In the field they were less aggressive toward full-siblings than half-siblings, and more likely to share a territory with a full-sister (Holmes & Sherman 1982). Phenotypic matching appears to allow them to behave differently to kin they have not previously met: in laboratory tests they are less aggressive to unfamiliar siblings than unfamiliar non-siblings (some genetic basis to the template odors is suggested because unfamiliar paternal half-siblings are preferred to unfamiliar non-siblings, Holmes 1986).

Relatively few field tests of phenotypic matching have been made but studies of beavers, *Castor canadensis*, support this mechanism (Sun & Müller-Schwarze 1997). As they disperse, young beavers might encounter unfamiliar siblings from litters dating from before they were born. The young beavers reacted less strongly toward anal gland secretions from unfamiliar siblings than those from unfamiliar non-relatives.

6.1.2.3 Clan or group odors and social mammals

As in beavers, many other social mammals such as ring-tailed lemurs, dwarf mongooses, hyenas, aardwolves, badgers, prairie voles, and mole rats are characterized by high levels of social behavior and exclusion of strangers from the group: behavior that relies on mutual recognition of group members. Odor cues dominate this behavior in mammals.

Depending on the social structure and juvenile dispersal pattern of the species, in many mammal species newly adult members will join the social group at intervals. As these animals come from outside they will not be closely related to the rest of the group they join. So, while there may be kin-related distinctions in behavior being made within the group, if the group works as a social unit, animals will need to recognize non-kin animals as group members.

In many social mammals there is a great deal of allomarking – marking of each other – often by the dominant individuals (Archie & Theis 2011). For example, European badgers, *Meles meles*, squat mark each other with their subcaudal and anal glands (Kruuk 1989; Kruuk *et al.* 1984; Roper 2010). All members of the clan then share a common odor identity – dominated by the odor of the dominant male (an effect like the queen's secretions in some ant species, Section 6.1.1). The fermenting bacterial flora of carnivore anal glands change over time (Albone 1984; Archie & Theis 2011; Macdonald 1985b) leading to changes in odors, but the regular marking of each other keeps the clan odor "up to date."

As in badgers, brown hyenas, *Hyaena brunea*, can recognize foreign conspecifics by smell: they invariably paste mark on top of marks from other groups (Chapter 5) (Mills *et al.* 1980). However, rather than being a demonstration of a distinctive shared "group" or clan odor, theoretically this response could be based on knowing all the individuals in the group and recognizing the marks as coming from unfamiliar individuals (and thus outside the group) (Macdonald 1985b).

Eusocial naked mole rats, *Heterocephalus glaber*, are highly xenophobic, even to closely related conspecifics from outside their colony (O'Riain & Jarvis 1997). As in social insects, the principal mechanism of recognition appears to be distinct colony odor labels, contributed by each colony member and distributed among, and learned by, all colony members. Differences in the mixture of these odors may provide even genetically similar colonies with a unique odor label.

6.1.2.4 Individual recognition

Other recognition events take place throughout life, with animals learning odor identities and recognizing individuals, so the perceiving animal can behave adaptively when it meets those individuals again. Discrimination of individuals by odor alone has been demonstrated by training experiments in the dwarf mongoose, *Helogale undulata*, and Indian mongoose, *Herpestes auropunctatus* (Gorman 1976; Rasa 1973), and for European badgers (Östborn 1976, cited in Macdonald 1985b). The mongooses could distinguish individual conspecifics by their anal gland secretions and badgers used subcaudal gland secretions.

It is surprisingly difficult to design experiments that test true individual recognition while eliminating the effects of familiarity and kinship (Halpin 1986; Thom & Hurst 2004). Many experiments would be better described as showing that animals can distinguish the odors from test subjects. Whether the subjects are recognized as individuals as opposed to simply being different from each other is usually unclear.

Many experimental designs rely on conditioning tests (for example using positive reinforcement) or habituation tests in which responses decrease as the same stimulus is presented repeatedly (Figure 2.2) (Chapter 2). An individual's odor that is perceived as different from the previous stimulus will elicit a greater response. Demonstrations of individual recognition by olfactory cues are usually in the context of an experiment and their importance in the social behavior of the animals is not always known (Thom & Hurst 2004; Wiley 2013).

In a more natural context, dominant mice seem to seek out particular subordinate individuals to attack if that subordinate (or experimenters on its behalf) urine mark a marking post reserved by the dominant male (Box 5.2) (Hurst 1993). Mate choice by female mice using scent matching to identify the territory owner seems to involve individual recognition (Chapter 3) (Hurst 2009, Hurst & Beynon 2013). Individual recognition is implicit in the "dear enemy" phenomenon (Chapter 5) as animals are presumably able to recognize particular neighbors by smell (so as to recognize them as familiar or not familiar) and they may be able to associate that individual with a particular expected location or direction. For example, when experimenters placed a scent mark from a neighbor on the territory of an aardwolf, *Proteles cristata*, it responded by going to the appropriate boundary and marking (Sliwa & Richardson 1998).

Individual recognition by odor is important for pair-bonding in monogamous species such as prairie voles (Carter & Roberts 1997). In the monogamous lizard, *Niveoscincus microlepidotus*, males follow the scent trails of their partners and this recognition may play a significant role in maintaining the prolonged partnership (Olsson & Shine 1998).

At a physiological level, the Bruce effect of pregnancy block caused by the odor of a strange male mouse relies on distinction between the signature mixture of the stud male she had mated with and the odor of any other male (Chapter 9). The odor cues responsible for the Bruce effect are different from those involved in mate choice, above (Hurst 2009). Reproductive suppression, which is achieved in different ways in prairie voles and common marmosets (Section 6.2.2), may also involve recognition of signature mixtures. An intriguing phenomenon is that within a species, some glands are more variable than others. For example, individual dwarf mongooses could be distinguished by their anal gland secretions but not by their cheek gland secretions (Rasa 1973). In habituation tests, golden hamster males treated some odors (flank, ear, vaginal, urine, feces) as different between individuals, but not others (for example saliva, secretions from the feet, behind the ear, gland-removed flanks) (Johnston 2003, 2008;

Johnston *et al.* 1993). Some glands might be specialized for individual variability (see Chapter 3).

If individuals are actually recognized, animals should have integrated multifactorial representations of these individuals (for reviews see Johnston 2008; Johnston & Jernigan 1994). A male golden hamster familiar with two females was habituated to vaginal odor from one female (A) then tested with flank odor from one or the other female. He treated the flank odor of the female (A) as if it were familiar (the habituation to her vaginal odor "carried over"). This was not a result of "generalization" between the two odors, however, as this effect was only shown if the male was familiar with the female and had learned her various odors (vaginal and flank odors were not responded to in this way if the female was a stranger).

6.2 Pheromones and reproduction in social groups: control or co-operative signaling?

Pheromones are important for the many species of mammal that live in social groups with animals sharing and defending the same territory (Chapter 5) and for social insects that live in colonies (Box 6.1). In some societies of mammal and social insect, co-operation is taken a stage further, with a reproductive division of labor: some individuals (helpers or workers) do not themselves reproduce but instead help to rear the offspring of other group members, usually the helpers' sisters or mothers. The reason that such altruistic reproductive behavior (helpers not breeding) can persist is kin selection, which allows the helpers to gain their inclusive fitness indirectly by rearing copies of their genes in their brothers and sisters (Bourke 2011a,b; Queller & Strassmann 2010; West & Gardner 2010). Hamilton's rule for kin selection predicts that altruistic behavior will be more likely to be selected for if the individuals are closely related (and thus more likely to share the helper's gene for helping) and if the decrease in the actor's personal fitness is relatively small compared with the increase in the recipient's fitness (Hamilton 1964).

Box 6.1 Social insect pheromones

The co-ordination and integration of colony activities, in particular recruitment for foraging and defense (see Chapters 7 and 8), has been an essential contribution to the success of social insects: the road to sociality was paved with pheromones (Blum 1974). Morgan (2008) suggests pheromones offer "chemical sorcery for sociality." Pheromones play a central role in recruitment, as markers (of caste, sex, and species), and spreading global information to enable colony homeostasis, through mediating social conflicts and integrating functions including provisioning, caste determination, and colony reproduction (Alaux *et al.* 2010).

In honeybees more than 50 chemical substances are known to be essential to the functioning of the society and more are still being found (Grozinger 2013; Kocher & Grozinger 2011; Slessor *et al.* 2005). Different, overlapping subsets of these molecules give different messages (Box 6.3) (I wonder if this is like playing different chords of notes on a musical instrument?). The meaning of the messages also depends on the receiver (for example young or old worker) and context.

It is difficult to imagine how the integration of colony functions could be made without pheromones transferring chemical messages through the colony, as there may be several million individuals in colonies of some ant species (Hölldobler & Wilson 2009). In addition to the pheromones there are the variable colony odors, used for nestmate recognition (Chapter 1) (Section 6.1.1). Termites show many features of convergent evolution of chemical communication with the eusocial Hymenoptera (Box 6.2).

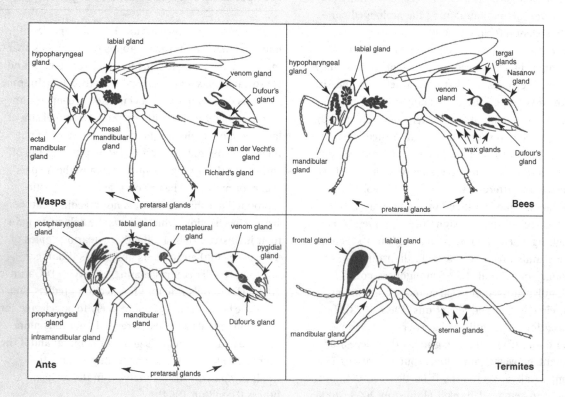

Box 6.1 (cont.)

Schematic profile drawings of the exocrine glands commonly found in wasps, bees, ants, and termites. Across the social insects more than 100 different glands have been found. Some glands produce largely one molecule, some produce many (Billen 2006; Morgan 2008). Pheromones may combine molecules from more than one gland, as for example queen retinue pheromone in honeybees (Box 6.3). Different castes may produce different molecules from the same gland, for example the mandibular gland of the ant, *Atta*, and in the mandibular gland of honeybee workers and queens due to the expression of different genes (Malka *et al.* 2009). Figure from Billen and Morgan (1998).

There is an enormous diversity of glands and gland products in social insects. While hydrocarbons are much discussed in the literature on social Hymenoptera and termites, a wide range of classes of organic molecules beyond the hydrocarbons have been adopted as pheromones (e.g., Keeling *et al.* 2004; Morgan 2008).

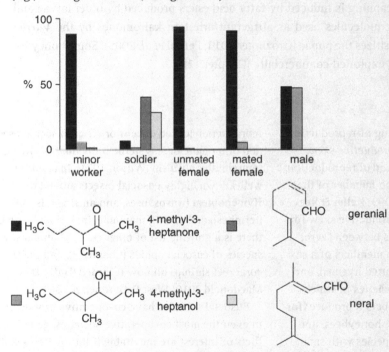

Caste differences in mandibular gland secretions of the leaf-cutting ant, *Atta*. Data from do Nascimento *et al.* (1993) in a figure from Chapman (1998). Many examples of different pheromone gland production in different castes of the same species of social insect are given in Robinson (2009). Differences in role are also reflected in large differences in perception, with specialized and enlarged brain glomeruli for trail pheromones in some *Atta* workers (Kuebler *et al.* 2010) (see Chapter 9).

Different castes in the same species may produce different pheromones from the same gland, as for example in the mandibular gland in worker and queen honeybees, and ants such as *Atta*. Worker honeybee mandibular gland secretions have 10-hydroxydec-2-enoic acid (10-HDA) rather than the queen's 9-HDA, due to different gene expression for enzymes in the two castes (Malka *et al.* 2009; Plettner *et al.* 1996). Different responses to pheromones depending on the caste of the receiver, combined with context-specific responses, create an enormously complex chemical signal environment inside and outside the nest (see figure, Box 6.3). We are just beginning to decode these messages. Multiple effects are common (Grozinger 2013; Slessor *et al.* 2005). For example in honeybees, the queen mandibular pheromone (QMP) is a sex pheromone, which attracts males and also affects worker reproduction (Box 6.3) (Section 6.2.1).

Box 6.1 (cont.)

It also has many other effects on worker behavior and physiology, probably by affecting juvenile hormone levels (Chapter 9) (Le Conte & Hefetz 2008; Robinson *et al.* 2005b). These include influences on swarming, stimulation of foraging behavior, and an increase in the age workers start to move from working in the hive to foraging outside (see figure, Box 6.3). In honeybees, pheromones produced by the workers and the brood (eggs, larvae, and pupae) help to co-ordinate activity to match the needs of the whole colony (Chapter 9) (Alaux *et al.* 2010; Grozinger 2013; Le Conte & Hefetz 2008). Brood pheromones, a complex mixture of fatty acid esters, stimulate worker brood-food glands, promote larval feeding and cell capping, and, like QMP, suppress worker reproduction. Cell capping is induced by fatty acid esters produced by older larvae and these esters are among the molecules used as attractant/arrestant kairomones by the *Varroa destructor* mite, which parasitizes the pupae (Grozinger 2013; Pernal *et al.* 2005). Some honeybee pheromones are now being exploited commercially (Chapter 12).

One way of describing the sharing of reproduction in social groups is by the term *reproductive skew*, which describes how much the spread of reproduction differs from an equal share for each member of that sex in the group (Hager & Jones 2009; Keller & Reeve 1994; Nonacs & Hager 2011). Reproductive skew for males or females in a group ranges between "zero," with equal shares (where all group members of a sex reproduce, for example female spotted hyenas), and "one," in highly skewed animal societies in which effectively only one or a few members reproduce (for example, females in termites, ants, honeybees, and the naked mole rat) (Figure 6.1). Species with small colonies can be highly reproductively skewed, for example bumblebees, paper wasps, and the common marmoset.

Using reproductive skew as a measure, one can envisage social animals of all kinds placed on a euso-ciality continuum from no skew to high skew (Figure 6.1) (Crespi 2005; Lacey & Sherman 2005). At the high-skew end, eusocial species show co-operative care of the brood, overlap of adult generations (with offspring helping parents) together with reproductive division of labor, with some individuals specialized for reproduction (called queens or kings in social insects), and other more or less sterile individuals showing reproductive altruism (Wilson 1971). In addition to the well-known highly eusocial insects among the Hymenoptera (wasps, bees, and ants) and Isoptera (termites), and among mammals (naked mole rats), there is a growing list of other eusocial animals, with species of eusocial aphids (Chapter 8), gall thrips, and coral reef shrimps all now recorded (Duffy & Macdonald 2010; West & Gardner 2010).

Eusocial societies have co-operative broodcare but in even the most co-operative societies, genetic conflicts of interest are inevitable (Clutton-Brock 2009; Ratnieks *et al.* 2006). In particular, group members will compete over who gets to reproduce. In most mammalian societies, and in those social insect species in which almost all individuals could potentially reproduce, fierce fighting determines who reproduces. Perhaps surprisingly at first sight, in some of the most skewed insect societies, with the greatest morphological differences between the queen and workers, pheromones produced by the dominant female "settle the dispute," by having the effect of stopping subordinate females from reproducing. However, it

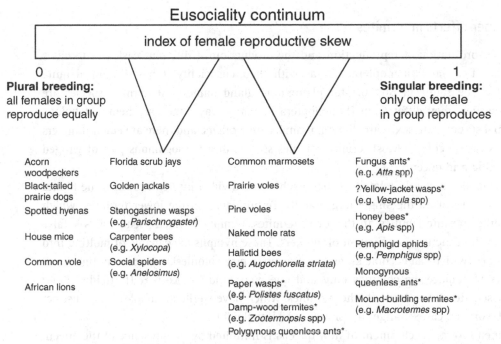

Figure 6.1 Co-operatively breeding vertebrate and invertebrate species can be arranged along an axis, "the eusociality continuum," which reflects how evenly reproduction is shared among females in a social group (Crespi 2005; Lacey & Sherman 2005; Sherman *et al.* 1995).The axis ranges from a skew of "0" when female lifetime success is equal among group members to "1" when reproduction is limited to one female in a group. Systems where pheromones have been identified as a mechanism for interaction between the breeding female and subordinates are starred* (see text). After Abbott *et al.* (1998), itself modified from Sherman *et al.* (1995).

appears that rather than being controlled by chemical manipulation, the queen's pheromones are honest signals (Chapter 1).

6.2.1 Social insect queen pheromones

Only one or a few individuals reproduce in eusocial Hymenopteran colonies. Since most social insect colonies are families not clones, there is potential conflict within colonies between workers, queens and workers, and queens over the level and timing of resources put into rearing reproductives, their sex ratio, and egg laying by workers (Ratnieks *et al.* 2006).

Despite the potential conflict, dominance with open aggression by the queen is virtually absent in more advanced insect societies, which have effectively sterile, morphologically distinct, worker castes (Hölldobler & Wilson 2009; Wilson 1971). As colonies increase in size, it is hard to see how physical domination could work for more than a few tens of animals let alone the 500,000 individuals in a weaver-ant colony, controlled by a single queen (Hölldobler & Wilson 1977). Instead, in many insect societies, pheromones take the place of fights. Queen pheromone influence on worker reproduction within social insect colonies is clear in many ants, wasps, and bees: in the presence of the queen pheromone, workers' ovaries do not develop and workers do not lay eggs (Gadagkar 2009; Heinze & d'Ettorre 2009; Kocher & Grozinger 2011; Le Conte & Hefetz 2008; Peeters & Liebig 2009). Pheromones may play similar roles in termites (Box 6.2) (Matsuura 2012).

Box 6.2 Primer effects in termites

Pheromone co-ordination of reproduction and the production of different castes in termites parallels that of social Hymenoptera, but as with their eusociality it has developed quite independently (see figure in Box 6.1 for pheromone gland names). All termite species are eusocial. However, they differ from Hymenoptera in many ways: they are hemimetabolous (with no pupal stage), both sexes are diploid, nymphs are workers, and there are equal numbers of both sexes in the colony. Most termite colonies start with a monogamous pair of winged sexuals, the king and queen.

Termites, which are closely related to cockroaches, are divided into two groups, the higher termites, which are all in the single large family Termitidae, and the lower termites, which include all other termite families. In the lower termites, nymphs (termed pseudogates = false workers) form the functional equivalent of workers. These nymphs are capable of molting into winged imagoes (which can become kings and queens of new colonies), replacement (neotenic) reproductives to replace the colony's king and queen, or soldiers (Korb & Hartfelder 2008; Vargo & Husseneder 2009). In the higher termites, which have sterile morphologically distinct soldiers and workers, caste changes are simpler and less flexible.

As in Hymenoptera, development of new queens is inhibited by the presence of the queen. The control of development of new termite queens is reviewed by Matsuura (2012) and Simpson *et al.* (2011). In termite colonies there is also a king who has a similar effect on the development of new kings. In the absence of the queen, the king produces pheromones that encourage the development of replacement queens.

Might queen (and king) influence on nymph sterility work in a similar way to Hymenoptera, with a pheromone fertility signal rather than pheromone control (Keller & Nonacs 1993)? The cue might similarly be CHCs: in the lower termite *Cryptotermes secundus*, queens had markedly different CHC profiles from nymphs (Weil *et al.* 2009) and in *Zootermopsis nevadensis*, CHC differences from nymphs in both queens and kings were related to fertility (eggs and sperm) but no differences were found between the sexes (Liebig *et al.* 2009). Cues that differ between the sexes, that are required to explain the differential effects of kings and queens, have been found in a different class of molecule, polar proteinaceous secretions, in species from three genera (*Prorhinotermes simplex*, *Reticulitermes santonensis*, and *Kalotermes flavicollis*) (Hanus *et al.* 2010). The proteinaceous secretions would fit with data from the last 50 years, which suggest that the pheromones inhibiting differentiation of the pseudogates come from secretions on the cuticular surface or from the anus of the queen, taken up by proctodeal trophallaxis (anal excretion feeding) and passed around the colony from worker to worker by oral trophallaxis (from mouth to mouth) or by mutual grooming. No effects have been demonstrated yet.

Box 6.2 (cont.)

However, in another lower termite species, *R. speratus*, a volatile multicomponent pheromone has been identified from fertile female neotenics (queens), which inhibits the development of new female neotenics and inhibits egg production in other queens (Matsuura 2012; Matsuura *et al.* 2010; Yamamoto & Matsuura 2011). The pheromone, *n*-butyl-*n*-butyrate and 2-methyl-1-butanol, is also produced by the eggs themselves. In a further species, *Cryptotermes secundus*, a gene (*Neofem2*), over-expressed in queens but not kings and workers, may be important for inhibition of workers in that system (Korb *et al.* 2009). In the higher termite *Nasutitermes takasagoensis*, a queen-specific volatile molecule, phenylethanol, has been identified but its function is not known (Himuro *et al.* 2011; Matsuura 2012).

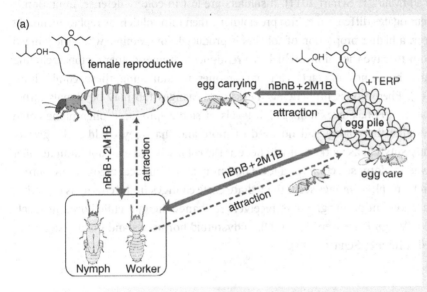

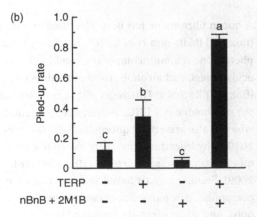

Box 6.2 (cont.)

Proposed functions of a volatile multicomponent pheromone n-butyl-n-butyrate (nBnB) and 2-methyl-1-butanol (2M1B) (see text) from queens (female reproductives) and eggs of the termite, *Reticulitermes speratus*. (a) These volatiles act to inhibit (bold arrows) the differentiation of female neotenics (replacement reproductives) from nymphs and workers. The volatiles also serve as an attractant (dashed lines) to locate eggs (right). On contact, eggs are distinguished by a surface "termite egg recognition pheromone" (TERP) consisting of an antibacterial protein lysozyme and a digestive β-glucosidase enzyme (Matsuura 2012). (b) Role of egg volatiles as an attractant pheromone. Addition of nBnB and 2M1B to TERP (consisting of lysozyme and β-glucosidase) significantly increased the rate at which dummy eggs were piled up. Dummy eggs coated only with nBnB and 2M1B were not carried into egg piles. Different letters indicate significant differences among treatments ($P < 0.05$, Tukey's HSD). Figure from Matsuura *et al.* (2010).

In some species, soldiers inhibit the production of more soldiers so homeostasis of numbers is maintained (Chapter 9) (Miura & Scharf 2011). If soldiers are lost in colony defense, inhibition is relaxed and more pseudogates differentiate into pre-soldiers, then into soldiers as replacements. At some times of the year, a higher proportion of soldiers is produced, for example when the winged reproductives are about to leave the colony, which is a reminder that many other factors including nutrition, season, and hormones interact with pheromones in managing these equilibriums (Miura & Scharf 2011). Pheromones influencing these caste changes too may be signals rather control (Keller & Nonacs 1993). For example, high levels of soldier-produced pheromone could signal to pseudogates that the colony had no need of more and that they could make greater contributions to colony survival as workers. It will benefit the colony to have an optimum number of soldiers, with between-colony selection for a local optimum based on local predator pressure.

Among the ways termite pheromones seem to mediate caste changes in the colony is by acting on the corpora allata (a key endocrine gland) of receivers to change their juvenile hormone levels (Miura & Scharf 2011; Vargo & Hussenender 2009). Ecdysteroid hormones and insulin signaling may also be involved (Miura & Scharf 2011).

Queen pheromones are molecules that are produced exclusively, or in greatest quantity, by queens and have the effect of "suppressing" ovary development in workers (Figure 6.2). If queen pheromone is a fertility signal to say that "I am laying eggs" then one should expect the time course of pheromone production to match egg laying and to correlate with fecundity, which it largely does in bees, ants, and wasps (Heinze & d'Ettorre 2009; Le Conte & Hefetz 2008; Peeters & Liebig 2009). At the time of writing only two queen pheromones with these effects on worker fertility have been identified in Hymenoptera, for the honeybee (Box 6.3) and for the ant, *Lasius niger* (Holman *et al.* 2010b).

A queen pheromone has been identified in termites (Box 6.2) (Matsuura *et al.* 2010). The honeybee queen pheromone is a multicomponent combination of fatty acids, esters, and alcohols, from more than one gland (Box 6.3)(Kocher & Grozinger 2011). In *Lasius niger*, the pheromone is a CHC, 3-methylhentriacontane, which is also present on queen-laid eggs (Holman *et al.* 2010b); hydrocarbons are likely queen pheromones in other ants and wasps (Liebig 2010; Peeters & Liebig 2009). Fertility signals found so far can be a single compound, or a particular combination of hydrocarbons, and while some are complex branched hydrocarbons, some are simple *n*-alkanes (Liebig 2010).

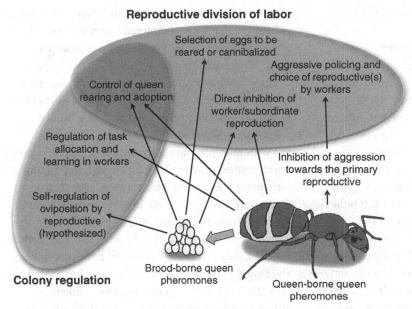

Figure 6.2 Known and hypothesized functions of queen pheromones produced by the queens (or other reproductive individuals) of social insects and carried on their brood. This illustration features a queen ant, but it could be a queen bee, wasp, or termite. Figure from Holman (2010), who gives additional examples of each of the phenomena, many of which are described in this chapter.

There is no need for them to be physiologically costly to produce (see Chapter 1, Section 1.6 signal honesty).

The physiological effects of the queen pheromone on worker reproduction have been demonstrated in the honeybee: synthetic queen mandibular pheromone is sufficient to stop ovary development in workers (Hoover *et al.* 2003) (brood pheromones produced by larvae also stop ovary development) (Box 6.3) (Chapter 9). Similarly, the queen pheromone of the ant, *Lasius niger*, the hydrocarbon 3-methylhentriacontane (3-MeC$_{31}$), is sufficient to prevent worker ovaries developing (Holman *et al.* 2010b).

News of the health of the queen, in the form of her pheromone, is continually spread throughout a colony of social Hymenoptera, in the case of honeybees by messenger bees (Box 6.3) and in the ant, *Camponotus floridanus*, on the eggs she lays (Endler *et al.* 2004). Queen pheromone effects can be dramatically demonstrated by removing the queen and seeing the rapid

changes in worker behavior and physiology; in honeybees, *Apis mellifera*, these can start in as little as 30 minutes (Box 6.3). Without the queen pheromone, worker honeybees start to rear new queens and the workers' ovaries start to develop.

Is the queen's pheromone a co-operative signal or queen control? The consensus is that workers are using the queen pheromone as an honest signal that the queen is present and the workers, her daughters, respond to it in their own best interests (Keller & Nonacs 1993; for review see Kocher & Grozinger 2011; Seeley 1985, p. 30). In the Hymenoptera, which are haplodiploid, an individual worker gains more by helping to rear the queen's eggs than by laying its own (the worker is more related to a sister (0.75) than to a son of her own (0.5), though see below) (Bourke 2011b, p. 32 ff.). The workers refrain from reproducing and also prevent nestmate workers from reproducing (worker policing, see below). For a discussion of signal honesty and reliability see Chapter 1 and Box 1.6.

What is the evidence that it is not queen control? First, manipulating the physiology of the workers against their interests is unlikely to be evolutionarily stable (Keller & Nonacs 1993). If it were not in their interest to respond, workers would evolve to ignore queen pheromones and ensuing arms races between queens and workers would decrease overall colony performance and be selected against. Genetic variation in worker sensitivity to queen pheromones, on which selection could act, has been demonstrated in honeybees and in "anarchic" bee colonies, mutant workers no longer respond to the queen's pheromones (see Le Conte & Hefetz 2008). Second, in different social insect species, the proportion of workers that develop active ovaries in the absence of the queen is a close match to the proportion predicted by workers' interests based on the relatedness of workers in that species (see below)

(Ratnieks & Wenseleers 2008). Third, there is worker policing in many hymenopteran insect societies that prevents workers from reproducing (see next paragraph, Chapter 1, and Figure 1.2).

The relatedness values between queens and workers depend on how often their mother (the queen) has mated. In hymenopteran insect societies headed by a single, once-mated queen, sisters are related by 0.75, but with multiple mating by the queen, actual levels of relatedness may not be high enough to account for worker sterility (Ratnieks & Wenseleers 2005, 2008; Ratnieks et al. 2006). In colonies with a multiply mated queen, workers might be more related to their own sons and should be selected to lay their own eggs but they do not. For example, honeybee queens mate with about ten males and workers are related by only 0.3, yet fewer than one in 1,000 workers attempts to lay eggs and only 1% of male eggs are derived from workers. The equilibrium proportion of

Box 6.3 Queen and brood pheromone of the honeybee

Some of the key pheromone molecules integrating the life of the honeybee colony are produced by the queen and the larvae (brood) (Alaux et al. 2010; Grozinger 2013; Kocher & Grozinger 2011; Le Conte & Hefetz 2008).

The pheromone blend of the queen is complex, and is produced from multiple glands, including the mandibular gland, tergal glands, and Dufour's glands (Grozinger 2013). The queen mandibular pheromone (QMP) is composed of five molecules (see figure opposite). The queen retinue pheromone (QRP) consists of the QMP plus four additional components from other glands (Slessor et al. 2005). The QRP attracts more workers than the QMP alone but the queen pheromone primer effect (Section 6.2.1) on worker ovaries is due to the five QMP components acting synergistically (Hoover et al. 2003). Some combinations of subsets of the QMP component molecules have their own releaser and primer effects. For example, the major queen ester, ethyl oleate, is involved in regulating the ontogeny (development) of the young workers who attend and feed the queen (Chapter 9) (Slessor et al. 2005). The QMP component homovanillyl alcohol (HVA) is structurally similar to the transmitter dopamine and HVA does activate the honeybee worker dopamine receptor AmDOP3 though the effects remain to be explored (Beggs & Mercer 2009). The many different effects of QMP and its components appear to act by different pathways (Chapter 9) (Grozinger et al. 2007a).

Box 6.3 (cont.)

(a) A queen honeybee, *Apis mellifera*, marked with a dot, surrounded by her retinue. The attendants collect queen retinue pheromone (QRP) and then distribute it around their nestmates. Photo by Scott Bauer, USDA ARS photo-library. (b) The molecules that make up queen mandibular pheromone (QMP) are a subset of the QRP. QMP: (*E*)-9-oxodec-2-enoic acid (9-ODA), both enantiomers of 9-hydroxydec-2-enoic acid (9-HDA), methyl *p*-hydroxybenzoate (HOB) and 4-hydroxy-3-methoxyphenylethanol (homovanillyl alcohol, HVA). In addition QRP consists of methyl oleate, hexadecan-1-ol, linolenic acid, and (2*E*)-3-(4-hydroxy-3-methoxy)-prop-2-en-1-ol.

If the queen is removed, the orphaned bees respond quickly to the sudden loss of QRP. Within 24 hours the workers start to rear new queens, enlarging cells and feeding some of the youngest larvae, previously destined to be workers, with royal jelly. The workers must act quickly as developmental paths for larvae are fixed at six days after egg laying (Chapter 9) and without a new queen the colony will die.

The relatively involatile QRP is spread from the queen around the colony by messenger bees. The QRP strongly attracts young workers and stimulates queen-tending (feeding, licking, and antennating the queen) leading to the queen being constantly surrounded by a changing retinue of eight to ten workers. After picking up the pheromone from the queen, these young workers groom themselves and then act as messengers by running through the rest of the colony for about 30 minutes, making frequent reciprocal antennal contacts with other workers, and passing on the QRP by contact, as if playing chemical tag (Seeley 1979).Tracking of radiolabeled (tritiated) QMP components demonstrated that the pheromone is carried from bee to bee as a unit and can travel to the edge of the colony in 15 minutes (Naumann *et al.* 1992, 1993).

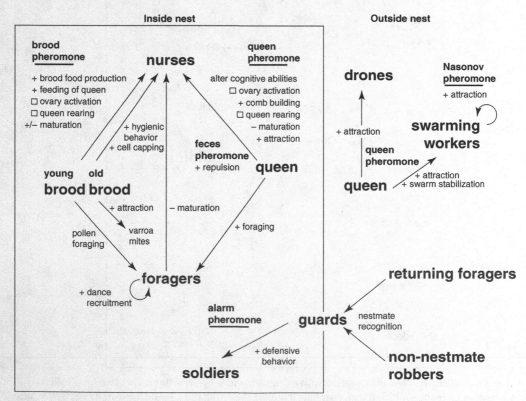

Pheromones of the honeybee, *Apis mellifera*, inside and outside the nest. The main pheromones are brood, alarm, Nasonov, queen, and feces pheromones. See Box 6.1 for the positions of the various glands. Royalactin, fed by nurse workers to larvae could be added as an allohormone pheromone (see Chapter 9). Diagram by P. Harland, from Grozinger (2013) in which these are all discussed in more detail.

The queen produces approximately 300 µg of QMP (the quantity in her mandibular gland, termed one queen equivalent, Q_{eq}) during each 24-hour period. The workers are extremely sensitive to QMP, down to a concentration of about $10^{-7}\ Q_{eq}$. However, a high percentage of the QMP produced is ingested by each bee that takes up the message, causing the pheromone to disappear as it is passed on. Constant production and rapid removal of the pheromone message means that if the queen dies and production of QMP ceases, the extinction of the message is almost immediate. Another inherent feature of pheromone message removal as it passes from worker to worker is the lessening of its effects as colony numbers increase, perhaps signaling to the workers that the size of the colony is reaching the optimum for fission (Keller & Nonacs 1993), which may explain why workers start building new queen cells as the colony size moves toward swarming size, despite the presence of an active queen secreting QMP.

Brood pheromone

Another pheromone that causes workers not to reproduce is "brood (ester) pheromone" produced by the larvae (brood) (Alaux *et al.* 2010; Grozinger 2013; Le Conte & Hefetz 2008). About ten fatty esters are produced by the bee larvae from their salivary glands. A subset including ethyl palmitate and methyl linolenate have the inhibiting effect on worker ovaries whereas methyl palmitate or ethyl oleate stimulates protein biosynthesis in worker hypopharyngeal glands, used to feed the larvae. Some components of brood pheromone also have releaser effects, for example methyl oleate and other esters together induce cell capping. An additional volatile pheromone E-β-ocimene also partially inhibits ovary development and prompts nest workers to start foraging earlier in life (Maisonnasse *et al.* 2010).

Le Conte and Hefetz (2008) note that ethyl palmitate is produced by both larvae and queen, to the same primer effect on worker reproduction. Further subtleties of pheromone integration of honeybee societies at behavioral, physiological, and gene expression levels are explored in Alaux *et al.* (2010), Le Conte and Hefetz (2008) and Grozinger (2013) (Chapter 9). Among the effects of the pheromones from a young brood are increased foraging for pollen.

workers that should reproduce, by Hamilton's rule for relatedness, is 14% for single-mated species, such as stingless bees or bumblebees, and 54% for the honeybee (Ratnieks & Wenseleers 2008). The high levels of observed worker altruism (sterility) can be achieved by coercion, with enforced altruism in the form of "worker policing": any eggs that are produced by workers are destroyed by other workers (Ratnieks *et al.* 2006). Theory suggests such policing can reduce the incentive for workers to lay eggs to such an extent that few, or none, should do so; this is supported by the observation that fewer workers lay eggs in species with more effective worker policing (Ratnieks & Wenseleers 2008).

Pheromones are involved in "worker policing" in two ways: first, workers only police if they detect the queen's pheromone signal (policing breaks down in the absence of the queen) (Figure 6.2). Second, worker policing is possible because the queen's eggs can be recognized by a pheromone mark (Liebig 2010;

Peeters & Liebig 2009). The queen marking of eggs may be by the same pheromone(s) as have the effect on worker sterility, as for example in the ant *Lasius niger* (Holman *et al.* 2010b), but in the case of the honeybee, the queen marking pheromone is not known, though it does not seem to be the queen mandibular pheromone (Martin *et al.* 2005).

6.2.2 Pheromone primer effects and reproduction in social mammals

It is in social mammals, those living in groups on shared territories and especially those breeding co-operatively, that mammal pheromone primer effects seem most developed, though probably not for reproductive suppression (see below). It has been suggested that pheromone stimuli in social mammals can induce hormonal changes, affect the success of pregnancy, alter the course of puberty, modulate cycles of female ovulation, and modulate reproductive behavior and aggression. These physiological effects include the Bruce and Whitten effects in mice (the hormonal basis of these and other primer pheromones is discussed in Chapter 9 along with doubts about the strength of the ecological evidence).

Co-operative breeding, with alloparental care in which non-breeding members of the social group assist in rearing young that are not their own, is rare but well developed in a limited number of mammalian taxa: rodents (especially mole rats), marmosets, and tamarins (Callitrichidae), dogs (Canidae), and diurnal mongooses (Herpestidae) (Figure 6.1) (Clutton-Brock 2009). Co-operative breeding can be "plural" (all females breed, although not all males) or "singular" (typically only one dominant female breeds). In rodents, co-operative or communal nesting and care of young have been reported for 35 species, representing nine out of 30 rodent families (Solomon & Getz 1997). Most of the pioneering work on mammal pheromone primer effects was on social rodents such as house mice, which are plural co-operative breeders. Female house

mice suckle each other's young and co-operatively defend the nest. An interplay of dominance (in particular between males), sex, and population density is characteristic of these societies.

Reproductive suppression of subordinates is common in singular co-operatively breeding mammal species (Figure 6.1) (Clutton-Brock 2009). As in many social insects, the subordinate females are often her daughters, and in mammals, as in social insects, suppression mechanisms affecting the reproduction of subordinates vary across species.

However, unlike advanced social insects in large colonies, mammal social groups do not seem to use pheromones in suppression of subordinate reproduction. For example, in the most eusocial mammals, naked mole rats *Heterocephalus glaber*, with colonies of up to 300 non-breeding workers, the complete suppression of worker fertility by the queen is not due to a pheromone (Figure 6.3) (Faulkes & Abbott 1993). Her odors are not enough to maintain the suppression in workers put in isolation. Instead, the queen, which is larger than other colony members, exerts her reproductive suppression on the non-breeding workers by physical dominance, "shoving" and pushing subordinates down the tunnels (for review see Faulkes & Bennett 2009). This is similar to the physical dominance mechanisms in social insects such as *Polistes* wasps, which form small colonies.

Olfactory cues for recognition (remembered signature mixtures), but not pheromones, may be important for reproductive suppression in the common marmoset, *Callithrix jacchus*. It is a small (approximately 300 to 400 g) South American primate, which shows singular co-operative breeding. Only the dominant female breeds. The young of the common marmoset stay within their natal group into adulthood. All group members, of both sexes, contribute to infant care (alloparenting), and may groom, tend (babysit), and transport young, but in addition may help with post-weaning feeding of infants (French 1997; Yamamoto *et al.* 2010). The initial benefits of alloparenting might originally have been increased survival of young, but once set on the path of helping, it has become a

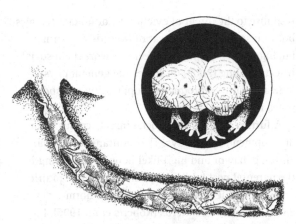

Figure 6.3 Naked mole rats are eusocial mammals with a worker caste, here shown digging a tunnel together. Although only one female reproduces, pheromones do not seem to be important in causing the workers not to reproduce (Faulkes & Abbott 1993; Faulkes & Bennett 2009). From a photograph by David Curl. Drawing by Priscilla Barrett in Manning and Dawkins (1998).

requirement as the energy costs of breeding are so high that a lone pair of common marmosets is effectively incapable of reproducing successfully (French 1997; Yamamoto *et al.* 2010). Suppression of ovulation in subordinate common marmoset females is by a combination of olfactory, visual, and behavioral cues but once reproductively suppressed, this can be extended by odor alone: a subordinate female isolated from the group will restart her ovarian cycle but this is delayed by about 20 more days if she is exposed to the individual scent of the familiar dominant female. The odor of an unfamiliar dominant female had no effect on ovulatory function in separated subordinates (Saltzman *et al.* 2009, citing T. E. Smith, unpublished thesis) so the effect is not due to a pheromone common to dominant females but instead relies on individual recognition by signature mixture (Chapter 1).

In the prairie vole, *Microtus ochrogaster*, monogamous pairs and their non-breeding offspring form the core of a communal breeding group (Carter & Roberts 1997; Solomon & Getz 1997). As in the common marmoset (above), "suppression" of reproduction in helping offspring relies on remembered

individual signature mixtures, in this case of familiar males. Young females delay puberty as they do not do nasogenital grooming (which would expose them to male pheromones) with familiar males (see Chapter 9). It is not likely to be caused by an inhibitory pheromone from the mother, as was once thought (Wolff *et al.* 2001).

The importance of ecological factors for the fine balance of benefits and costs to helpers is shown by the patchy distribution of singular and plural co-operative breeding across related genera. Even in the same genus there may be species that are singular breeders and others that are plural breeders, for example prairie voles and common voles, respectively. Populations of the same species in different places, for example prairie voles (Roberts *et al.* 1998), may show more or less alloparental care according to local ecological conditions.

6.2.3 Parallels between social mammals and social insects

Is the lack of confirmed examples of the involvement of pheromones in suppression of breeding in mammals because their groups are smaller than social insect ones? Perhaps even in the largest groups of naked mole rats, the most eusocial mammal, the queen can physically dominate members of the colony. Or does the lack reflect the taxonomic distance between mammals and social insects?

The suppression of reproduction by subordinate marmosets or prairie voles may be an adaptive response to signals from the principal female, analogous to the worker responses to signaling by social insect queen pheromones (Section 6.2.1) (Keller & Nonacs 1993). Subordinates in marmosets, prairie voles, and social insects may have evolved specific, adaptive responses to signs of subordinate status that lead them to respond with alloparental and other behavior that increases their inclusive fitness by helping the society or family group (Abbott *et al.* 1998). Like workers in social insect colonies, subordinate female marmosets show many behavioral, neuroendocrinological, and physiological differences from dominant females. The behavior and physiology of subordinates seem to be a stable alternative to dominant status, not a state of generalized stress imposed by the dominant female and endured by the subordinates to their physiological detriment (there is no elevation in the circulating hormones, cortisol or prolactin, associated with stress) (Abbott *et al.* 1998). Saltzman *et al.* (2009) conclude that the pre-conception suppression of reproduction in common marmosets is not directly imposed on subordinate females by dominants, but is instead self-imposed by most subordinates, consistent with restraint models of reproductive skew. In contrast to restraint models, however, this self-suppression probably evolved not in

response to the threat of eviction by dominant females but in response to the threat of infanticide, perhaps analogous to policing by queens or workers in social insects. Young (2009) discusses the general physiological basis of reproductive suppression of subordinates in mammal groups.

A further parallel with social insects comes from developmental pathways. The spontaneous alloparenting behavior and high likelihood of remaining in the parental nest (philopatry) of subordinate prairie voles are influenced by their pre-natal hormonal environment in the uterus (Roberts *et al.* 1996). I wonder how different this is from developmental influences on social insect larvae (Chapter 9) as they are directed to worker or queen roles?

Summary

The individuals in animal societies interact via a complex web of semiochemical signals. Odor recognition is also important for kin recognition for mate choice and other behavior in less social species. A common feature of recognition is the learning of odor cues at a certain life stage. Social insects, such as ants, bees, wasps, and termites use hydrocarbon odor cues on their cuticles. Colony odors are usually the combination of secretions under genetic control and odors gained from the environment. The mixture of odors may serve to mask the underlying genetic heterogeneity of a nest of honey bees, which share a mother but have different fathers. Social mammals use mixtures of odors that are just as complex. In both insects and mammals, the odor signatures are the passport to acceptance by the group. Eusocial species of social insects and social mammals are characterized by reproductive division of labor. In some species, group members fight to establish which animals will reproduce. Pheromones in social insects are commonly used to signal queen status or to suppress reproduction in subordinates, but this does not seem to occur in equivalent mammal societies.

Further reading

Bourke (2011b), West and Gardner (2010), and Davies *et al.* (2012) provide readable introductions to the evolution of social behavior including inclusive fitness and Hamilton's rule. Le Conte and Hefetz (2008) and Kocher and Grozinger (2011) review pheromone primer effects in social Hymenoptera. Chapters in Gadau and Fewell (2009) cover evolutionary aspects of social insect societies, including chemical communication. Chapters in Blomquist and Bagnéres (2010) cover the CHCs so important for both colony odors and fertility signals in many social insects. Ratnieks *et al*. (2006) review intra-colony conflict in social insects and its resolution. Chapters in Hager and Jones (2009) cover vertebrate reproductive skew and mammalian co-operative breeding.

Be aware when reading the past and current literature that the term "pheromone" is often used ambiguously and may be used in contexts where "signature mixture" or "olfactory cues" might be more accurate or helpful.

For molecular structures see sites such as www.chemspider.com, which allows you to search by common name and shows synonyms as well as the systematic names. See also the Appendix for a short guide to the terminology (available for free download from the website associated with this book). Many pheromones are illustrated on Pherobase www.pherobase.com (El-Sayed 2013).

Pheromones and recruitment communication

The ability to recruit group members to new sources of food, to defend the territory, or to protect the group against enemies is crucial to the success of social species across the animal kingdom. This ability is one of the key factors behind the extraordinary ecological dominance of social insects in so many habitats (Hölldobler & Wilson 2009). Recruitment brings nestmates to the place and task required.

Recruitment signals are commonly pheromones, in part because the taxa that show the most development of recruitment behaviors are those most reliant on pheromones, but also because pheromones enable mass recruitment to tasks. "Call to arms" pheromones for collective defense are also common in social insects (Chapter 8). New nest site finding and colony moving in many ant and wasp species involves phero- mones, as does the honeybee swarm's entry into its new nest (Bruschini *et al.* 2010; Hölldobler & Wilson 2009; Seeley 2010, pp. 184 ff.). Even elaborate nest building itself may be organized using pheromones (Section 7.2.2).

The co-ordinated activity of social insect colonies puts them at a pinnacle of biological complexity, but although hundreds of thousands, even millions, of individuals may be involved, activity is not commanded from the center: perhaps the complexity is possible precisely because of this. Instead, the social integration and assembly of colony-level patterns come from self-organization result- ing from simple interactions between individuals and from individual responses to local conditions, largely mediated by pheromones (Section 7.2) (Camazine *et al.* 2001; Sumpter 2010; Traniello & Robson 1995).

The range of recruitment tasks mediated by phero- mones in social insects is illustrated by African weaver ants, *Oecophylla longinoda*, which may have the most complex set of signals yet identified in ants (Hölldobler &

Wilson 1978, 2009). These ants have five different recruitment systems, which use combinations of differ- ent tactile signals and pheromones from two glands (Table 7.1) (see Box 6.1 for positions of social insect pheromone glands).

7.1 Foraging ecology and evolution of recruitment communication

Eusocial insects (ants, wasps, bees, and termites), some social caterpillars, and the naked mole rat are central- place foragers, going out from the nest in search of food. While all termites show foraging recruitment behavior, the development of such behaviors in ants, bees, and wasps varies greatly between species. Only a minority of social caterpillar species show recruitment (Box 7.1). In ants, different recruitment mechanisms include tandem running in which the scout ant leads one nestmate to the resource; group recruitment, which recruits tens of nestmates; and mass commu- nication, which uses pheromones alone to recruit large numbers of nestmates (Box 7.2).

Which recruitment mechanism is used depends as much on the ecology of the species as on its taxonomic position. Comparing closely related species with differ- ent foraging strategies reveals that key factors which select for recruitment behavior are clumped, patchy food resources (Traniello & Robson 1995). For example, the ant *Pachycondyla obscuricornis* hunts small arthro- pods, and it has no need for foraging recruitment as each prey item can be carried back to the nest by one forager, the finder. However, for nest moving this spe- cies does use tandem running (Box 7.2), facilitated by pygidial gland pheromones. The congener *P. (Termitopone) laevigata* is a specialist predator of

Table 7.1   Basic properties of the five recruitment systems of the African weaver ant, *Oecophylla longinoda*. The recruitment systems are directed respectively to food, potential new territory, new nest sites, organization of emigration, and responses to enemies at short range and at long range. The weaver ants are strongly territorial and defend their trees against conspecifics and other ant species (Box 5.1).

System	Chemical signals	Tactile signals	Pattern of movement	Apparent function
Recruitment to food	Odor trail from rectal gland; regurgitation of liquid crop contents	Antennation; head waving; mandible opening associated with food offering	Occasional signpost marking with looping trails laid around food source; main trail directly to nest	Recruitment of major workers to immobile food source, especially sugary materials
Recruitment to new terrain	Odor trail from rectal gland	Antennation; occasional body jerking	Main trail directly to nest; broad looping movements resembling signpost marking, but only after foragers physically contact terrain; increase in frequency of anal spotting	Recruitment of major workers to new terrain
Emigration	Odor trail from rectal gland	Antennation; physical transport of nestmates and tactile invitation of signals leading to transport	Main trail directly to nest site; no signpost marking; predictable sequence of categories of nestmates carried	Emigration of colony to new nest site
Short-range recruitment to enemies	Short looping trails from sternal gland and exposure of gland surface with abdomen lifted in air	None	Trails short, looping and limited to vicinity of contact with enemy	Attraction and arrest of movement of nestmates; inducement of clumping and quicker capture of invaders and prey
Long-range recruitment to enemies	Odor trail from rectal gland	Antennation; at higher intensities, body jerking	Main trail directly to nest; no signposts laid	Attraction of major workers to vicinity of invaders and prey operation in conjunction with short-range recruitment; especially intense during territorial wars with conspecifics

From Hölldobler & Wilson (1978). See also Crozier *et al.* (2010).

Box 7.1 Tent caterpillars

Caterpillars of some moth and butterfly species live in large groups and use pheromone trails. Social caterpillars living in silk tents have enhanced predator defense, thermoregulation, and foraging efficiency (Fitzgerald 1995). The most sophisticated trail communication has developed in species such as the eastern tent caterpillar, *Malacosoma americanum*, which forages from a fixed base on a patchily distributed resource. *M. americanum* lives colonially, with some 50 to 300 sibling individuals from a single egg mass in a silk tent in the branches of cherry or apple trees (see figure below). Hungry caterpillars forage in groups, leaving the tent in search of food every six hours or so. They travel to fresh areas, ignoring previously visited, now defoliated, sites. The nest serves as an information center: animals that were not successful in finding a good feeding site of their own will find productive areas if they follow the trails of successful conspecifics.

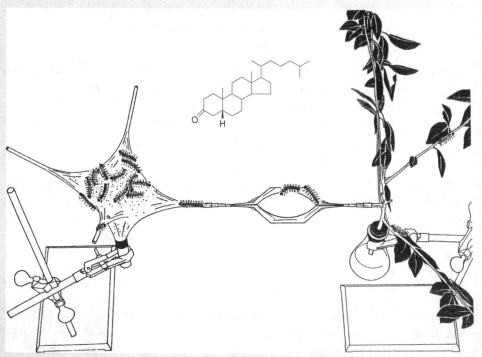

The eastern tent caterpillar, *Malacosoma americanum*, lives socially in a silk nest (left) and caterpillars travel to feed on leaves (right). Caterpillars returning from good feeding sites mark the trail with a relatively involatile steroid pheromone, 5β-cholestan-3-one, identified using this Y-maze bioassay. Figure from Fitzgerald and Gallagher (1976).

Box 7.2 Mechanisms of recruitment

Ants show the whole range of recruitment behavior from none through to mass communication by pheromones alone (see reviews by Hölldobler & Wilson 1990, 2009; Jackson & Ratnieks 2006; Vander Meer & Alonso 1998).

One of the more primitive forms of ant recruitment is **tandem running**: in ants such as *Leptothorax acervorum* a single nestmate is recruited by the returning scout and led to the new nest or food source. The follower touches the leader's abdomen with her antennae as they run. This behavior may have evolved from tandem running in ant courtship. The scout's recruitment signals often include motor displays (which are ritualized, stereotyped movements such as tugging) and in some species are accompanied by pheromone release to attract workers. Recruitment in tandem-running species is relatively slow as each scout only brings back one nestmate with her.

More rapid recruitment is accomplished by species that use **group recruitment** to excite tens of nestmates at a time and then lead them out to the site. In the ant *Camponotus socius*, scouts leave chemical signposts round new food sources and lay a hindgut contents trail back to the nest. More than the trail is needed to recruit the nestmates: the recruiting ant performs a waggle display when facing nestmates head-on. Workers then follow the scout along the trail. Stingless bees have a similar behavioral display that excites nestmates to follow the pheromone trail. Naked mole rats may use group recruitment.

Mass communication, with recruitment by chemical signals alone, is the only method of termite and tent caterpillar recruitment communication and is characteristic of many species in the majority of ant subfamilies. Recruitment by pheromones allows rapid recruitment of large numbers of nestmates as no physical contact between individuals is required, for example in the fire ant, *Solenopsis invicta*, so that the signal can travel quickly throughout the colony.

termites: when a scout ant finds a single termite, she returns to her nest, laying a pygidial gland trail (Box 7.3). The pheromone acts as a mass recruitment signal stimulating nestmates to leave the nest and follow the trail to attack the termite colony. Clumped food resources (termite nests) have selected for changes in foraging behavior mediated by (presumably) relatively small changes in pygidial gland chemistry (Traniello & Robson 1995).

Species that can recruit many workers to bring back large prey items can expand the diet of the colony to include items that greatly exceed the ability of any single worker (Traniello & Robson 1995). For example, foragers of the ant *Lasius neoniger* are all small. Co-operative foraging increases the diet breadth of the colony 30-fold.

Foraging communication recruits workers but also allows the colony to regulate total foraging activity, to retain a memory of previously rewarding locations, and to select between locations of different profitability (Dussutour & Beekman 2009; Jackson & Ratnieks 2006).

Box 7.3 **Trails in and out**

Trails will be familiar to anyone who has watched a line of ants running across a wall or path, appearing to follow a strong invisible "thread." In most ant species, the trail, which can be erased by rubbing a finger across it, is a pheromone laid down by the workers as they travel back from the food source. Similarly, tent caterpillars mark with a recruitment trail on return trips if they have encountered a good food resource (Box 7.1) (Fitzgerald 1995). Termites, ants, and tent caterpillars show similar behavior when laying trails, touching the ground with the sternal or other trail gland (see below). Conspecifics follow the trails to the food. Most ant species rely on visual cues (e.g., skylight compass or landmarks) and even general olfactory (smell) landmarks for return to the nest and do not lay trails outbound (Hölldobler & Wilson 2009; Kathrin 2012). However, some social insects lay trails on both the outward trip, to guide them back to the nest, and on return trips (if they have found food). These include tent caterpillars, most termites (which are blind), and blind ants such as *Eciton* army ants (Section 7.2.1).

A *Zootermopsis nevadensis* termite nymph laying a trail from its sternal gland on the underside of its abdomen. Figure from Stuart (1969).

The great sensitivity of worker ants to trail pheromone, with a threshold of 0.08 pg/cm (3.5×10^8 molecules/cm) found for the leaf-cutter ant, *Atta texana*, means that only minute quantities are needed (for trail-following behavior, see Chapter 10) (Tumlinson *et al.* 1971: at any one time, each worker might contain as little as 0.3 ng of the main component, methyl 4-methylpyrrole-2-carboxylate). The total amount of trail pheromone in one nest of *Atta texana* (about 1 mg) could theoretically lay a trail leading a column of ants round the world three times. One milligram of the trail pheromone of the leafcutter ant, *Atta vollenweideri*, could lead a worker 60 times round the world (Hölldobler & Wilson 2009, p. 206; Kleineidam *et al.* 2007).

7.1.1 **Stingless bees, bumblebees, honeybees, and rodents**

Group recruitment mediated by pheromones, found in many ants (Box 7.2), also takes place in different bees and a very few mammals.

Many species of neotropical meliponine stingless bees spit "dots" of pheromone as "odor beacons," rather than a line on the ground. These guide flying foragers through the rainforest to gather quickly and precisely at a particular food source, easily competing with the recruitment efficiency of honeybees (Figure 7.1) (Barth *et al.* 2008; Jarau 2009). On her return trip from a good nectar source, a worker lays down marks from her labial gland (rather than the mandibular gland as previously thought) on vegetation around the source and then at intervals back to the nest. Back at the nest, in some species the scout, after stimulating other workers with jostling, flies back to the source, perhaps releasing scent from her labial glands as she goes, followed by other workers. In *Trigona recursa* and *T. spinipes* the major (or sole) components of their trail pheromones are hexyl decanoate and octyl decanoate, respectively (Barth *et al.* 2008; Jarau 2009). For directing foragers in three dimensions, *Trigona*'s pheromone marks may be more effective than the honeybee's waggle dance, which only gives the distance and direction of resources. The trail-marking secretions from different colonies of a third species, *T. corvina*, differ in relative quantities of the constituents (John *et al.* 2012). By choosing to follow only their own colony's marks, *T. corvina* may avoid aggressive encounters with other colonies at food resources. It seems there is a core combination of molecules representing a shared trail pheromone together with variable amounts of other molecules making up a colony-specific odor mixture, which is learned (John *et al.* 2012; Reichle *et al.* 2013).

Bumblebee foragers may signal to nestmates. On return to the colony after a rewarding trip, they make excited runs with bouts of wing fanning while releasing an alerting pheromone, eucalyptol, from their tergal glands (Granero *et al.* 2005). The foragers do not communicate spatial co-ordinates of food sources but they do inform nestmates about the general availability and the scent of rewarding food sources. Prompted by the pheromone, nestmates are activated and leave the nest in search of food. The pheromone may have applications for managing bumblebees used for commercial pollination (Chapter 12) (Molet *et al.* 2009).

Honeybees, most famous for their dance communication of food location, also use recruitment pheromones. Waggle-dancing honeybees release volatile cuticular hydrocarbon (CHC) pheromones from their abdominal cuticles (Thom *et al.* 2007). Release of the molecules (two alkanes, tricosane and pentacosane, and two alkenes, *Z*-(9)-tricosene and *Z*-(9)-pentacosene) increases the number of bees that exit the hive to forage

(Gilley *et al.* 2012; Thom *et al.* 2007). The chemical nature and source of the compounds of the waggle-dance scent differ from those of the bumblebee foraging recruitment pheromone so it is likely that these have evolved independently.

Naked mole rats, *Heterocephalus glaber*, are eusocial, burrowing rodents that inhabit arid regions of north-eastern Africa (Figure 6.3) (Faulkes & Bennett 2009). They forage from a central nest to find concentrated food sources, such as plant tubers, that are patchily distributed – bigger colonies can dig more tunnels, increasing the chance of coming across tubers. Successful scouts return directly to the nest, making "chirp" sounds (not given by unsuccessful scouts or later recruits) and, back at the nest, wave the new food around (Judd & Sherman 1996). Nestmates followed the individual odor trail (perhaps urine?) left by the particular mole rat that had found a new food source.

Norway rats, *Rattus norvegicus*, moving between foraging and harborage sites tend to follow trails made by fellow group members. Further, laboratory tests suggest that Norway rats lay trails back from good food sources. Using a T-maze with a cleanable floor, Galef and Buckley (1996) showed that a second rat would choose the return path of a previous rat that had found food. It was not simply a matter of following any trail: outward trails or trails of rats returning unfed were not followed.

7.1.2 Trail specificity in termites and ants

There seems to be little variation in trail pheromones in termites, with only eight different molecules for the 60 species investigated so far (Bordereau & Pasteels 2011). Some of these molecules are nonetheless characteristic of particular taxa and the molecules used by basal termites (branched C_{13}, C_{14} alcohols or C_{14}, C_{18} aldehydes) are different from more derived termites (unbranched C_{12} alcohols or a C_{20} diterpene). In all termites, the trail pheromones are secreted from the sternal gland. Trail-following pheromones of termites have a double role of orientation and recruitment but in most species appear to be composed of only one compound. However, in some species, especially in

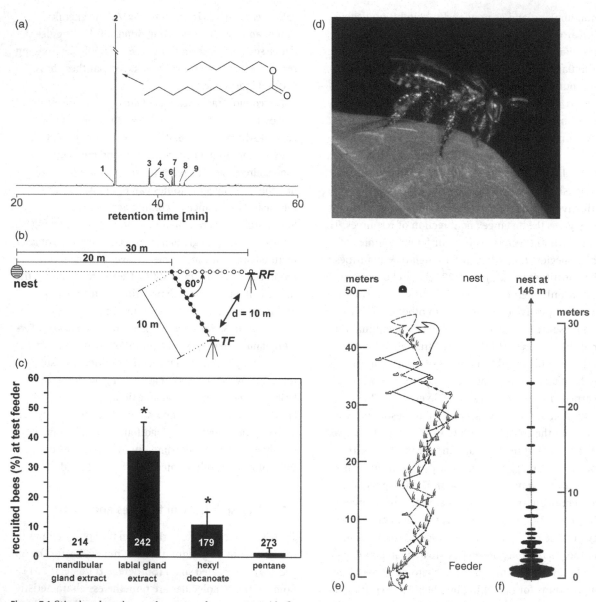

Figure 7.1 Stingless bees leave pheromone beacons to guide foragers.

(a) Gas chromatogram of a labial gland extract taken from a *Trigona recursa* forager. The main compound, peak 2, is hexyl decanoate (see Jarau *et al.* 2006 for other numbered peaks).

(b) Behavioral assay of recruited *T. recursa* foragers to artificial scent trails. *Filled circles* represent baited, *open circles* unbaited short branches taken from a Pitanga tree, *Eugenia uniflora*. *RF* recruitment feeder, *TF* test feeder; distance between the two feeders 10 m.

(c) Response to artificial scent trails measured as the percentage of newcomers that reached the test feeder (100% = newcomers at recruitment and at test feeder). Mean (± standard deviation). Number of individuals tested given in or above bar. Asterisks indicate significant difference from the pentane (solvent control) trail. (Jarau *et al.* 2006).

(d) Forager *Trigona recursa* bee depositing a scent mark on a leaf with its extended proboscis. Photograph courtesy Stefan Jarau.

Nasutitermitinae, trail pheromones have two components, neocembrene[1] (see Appendix) with either dodecadienol ((Z,Z)-dodeca-3,6-dien-1-ol) or dodecatrienol ((3Z,6Z,8E)-dodeca-3,6,8-trien-1-ol), and minor components may be present in other species (Bordereau & Pasteels 2011). In many species the trail-following molecule is also used by reproductive castes as a sex-pairing pheromone, but in quantities about 1,000 times greater than those in workers.

In contrast, among the ants there is enormous variation and trail communication appears to have evolved many times: some ten different glands are sources of trail pheromone and there is great diversity in the glands and compounds used in different ant genera (Hölldobler & Wilson 1990, 2009; Morgan 2009). Among the compounds are a variety of terpenes, pyrazines, coumarins, fatty acids, alcohols, esters, and aldehydes. Ant trail pheromones may be a single compound or, in one species, *Leptogenys peuqueti*, consist of a blend of as many as 14 compounds (Morgan 2009). The pheromones may come from a single gland, or a combination of glands. They may be peculiar to a single species or may be shared by a number of species. However, although different species in a genus may share the main components of their trail pheromones, as for example in the sympatric fire ant *Solenopsis* spp., workers will only follow those of their own species. Similarly, the recruitment pheromone seems to be the same in all species of the harvester ant, *Pogonomyrmex*, but the Dufour's gland secretion, used to mark persistent trunk routes, has a hydrocarbon blend that varies between species (and other markers may even differ between colonies, and, in some species, between individuals) (see Chapters 1 and 6) (Hölldobler & Wilson 1990; Morgan 2009). As described below, there may be more than one trail pheromone used by a species depending on context and some may have different roles such as recruitment or orientation.

7.1.3 Recruitment and competition in ants

Recruitment is also important to secure resources against intra- and inter-specific competition with other ants, particularly as most species do not have exclusive territories so ownership of any prey item could be disputed (Chapter 5) (Hölldobler & Wilson 1990; Traniello & Robson 1995). When a scout of the large desert ant, *Aphaenogaster (= Novomessor)*, finds prey that is too big to be carried unaided, such as a dead insect, it recruits colony mates from up to two meters away by releasing poison-gland pheromone into the air and stridulating. As soon as enough colony mates arrive, they gang-carry the prey back to the nest (Hölldobler *et al.* 1978). Speed is of the essence because they must move the prey to safety before formidable but slower moving ants, including fire ants, can claim the prize. Small ant species such as the tiny North American *Monomorium minimum*, which are unable to carry large food items, can monopolize these by rapid mass recruitment and chemical repellents to deter other species (Adams & Traniello 1981).

Figure 7.1 (cont.)

(e), (f) Spatial distribution of pheromone marks deposited (e) by *Scaptotrigona postica* foragers on stones and ground-proximate vegetation and (f) by *Trigona hyalinata* foragers along a rope with leaves fastened to it at 1 m intervals, which extended for 100 m from the food source toward the nest (the leaf-bedecked rope was the only possibility for bees to deposit scent marks during the experiment). Recruiting foragers of *S. postica* deposit pheromone marks along most of the distance covered by their return flight to the nest (two forager flights shown; dots denote a scent mark, arrows indicate the flight direction). By contrast, *T. hyalinata* foragers deposit most scent marks at the food source and along the first few meters toward the nest (area of ellipses correspond to the number of scent marks deposited, the smallest ones being equal to one mark; in sum, 103 marks are shown). The food sources (sugar solution feeding tables) were located at 0 m; note the different scales and distances to the nest in (e) and (f). The black hive symbol in (e) denotes the location of the bees' nest. (e) after Lindauer and Kerr (1958); (f) after Nieh *et al.* (2003). Figure and caption from Jarau (2009).

---

[1] (1*E*,5*E*,9*E*,12*R*)-1,5,9-trimethyl-12-(1-methylethenyl)-1,5,9-cyclotetradecatriene

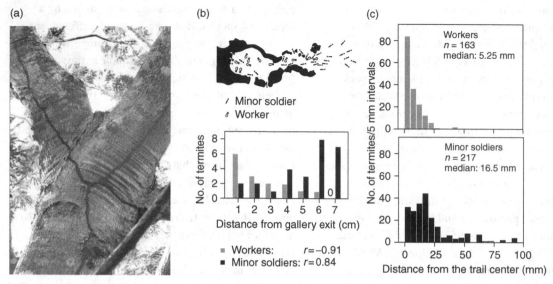

Figure 7.2 Termites avoid exposure to the drying air and predators by travelling to foraging sites in shelter tubes or galleries, seen here (a) on a tree in Mexico. The "carton" shelter tubes are made from soil and plant matter glued together with saliva and feces (Photograph Jim Conrad, Creative Commons). (b) and (c) Termites are in constant risk of attack by ants. The vulnerable *Schedorhinotermes lamanianus* workers are always protected by a wall of soldiers, which travel further from the exit of foraging galleries (b) and the center of pheromone trails (c). Figure from Kaib (1999).

7.1.4 Foraging patterns in termites and ants

Just as the kind of food drives the evolution of different recruitment communication mechanisms, how long the food supply lasts has selected for the longevity of pheromones used for trails and the form of foraging. Ants such as *Pheidole oxyops*, recruiting nestmates to collectively retrieve a single prey item, lay transient trails that fade in minutes (Czaczkes & Ratnieks 2012). The short life of the pheromone ensures that the trail remains relevant: if not renewed, it fades (Chapter 10). Species with longer lasting food supplies such as trees, for example *Atta* leafcutter ants that collect leaves for their fungus garden (Chapter 11), have long-lasting trunk trails. Ants that collect a renewable resource, honeydew from aphids, have trails between trees where they tend the aphids and the nest that follow the same routes for years (and individual ants may follow particular trails for more than one season) (Quinet & Pasteels 1996). Some termite species have trail pheromones still active after months or years (Bordereau & Pasteels 2011).

Termites feed mostly on dead vegetation or wood so foraging is more like slowly mining a coal face (see review by Bordereau & Pasteels 2011). It is quite rare to see termites on the surface as they are usually hidden in their covered walkways, roofed over with "carton" made of earth or wood mixed with feces and saliva (Figure 7.2). In some species, rather than sending individual scouts foraging far from the nest, soldiers go a short distance from the end of the covered trails and, using sternal gland secretions, recruit workers to any food they find (Figure 7.2). Once there, workers gnawing food release a persistent, phagostimulating pheromone from their labial glands that causes other workers to gnaw there (Figure 7.3). In the Australian termite species *Mastotermes darwiniensis*, the labial gland pheromone is hydroquinone (1,4-dihydroxybenzene) (Reinhard *et al.* 2002), which may also have this activity in many, but not all, termite species (Bordereau & Pasteels 2011).

Patterns of trails can adjust searching to match areas of high productivity, adjust movement of workers to avoid conflict with neighboring nests, and direct foragers to sites needing recruitment (Figure 7.4) (Hölldobler & Wilson 1990; Jackson & Ratnieks 2006; Traniello & Robson 1995). For example, colonies of the Mediterranean harvester ant, *Messor barbarous*, cut smooth trunk trails cleared of vegetation, that lead from the nest to foraging areas with seeds (Lopez *et al.*

Figure 7.3 Food collection by workers of the termite *Schedorhinotermes lamanianus* is organized by pheromones. Gnawing is stimulated by long-lasting labial gland pheromone, hydroquinone, laid down by workers (Reinhard *et al.* 2002). Other workers detect this with contact receptors on their antennae, which they sweep from side to side. The workers do not gnaw if they detect sternal gland trail pheromone, so trails are kept clear of gnawers. The result is a flexible division of the "mining zone" into feeding and trail areas. Figure from Reinhard and Kaib (1995).

1994). Where there is a low density of seeds the trails are long and have few branches. In areas richer in seeds, they have shorter trunk trails leading from many nest entrances of the same colony (thus reducing foraging transport costs). The proximate cue that leads workers of the ant *Pheidole pallidulato* to rapidly mass recruit other workers to bring home large prey items species appears to be the "tractive resistance" of the prey. If the scout cannot move the prey it gives up quickly and goes back to the nest, laying a more continuous pheromone trail than if it finds a small prey item that it can carry by itself (Detrain & Deneubourg 1997). Many aspects of ant foraging behavior seem to emerge from self-organization and these are discussed in Section 7.2. How ants know which direction to travel on trails is discussed in Chapter 10.

Many species have multiple pheromones – or molecules used in different combinations – which allow flexible and complex uses for trail signaling (Jackson & Ratnieks 2006). For example, in *Solenopsis* fire ants, the six or so individual trail pheromone components may have different roles, some being involved in recruitment (alerting) and others in orientation (the tendency to follow a trail once alerted)

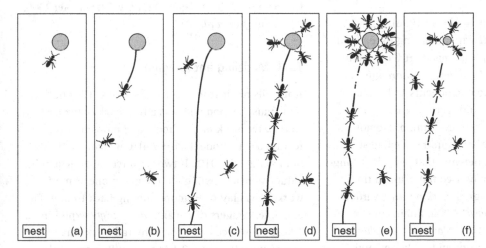

Figure 7.4 A scout worker that has discovered a food source (gray circle) lays a pheromone trail back to the nest (the line represents the pheromone trail – the thicker the line, the fresher the odor trail); interrupted lines show evaporating older trails. Other ants follow the trail from the nest, reinforcing it with pheromone as they return full of food. As the food diminishes (f), the returning ants no longer lay pheromone and the trail disappears. These behaviors can be modeled (e.g., Figure 7.6). Figure from Billen (2006).

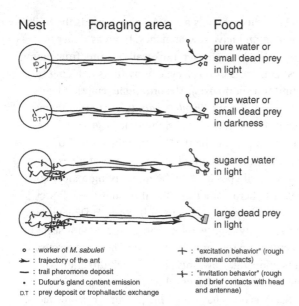

Nest   Foraging area   Food

pure water or
small dead prey
in light

pure water or
small dead prey
in darkness

sugared water
in light

large dead prey
in light

o : worker of *M. sabuleti*
➤ : trajectory of the ant
— : trail pheromone deposit
• : Dufour's gland content emission
D.T : prey deposit or trophallactic exchange

✚ : "excitation behavior" (rough
antennal contacts)

✚ : "invitation behavior" (rough
and brief contacts with head
and antennae)

Figure 7.5 Signals in the ant *Myrmica sabuleti* vary according to the prey they encounter. In the light they appeared to lay less pheromone. In the dark, the ants lay poison-gland pheromone trails. If a worker finds sugar water or carryable insect prey (a fruit fly for example) she then lays a trail with Dufour's gland secretions just outside the nest entrance. If the prey is a large insect, scouts deposit Dufour's gland secretions all along the trail. Figure from Cammaerts and Cammaerts (1980).

(Chapter 2) (Vander Meer & Alonso 1998). Signals in the ant *Myrmica sabuleti* vary according to the food encountered (Figure 7.5) (Cammaerts & Cammaerts 1980). Pharaoh's ants, *Monomorium pharaonis*, have at least seven pheromone components, delivered from the Dufour's and sting gland. These are used to create distinct short-lived (~33 minutes) trail pheromone to guide foragers to currently rewarding feeding sites, a longer lived (~days) trail network that provides a long-term memory of previously used trails, and a third, short-lived, "no entry" signal marking unrewarding trail branches (Jackson *et al.* 2006; Robinson *et al.* 2005a) (see also Figure 10.2). *Monomorium pharaonis* also has worker specialization in trail laying and detection: "pathfinders," probably a specialized sub-caste of workers, walk with their antennae in contact

with the substrate, detecting and renewing the long-lived trail pheromone (Jackson *et al.* 2007). The ant *Lasius niger* appears to adjust its marking behavior in the presence of its colony's home-range markings (CHCs secreted from tarsal glands on the feet and passively deposited on surfaces that ants walk over) (Czaczkes *et al.* 2012). Contact with the CHCs characteristic of scouts coming back into the nest prompts workers of the harvester ant, *Pogonomyrmex barbatus*, to switch from inside-nest tasks to forage themselves (Greene 2010; Greene & Gordon 2003).

7.2 Social insects as self-organizing systems

Self-organization models offer promising ways of understanding how colonies organize foraging effort so that it matches food profitability and how colonies construct nests: in both cases without central direction but instead by using simple rules in response to local situations. Complex, characteristic colony-level foraging or nest-building patterns, for example, can be generated from simple, local interactions between many simple units (individual workers) each with simple pheromone trail-laying and trail-following behavior (see reviews by Camazine *et al.* 2001; Jeanson *et al.* 2012; Sumpter 2006, 2010; Sumpter *et al.* 2012).

7.2.1 Modeling ant foraging

Ant trails result from positive feedback. The ant that first finds the food and lays a trail back to the nest starts a feedback loop as more and more ants are recruited to the food (Figure 7.4) (Camazine *et al.* 2001; Sumpter 2006, 2010). If two resources are unequal in value, more ants will go to the richer one as more workers will lay pheromone coming back from it. For example, foragers of the ant *Lasius niger* exploiting a 1 M sugar source laid 43% more trail marks than those exploiting a 0.05 or a 0.1 M source (Beckers *et al.* 1993). The greater trail laying is enough to account for choice of the richer source both in experiments and in models of trail recruitment.

The finding of new sources (and shorter routes to get there) may in part be stochastic – by the chance mistakes ants make in following trails. Pasteels *et al.* (1987) included an element of chance in their models. They studied the finding and exploitation of new food sources by the mass-recruiting ant *Tetramorium caespitum*. When presented simultaneously with two non-depleting food sources, many mass-recruiting ant species initially recruit in equal numbers but then effort switches to one source only. The switch of foraging to one source can be explained by the responses of individuals responding to the trails: if there is a stochastic (by chance), initial difference in pheromone concentration between the trails, the one with more will become amplified by positive feedback as more individuals will follow it and lay down pheromone, whereas even though both are of equal food value the other trail will diminish (an effect also seen in foraging social caterpillars and other animals, Sumpter 2010). Observations and models also reveal how ants optimize traffic flows in two directions along foraging trails (Fourcassié *et al.* 2010).

The principles of modeling ant colony foraging in the laboratory can be extended to army ant raids.

7.2.1.1 Army ants

The swarm raids of army ants, one of the wonders of the natural world, provide one of the best examples of self-organized, decentralized control (Camazine *et al.* 2001). The hundreds of thousands of ants forming an *Eciton burchelli* swarm raid are virtually blind and the complex structure of the raid system is generated by the blind leading the blind (Camazine *et al.* 2001; Franks *et al.* 1991).

Deneubourg *et al.* (1989) used a computer simulation model to investigate the way different food resource patterns may help explain the raiding patterns of different army ant species (Figure 7.6, top). Their model has virtual "ants" moving across a

two-dimensional landscape: at each point the "ant" can move ahead to the left or right (Figure 7.6, inset bottom left). Like a real foraging army ant moving forwards, the model ant adds pheromone at the point chosen. The pheromone left by each ant that passes a point increases the probability that a later ant will follow the same path so, although the movement starts as a random walk, a positively reinforcing "trail" rapidly forms. Like real pheromone, the virtual "pheromone" evaporates over time. Ants capturing food then return from the swarm front, using similar rules (but laying more pheromone). Keeping the model rules the same, but varying the food densities at the swarm front, was enough to produce different colony-level foraging patterns (Figure 7.6, bottom). When each point has a fixed probability of containing one non-renewable food item, transportable by one ant, trails from that are roughly equivalent to those of *E. burchelli* (which feeds mostly on scattered single arthropods). With rare concentrated food densities, trails similar to those of *E. hamatum* form (this species feeds on concentrated prey – social wasp nests). In field experiments, Franks *et al.* (1991) were able to change the foraging behavior of *E. burchelli* colonies by placing large bags of dead crickets (mimicking concentrated wasp nests) in the path of the swarm: the foraging trails became similar to those of *E. hamatum*. With intermediate densities and probabilities, trails similar to those of *E. rapax* formed. Camazine *et al.* (2001) discuss the models in detail.

7.2.2 Building a termite nest

Termites are among the most impressive builders in nature, constructing elaborate nest structures up to 30 m in diameter, on a scale some 10 million times bigger than the individual workers (Bonabeau *et al.* 1998; Camazine *et al.* 2001). The five-meter high, air-conditioned mounds of *Macrotermes* termites have a complex internal structure of air spaces with an intricate system of air ventilation ducts in

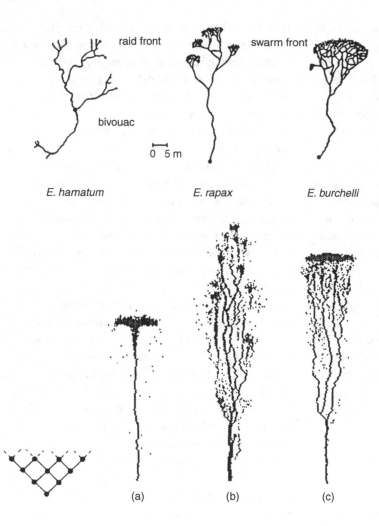

Figure 7.6 (Top) Foraging patterns of three army ant species: *Eciton hamatum*, *E. rapax*, and *E. burchelli* (Burton & Franks 1985). (Bottom) The foraging of army ants can be modeled as ants moving across a 2-D landscape; the inset on the left represents the network of points in 2-D space for a virtual "ant." Three distinct foraging patterns develop in simulations using the same model with three different food distributions (Deneubourg *et al.* 1989). (a) no food; (b) each point has a 1/100 probability of containing 400 food items (as for an army ant such as *E. rapax* specializing on social insect nests); (c) each point has a 0.5 probability of containing one food item (as might be the case for a generalist predator such as *E. burchelli*). Top redrawn from Burton & Franks (1985) and *E. hamatum* and *E. burchelli* after Rettenmeyer (1963). Lower figures after Deneubourg *et al.* (1989).

the walls for gas exchange with the outside air, brood chambers for the young termites and fungus gardens, and a royal chamber for the queen and king. However, there is no termite architect with a plan directing the half million workers, who must in any case work literally and figuratively in the dark. Instead, the structure emerges from simple responses of individual termites to local cues. The termite *Macrotermes* uses soil pellets impregnated with cement pheromone to build pillars. In the first phase, pellets are deposited at random. If enough pellets collect, the termites respond by building a pillar or

strip (a feedback effect called "stigmergy" by Grassé 1959) (Camazine *et al.* 2001, pp. 56 ff.). The process can be explained by mathematical models of pheromone concentrations over time and space that incorporate features such as the attraction of the cement pheromone to workers to lay more pellets, how the pheromones diffuse and fade, and thresholds for laying pellets at different trail pheromone concentrations (Bonabeau *et al.* 1998; Camazine *et al.* 2001). The models suggest that the same simple responses by individual workers are enough to create the different complex elements of the

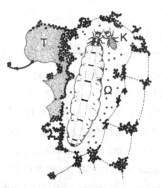

Figure 7.7 A *Macrotermes subhyalinus* termite queen (Q) and king (K) in their royal chamber. The queen seems to produce a pheromone, as yet unidentified, which diffuses from her. Workers respond to this "chemical template" and drop their soil pellets where local pheromone concentrations are between particular thresholds. This results in the chamber being built with the walls and roof at a certain distance around her. The black dots correspond to pillars and vaults; T (the stippled area) is a portion of the roof; the white circles represent material deposited at the beginning of the building activity, which has subsequently lost its attractiveness. Tactile stimuli and other pheromones, such as cement and trail pheromones, also facilitate recruitment and orientation of workers. Drawing from Grassé (1984). Caption adapted from Camazine *et al.* (2001).

termite mound. The pheromone cues are influenced by, for example, responses to the structures already built, the air currents that these create, and additional sources of pheromone such as the queen (Figure 7.7) (Bonabeau *et al.* 1997; Camazine *et al.* 2001).

7.2.3 Self-organization, natural selection, and "simple units"

Self-organization is not an alternative to natural selection – on the contrary, the two are intimately linked as natural selection molds the rules of interaction between the components of a living

system (Camazine *et al.* 2001, p. 89). An adaptive biological pattern may have to be within a particular range of parameter values; natural selection is likely to have tuned the behavior of the individual ants, for example, to achieve these adaptive patterns (Camazine *et al.* 2001). It is also worth noting that no particular hypothetical self-organized system is necessarily the optimum solution for a given problem (Reeve & Sherman 1993; Traniello & Robson 1995).

While studies of self-organization show that "simple units" responding to local conditions can become more than the sum of their parts, if the units are social insects they nonetheless have great behavioral complexity and cognitive abilities (Jeanson *et al.* 2012; Seeley 2002; Sumpter 2006). Detrain and Deneubourg (2006) remind us that each ant individual, with its 10^6 neurons and numerous sensors, is a sensitive unit that can process a lot of information. As seen in the rest of this book, the behavior of individual social insects is complex and subtle, with sensitive responses to pheromones affected by genes, experience, and context. Each individual follows decision rules, modulates its interactions with nestmates, tunes emitted signals or exhibits variations in its response thresholds in order to cope with complex and changing situations (Detrain & Deneubourg 2006; Jeanson *et al.* 2012).Thus, self-organization gives us, through mathematical models, a means of predicting the consequences of certain interactions between individuals, but it is by no means an exclusive principle by which insect societies are organized (Jeanson *et al.* 2012; Sumpter 2006). Duarte *et al.* (2011) explore how self-organization models might be integrated with the evolution of the division of labor in social insects (between queen and workers, and different worker castes, Chapter 6).

Summary

The ability to recruit group members is one of the most important factors behind the extra-ordinary ecological dominance of social insects in so many habitats. Recruitment signals are commonly pheromones. Eusocial insects (ants, wasps, bees, and termites) and the naked mole rat are central-place foragers, going out from the nest in search of food. While all termites show foraging recruitment behavior, the development of such behaviors in ants, bees, and wasps varies greatly between species and only a minority of species of social caterpillar show recruitment. In ants, different recruitment mechanisms include tandem running, in which the scout ant leads one nestmate to the resource; group recruitment, which recruits tens of nestmates; and mass communication, which uses pheromones alone to recruit large numbers of nestmates. One factor that selects for recruitment behavior is a clumped, patchy food resource. Despite the overall size and complexity of the colony, the social integration and assembly of colony-level patterns come from self-organized behaviors, emerging from simple interactions between individuals and individual responses to local conditions. Most of these interactions and responses are mediated by pheromones.

Further reading

Camazine *et al.* (2001), Jeanson *et al.* (2012), and Sumpter (2010) describe collective behavior and self-organization in a wide range of animals, including many of the examples in this chapter. Hölldobler and Wilson (1990, 2009), Vander Meer and Alonso (1998) and Morgan (2009) discuss trail recruitment and pheromones in ants. Bordereau and Pasteels (2011) review recruitment behavior in termites. Duarte *et al.* (2011) consider integration of self-organization models and the division of labor in social insects.

For molecular structures see sites such as www.chemspider.com, which allows you to search by common name and shows synonyms as well as the systematic names. See also the Appendix for a short guide to the terminology (available for free download from the website associated with this book). Many pheromones are illustrated on Pherobase www.pherobase.com (El-Sayed 2013).

8

Fight or flight: alarm pheromones and cues

Alarm signals offer perhaps one of the greatest challenges to evolutionary explanations of behavior (Bradbury & Vehrencamp 2011). Why should animals make alarm signals, at potential risk to themselves, to warn conspecifics of danger? For many animal species that release alarm pheromones the answer is likely to be the benefits to kin, either direct family members or fellow members of a clone or social insect colony.

In social insects alarm pheromones are the most commonly produced class of chemical signal, after sex pheromones, and have evolved independently within all major taxa (Blum 1985). Like other pheromones, alarm pheromones are likely to have evolved as secondary or modified uses of existing compounds (Chapter 1). Two common evolutionary routes seem to be from either chemicals used in defense or those released by injury, both events being linked to predation.

A very different phenomenon, discussed in Box 8.1, is the evolution of alarm responses to cues of injury of unrelated conspecifics and even other species, found in a wide variety of aquatic organisms of all kinds (Ferrari et al. 2010; Wisenden 2014). These cues, sometimes called "public chemical information," are not evolved signals (pheromones).

8.1 Evolution of alarm signals by kin selection

When already in the predator's jaws, giving an alarm signal might not save the victim, but the victim's signal could evolve by kin selection if it can save the lives of relatives such as offspring, siblings, or clones, by alerting them to danger.

Along the continuum of relatedness, clones (with all individuals genetically identical) ought to offer the most extreme cases of altruism (helping others at direct personal cost), and altruism should be greatest where

clonal individuals stay close to each other (Hamilton 1987), as in the sea anemones and aphids discussed below. We might also expect altruism, including alarm signals, in eusocial insects with a high degree of relatedness within the colony. We start with alarm pheromones evolved in species with family groups (subsocial insects and mammals).

8.2 Subsocial insect families

The female of the subsocial lace bug, *Gargaphia solani*, stays with her eggs and nymphs. She runs toward any predator, fanning her wings, an effective defense against predators such as coccinelid beetles or small spiders (Tallamy 2005 Tallamy & Denno 1982). If caught by a predator, the larva emits an alarm pheromone, geraniol (3,7-dimethylocta-2,6-dien-1-ol), from its dorsal glands (Aldrich et al. 1991). Nearby nymphs stop feeding and run.

8.3 Alert signals in deer family groups

Many mammals release odors when alarmed or pursued, but these "alert" alarm signals are more like alarm calls in birds, not given on death but given when a predator is detected. Alert signals by prey may be directed at the predator itself, an "I've seen you" signal (Bradbury & Vehrencamp 2011). The signaler may benefit if the predator is less likely to attack once it has lost the element of surprise. Alternatively, alert signals may be directed at the signaler's kin.

Alert signals in family groups of black-tailed deer, *Odocoieleus hemionus columbianus*, may have evolved in response to both selection pressures. When disturbed, black-tailed deer expose the metatarsal gland

Box 8.1 Evolution of responses to alarm cues in unrelated individuals: cues not pheromones

Aquatic prey animals respond to the chemical cues released by predators (kairomones, Chapter 11) and also disturbance cues, possibly a pulse release of urine, as prey nearby are startled or disturbed (Chivers *et al.* 2012; Ferrari *et al.* 2010). There are also many examples of alarm responses by aquatic animals to molecules leaking from conspecifics or other animals injured by a predator. The responding organisms range from spiny lobsters and sea urchins to freshwater mosquito and amphibian larvae, and fish (Aggio & Derby 2011; Chivers *et al.* 2012; Ferrari *et al.* 2010). Their responses may be conspicuous behavioral reactions such as freezing, fleeing, or tighter shoaling, and can sometimes involve morphological or life-history changes. A number of species respond to alarm cues from other species, perhaps because of shared molecules in related taxa or learning (see below).

However, the injury-related molecules that elicit alarm responses seem to be public chemical information cues rather than evolved pheromone signals (Ferrari *et al.* 2010). This is because aquatic organisms with alarm responses do not seem to live in kin groups, so it is hard to see how alarm pheromones could evolve in these animals as the responders are not related to the injured animal (clonal sea anemones are an exception). Instead, there is a highly evolved response but not an evolved signal. Most studies of fish, tadpoles, and other animals show survival benefits for animals that respond to alarm cues (albeit mostly in laboratory experiments though fish responses have been observed in the field) (Chivers *et al.* 2012; Ferrari *et al.* 2010).

One of the best studied systems is the alarm response of ostariophysan fish to the alarm cue, Karl von Frisch's "Schreckstoff" (fear substance), released when the skin is damaged by a predator (Chivers *et al.* 2012; Ferrari *et al.* 2010; Wisenden 2014). The chemical nature of the alarm cues has not been fully characterized, though it may include a glycosaminoglycan, chondroitin (Mathuru *et al.* 2012). The much discussed epidermal club cells of Ostariophysi and Percid fishes do not seem to have evolved to produce alarm cues and may instead be related to immune response to pathogens or parasites (Chivers *et al.* 2012; Ferrari *et al.* 2010; Wisenden 2014).

Fish are highly sensitive to the alarm cues and the olfactory neuron responses may be highly specific to particular molecules, with identifiable brain circuits for alarm (Chapter 9) (Døving & Lastein 2009; Lastein *et al.* 2014; Mathuru *et al.* 2012). The evolved sensitivity of receivers to alarm cues is similar to the high sensitivity and specificity of the mosquito olfactory receptors evolved to detect the molecules given off involuntarily by their human hosts (Carey & Carlson 2011; Smallegange *et al.* 2011). Humans do not release these molecules to attract mosquitoes. The benefit is to the receiver.

Box 8.1 (cont.)

The fright reaction to the alarm cue itself seems to be innate, but many prey species can learn to associate any predator odor with danger when they sense these with the alarm cue (Ferrari *et al.* 2010; Wisenden 2014). It only needs a single reinforcing event to form a virtually permanent association in many prey animals. Later they will react with an alarm response to the predator odor alone. Learning gives flexibility because prey can learn to respond to the predators found in their local habitat. For example, field populations of predator-naïve fathead minnows learned to respond to first the odor and then sight of predators within days after the natural predatory fish (the pike, *Esox lucius*) were added to a pond (Brown *et al.* 1997). Young coral reef fish can quickly learn the predators on whichever reef they migrate to (Mitchell *et al.* 2011). Even as embryos, tadpoles can learn the odor cues of local predators and even the dangerous times of day (Ferrari *et al.* 2010).

on the hind leg, releasing a strong garlic-like odor, which causes other deer to lift their heads and scan the surroundings (Müller-Schwarze *et al.* 1984). The recipients of the signal are likely to be kin, as mothers, their female yearlings, and fawns tend to associate together.

In black-tailed deer, the alert signal also includes an erect tail, raising of anal hair, cocking of the ears, hissing, and foot stamping. The redundancy of signal, with simultaneous visual, sound, and odor signals, and the subtlety of receiver responses, helps to explain why mammal alarm pheromones are so difficult to study.

8.4 Clonal sea anemones

The sea anemone *Anthopleura elegantissima* lives on rocky shores along the Pacific coast of North America. It forms clonal groups of genetically identical individuals produced by asexual division (binary fission). The anemone shows a characteristic rapid response to the wounding of nearby individuals: first, it gives a rapid shake of the tentacles; second, it withdraws the vulnerable tentacles and oral disc; and third, the whole anemone contracts, all within three seconds. The response is to a pheromone, anthopleurine,

released by damaged individuals (Howe & Sheik 1975). The likely receivers of the pheromone would be members of that clone (although it would benefit the receiver to respond to withdraw vulnerable parts whether or not it was related to the signaler).

The behavior of one of the anemone's specialist predators, the nudibranch seaslug, *Aeolidia papillosa*, suggests that anthopleurine evolved originally for defense: the predator prefers to feed on the parts of the anemone with the least anthopleurine. Anthopleurine remains in the tissue of the predator, days after eating anemones, sufficient to diffuse out and to alarm anemones ahead of it (Howe & Harris 1978).

8.5 Aphids

Aphids are another clonal organism with separate individuals. For much of the year parthenogenetic (asexually reproducing) wingless mother aphids are surrounded by their clonal offspring. As expected, aphids show a well developed alarm pheromone system (Dewhirst *et al.* 2010). If an aphid is attacked by a predator or parasitoid wasp, alarm pheromone is released in cornicle secretions (Figure 8.1), which can

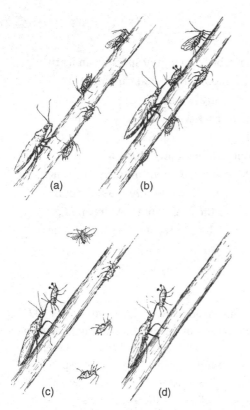

Figure 8.1 (a) A predatory insect (nabid) attacking a colony of aphids feeding on a plant stem. (b) The nabid has seized its prey and the aphid has released the alarm pheromone from its cornicles. (c) Other aphids detect the alarm pheromone and scatter by flying or dropping off the plant. (d) Although the nabid continues to eat its prey, other aphids in the vicinity have departed. Figure from Nault (1973).

themselves glue predator mouth-parts together (as well as "belling the cat" by marking the predator with the pheromone [Mondor & Roitberg 2004]). In response to the alarm pheromone, nearby aphids stop feeding abruptly, pulling their mouthparts from the leaf, back away, run away, or even jump or drop off the leaf to escape. The result is an aphid-free halo around the predator. The dropping behavior can have a real effect on coccinellid predator foraging success as shown by direct behavioral observations (Francke et al. 2008) and indirectly by showing that aphids responding to the alarm pheromone had lower mortality from predators than aphids habituated and

thus non-responsive to alarm pheromone (de Vos et al. 2010). Signaling aphids benefit by kin selection as the majority of receivers will be clone mates.

The main component of the alarm pheromone of many aphids is the sesquiterpene (E)-β-farnesene (Dewhirst et al. 2010). The response to alarm pheromone can depend on context. For example, aphid species attended by ants respond less to alarm pheromone, particularly if ants are present (Chapter 11). Some conditional factors may have evolved to reduce the chance of false alarms (especially as plants also produce some of the compounds). For example, in the turnip aphid, *Lipaphis (Hyadaphis) erysimi*, individuals do not respond unless (E)-β-farnesene is combined with odors produced by aphids feeding nearby on the plant. In some species, aphids respond more to alarm pheromone after the plant has been vibrated in the same way that a hunting predator would (Clegg & Barlow 1982).

Responding to alarm pheromone has its costs. For example, the costs of dropping off the plant include potential lost feeding time, the chance of not finding the host plant again, or death. The costs will vary according to the value of the feeding site the aphid left: aphids feeding on nutritionally richer artificial diets are less likely to move in response to alarm pheromone than those on poorer diets (Dill et al. 1990). Costs also vary by habitat. In hot, dry habitats there is higher mortality of aphids on the ground due to desiccation: pea aphid, *Acyrthosiphon pisum*, clones from hot, dry inland sites in Canada, where soil temperatures reach up to 42°C, were less likely to drop in response to alarm pheromone than those from moister, cooler, coastal sites (Dill et al. 1990). The proportion of a clone that drops in response to alarm pheromone can be quickly changed by selection experiments (Andrade & Roitberg 1995).

Parasitism may also affect responses to alarm pheromone. In the hot inland populations of *A. pisum*, aphids parasitized by a wasp are more likely to drop off the plant than unparasitized individuals if approached by a predator or stimulated by alarm pheromone

(McAllister & Roitberg 1987). This could be an adaptive host-suicide response by the parasitized aphids – they are more likely to die on the ground than by staying on the leaf with the predator (but in that case why don't aphids simply drop as soon as they are parasitized?).

The alarm pheromone (*E*)-β-farnesene also induces the pea aphid, *A. pisum*, to give birth to winged, rather than wingless, morphs, which leave their host plant to disperse to enemy-free space (Kunert *et al.* 2005; Podjasek *et al.* 2005). However, the alarm pheromone only has this effect with groups of aphids not single aphids on their own, suggesting that the alarm pheromone works indirectly by causing more aphid movement leading to more physical contacts, as occurs when aphids are crowded (Kunert *et al.* 2005).

Aphid alarm pheromones have been exploited deceitfully by plants to repel boarders and eavesdropped by predators as kairomones (Chapter 11), and might have application in aphid pest control in crops (Chapter 12) (Dewhirst *et al.* 2010).

8.6 Social aphids

Altruism is taken one step further in social aphid species with soldiers, first investigated by Aoki (1977). Soldier castes have evolved repeatedly across the aphids and some 60 species have soldiers (Pike & Foster 2008; Stern & Foster 1996). The soldiers, often first instar nymphs, attack intruders. In many species they have enlarged mouthparts, legs, or horns, which they use to stab invertebrate enemies (Figure 8.2). The aphid soldiers may die defending their clone. In the Japanese social aphid, *Ceratovacuna lanigera*, though all stages can secrete alarm pheromone, it elicits attack in soldiers but escape in adults and other non-soldiers (Arakaki 1989).

8.7 Termites and social Hymenoptera

Social insects are not clones, but they do form groups of related individuals. Eusocial species show co-operative care of the brood, overlap of adult generations (with

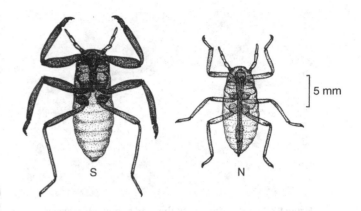

Figure 8.2 (a) The first instar aphid soldier (S) of *Colophina monstrifica* is much larger than the normal larva (N) and has a shorter beak and thickened cuticle with enlarged horns and mouthparts. Figure from Stern and Foster (1996). (b) First instar soldiers of the social aphid *Ceratovacuna lanigera* attacking a predatory syrphid larva, which they have marked with alarm pheromone (Arakaki 1990). Photograph by N. Arakaki.

offspring helping parents) together with reproductive division of labor, with some individuals specialized for reproduction (Chapter 6) (Wilson 1971). Most termite colonies start with a monogamous pair of winged sexuals, the king and queen. In social Hymenoptera, individuals are more closely related than usual diploid animals such as termites, because of their unusual haplodiploid genetics (Chapter 6). Like social aphids, these animals show "suicidal" responses in defense of the nest, on behalf of kin (Figure 8.3). Indeed, eusocial wasps, bees, ants, and termites (Chapter 6) show some of

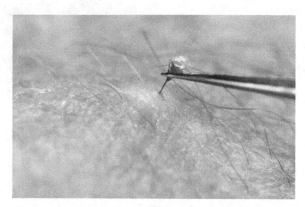

Figure 8.3 In a suicidal attack the worker honeybee leaves its autotomized sting in the vertebrate's skin while simultaneously releasing alarm pheromone and injecting venom into the victim. The sting continues to release pheromone and acts as a beacon to attract other bees to sting the intruder (which is why it is worth removing the sting as in the photograph) (Breed *et al.* 2004b; Schmidt 1998). In the European honeybee, isopentyl acetate is the principal active component of the alarm pheromone blend. This molecule is responsible for the majority of sting-releasing activity but many other molecules may also be involved as some 30 volatile compounds are found in extracts of the sting apparatus (Breed *et al.* 2004b; Grozinger 2013). Photo by Scott Bauer, USDA ARS photo-library.

the most highly developed alarm pheromone systems of any animal (Blum 1985).

Alarm pheromones are typically small, volatile molecules such as terpenoids (Keeling *et al.* 2004; Morgan 2008; Šobotník *et al.* 2010). In ants, alarm systems have been found in every species tested (Hölldobler & Wilson 1990). The pattern and range of compounds used as alarm pheromones, and the great variety of different glands they are secreted from even within a group such as the ants, strongly suggest multiple independent evolution across the social insects (Chapter 1) (Blum 1985; Morgan 2008). Many ant alarm pheromones appear to have evolved from molecules used in defense by that species; so for example, in *Formica* wood ants formic acid has both roles (Chapter 1). Most alarm pheromones are multi-component, with two or more compounds (which can come from different glands) (Bruschini *et al.* 2010;

Keeling *et al.* 2004; Slessor *et al.* 2005). The compounds may simultaneously alert, attract, and evoke aggression (Chapter 10) (Breed *et al.* 2004a; Hölldobler & Wilson 1990, 2009).

For eusocial species, alarm pheromones are a key evolutionary development that enables the colony to recruit its collective force to defend the nest against attack (Blum 1985). Defense is needed because colonies represent rich concentrated resources of brood and, in honeybees, wax, pollen, and stored honey (Winston 1987). Predators of honeybee nests include humans and other vertebrates such as bears, honey badgers, anteaters, and birds. Invertebrate threats include other social Hymenoptera and even conspecifics from other nests (see Chapter 11).

The effectiveness of colony defense is illustrated by the behavior of the Japanese honeybee, *Apis cerana japonica*, when threatened by the giant hornets, *Vespa mandarinia japonica* (Ono *et al.* 1995). If the giant hornets get into the nest, they will destroy it, taking the brood to feed their own larvae. An attack starts when a hornet scout finds the nest and marks the site with its sternal van der Vecht gland. Soon, nestmate hornets flying in the same area congregate and attack the bee nest together.

Japanese honeybees have evolved a co-ordinated response triggered by the hornet's marking pheromone. When guard honeybees detect the hornet marking pheromone, they signal alarm by releasing their own pheromones and bees in the nest entrance retreat into the nest. Meanwhile, more than 1,000 workers leave the comb and wait just inside the entrance. If a foraging hornet tries to enter, a buzzing ball of more than 500 bees forms around it (Figure 8.4). Pheromones in the bee venom and mandibular secretions attract yet more workers and focus attack on the hornet, which is now chemically marked like a beacon. Remarkably, the hornet is killed by the combined effect of the heat generated by the swarm, which can reach about 47°C at the center, and the raised carbon dioxide levels in the ball (Sugahara & Sakamoto 2009). The bees survive as they have higher tolerances than the hornet for both stresses. Using pheromones to gather assistance, the Japanese honeybees can collectively

overcome a predator many times bigger and more powerful than any individual bee. European honeybees, *Apis mellifera*, introduced to Japan are helpless when attacked by the hornets as they have not evolved a similar response.

The cascade and escalation of response through recruitment is characteristic of eusocial bees, wasps, ants, and termites. The wasp *Vespula vulgaris* marks attackers with venom from the sting, attracting more defending wasps as the venom contains the alarm pheromone (Bruschini *et al.* 2010). Honeybees leave their sting in the attacker's skin, pumping out more alarm pheromone (Figure 8.3) (Grozinger 2013). Similarly, *Nasutitermes* termite soldiers spray alarm pheromone and defensive secretions from their enormously enlarged frontal gland via a head nozzle at intruders, recruiting more soldiers (Šobotník *et al.* 2010). Termites also use vibration signals as part of their alarm signal.

However, not all social insects have the aggressive alarm responses shown by the examples above (Hölldobler & Wilson 1990). Some species of ant, for example *Lasius alienus*, show a panic and escape response to their own alarm pheromone: the colony as a whole flees from the stimulus or dashes around in erratic patterns (Regnier & Wilson 1969). If disturbed enough, individuals may even evacuate the nest. This may be an adaptive response for species of small ants that form small colonies with little possibility of realistic defense.

Responses within a species of social Hymenoptera differ not only between castes and age classes (with major, or older, workers and soldiers more active in attack of intruders) (Hölldobler & Wilson 2009, p. 230). Termites show a similar range of caste-specific responses to alarm pheromones (Šobotník *et al.* 2010). Older termite *Neocapritermes taracua* workers have "explosive back-pack" glands, which they use in suicidal defense of the colony (Šobotník *et al.* 2012). Some otherwise aggressive ant species may give hair-trigger

Figure 8.4 The Japanese honeybee, *Apis cerana japonica*, has a mass defense of the nest, co-ordinated by responses to alarm pheromone, when (a) the pheromones left by scouts of the predatory giant hornet, *Vespa mandarinia japonica*, are detected. (b) Hundreds of tightly aggregated bees surround the giant hornet and (c) their combined heat and carbon

Figure 8.4 (cont.)
dioxide, up to 47°C and ~3% CO_2, kills the hornet (Ono *et al.* 1995; Sugahara & Sakamoto 2009). Photographs: © M. Ono, Tamagawa University.

responses, including panic, to other species especially dangerous to them, termed enemy specification by Hölldobler and Wilson (1990, p. 261). For example, a single fire ant, *Solenopsis invicta*, approaching the nest is enough to cause minor workers of the ant *Pheidole dentata* to recruit majors to start nest defense. The powerful effects of alarm pheromones on social insect colonies can be exploited as "propaganda" by other species, to devastating effect (Chapter 11).

Other factors including context, experience, and genetics may influence the type of response of social insects to their alarm pheromones (Breed *et al.* 2004a;

Grozinger 2013). In many social insects, individuals that would react very aggressively to alarm pheromone near the nest, respond away from the nest by showing escape behavior (Hölldobler & Wilson 2009, p. 230). Recent experience also affects the sensitivity of response: honeybee workers are more alert and aggressive for several hours after exposure to alarm pheromones (Alaux & Robinson 2007). The effects of recent exposure, genetics, and age on the response of honeybees to alarm pheromone (Breed *et al.* 2004a) are reflected in patterns of gene expression in the honeybee brain (Chapter 9) (Alaux *et al.* 2009b).

Summary

Alarm signals offer perhaps one of the greatest challenges to evolutionary explanations of behavior. Benefits to kin probably account for the evolution of alarm signals in most species. These include alarm pheromones released between siblings, offspring, and parents in subsocial insects and in family groups of deer. Alarm pheromones are found in clonal animals such as aphids and sea anemones, in particular in social aphid species with soldiers. In social insects, after sex pheromones, alarm pheromones may be the most commonly produced class of chemical signal, and have evolved independently within all major taxa. In ants, bees, wasps, and termites alarm pheromones co-ordinate attack on enemies threatening the nest. Some species show a panic and escape response to situations where defense is not realistic.

Where victims are unlikely to be surrounded by kin, the alarm responses may be adaptive to the responder rather than an evolved signal from the sender. There are many examples of responses to injury-related cues among a wide range of aquatic animals.

Further reading

Šobotník *et al.* (2010) review alarm and defense in termites and Bruschini *et al.* (2010) those in social wasps. Hölldobler and Wilson (1990, 2009) cover the ants. Ferrari *et al.* (2010) and Wisenden (2014) review aquatic alarm cues in fish and other animals. For a good introduction to kin selection, see Davies *et al.* (2012).

For molecular structures see sites such as www.chemspider.com, which allows you to search by common name and shows synonyms as well as the systematic names. See also the Appendix for a short guide to the terminology (available for free download from the website associated with this book). Many pheromones are illustrated on Pherobase www.pherobase.com (El-Sayed 2013).

9

Perception and response to chemical communication: from chemosensory receptors to brains, behavior, and development

All animals detect and react to molecules in the external environment, including pheromones, signature mixtures of other individuals, and the chemical cues that indicate food, shelter, pathogens or predators. Despite the great external diversity in appearance of antennae and noses, animals of all kinds perceive these chemical stimuli in the same way (Hildebrand & Shepherd 1997; Su *et al.* 2009; Touhara & Vosshall 2009). Chemosensory receptor proteins are exposed to the outside world in the membrane of chemosensory nerve cells, often through a "window" in an otherwise impermeable skin or cuticle (Figure 9.1). The arrival of an odor molecule (odorant) is converted into a signal to the brain by first binding to a chemosensory receptor protein. Odorants can be any kind of molecule, as large as a protein or as small as formic acid.

Whether in air or water, olfaction is the key sense used to detect most chemical cues, rather than taste (gustation) (see Box 9.1 for the differences between these senses). All signature mixtures (the molecules learned highly varied individual or colony odors; Chapters 1 and 6; Sections 9.8 and 9.9), and probably most pheromones, whatever the size of molecules, are detected by olfaction. However, some pheromones in invertebrates are detected by gustation, including some in *Drosophila* (Section 9.5.2). Some pheromones in both vertebrates and invertebrates may act directly on the brain or other organs (Sections 9.5.2 and 9.7.3.2).

This chapter starts with its two main themes: how olfaction works and how the olfactory receptors themselves evolve, in enormous variety.

Vertebrates and invertebrates are similar in the way they detect and process chemical cues, by combining

inputs from neurons carrying different olfactory receptors, but they achieve this with quite different receptor families, which evolved independently.

I then discuss the chemosensory subsystems in mammals and insects, brain circuits for sex pheromones, what affects responses to pheromones, primer effects of pheromones on reproduction and development, learning of signature mixtures for recognition, and how responses to pheromones are affected by signature mixtures (and vice versa). Which "nose" mammals use to process pheromones was controversial for some years, so the resolution gets its own section (9.4).

Mice, moths, honeybees, *Drosophila*, and *C. elegans* rather dominate this chapter. This reflects the power of model systems, in particular ones that can be manipulated genetically with gene "knockouts" for example. The general principles emerging from these studies do, however, apply to other animals. If you work on animal behavior, ecology, and evolution, this chapter aims to give you an understanding of the sensory systems that selection is acting on.

9.1 How olfaction works: combinatorial processing of odorants including pheromones

The combinatorial olfactory systems of most animal species have the ability to detect, discriminate, and distinguish innumerable different molecules as different odors, including ones they have never encountered before. The ability is given by these features of olfactory systems: diverse chemosensory receptor proteins with broad but overlapping specificities, expression of

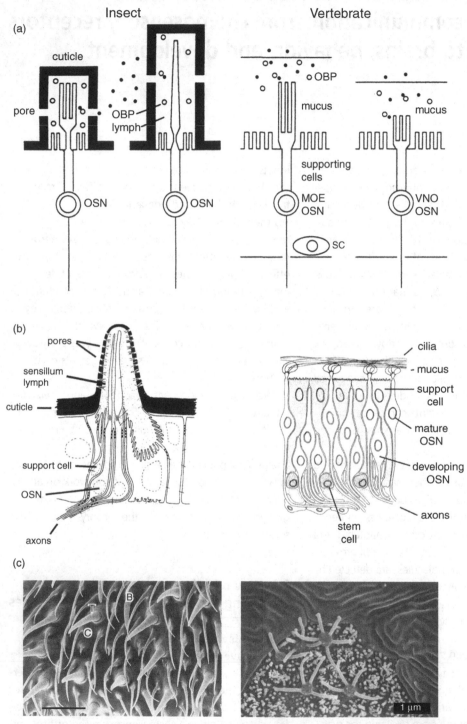

Figure 9.1 Olfactory sensory neurons (OSNs) are functionally similar across the animal kingdom, from insects (left) to vertebrates (right). Some authors call these olfactory receptor neurons (ORNs).

(a) In these schematic diagrams, odor molecules (black dots) diffuse through pores in the cuticle of arthropod sensory hairs (left) or through the mucus protecting vertebrate olfactory tissue (right). The odorants may be carried to the receptors by odorant

Box 9.1 Taste and smell contrasted

The difference between taste (gustation) and smell (olfaction) is not the medium, air or water, or even the distance between signaler and receiver. Chemosensory stimulus of olfactory sensory neurons at short range is still olfaction – for example the touch (or near touch) of an ant's antennae on the cuticle of an ant, while often called "contact" chemoreception in the insect literature, is olfaction (Brandstaetter *et al.* 2008); it is not a question of volatility or distance of communication.

Instead, what sets gustation apart is that gustatory cells link to the brain in different, simpler ways from the olfactory pathways and, in both invertebrates and vertebrates, gustatory receptors come from different families of receptors from olfactory receptors (Montell 2009; Nei *et al.* 2008; Yarmolinsky *et al.* 2009). Whereas olfactory neurons have axons that go to the brain, taste cells are secondary sensory cells that do not themselves extend into the brain but instead pass on their signal to a sensory nerve fiber. Taste cells can be anywhere on the surface of the animal, but regardless of location, taste cells always connect to nerves leading to a different area of the brain from the olfactory system, and in a much simpler manner, without the integration of information and coding found in olfactory systems. The number of different receptors (except in *Drosophila*), sensory cells, ranges of substances that stimulate them, and qualities or categories of stimuli that can be discriminated, are all smaller for taste than for olfaction. The enormous range of flavors that we can distinguish in food, for example, is almost entirely by olfaction not taste (Shepherd 2006, 2012).

When considering invertebrates it may be better to use "chemoreception" or "gustation" rather than "taste" (Derby & Sorensen 2008; Schmidt & Mellon 2011). Some pheromones are detected by gustatory receptors, for example in *Drosophila* (Section 9.5.2). Male crayfish, *Orconectes rusticus*, detect female odors with their main chelae (claws) using their non-glomerular distributed chemosensory system (Belanger & Moore 2006).

Figure 9.1 (cont.)

binding proteins (OBPs) (open circles) (see Section 9.3.2.1). The binding of the odor molecule to an olfactory receptor protein in the membrane initiates a signal sent down the nerve axon to the brain. Other abbreviations: MOE, main olfactory epithelium; SC, stem cell; VNO, vomeronasal organ. Figure after Hildebrand & Shepherd (1997).

(b) Schematic diagrams of: (left) an insect olfactory sensillum with two OSNs, based on electron micrographs of moth sensilla trichodea. Figure from Kaissling (1998); (right) vertebrate OSNs in olfactory epithelium. Unlike other vertebrate neurons, OSNs are short lived and continually replaced from stem cells in the adult, every 30 days or so. Figure from Farbman (1992).

(c) Scanning electron micrographs (SEMs) of olfactory sensory organs and neurons. (Left) *Drosophila melanogaster* third antennal segment, showing three morphological types of olfactory sensilla: coeloconic (C), basiconic (B), trichoid (T). Scale bar = 10 μm. Photo by C. Bock, Seewiesen from Shanbhag *et al.* (1999). (Right) Vertebrate olfactory epithelium. In terrestrial vertebrates the two kinds of olfactory sensory cell (ciliated and microvillous sensory neurons) occur in different organs, in the MOE and VNO, respectively. However, both cell types are mixed in the same epithelium in many fish and salamanders. This is illustrated here by the SEM of the olfactory epithelium of the catfish, *Plotsus lineatus*. Photo by B. Theisen from Theisen *et al.* (1991).

As the only part of the central nervous system exposed directly to the outside world, olfactory tissue has a particular need for ways to deactivate the wide range of molecules, biologically important or not, that it meets. Enzymes found in both insect sensory lymph and mammal mucus break down odorant and other molecules on the sensory epithelium (Durand *et al.* 2010; Watelet *et al.* 2009). In insects, some enzymes may specifically degrade pheromones (Leal 2013).

one receptor type per olfactory sensory neuron (OSN), and targeting the OSNs sharing the same receptor to a collection point in the brain, the glomerulus, one for each receptor type (Axel 2005; Buck 2005). Odorant (or olfactory) receptor proteins (ORs) are activated by multiple odorants and odorants activate multiple ORs in different combinations (see below).

The OR types (~1,000 in mouse, ~60 in *Drosophila*) are far outnumbered by the enormous number of possible odorants, but the diversity of receptors and the combinatorial process allows a huge "odor space" to be covered. Whereas vision depends on a handful of related receptors (opsins), olfaction relies on large numbers of receptors belonging to diverse, multiple families (Malnic *et al.* 2010; Spehr & Munger 2009; Su *et al.* 2009).

In both invertebrates and vertebrates most ORs are broadly tuned so any odor molecule will stimulate a distinct subset of receptors and their associated

glomeruli, giving a combinatorial code or "odor map" characteristic for that odor molecule (Figure 9.2) (Hallem & Carlson 2006; Kaupp 2010; Malnic *et al.* 1999, 2010). So for example, hypothetically, a molecule might stimulate the ORs on OSN types 1 and 2, and thus their associated glomeruli. A different molecule might stimulate OSN types 1 and 3 (Figure 9.2). The brain builds up an olfactory picture of the world from these, integrating the responses across the different glomeruli (Figure 9.3). In both insects and vertebrates, higher concentrations of the same odorant will recruit more glomeruli as even receptors only marginally sensitive to the molecule are activated (see below).

9.1.1 Olfactory sensory neurons converge on glomeruli

In the mammalian olfactory system, each OSN expresses only one olfactory receptor protein (OR)

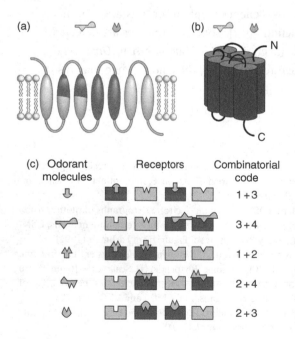

Figure 9.2 Olfactory receptors, odorants, and combinatorial codes. (a) A vertebrate olfactory receptor protein (OR) in the cell membrane. Darker regions denote regions of the vertebrate olfactory receptors, classic GPCRs, that are highly diverse between different receptors, suggesting a role in ligand (odorant) recognition. Small odor molecules reach the receptor from the extracellular environment (above)(Bargmann 2006a). (b) The usual barrel shape the GPCR takes in the membrane (Fleischer *et al.* 2009). (c) Diagram of hypothetical stimulatory molecules (odorants) with some shared as well as some different molecular features (left) and the different patterns of neural activity each molecule would elicit across a hypothetical population of four receptor cells each expressing one OR protein tuned to a particular molecular feature (right). (This schematic applies to vertebrate and invertebrate olfaction). The cells that are activated by each molecule are shaded. A single odorant is recognized by multiple receptors and a single OR recognizes multiple odorants. The identities of different odorants are encoded by different combinations of receptors. A given odorant sends a combinatorial code of receptor (and hence glomerular) activation up to the next levels in olfactory processing in the brain (Figure 9.3). The code is shown on the right. Given the immense number of possible combinations of ORs, this scheme could allow for the discrimination of an almost unlimited number and variety of different odorants. Adapted from Malnic *et al.* (1999).

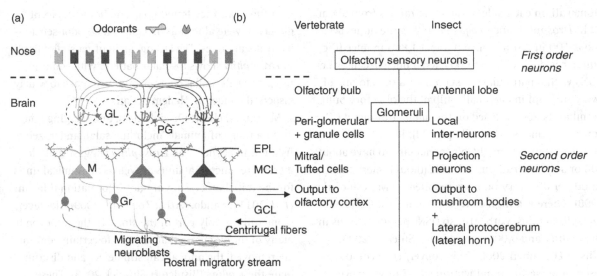

Figure 9.3 Functional parallels in the synaptic organization of the olfactory system in mammals (illustrated, (a)) and insects. The equivalent terms (b) are given for the analogous cells and structures in insects.

Olfactory sensory neurons (OSNs) expressing the same olfactory receptor send their axons to the same glomerulus (GL). Three different populations of OSNs are shown, with different shading and symbolic representations of their ORs, sending their axons to their respective glomeruli. The projection neurons, mitral and tufted cells, extend a single apical dendrite, which arborizes within one glomerulus, and several lateral dendrites in the external plexiform layer (EPL). In the glomerulus, OSN axons make excitatory synapses onto the apical dendrites of mitral/tufted (M) cells, as well as onto the peri-glomerular cell (PG) dendrites. Intra-glomerular circuits also include reciprocal dendrodendritic synapses between the mitral/tufted cell dendrites and the peri glomerular cell dendrites. The other primary population of inter-neurons, granule (Gr) cells, have their cell bodies located in the granule cell layer and an apical spiny dendrite that arborizes in the external plexiform layer, where they establish reciprocal dendrodendritic synapses with the secondary or lateral dendrites of mitral/tufted cells. See also Figure 9.14 for interactions between these cell types.

Inputs from centrifugal axons bring top–down influence from other parts of the brain. The rostral migratory stream provides a continuous addition of new neuroblasts, which differentiate into granule and periglomerular cells.

EPL, external plexiform layer; MCL, mitral cell layer; GCL, granule cell layer; PG, peri-glomerular cell; M, mitral cell; Gr, granule cell. Diagram (a) and caption text adapted from Whitman and Greer (2009).

from a large repertoire and expression is limited to either the maternal or paternal allele – a monogenic and mono-allelic mode of receptor gene choice known as the "one neuron – one receptor" rule (Imai *et al.* 2010). In insects too, with few exceptions, all OSNs express just one OR (though the mechanism for achieving this is different) (Section 9.2.4) (Imai *et al.* 2010). The OR that an OSN expresses will determine which odorants will activate it.

Each OSN sends a single unbranched axon into the brain. In mammals and in insects, the axons of OSNs expressing the same OR converge with great precision on the same glomerulus (neuropil) on that side of the brain (Figure 9.3) (Hildebrand & Shepherd 1997; Su *et al.* 2009; Touhara & Vosshall 2009; Zou *et al.* 2009). The number of glomeruli on each side of the bilaterally symmetrical brain is thus related to the number of types of ORs and corresponding number of different kinds of OSNs in that organism. The number of glomeruli on each side of the brain ranges from about 50 in *Drosophila melanogaster*, 160 in the honeybee, about 100 in zebra fish, and about 1,800 in mice (the number is double the ~1,000 ORs in mice because the OSNs with a particular OR send their axons to one of two glomeruli in each half-bulb of the olfactory bulb, Mombaerts 2006). Based on the number of intact OR genes in humans (~370, Chapter 13), if we follow the mouse pattern, we might expect humans to have about 800 or so glomeruli on each side (though more than 5,500 per olfactory bulb are reported by Maresh *et al.* 2008). There is a high convergence ratio of OSNs onto each glomerulus, with 50 or so OSNs per glomerulus in *Drosophila* and 5,000:1 in rodents (Su *et al.* 2009; Wilson & Mainen 2006). This convergence could represent a powerful amplification step. Convergence could also increase the signal-to-noise ratio for olfactory information (Nehring *et al.* 2014; Wilson & Mainen 2006). (The only notable exceptions to glomerular organization for olfaction, apart from those animals with no nervous system, are the nematodes (Bargmann 2006b; Eisthen 2002) and some homopteran insects (sucking bugs) including aphids (Kristoffersen *et al.* 2008)).

The glomeruli form a layer on the surface of the olfactory bulb (OB) in vertebrates and form the major part of the antennal lobe in insects. There seems to be a topographic map of the position of the glomeruli in the mammal OB and the insect antennal lobe, with positions mirrored between the left and right sides of the organ. In rats and mice a comparison of the position of glomeruli, identified by their unique response profiles to stimulation by hundreds of odors, showed that their layout varied little, by only 1 glomerulus in 1,000 between individuals (Soucy *et al.* 2009) (there might be more variation between out-bred animals with more OR variation, see later). However, it seems doubtful that odorants map out in a simple way onto a simple topographical map of the glomeruli (Restrepo *et al.* 2009; Zou *et al.* 2009). For example, in rats and mice, Soucy *et al.* (2009) found a loose chemotropic map at a 1 mm scale with some clustering of glomeruli responding to similar odorants, for example aliphatic aldehydes, but on a fine scale they found no chemotropy: adjacent glomeruli were almost as diverse in their odor sensitivity as distant ones. The arrangement of most glomeruli in *Drosophila* shows a similar pattern, with some receptors with similar odor sensitivity mapped to widely dispersed glomeruli (Hallem & Carlson 2006).

Molecular, electrophysiological, and imaging studies in a range of animals including salamander, zebra fish, rat, honeybee, and *Drosophila* suggest that glomeruli are functional units as well as anatomical units for the initial processing of odor information (Martin *et al.* 2011a; Sandoz 2011; Zou *et al.* 2009). However, glomeruli are only part of the story. In the rat, even if many of the glomeruli that respond to certain odorants are removed, the animal can still detect and discriminate these odors (Bisulco & Slotnick 2003). These counter-intuitive outcomes suggest that odorants may be coded by a highly distributed pattern of outputs. It is possible that rather than glomeruli being simply functional units they are a solution to the wiring problem of, in a mouse, linking several million OSNs to their correct places in the olfactory bulb (Section 9.2.4) (Zou *et al.* 2009). It is likely that much of the real work of processing is done by the interactions of the mitral/tufted cells working with other cells in the olfactory bulb (see below) (Figure 9.3).

Glomeruli in insects and mammals may represent the convergent evolution of a logical response to the demands of olfactory processing, rather than a shared feature of a common ancestor (Imai *et al.* 2010; Strausfeld 2009; Strausfeld & Hildebrand 1999). This is supported by the independent evolution of their receptor families and significant differences between the developmental mechanisms for the formation of glomeruli in insects and mammals. The question will perhaps be resolved when we find the developmental genes for neuropil (glomeruli) formation.

9.1.2 Receptor properties and combinatorial processing of odorants across glomeruli

A given odorant molecule will stimulate a characteristic selection of receptors and thus a unique combination of glomeruli. Which receptors are stimulated depends on the characteristics of the odorant molecule, including its functional groups, charge, and size as well as shape, hydrophobicity, and flexibility (Kato & Touhara 2009; Reisert & Restrepo 2009). The specificity of receptors can evolve (and can be changed by site-directed mutagenesis) (Section 9.2). Concentration greatly affects specificity: if the concentration of odorants is raised high enough, many olfactory sensory cells with receptors that are narrowly tuned at low concentrations will respond (Christensen 2005; Kreher et al. 2008; Leal 2013). This explains the importance of using biologically relevant concentrations in bioassays of putative pheromones, in order to avoid ambiguous results (Chapter 2).

In addition to activation, receptors exhibit another response: inhibition. It is found in both invertebrates and vertebrates (Reisert & Restrepo 2009). In *Drosophila* the same odorant might stimulate some receptors and inhibit others, and an individual receptor might be inhibited by some odorants and stimulated by others (Su et al. 2009). How inhibitors (antagonists) might affect the odor map is not understood yet, though they may also affect response duration (Su et al. 2011).

Some receptors are so narrowly tuned to a particular molecule that, at natural concentrations, just one glomerulus is stimulated, giving the impression of a so-called "labeled line" from antenna to glomerulus (Christensen 2005; Galizia & Rössler 2010). However, these OSNs may differ only in the specificity of the receptor in relation to other OSNs in the olfactory system (Christensen & Hildebrand 2002). There is no compelling evidence for a functional distinction of generalist versus specialist receptors in insects (Kaupp 2010). Results that seem to be non-combinatorial may occur when there is a response to one pheromone molecule acting singly, for example ESP1 in the mouse and *cis*-vaccenyl acetate (cVA) in *Drosophila* (Section 9.5).

Some of the most spectacular examples of narrowly tuned receptors are those in the male moth for individual components of female sex pheromones, but even these molecules are processed combinatorially. Each pheromone component of the female's multi-component pheromone is detected by an OSN with a narrowly tuned OR, which sends its axon into a specialized glomerulus for that component, in a particular part of the antennal lobe, the macroglomerular complex (MGC) (Figure 9.4) (de Bruyne & Baker 2008; Haupt et al. 2010; Martin et al. 2011a). However, despite the narrow tuning of the receptors, the processing of the information is combinatorial, with key projection neurons responding to the simultaneous stimulation of key glomeruli representing the key multicomponent blend (Christensen & Hildebrand 2002; Galizia & Rössler 2010; Sorensen et al. 1998). All the required components of the blend need to be present in the right ratio (Figure 9.4) (Cardé & Haynes 2004; de Bruyne & Baker 2008). Projection neurons provide this integration. The combinatorial processing creates the phenomenon of synergy (Chapter 1) because the conditional response only occurs if all molecules are present, and potentially only if in the right ratio or concentrations. When the glomeruli for the different pheromone components are stimulated at the same time, it indicates that the male's antenna has hit a pocket of pheromone-laden air that has come from an upwind female of his species (Chapter 10) (Martin et al. 2011a). He responds with brief upwind surge of flight. However, a male will not fly to calling females of related, sympatric species even though many of the pheromone components in a plume are the same (Chapter 3). If the pocket of wind contains the distinctive compounds of other sympatric species, the male will turn back, so that he does not waste time flying to a female of another species (Figure 9.4).

Goldfish also seem to have pheromone components detected by narrowly tuned receptors, stimulating particular glomeruli and then integration of this information for a species-specific response to sex pheromone (Hamdani & Døving 2007).

Combinatorial processes also occur in the mouse accessory olfactory bulb as well as the main olfactory bulb (Figure 9.5) (Chamero et al. 2012). The mouse pheromones dehydro-*exo*-brevicomin and thiazole are

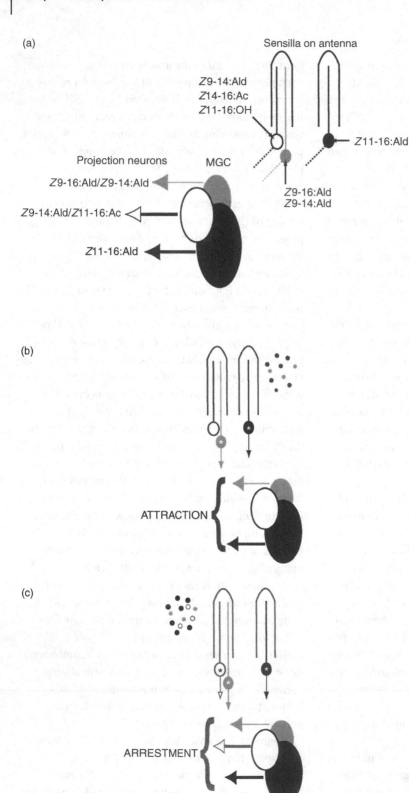

(a)

Sensilla on antenna

Z9-14:Ald
Z14-16:Ac
Z11-16:OH

Z11-16:Ald

Projection neurons   MGC

Z9-16:Ald/Z9-14:Ald

Z9-16:Ald
Z9-14:Ald

Z9-14:Ald/Z11-16:Ac

Z11-16:Ald

(b)

ATTRACTION

(c)

ARRESTMENT

Figure 9.4 Male moths use a combinatorial process to make split-second orientation responses to the temporal structure of a pheromone plume and the chemical content of individual filaments in the plume (Chapter 10). (a) The moth *Helicoverpa zea* uses a multicomponent pheromone, with the major component Z11–16:Ald detected by the black olfactory sensory neuron (OSN). The secondary component, Z9–16:Ald stimulates the gray OSN (as does a non-pheromone molecule Z9–14:Ald). These sensilla also house an OSN sensitive to molecules that are important secondary components (Z9–14:Ald, Z11–16:Ac, Z11–16:OH) of pheromones used by related moth species that live in the same region of North America. (b) If the antennal OSNs for its own two pheromone components (shown as black and gray) are both stimulated (*) then the combinatorial outputs from their respective glomeruli in the macroglomerular complex (MGC) of his antennal lobe cause the male moth to fly forward ("ATTRACTION"). (c) If the specialized OSNs sensitive to the minor, characteristic antagonistic pheromone components (white circles) of sympatric species are stimulated (*) along with the black and gray OSNs then the combinatorial message is "ARRESTMENT" (stop) as this indicates the pheromone is from the wrong species. The *H. zea* males will turn back even though *H. zea* shares other pheromone components (black and gray). See Appendix for the naming of pheromones. Figure adapted from Vickers *et al.* (1998).

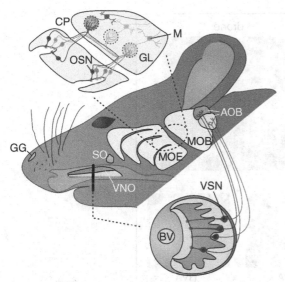

Figure 9.5 The mouse has parallel processing of overlapping sets of social chemosignals by its main and accessory olfactory systems. Schematic diagram showing the organization of the rodent nose into chemosensory subsystems with the main olfactory epithelium (MOE), vomeronasal organ (VNO), septal organ (SO), and Grueneberg ganglion (GG). Olfactory sensory neurons (OSN) in the MOE project through the cribiform plate (CP) to glomeruli (GL) in the main olfactory bulb (MOB). Olfactory sensory neurons expressing the same olfactory receptor genes converge on single glomeruli in the MOB. Main olfactory bulb mitral cells (M) extend an apical dendrite into adjacent GL and extend axons to the forebrain. Vomeronasal sensory neurons (VSNs) extend axons from the VNO to the accessory olfactory bulb (AOB). Vomeronasal sensory neurons in the apical zone (darker gray, left) of the VNO project to the rostral AOB; VSNs in the basal zone (lighter gray, right) of the VNO project to the caudal AOB. The trigeminal system is not shown. BV: blood vessel. Figure and caption adapted from Spehr et al. (2006a) with caption additions from Baum and Kelliher (2009).

detected by narrowly tuned vomeronasal sensory neurons (Leinders-Zufall et al. 2000). These compounds act together synergistically so presumably the outputs from the glomeruli in the accessory olfactory bulb (AOB) are integrated combinatorially, together with responses to other urine volatiles, which are also required. Studies of urine stimulation of VNO slices suggest that individual signature mixture information,

even allowing distinction of littermates, is encoded by the combinatorial activation of VNO neurons, presumably by very complex mixtures including other molecules as well as peptides (He et al. 2008). The vomeronasal sensory neurons (VSNs) detecting peptides show combinatorial activation with overlapping specificities (Chamero et al. 2012; Leinders-Zufall et al. 2009) .

9.1.2.1 The significance of the insect macroglomerular complex

Model systems are powerful ways to study phenomena but they may not be typical of all pheromone processing. For example, studies of male moths and their response to female sex pheromone show specialized receptors for pheromone and a dedicated brain area, the macroglomerular complex (MGC), for pheromone processing (Haupt et al. 2010; Martin et al. 2011a). However, other pheromone processing in insects may be by less specific receptors, without dedicated glomeruli in the brain. Narrowly tuned specialized receptors and dedicated glomeruli are not a pre-requisite for pheromone communication.

If sensitivity is important, there is selection for multiplication of OSNs for those molecules leading to a larger glomerulus. Male moths detect females from some distance and compete to be the first to arrive (Chapter 3). Males of many other insects with similar scramble competition for mates do have MGCs for the perception of female sex pheromones, for example, male honeybees (Figure 9.6) and male American cockroaches. The selective pressure for signal sensitivity on males in scramble competition leads to enormous numbers of OSNs for each of the pheromone components (Box 9.2). A large glomerulus may simply reflect the large number of OSNs carrying a particular receptor converging on it. *Drosophila* males do have three glomeruli that are larger than in the female but the size difference is not great (Dickson 2008), which probably reflects the close range that chemical interactions occur (large numbers of sensilla are not needed) (Section 9.5.2).

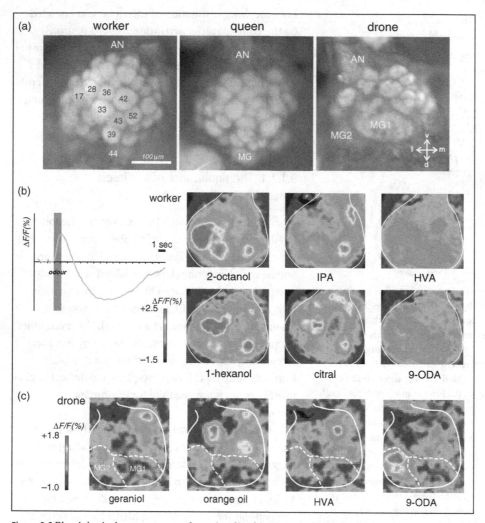

Figure 9.6 Physiological responses to odors visualized in the antennal lobe of honeybees.

(a) Anatomy of the antennal lobe in the three honeybee castes (worker, queen, and drone (male)). The lobes are shown in frontal view, in the position in which they can be accessed during calcium imaging recordings. About 25% of the antennal lobe can be seen (30 to 40 glomeruli in workers). AN, antennal nerve; MG, macroglomerulus; v, ventral; l, lateral; m, medial; d, dorsal. Numbers in workers refer to identified glomeruli of the T1 OSN tract.

(b) Worker bee. Calcium imaging recording of the antennal lobe of a worker bee. Upper left: odor delivery stimulates a biphasic fluorescence signal in active glomeruli, with a first fast positive component (max after ~1 second), followed by a slow – highly spatially correlated – negative component (minimum after 8 to 10 seconds). Right: odor activity maps, showing the amplitude of the biphasic signal for each pixel in a false-color code.

General odors (1-hexanol and 2-octanol) and social pheromone molecules (isopentyl acetate (IPA) and citral) elicit combinatorial activity in the imaged glomeruli. Note that the glomeruli activated by the pheromones may also be activated by general odors and vice versa. By contrast, no clear signals appeared with components of the queen mandibular pheromone (QMP): (E)-9-oxodec-2-enoic acid (9-ODA) and 4-hydroxy-3-methoxyphenylethanol (HVA). The glomeruli responsible for processing of these signals may be in other – as yet unimaged – parts of the antennal lobe.

A non-sex pheromone macroglomerulus is found in large workers of the leaf-cutter ant, *Atta vollenweideri*, which processes the releaser component of its trail pheromone (Kleineidam & Rossler 2009; Kuebler *et al.* 2010).

Other insects that use pheromones for communication do not seem to have "labeled lines" and highly specialized receptors leading to an MGC (Galizia & Rössler 2010). For example, most non-sex pheromones in honeybees and ants stimulate more general receptors and thus a number of glomeruli (which might also respond to floral or other odors). The activation of these glomeruli (visualized using calcium and other imaging) and activation of projection neurons (by calcium or direct recording) in response to pheromones is repeatable, for alarm pheromone in ants (Yamagata *et al.* 2007) and in honeybees (Sandoz *et al.* 2007) (Figure 9.6). Presumably, it is the integration by the projection inter-neurons that leads to the appropriate characteristic and stereotyped response to alarm pheromone. In one ant species, *Camponotus obscuripes*, the alarm pheromone seems to be processed in a particular group of glomeruli (Yamagata *et al.* 2006). However, in a related ant species, *Camponotus floridanus*, Zube *et al.* (2008) observed that the patterns of glomeruli activated by pheromonal and non-pheromonal odors were partly overlapping, indicating that processing of these odor classes is not spatially segregated within the antennal lobe. In honeybees the processing of alarm pheromone seems to be by non-specific glomeruli, in a similar and overlapping pattern with the processing of floral scents (Figure 9.6) (Wang *et al.* 2008b).

9.1.3 Olfactory receptor circuits to the higher brain

In both insects and vertebrates, the glomeruli form the first relay in the olfactory circuits, where the OSNs interact with the secondary neurons linking them, which in turn sends their axons to the next relay levels, in the primary olfactory cortex of vertebrates and protocerebrum of insects (Figure 9.3) (Martin *et al.* 2011a; Mori & Sakano 2011; Su *et al.* 2009; Touhara & Vosshall 2009). In vertebrates, the OSNs synapse in glomeruli with secondary neurons called mitral/tufted (M/T) cells, which project their axons to the olfactory cortex. In insects, the analogous secondary neurons synapsing in the glomeruli are the projection neurons, which project their axons to the higher brain regions of the mushroom body and the lateral horn of the protocerebrum (Carey & Carlson 2011; Martin *et al.* 2011a; Sandoz 2011). Projection neurons from glomeruli stimulated by general odors such as food seem to go to a different part of the lateral horn (lateral protocerebrum) from the sexually dimorphic targets of glomeruli processing pheromones (Jefferis *et al.* 2007).

Lateral interactions in the OB transform the precise array of OSN inputs: local interneurons interact with OSNs and output neurons (M/T) and projection neurons in vertebrates and insects respectively to modulate olfactory signals (reviewed in Wilson & Mainen 2006). In the mammalian OB these local inter-neurons include juxtaglomerular cells working within a glomerulus and granule cells forming inhibitory synapses on M/T cells in multiple glomeruli and mediating information transfer between glomeruli.

Figure 9.6 (cont.)

(c) Drone. Calcium imaging recordings of antennal lobe activity in a drone (male) bee. The odor activity maps are calculated as in (b). The position of the two accessible macroglomeruli is overlaid on the maps (white). General odors (here a complex blend, orange essential oil) and social pheromones (here geraniol) induce activity in ordinary sized glomeruli, i.e., on the medio-ventral side of the antennal lobe. The major component of the queen mandibular pheromone, 9-ODA, which is involved in the attraction of males toward queens during nuptial flight, is specifically detected by the largest macroglomerulus of the drone antennal lobe, MG2 (bottom left). By contrast, HVA, another QMP component whose role has only been shown in workers, induces activity in an ordinary-size glomerulus. Figure and caption adapted from Sandoz *et al.* (2007). For more information about QMP see text and Box 6.3.

Box 9.2 Functional design of olfactory organs: flow and scale

In most animals, olfactory sense cells are concentrated in organs such as antennae or nasal cavities, specialized for detecting chemical signals. Some animals also have chemosensory organs on their feet or ovipositors. Some get greater spatial separation between the sensory cells by having long antennae or widened heads like hammer headed sharks (Abel *et al.* 2010; Gardiner & Atema 2010) (other animals move their heads from side to side to achieve the same effect (Chapter 10). The designs of olfactory organs are, like other biological structures, an evolved compromise between conflicting selective forces. Olfactory organs must first, maximize exposure of the chemosensory neurons to the outside environment while protecting these relatively delicate cells; and second, maximize the effective capture of odor molecules within the biomechanical constraints of size, flow, and fluid dynamics (Atema 2012). Insect sensory neurons are protected inside cuticular "hairs" or sensilla and bathed in sensillum lymph (Figure 9.1).Terrestrial vertebrates enclose their delicate olfactory tissue in the nasal cavity, protected by a continuously secreted layer of mucus (Pelosi 1996). Scramble competition to be the first male to detect and locate an upwind female releasing pheromone has been a powerful selection pressure on male moths (Chapters 3 and 10)(Symonds *et al.* 2012). Such selection has resulted in antennae exquisitely sensitive to the specific pheromone and able to detect picogram quantities (de Bruyne & Baker 2008; Ishida & Leal 2005). A stimulus pulse delivering as little as one molecule per pore tubule is a very strong stimulus for a pheromone olfactory sensory neuron. The great sensitivity of male insects comes from the specificity of the olfactory receptor protein and the pheromone binding protein (present in large quantities) combined with the design of the antennae, which carry thousands of the specialized OSNs. For example the male silk moth, *Bombyx mori*, has some 17,000 hairs on each antenna, each containing the ends of two OSNs, one tuned to the sex pheromone bombykol (Kaissling 2009a) (see figure opposite). The combined area of the antennae is more than 170 mm^2. Radioactive-label studies suggest that most pheromone molecules from the airstream passing the cross-sectional area of the antennae are adsorbed, and that 80% of these are caught by the long hairs of the sensilla hairs (Kanaujia & Kaissling 1985).

However, scale is very important: structures this small are affected by the treacle-like viscous characteristics of airflow at these low Reynolds numbers (*Re*, a term used to describe characteristics of flow speed, viscosity, size, and shape) (see Box 10.2) (Koehl 2006). Direct measures of the giant silk moth, *Actias luna*, showed that only 8 to 18% of the air directly upstream passed through the antenna, while the rest went around (Vogel 1983) (in contrast to air, some 38% of a light beam passed through). More, finer, hairs increase the probability of pheromone molecules hitting the antenna but lead to less air flowing through. The *Re* of the fine hairs themselves is around 0.2. At these *Re* values, rows of hairs tend to function as paddles rather than as sieves, which helps to explain why so much of the air passes around the antenna (Koehl 2006). Close to the antennal surface there is an almost stationary boundary layer (Chapter 10); the lower the Reynolds number the thicker it is. The shape of the antenna will affect the thickness of the

Box 9.2 (cont.)

boundary layer, in turn affecting the perceived signal, including the relative importance of diffusion effects on pheromone encounters of the sensilla hairs (Koehl 2006) (see Box 9.3 for sampling by flicks of the antenna).

Vertebrates, such as dogs, which rely on their sense of smell for prey detection, increase their sensitivity to odors by having multiple copies of generalist olfactory sensory cells, in a large area of olfactory epithelium (~30,000 mm^2 compared with the olfactory area in humans of about 400 mm^2). Dogs have stable concentration thresholds of 1 to 2 parts per trillion (ppt), about 10,000- to 100,000-fold lower than for humans at parts per billion (ppb) for the same molecule n-amyl acetate (Walker *et al.* 2003, 2006) (see Box 9.3 for how this is achieved).

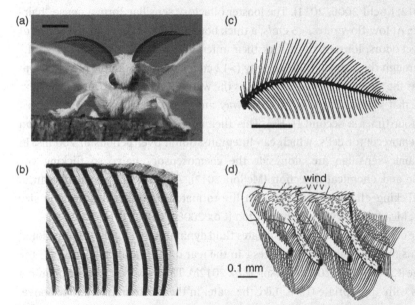

(a) Male silk moth, *Bombyx mori* (scale 2 mm). (b) Close-up of a male *B. mori*'s antennae showing the numerous side branches and sensillary hairs of the antennae, which form an effective olfactory sieve for molecules of the pheromone, bombykol. (c) The male antenna of the silk moth, *Antheraea polyphemus* (scale 5 mm). (d) Two segments of a male antenna of *A. polyphemus*. The long sensilla hairs are each innervated by one to three pheromone-sensitive cells and are regularly arranged to form an "odor sieve." Photographs (a) and (b) by R. A. Steinbrecht, (c) from Keil (1992). Antenna drawing (d) from Kaissling (1987).

The OB also receives top-down inputs: first, from reciprocal projections from most of the higher brain areas including the olfactory cortex, amygdala, and hippocampal formation, and second, from centrifugal ascending neurons releasing neuromodulators (norepinephrine, acetylcholine, and serotonin), for example from vaginocervical stimulation at giving birth (Section 9.8.1) (Restrepo *et al.* 2009; Wilson & Mainen 2006). (In moths, there is evidence of analogous effects in the antennal lobe; Section 9.6). The

Box 9.3 Sniff and flick: the behavior of smelling

For many animals, olfaction involves active sampling of the environment: "In the realm of odors, scents, perfumes, effluvia, stenches, and vaporous aphrodisiacs, mammals sniff, lobsters, insects, and snakes flick" (Dethier 1987). Sniffs, flicks, and pulses may be essential for animals to sense the world: with a continuous stimulus, sensory cells adapt and stop signaling, the central nervous system habituates, both responses being ways of reducing useless or irrelevant information (Dethier 1987). Interrupted sampling allows the sense cells to sense. Increasing the speed of the medium, air or water, during the sniff will also reduce the thickness of the boundary layer over the olfactory sensory cells (Chapter 10), speeding access of chemical signals.

The boundary layer is especially important for aquatic animals and flicking provides a solution (Chapter 10) (Atema 2012; Koehl 2006, 2011). The lobster olfactory sensillae form a dense "hairbrush" on the antennule. At low flow rates, <5 cm/s, a thick boundary layer prevents the access of odor molecules. To detect odors, lobsters must flick their antennules, effectively taking a "sniff." Flicking drives water through the hairs at high velocity (>12 cm/s), taking a sample of water kept with the sensory hairs as the slower return stroke retains the water (Koehl 2006). During the next rapid flick downstroke, that water sample is flushed away and replaced by a new one. Excited lobsters flick at up to four times a second (4 Hz). This flicking rate matches the flicker fusion frequency of lobster chemoreceptor cells, which can integrate stimuli over periods of 200 ms. In crustaceans, hydrodynamic sensillae are alongside the chemosensory hairs, so flicking coordinates hydrodynamic and chemical detection (Mellon 2012). Another marine crustacean, a stomatopod, alters its flicking characteristics through life to match its ten-fold growth in size (Mead & Caldwell 2011; Mead et al. 1999). Fish sniff too (Cox 2008; DeBose & Nevitt 2008).

When a dog sniffs, the design of its nasal cavity creates fluid dynamics that send a unidirectional airflow carrying odorants to a special olfactory recess, in the rear of the nasal cavity and off the main respiratory passage (Craven et al. 2010; Lawson et al. 2012b). The flow may sort the odorants too, depositing in zones with appropriate OSNs. Like the water in the lobster return stroke above, the air in the olfactory recess is undisturbed during the expiration cycle, leaving quiescent scent-laden air there, providing an additional residence time for enhanced odorant absorption. Other species with a specially developed sense of smell, including carnivores, rodents, ungulates, and marsupials appear to have a similar nasal design with an olfactory recess.

In mammals including humans, sniffing also stimulates patterns of neural activity throughout the olfactory system (Section 9.1.4) (Kepecs et al. 2006; Mainland & Sobel 2006; Wachowiak 2010).

effect of these top-down inputs means that mitral cell outputs are influenced by learning, behavior and context (Restrepo et al. 2009). Mammals also regulate input to the olfactory bulb by modifying their sniffing behavior (Box 9.3).

In both mice and *Drosophila* there are stereotypical patterns of links from the glomeruli up to the higher levels of the brain (Mori & Sakano 2011; Su et al. 2009; Touhara & Vosshall 2009). In *Drosophlia*, these are to the mushroom body and lateral horn of the protocerebrum; general odors such as food seem to go to a different part of the lateral horn from pheromones (Section 9.5.2) (Jefferis *et al.* 2007). In mammals, M/T cells send their axons via the lateral olfactory tract to particular parts of the olfactory cortex, thus transferring the olfactory stimuli from the glomeruli to higher parts of the brain. However, it seems that topography is not a prominent feature of projections from the olfactory bulb to higher brain areas in the mouse (Friedrich 2011). Projections to two cortical targets, the anterior olfactory nucleus (AON) and the cortical amygdala, are topographically organized at coarse, but not fine, spatial scales. No topography was found at all in projections to the piriform cortex, the largest target area and a zone important for learning and associations (Friedrich 2011). For processing and learning of signature mixtures see Section 9.8.

In fish, groups of glomeruli in particular parts of the olfactory bulb respond to different types of odors. Their mitral cells project to higher brain areas by different tracts: sex pheromones by the lateral medial olfactory tract (lMOT), alarm cues by the medial MOT (mMOT), and glomeruli stimulated by food odors via the lateral olfactory tract (LOT) (Hamdani & Døving 2007; Mathuru *et al.* 2012). The species specificity of response to sex pheromones suggests that fish are responding to a number of molecules (Levesque *et al.* 2011; Lim & Sorensen 2012), implying a combinatorial response, as in male moths.

In mammals, as in the *Drosophila* and fish examples above, within the main olfactory system there may be some kind of separation between general olfaction and responses to pheromones and other pre-disposed chemical stimuli that elicit an "innate" response (see Section 9.6) (Mori & Sakano 2011). For example, the response to urine volatiles in mice was localized to mitral cells in two clusters of OB glomeruli and within these zones there were mitral cells specifically responding to the male pheromone (methyl-thio)methanethiol (MTMT) (Lin *et al.* 2005). These specific responses may work in parallel with more general odor processing across many glomeruli (Lin *et al.* 2006).

An indication of how innate responses to particular molecules of importance might be processed in mammals comes from a study of the innate response to aversive, non-pheromone, molecules in the mouse (Mori & Sakano 2011). The dorsal zone of OSNs in the main olfactory epithelium (MOE) and their associated glomeruli, which include the glomeruli that process the predator odor trimethyl-thiazoline, can be ablated by selective expression of diphtheria toxin (Kobayakawa *et al.* 2007). This seemed to abolish the innate fear response to predator odor and the innate aversion to the odors of spoiled food, yet left the rest of the main olfactory system sufficiently intact for olfactory learning of general odors (Kobayakawa *et al.* 2007). These dorsal domains also include glomeruli that respond to the small molecule mouse pheromones such as 2-*sec*-butyl-4,5-dihydrothiazole and dehydro *exo*-brevicomin (Mori & Sakano 2011).

9.1.4 Spatiotemporal coding in brains

The activation of unique patterns of glomeruli by different odorants has provided strong evidence for the spatial arrangement of olfaction processing in the olfactory lobe. However, different odorants are coded not just by which output neurons fire but also the pattern of firing of those neurons in relation to others over millisecond periods during and after the stimulus (Bathellier *et al.* 2010; Kay *et al.* 2009; Martin *et al.* 2011a; Wilson & Mainen 2006). Among the best studied in vertebrates are gamma oscillations at ~70 Hz in rats and mice, evoked by olfactory stimulation and initiated at the end of the inhalation cycle or "sniff" (Box 9.3) (Kay *et al.* 2009; Wachowiak 2010). Homologous and analogous oscillations of neural activity (at a frequency of 10 to 30 Hz) have been recorded from olfactory systems in many other vertebrates including bony fishes, frogs, turtles, rabbits, cats, and humans as well as insects and mollusks. In insects, and perhaps other organisms, the synchronization of the firing seems to be an important feature (Martin *et al.* 2011a). The ubiquity of fast oscillations in olfactory systems suggests that they may be important in the higher neuronal processing of odor stimuli.

9.2 Evolution of chemoreceptors

The great flexibility of the olfactory system means that any kind of chemical can potentially evolve into a pheromone. A wide range of relatively non-specific chemosensory receptor proteins, each expressed on different chemosensory neurons linked to glomeruli (Section 9.1), gives evolutionary flexibility to olfaction and to the evolution of chemical signals (pheromones) (Chapter 1). Because of its wide range of broadly tuned receptors, the olfactory system can respond to the widest possible range of odors, even as entirely new compounds are created (in evolving biological systems such as the rise of flowering plants; Bargmann 2006b). Although a given odorant may be unlikely to fit any one receptor perfectly, it is likely to stimulate some. This allows the olfactory system to track a moving world of cues generated by other organisms, and under natural selection, test, refine, or discard variant receptor genes and coding strategies over evolutionary time (Bargmann 2006b; Bendesky & Bargmann 2011). Animals can thus evolve the chemosensory sensitivities needed to exploit new habitats and new feeding niches.

This is also the mechanism for the evolution of pheromone signals from molecules that turn out to have signal value to receivers (and senders), for example sex hormones "leaking out" of females and detected by male goldfish have become pheromones (Chapter 1). If these lead to greater reproductive success or survival, natural selection can work on improving selectivity and sensitivity in the receiver, and greater production in the sender. The result is the wonderfully diverse range of compounds used as pheromones by organisms across the animal kingdom, a chemical diversity that can seem weird and haphazard at first sight (Chapter 1).

The combinatorial flexibility of olfaction may be the reason that the great majority of pheromones are perceived by glomerular systems (Section 9.1). The glomerular organization may facilitate less catastrophic changes in receptors, allowing diversification (Section 9.2.4). Evolution of gustatory receptors of course also occurs, though their variety and number in vertebrates is much lower than ORs and vomeronasal receptors (VRs) perhaps because their neurons lack a glomerular organization.

The evolution of animal chemoreception raises at least five questions: first, which receptors become co-opted as chemosensory receptors for olfaction or gustation; second, how are such large numbers of different olfactory receptor genes generated (forming 1% of the human genome for example); third, how do changes in receptors' binding sites change their response to odorants; fourth, how might glomerular organization facilitate the evolution of new receptors, particularly in vertebrates; and lastly, how do individuals vary in their chemosensory receptors, and thus the molecules they perceive?

9.2.1 Co-option of receptors for chemoreception

A characteristic of chemoreception is the opportunistic co-option of cell surface receptors of various kinds, whether originally functioning for example as a receptor for hormones or neurotransmitters, so long as it is in the cell membrane and can interact with external odorants important for that animal taxon. For example, vertebrate formyl peptide receptors in the VNO may have been co-opted from the immune system (Liberles *et al.* 2009; Rivière *et al.* 2009).

The opportunistic nature of chemoreception explains how the chemosensory system relies on completely independent families of receptor genes in different animal taxa. The chemoreceptors of vertebrates and insects seem to have evolved completely independently (Section 9.3). Vertebrates and insects each have a variety of receptor types, some of which are unrelated to the others. Nematodes show a similar independently evolved range of receptors including diverse GPCRs and non-GPCRs such as guanylate cyclases (Bargmann 2006a,b).

9.2.2 Birth and death of chemoreceptor genes

The number of chemosensory receptor genes varies extensively among different vertebrate or insect species, owing to repeated gene duplication and deletion or inactivation as pseudogenes (Nei *et al.* 2008). The process is "birth-and-death" evolution. Most multigene families are subject to this mode of evolution, but vertebrate OR and VR genes represent extreme cases with dramatic gene duplication and diversification.

There was an enormous expansion of OR genes in terrestrial vertebrates when they evolved onto land from our common ancestor with fish (Nei *et al.* 2008). Fish have around 100 OR genes but that is much fewer than the 400 to 2,100 OR genes in tetrapod vertebrates including amphibians, birds, and mammals. The expansion in terrestrial vertebrates for ORs to detect airborne odorants was from just two of the nine groups of OR genes that our common ancestor with fish had. The OR gene repertoire in birds is similar to that in mammals (Steiger *et al.* 2008b).

There is less variation in the number of olfactory or taste receptor genes in insects than in vertebrates (Nei *et al.* 2008). One of the exceptions is the honeybee, which has 163 intact OR genes but just 10 gustatory receptor (GR) genes (compared with *Drosophila*'s 60 OR and 68 GR genes) (Robertson & Wanner 2006). Almost all of the expansion in honeybee ORs has been in one honeybee-specific subfamily of ORs and the expansion may reflect the importance of pheromones for social organization, recognition of colony signature mixtures, and diverse floral odors when foraging (Chapter 6) (Sections 9.7.3 and 9.8).

The "birth" of new genes comes largely from unequal crossing over, generating tandem gene duplications of chemosensory genes (Nei *et al.* 2008). This could be the principal mechanism for receptor repertoire expansion as there is an element of positive feedback to this process: duplicated genes increase the likelihood of additional allelic mispairings and further gene duplication events (Ramdya & Benton 2010). Duplicated chemoreceptor genes may later get separated within the genome. Chemosensory gene losses occur by deletion or from pseudogenization by nonsense or frameshift mutations.

New receptors are likely to be closely related, if not identical, in amino acid sequence to their parental receptors and often – though not necessarily – in their expression pattern (Ramdya & Benton 2010). Commonly one copy may degenerate, indicated by the presence of many pseudogenes in tandem arrays of recently duplicated OR genes. In other cases a duplicated gene, free of selective pressures to maintain a redundant function, may mutate and acquire new properties. If advantageous, perhaps by providing new odorant detection capabilities, this receptor gene could be subject to unique selective pressures and maintained in the genome (Nei *et al.* 2008; Ramdya & Benton 2010). Evidence for functional divergence of recently duplicated receptors comes from comparative studies in drosophilids (Ramdya & Benton 2010).

Genomic drift from almost random gene-duplication and inactivation events plays a significant role in creating the diversity of the OR system, on which natural selection might act (Nei *et al.* 2008). The importance of non-random changes to the genome is shown by a detailed comparison of the functional OR genes in three independent lineages of aquatic and semi-aquatic mammals including whales, manatees, and carnivores such as otters, compared to terrestrial mammals (Hayden *et al.* 2010). All three aquatic and semi-aquatic lineages showed OR gene losses, yet they have convergently and selectively retained similar functional OR families. As it was possible to assign the animals to their correct ecotype (aquatic for example) based on their functional OR gene repertoire rather than phylogenetic relatedness, natural selection on ecological niche seemed to shape their OR gene repertoire more than shared evolutionary history and chance (Hayden *et al.* 2010). Adaptive changes fostered by environmental shifts are similarly seen in the evolution of chemosensory receptor families in insects and are likely to involve reproductive, ecological, or behavioral traits (Sanchez-Gracia *et al.* 2009).

An animal's ecology affects its chemoreceptor repertoire. *Drosophila* species, such as *D. sechellia*, which have specialized on one species of fruit show an increased rate of OR gene loss relative to other drosophilids, suggesting that *D. sechellia* has lost sensory inputs no longer relevant to its restricted food-seeking behaviors (see Hansson & Stensmyr 2011; Ramdya & Benton 2010). In the nematode *C. elegans*, culturing at high densities in the laboratory selected for deletions that disrupted two pheromone receptor genes, making the worms insensitive to the pheromone (Section 9.7.2) (McGrath *et al.* 2011). Both pheromone receptor genes were disabled independently in two strains grown at a high density. A similar pheromone receptor gene was disabled in a different nematode species, *C. briggsae*, grown under these conditions. However, the suggestion that the high rate (~50%) of OR pseudogenes in humans stems from the lesser importance of olfaction after the acquisition of trichromatic color vision is not supported by more recent primate genome comparisons (Chapter 13) (Matsui *et al.* 2010).

9.2.3 Chemosensory receptor binding sites and response

Characteristics of binding sites are emerging for vertebrate ORs, which are seven transmembrane-spanning GPCRs (though less is known about insect ORs and IRs) (Ramdya & Benton 2010; Silbering *et al.* 2011). As might be expected for a family of proteins that binds such diverse ligands (odorants), different GPCR ORs differ greatly in sequences of the transmembrane helices that come together to form a barrel-shaped odorant-binding site (Figure 9.2), in particular on the outward facing side (Kato & Touhara 2009; Reisert & Restrepo 2009). Data mining of OR sequences between and within species, molecular modeling, site-directed mutagenesis, and heterologous expression (in another species) of ORs are used to study the binding site (Malnic *et al.* 2010; Nei *et al.* 2008; Reisert & Restrepo 2009). Such studies have revealed, for example, that the majority of onefamily of vertebrate GPCRs share a similar ligand-binding site located deep in four transmembrane regions (Kato & Touhara 2009). Studies of a number of invertebrate and vertebrate ORs show that single-point mutations can change the ligand specificity (Kato & Touhara 2009; Leary *et al.* 2012; Reisert & Restrepo 2009) (as also shown in the androstenone OR, Chapter 13). For aquatic organisms, these sensitive binding interactions are likely to be adversely affected by increasing acidity resulting from rising CO_2 levels (Chapter 1).

9.2.4 Glomerular development and receptor evolution

How does the animal olfactory bulb or antennal lobe adapt to the birth and death of olfactory receptors at such a fast rate compared with other evolutionary rates? New receptors will need new glomeruli (Section 9.1). The glomerular organization and how it develops in the embryo provide the answer, at least in mammals.

In mammals, ORs themselves help guide axons to their glomeruli. This provides a neat mechanism for incorporating new ORs as well as solving the developmental problem of wiring up several millions of OSNs, each expressing one of up to 1,000 ORs, to up to 2,000 glomeruli in the mammal olfactory bulb (Imai *et al.* 2010; Malnic *et al.* 2010; Mombaerts 2006; Mori & Sakano 2011; Zou *et al.* 2009). This mechanism simplifies the assignment of glomeruli if a new OR gene emerges (Mombaerts 2006). If the new OR is substantially different from any other OR in the repertoire, the glomerulus for the new OR is likely to be novel and must be squeezed somewhere within the glomerular array. Even slight variations in the amino acid sequence of an OR can result in new glomeruli (Zou *et al.* 2009). As OR genes become pseudogenized over evolutionary time, and thus not expressed, their glomeruli will no longer form. In this way, the olfactory system can adapt to changes in the OR gene repertoire whether small, or as large as the OR expansion when tetrapods came on to land.

The stochastic (chance) mechanisms used in mammal OSNs for choosing which single OR will be expressed also facilitate the incorporation of new ORs as these, like long existing OR, also have a chance of being expressed. Both the guidance and stochastic mechanisms contribute to the "evolvability" of the mammalian olfactory system, the ability to respond to selective pressures with adaptive genetic changes (Bendesky & Bargmann 2011). In the adult mammal, OSNs are continuously replaced in the olfactory epithelium and new axons use these or similar mechanisms to find the proper glomerulus and synaptic targets (Whitman & Greer 2009).

Insects do it differently. The insect antennal lobe is patterned by hard-wired genetically determined mechanisms and the ORs themselves have no influence over map formation (Imai et al. 2010; Ramdya & Benton 2010). In Drosophila individual sensilla develop under precise genetic control resulting in an almost completely stereotypical number of OSNs of defined OR identity (not stochastically) in a characteristic position on the antennal surface. The prescription continues in the antennal lobe with the ~50 Drosophila glomeruli having stereotyped shapes, sizes, and positions (a great advantage for scientists studying the circuits).

Are the deterministic receptor gene choice and prescriptive wiring development among the reasons that insects have so few different chemoreceptors, compared with the thousands in mammals? Insects may need changes in their genetic control mechanisms to expose new ORs. Some of the ways new neural circuits might evolve in response to or incorporate inputs from new receptors are explored by Ramdya and Benton (2010). Conceptualizing the integration of novel glomeruli into circuits beyond the antennal lobe is currently difficult.

9.2.5 Chemoreceptor variation between individuals: polymorphisms

While most accounts of chemoreception in this chapter downplay individual variation in response, chemoreceptor genes vary between individuals as much, if not more, than other parts of the genome. This is the variation that natural selection works on. For example, genomic analysis of humans has shown that each of us smells a slightly different world depending on which functioning ORs and variant OR alleles we have (Chapter 13) (Krautwurst 2008; Olender et al. 2012). Human OR genes carry more than 6,000 genetic variations, including single nucleotide polymorphisms (SNPs) and/or copy number variants (duplication and deletions of OR genes) in 413 OR genes (Olender et al. 2012). These can either turn the gene into a non-functioning pseudogene or change its sensitivity or interaction with odorants. Each of us has different combinations of these alleles. If we lose a functioning receptor that has a narrow specificity, a "specific anosmia" can result (Chapter 13). Differences in human sensitivity have been linked to OR amino acid sequence variations for the high-affinity receptor for isovaleric acid and the receptor for androstenone (Chapter 13) (Keller et al. 2007; Menashe et al. 2007). Mouse VRs are also likely to be highly variable between individuals (Wynn et al. 2012).

Natural polymorphisms in ORs have been revealed within a population of wild Drosophila melanogaster (Richgels & Rollmann 2012; Rollmann et al. 2010). The OR genes contained multiple SNPs that were associated with individual variation in behavioral responses to one or both of two odorants. Many of the single nucleotide polymorphisms (SNPs) associated with variation in behavioral responses were in non-coding regions, which could change gene expression levels or mRNA stability. This is a reminder that mutations affecting olfaction are not limited to OR proteins themselves.

However, the combinatorial nature of smell and broad tuning of most receptors means that the mutation of a single OR might have a small impact on overall perception and behavior (Ramdya & Benton 2010; Reisert & Restrepo 2009), as shown for some ORs in Drosophila (Keller & Vosshall 2007). If other receptors bind the odorant with an affinity similar to the mutated OR then the effect might not be noticeable (Chapter 13) (Reisert & Restrepo 2009). The way combinatorial processing can damp the effect of mutating an individual olfactory receptor or changing an inter-neuronal circuit could facilitate receptor diversification by making the effects of changes less dramatic.

There is some individual variation in male moth pheromone receptors, which allows speciation to occur if the females' pheromone blend changes (Chapter 3), but the variation is constrained by stabilizing selection as males with unresponsive receptors are strongly selected against.

9.3 The many chemosensory subsystems in mammals and insects

This section describes the detailed organization and variety of chemosensory subsystems used by mammals and insects. You can skip this section if you wish and return to it for reference. For integration of the mammal subsystems, see Section 9.4; for sexual circuits in insects and mammals see Section 9.5; and for more on primer effects, mediated by these subsystems, see Section 9.7.

In a given animal type, these chemosensory systems often differ in body location, receptor types, and the targets of their neurons within the central nervous system (Su et al. 2009; Touhara & Vosshall 2009). Animals typically integrate information from the different subsystems.

9.3.1 Mammal chemosensory subsystems

The mammalian chemosensory system is composed of multiple subsystems, some well known and others only recently characterized (Figure 9.5) (Chamero et al. 2012; Ma 2010; Mucignat-Caretta et al. 2012; Tirindelli et al. 2009). In mammals, the principal subsystems are the main olfactory system (MOS) and, in those that have it, the accessory olfactory system (AOS). Depending on the species, pheromones and

individual recognition cues (signature mixtures) may be detected by the MOS or the AOS, or both; the integration of the inputs from both systems higher in the brain is also now well established (Section 9.4). Both systems are highly sensitive, with thresholds in the sub-nanomolar range (Spehr et al. 2006b). Across the mammalian chemosensory system, five types of chemosensory GPCRs have been identified so far: odorant (olfactory) receptors (ORs) and trace amine-associated receptors (TAARs) expressed in the main olfactory epithelium, and in the vomeronasal organ, two distinct vomeronasal receptor families (V1Rs and V2Rs) and formyl peptide receptors (FPRs) (Kaupp 2010; Spehr & Munger 2009). Vertebrate gustatory receptors are also GPCRs.

There are similarities between both AOS and MOS in many of the neural mechanisms involved, such as adult neurogenesis (the generation of new neurons) (Lledo et al. 2006; Whitman & Greer 2009). The olfactory system is one of only two areas of the adult mammalian brain to exhibit continuing neurogenesis (the other is the dentate gyrus of the hippocampal formation). Neural stem cells located in the subventricular zone of the lateral ventricles give birth to neuroblasts that migrate along the rostral migratory stream toward the olfactory bulb. Once in the OBs, these new cells acquire their role, mainly as inhibitory inter-neurons (granule cells), either in the MOB or AOB, and participate in olfactory processing (Lledo et al. 2006; Whitman & Greer 2009). In the AOB, some of these new cells are involved in the formation of the female's memory of her mating partner's signature mixture in the Bruce effect (Section 9.9.3) (Oboti et al. 2011). In the olfactory epithelium, OSNs are continuously replaced and new axons must find the proper glomerulus and synaptic target (Section 9.2.4) (Whitman & Greer 2009).

9.3.1.1 Main olfactory system

The main olfactory system (MOS) consists of the main olfactory epithelium in the dorsal nasal cavity, containing olfactory sensory neurons (OSNs), which send their axons to the glomeruli of the main olfactory bulb (MOB), and the higher olfactory centers, which receive directly or indirectly information from the MOB (Figure 9.5) (Munger *et al.* 2009). The MOB's projections include the anterior medial amygdala where signals may be integrated with inputs from the vomeronasal system (Section 9.4). The odorant (or olfactory) receptor proteins (ORs) of vertebrates are encoded by a large multigene family, in rats and mice estimated to be about 1,000 genes or 1% of the genome, by far the largest family of genes in the genome of any species (Malnic *et al.* 2010; Zhang & Firestein 2002). Humans have about 370 functioning ORs, which nonetheless probably give us a similar odor or olfactory space to a mouse (Chapter 13) (Shepherd 2004, 2010). Aquatic vertebrates tend to have the more ancient class I ORs, generally tuned to water-soluble odorants, whereas terrestrial vertebrates have both these and class II receptors tuned to airborne hydrophobic odors (Section 9.2) (Malnic *et al.* 2010; Nei *et al.* 2008).

Some ORs in the MOE recognize pheromones. For example, small molecule pheromone components found in mouse urine (2,5-dimethylpyrazine, 2-*sec*-butyl-4,5-dihydrothiazole, dehydro-*exo*-brevicomin, and 6-hydroxy-6-methyl-3-heptanone, and 2-heptanone) stimulate the MOE at nanomolar concentrations and the likely OR (OR912–93) for one of them, 2-heptanone, has been identified in the mouse (Gaillard *et al.* 2002; Spehr *et al.* 2006a,b). Pheromone detection by the MOS is discussed further in Section 9.4. A minority of OSNs have different transduction molecules associated with their OR. Mouse pheromones of various kinds have been shown to stimulate both this minority and ORs with the majority transduction mechanism (see Figure 9.7) (Baum & Kelliher 2009; Tirindelli *et al.* 2009).

Pheromones might also be detected by a subset of OSNs in the main olfactory epithelium, which have a different kind of receptor from the others: their receptors are trace amine-associated receptors (TAARs) (Liberles & Buck 2006; Ma

2010). TAAR 5 is a receptor for the male mouse pheromone trimethylamine (Li *et al.* 2013). It is not known if TAARs share the same properties of glomerular convergence as ORs. Some TAARs might be sensitive to characteristic odors produced by predators (Ferrero *et al.* 2011) (see Section 11.1.2).

9.3.1.2 Accessory olfactory system

In addition to the MOS, many mammals and other terrestrial vertebrates, including amphibians and reptiles (though not birds), have an accessory olfactory system (AOS) consisting of the vomeronasal organ (VNO = Jacobson's organ), which has vomeronasal sensory neurons (VSNs), the accessory olfactory bulb (AOB) to which the VSNs project, and the higher olfactory centers that receive information from the AOB (Figure 9.5) (Brennan 2010; Chamero *et al.* 2012; Halpern & Martinez-Marcos 2003; Munger *et al.* 2009). The range of molecules detected by the AOS shows considerable overlap with those detected by the MOS: both detect volatile and non-volatile molecules (Section 9.4). The VNO responds to non-pheromonal molecules such as those from prey or predators as well as pheromones. Some mammals such as male antelopes, other ungulates, and felids such as lions and the domestic cat, sniff and lick the urine of estrous females, often showing a characteristic lip-curling "flehmen" behavior thought to aid access of chemo-signals to the VNO. Entry of large and small molecules may also be facilitated by pumping (Brennan 2010; Halpern & Martinez-Marcos 2003). In reptiles, such as snakes, the VNO openings are in the roof of the mouth and stimuli are delivered by the tips of the tongue (Chapter 10).

In mammals, the VNO has two distinct families of chemoreceptor GPCRs, vomeronasal receptors 1 and 2 (V1Rs and V2Rs), which evolved independently from the ORs of the main olfactory epithelium (and from each other) (Chamero *et al.* 2012; Tirindelli *et al.* 2009). There are fewer VRs in the mouse than ORs: 187 V1Rs and ~120 V2Rs. V1Rs are expressed by the VSNs in

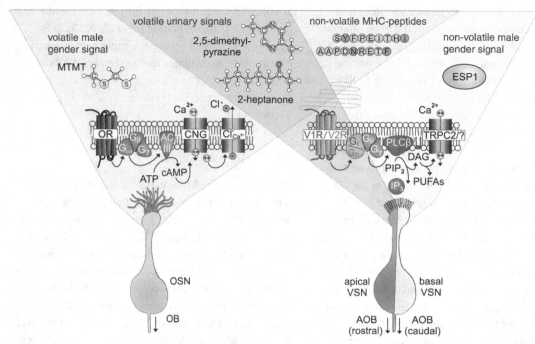

Figure 9.7 In mice, partially overlapping sets of social chemosignals are detected by the main and accessory olfactory systems. Just a few of the many molecules are shown here. Both volatile (e.g., 2-heptanone, 2,5-dimethylpyrazine) and non-volatile social signals (e.g., MHC peptides) are potent sensory stimuli for olfactory sensory neurons (OSNs) and vomerosensory neurons (VSNs). The diagram also shows two male-specific signals, a urinary volatile, (methylthio)methanethiol (MTMT), and a 7-kDa peptide named ESP1 (exocrine gland-secreting peptide 1) that is found in mouse tears and activates basal VSNs (Figure 9.12). So far as I know, the effects of these two molecules have not been compared directly in both systems.

OSNs and VSNs employ distinct signaling mechanisms with different ion channels. Knockouts of the ion channels CNGA and TRPC2, knocking out most of the MOE or VNO respectively, have been used to explore which olfactory subsystem is being stimulated by different molecules. However, the knockout effects may not be limited to the olfactory system (see Chapter 2 for more comment). In the main population of OSNs, olfactory receptors (ORs) as well as other canonical signaling proteins are densely packed in apical ciliary membranes. Ligand binding in ORs activates type III adenylate cyclase (AC III) via the G protein $G\alpha_{olf}$. In turn, an increase in ciliary cAMP opens a cyclic nucleotide-gated (CNG) channel consisting of three subunits (CNGA2, CNGA4, and CNGB1b). Cation influx and successive activation of Ca^{2+}-gated Cl^- channels (Cl^-Ca^{2+}) result in a depolarizing receptor current.

By contrast, apical VSNs co-express members of a multigene GPCR family, the V1Rs, with $G\alpha_{i2}$, whereas unrelated V2Rs and $G\alpha_o$ are found in basal VSNs. Downstream products of phospholipase C (PLC) such as inositol 1,4,5-trisphosphate (IP_3), diacylglycerol (DAG), and polyunsaturated fatty acids (PUFAs) have all been implicated in VSN signaling. The primary transduction channels are formed, in part, by the transient receptor potential channel TRPC2. Caption and figure after Spehr *et al.* (2006a).

the apical part of the VNO. They interact with smaller, volatile molecules including the small molecule pheromones such as 2-heptanone, also detected by the MOE (above), as well as other small odorants. V2Rs are expressed by the VSNs in the basal layer of the VNO. V2Rs are stimulated by water-soluble molecules such as the peptide pheromone, exocrine-gland secreting peptide (ESP1) (Section 9.5.3), likely the major urinary protein (MUP) darcin, and also other ESPs, MHC peptides, and MUPs important for signature mixtures (Sections 9.8 and 9.9). The apical and basal VSNs project to separate parts of the AOB.

Vomeronasal sensory neurons expressing a V1R show the same single-receptor type expression as OSNs. A small subfamily of V2R genes, the V2R2 genes, are an exception to the one neuron-one receptor rule as they co-express with other V2Rs (Touhara & Vosshall 2009). More molecules that stimulate the VSNs are being identified and linked to their VRs. These include sulfated steroids in mouse urine and molecules from predators and other rodent species (Chamero *et al.* 2012; Isogai *et al.* 2011; Nodari *et al.* 2008). Some of the molecules stimulating the VNO might also stimulate the MOE, but this is rarely tested (see Section 9.4).

The wiring logic of the connection of VSNs to their glomeruli differs fundamentally from the pattern in the main olfactory system (Tirindelli *et al.* 2009). The VSNs expressing a particular receptor protein connect to 10 to 30 distinct glomeruli, albeit in a particular zone of the accessory olfactory bulb, and each glomerulus seems to receive input from more than one type of sensory neuron. It is not yet clear what the implications of these, and other features of neuronal organization in the AOB, are for the processing of stimuli.

A subset of rodent VSNs express one of five formyl peptide receptors instead of a V1R or V2R (Liberles *et al.* 2009; Rivière *et al.* 2009). These receptors may allow detection of molecules characteristic of bacterially spoiled food or infected conspecifics.

9.3.1.3 Other mammalian chemosensory systems

In addition to the MOE and VNO, rats and mice have other chemosensory systems in the nose, which include the septal organ of Masera, Grueneberg ganglion, and trigeminal system (Figure 9.5) (Ma 2010).

The septal organ of Masera is a small isolated patch of sensory epithelium with OSNs that each present one of a subset of the MOE's ORs (Ma 2010; Munger *et al.* 2009). Their axons project to glomeruli in part of the MOB. As well as responding to volatiles, the septal organ OSNs may, like those of the MOE, respond to the air pressure of a sniff (Box 9.3) (Grosmaitre *et al.* 2007). This mechanoreceptor response may modulate sensory responses of OSNs in relation to airflow and perhaps synchronize rhythmic activity in the olfactory bulb (Section 9.1.4).

The recently rediscovered Grueneberg (or Grüneberg) ganglion is another chemosensory organ, in the apical region of the nose, which appears to respond to odors released by stressed mice and possibly also to cool ambient temperatures as experienced by abandoned pups (Brechbühl *et al.* 2008; Fleischer & Breer 2010). Its OSNs each present one of a limited number of VR or TAAR receptors and send their axons to ten glomeruli in a particular area of the OB, possibly part of what is called the "necklace" olfactory system (Ma 2010).

The trigeminal system responds to noxious or chemically irritating stimuli and triggers protective reflexes such as sneezing (Ma 2010; Munger *et al.* 2009).

In addition, outside the nose, there are gustatory receptors for taste but these do not seem to be involved in communication in mammals so are not covered here.

9.3.2 Insect chemosensory subsystems

Insects also have multiple distinct systems for chemoreception: antennae, maxillary palps, and gustatory sensory cells (on proboscis, legs, wings, and genitalia) (Figure 9.8) (Hansson & Stensmyr 2011; Su *et al.* 2009; Touhara & Vosshall 2009). Sensory hairs (sensilla) on the antennae and palps each contain up to four OSNs in *Drosophila* (and up to 30 or more in other insects). The OSNs from the antennae and maxillary palps go to glomeruli in different parts of the antennal lobe (AL) though they respond to overlapping sets of odorants. Inter-neurons go from the AL to the mushroom body and lateral horn of the protocerebrum. There are three types of sensilla (basiconic, trichoid, and coeloconic) (Figure 9.1c), which may have different functions. Basiconic sensillae on the antennae and

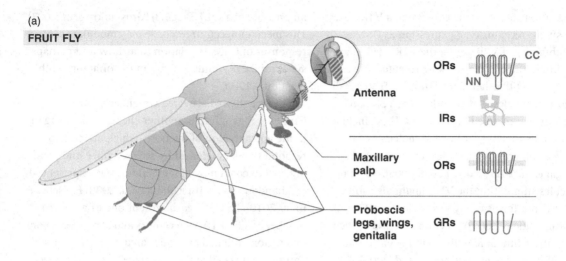

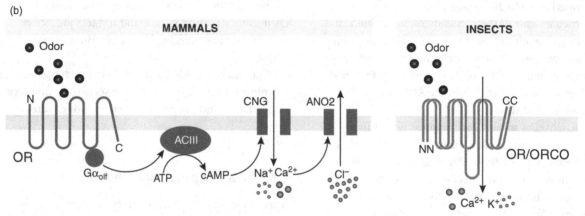

Figure 9.8 The chemosensory receptors of insects evolved independently of vertebrate ones. (a) Olfactory sensory neurons expressing odorant receptors (ORs) and ionotropic receptors (IRs) link to glomeruli in the brain. The gustatory sensory neurons expressing gustatory receptors (GRs) link to the suboesophageal ganglion. (b) (right) The insect ORs are dimers formed by two proteins, an OR and ORCO. Insect chemosensory receptors are ionotropic: the pore opens for ion flow when the odorant binds. Contrast with the metabolic cascade in vertebrate chemoreceptors, left and Figure 9.7. Note also the opposite orientation in the membrane: insects have the N-terminal inside the cell, compared with vertebrate GPCRs (left, and Figure 9.2). Figure adapted from Silbering and Benton (2010).

maxillary palps respond to general odorants. Trichoid sensillae are found on the antennae; some respond to pheromones including the pheromone *cis*-vaccenyl acetate (cVA) in *Drosophila*. These are also the pheromone-sensitive sensillae in moths but some beetles, for example, use other sensilla types for pheromone detection (de Bruyne & Baker 2008).

Coeloconic sensillae, on the antennae, respond to amines and carboxylic acids.

Three key kinds of insect chemosensory receptor proteins are now known: odorant receptors (ORs), ionotropic receptors (IRs), and gustatory receptors (GRs). None are GPCRs and instead all seem to use ionotropic signaling pathways. Though called ORs, the insect odorant receptors are unrelated to vertebrate ORs (Su *et al.* 2009; Touhara & Vosshall 2009). This explains why it took so long to identify insect chemosensory receptors. While the putative ORs of vertebrates have been known since Buck and Axel's paper in (1991) which led to their Nobel prize in 2004, ORs were only identified in insects in 1999, in the fruit fly *Drosophila melanogaster* (Clyne *et al.* 1999; Gao & Chess 1999; Vosshall *et al.* 1999).

The ionotropic signaling pathways used by insects are much faster than the metabotropic pathways used by vertebrate GPCRs: ionotropic signaling is on a millisecond (ms) to sub-ms time scale, as ligand binding directly gates the ion channel (Nakagawa & Vosshall 2009; Silbering & Benton 2010). By contrast, metabotropic receptors have a longer latency, from a few tens to several hundred milliseconds, the time needed to produce second messengers and activate secondary effectors. This speed advantage may be important for tracking dynamic odor plumes on a millisecond time scale (Chapter 10) (Silbering & Benton 2010). There may be modulation by G-proteins.

Insect ORs form ligand-gated ion channels by assembling a ligand-selective OR together with a universal olfactory co-receptor (Nakagawa & Vosshall 2009; Touhara & Vosshall 2009). The co-receptor, highly conserved across insects, is now called ORCO, short for "olfactory receptor co-receptor" (formerly known in different species as OR83b, OR1, OR2, and OR7) (Vosshall & Hansson 2011). The ORCOs were found independently by a number of research groups (see Vosshall & Hansson 2011). There is still much to discover about insect olfactory receptors (see in Nakgawa & Vosshall 2009 and Nakagawa *et al.* 2012 for discussion of some of the unresolved questions).

A second family of ~60 insect chemoreceptors, the ionotropic receptors (IRs), again not GPCRs, has been identified in *Drosophila* (Benton *et al.* 2009). Ionotropic receptors are homologous to the ionotropic glutamate receptor family of ligand-gated ion channels in vertebrates. Ionotropic receptors are found in coeloconic sensilla on the antennae. Ionotropic receptor and OR sensory neurons in *Drosophila* are tuned to complementary chemical classes of odorants: IRs for specific amines and acids and ORs broadly tuned for esters and alcohols (Silbering *et al.* 2011).

Gustatory sensory neurons (GSNs) are found in many places including the labial palps of the proboscis, legs, wings, and genitalia (Montell 2009). They express gustatory receptors (GRs), which are distantly related to the insect ORs. Gustatory sensory neurons located in different organs and body appendages all send their outputs to the suboesophageal ganglion (SOG), which is located slightly behind and ventral to the brain proper (Vosshall & Stocker 2007) (exceptionally, at least four are expressed in antennal neurons, including two CO_2 detector GRs, which project to antennal lobe glomeruli (Montell 2009)). Unlike the antennal lobe, the SOG does not appear to have any organization analogous to glomeruli. Some GRs may respond to CHC pheromones important in sex and species recognition in *Drosophila* courtship (Section 9.5.2).

9.3.2.1 Odorant- and pheromone-binding proteins

Many pheromones used by terrestrial animals are hydrophobic molecules yet they have to reach the OSN membrane through the watery sensillum lymph of chemosensilla in insects (or, in vertebrates, a protective layer of watery nasal mucus) (Figure 9.1) (Leal 2005, 2013; Pelosi *et al.* 2006; Zhou 2010). Soluble odorant-binding proteins (OBPs) reversibly bind odorant molecules (ligands) in a protective cleft and ferry them to the OSN membrane. (Though they share the same convergent functionality, the OBPs of vertebrates and insects have different amino acid sequences and three-dimensional structures; Pelosi *et al.* 2006). In both insects and in mammals, proteins in OBP families have non-olfactory transporting roles in other parts of the body (Pelosi *et al.* 2006). For example, in humans, lipocalins transport molecules to the sweat glands (Chapter 13) and lipocalin urinary proteins carry small molecule pheromones secreted by male elephants and rodents (one protein, darcin, is itself a mouse pheromone). In moths, some other OBPs may be involved in the secretion of moth female sex pheromones (e.g., Dani *et al.* 2011).

In insects, a subclass of OBPs, pheromone-binding proteins (PBPs) bind with high specificity to the pheromones of that species, providing a first level of selectivity (Leal 2005). Different OBPs may be expressed in different sensillum types. As each insect sensillum hair is isolated from others, its responses are a product of its OBPs and the OR on each of the OSNs in the sensillum, giving the idea of a "compound nose." For some PBPs, a fast conformational change in the insect PBP–pheromone complex releases the pheromone when it reaches the negatively charged zone near the OSN membrane (Leal 2005, 2013; Leal *et al.* 2005). The PBP also protects the pheromone from pheromone-degrading enzymes before the signal is delivered. There is one example, perhaps unusual, of a particular PBP being required for activity: the *Drosophila* OSNs that express the receptor OR67 to detect the pheromone cVA (Section 9.5.2) also need the pheromone-binding protein LUSH (OBP76a), and sensory neuron membrane protein (SNMP), which both play critical roles (Laughlin *et al.* 2008; Leal 2013) (though see Gomez-Diaz *et al.* 2013 for an alternative view).

Odorant-binding proteins are also involved in gustation and can affect the selectivity of response. *Drosophila sechellia* is a species that has specialized on the *Morinda citrifolia* fruit, repellent to other *Drosophila* including its cousin species, *D. simulans* (Hansson & Stensmyr 2011; Matsuo *et al.* 2007). In other *Drosophila* species, two OBPs expressed in tarsal (leg) gustatory neurons are involved in detection of characteristic *Morinda* molecules, hexanoic and octanoic acid, leading to avoidance. In *D. sechellia*, a mutation in the promoter sequence of the OBPs has altered their expression and, with that, the gustatory sensitivity to the two fatty acids, meaning *D. sechellia* is attracted not repelled (Matsuo *et al.* 2007).

9.4 The overlapping roles and integration of the accessory and main olfactory systems in mammals

Classical behavioral and neuroscience studies had always suggested that mammals use the main olfactory system, the accessory olfactory system, or both working together, to detect pheromones and signature mixtures depending on species, signal, and previous experience (Baxi *et al.* 2006; Johnston 1998; Meredith 1998). Such studies on model species, including the mouse and hamster, also showed that the inputs from the VNO-AOS and the MOE-OB were integrated at higher brain levels such as the amygdala. These conclusions are now confirmed by new evidence including genetic ablations, molecular markers, and electrophysiological recordings, which show: (i) pheromones are detected by both the VNO and the main olfactory system, depending on species and pheromone; (ii) the VNO also responds to odorants apart from pheromones; and (iii) there is extensive higher brain integration of inputs from the two olfactory systems (Figures 9.7 and 9.9) (reviews by Baum & Kelliher 2009; Brennan & Kendrick 2006; Keller *et al.* 2009; Mucignat-Caretta *et al.* 2012; Munger *et al.* 2009; Tirindelli *et al.* 2009). Neither system has an exclusive role for pheromone perception. Both detect small molecule non-pheromone odorants. Depending on species, either olfactory system can be involved in chemical recognition of individuals by their signature mixtures (Section 9.8).

In many mammals, the main olfactory system is the principal route for detection of pheromones and/or signature mixtures. For example, the rabbit mammary pheromone is perceived by the rabbit pup's MOE (Chapter 1) (Charra *et al.* 2012). In the domestic pig, *Sus scrofa*, the male steroid pheromone androstenone is detected by the female's MOE (Dorries *et al.* 1997). In sheep the luteinizing hormone (LH) surge in ewes elicited by male odors (the "male effect") and recognition of lamb signature mixtures is mediated via the MOE (see below and Sections 9.8.1 and 9.9.4).

In mice, some releaser effects such as male responses to the male MUP pheromone (likely to be darcin) and female responses to darcin are mediated via the VNO (Chamero *et al.* 2007; Roberts *et al.* 2010) (Section 9.9.1) but female attraction to the small molecule pheromone (methylthio) methanethiol (MTMT) is mediated by the MOE (Lin *et al.* 2005). While some work using mutant mice with VNO receptor transduction knocked out (TRP2C null, see Figure 9.7) suggests that the VNO is used in distinguishing the sexes, other experiments suggest the MOE is the more likely critical first step in mate discrimination (Baum 2012; Baum & Kelliher 2009).

In mice, and likely other mammals, both olfactory systems are highly sensitive and overlapping, with sub-nanomolar thresholds for many molecules (Figure 9.7) (Spehr *et al.* 2006b and reviews above). For example 2-heptanone, with primer effects on female estrus (Section 9.7.1), is detected by both the MOE and VNO at similar concentrations (Spehr *et al.* 2006a) but by different receptors (OR912-93 and V1Rb2 respectively) (Gaillard *et al.* 2002; Spehr *et al.* 2006b). Using fMRI to look simultaneously at activation of the MOB and AOB in a mouse, Xu *et al.* (2005) found that both responded to 2-heptanone and to urine odors. Major histocompatibility complex peptides are similarly detected by both MOE and VNO at the same low concentrations (Spehr *et al.* 2006a). These and other non-volatile molecules probably gain access to the MOE during the close contact of mouse social interactions. However, though both systems detect and respond to an overlapping set of molecules, different information may be gained from the same stimuli: first, because the tuning characteristics of MOE ORs and VNO VRs differ and the wiring logic of the OB and AOB is different; second, because the OB and AOB project to some different brain areas, as well as overlapping in other parts of the brain, notably the amygdala (Figure 9.9) (Spehr *et al.* 2006b).

9.4.1 Pheromone-activated hormonal systems in mammals

Projections from both the main olfactory and accessory olfactory systems converge on the amygdala, enabling extensive integration of their chemosensory inputs (Figure 9.9) (Brennan & Kendrick 2006; Gutiérrez-Castellanos *et al.* 2010; Mucignat-Caretta *et al.* 2012; Swann *et al.* 2009). Some neurons in the amygdala are stimulated by both systems (Meredith 1998). As well as pheromone inputs, MOE and VNO inputs for signature mixtures and social learning are also integrated in various parts of the amygdala (Section 9.8) (Brennan & Kendrick 2006; Brennan & Zufall 2006). Learning of olfactory cues prompted by combined inputs to the amygdala explains how sexually experienced hamster males, unlike naïve males, no longer need inputs from the VNO to stimulate copulation in response to vaginal fluid, but instead can respond to learned olfactory cues, perceived through the MOE (Section 9.6) (Meredith 1998).

Outputs from the medial amygdala, influenced by both main olfactory and accessory olfactory systems, lead to pheromone primer effects by activating particular neurons in the hypothalamus, the brain's chief hormonal, or endocrine, control center (Figure 9.9) (Simerly 2002). These are the GnRH neurons, in the medial pre-optic area (MPA) and ventromedial (VMH) parts of the hypothalamus. The ~800 GnRH neurons secrete gonadotropin-releasing hormone (GnRH, also known as luteinizing hormone-releasing hormone, LHRH), the key regulator of reproductive function and sex hormones in vertebrates. Gonadotropin-releasing hormone stimulates the anterior pituitary gland to release two gonadotrophins: luteinizing hormone (LH) and the follicle-stimulating hormone (FSH), which in turn control release of gonadal steroids affecting reproductive physiology and behavior (Section 9.6). GnRH neurons also affect behavior directly by their synapses in brain areas such as the medial pre-optic nucleus that, for example, controls male sexual behavior.

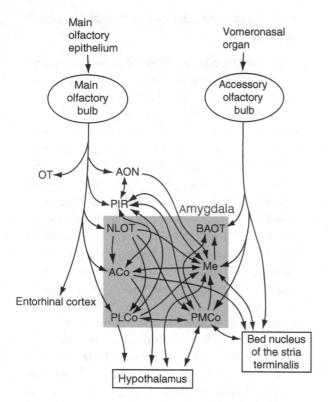

Figure 9.9 Schematic of the major projections of the main olfactory system and the vomeronasal system in the rat. Selected second order connections are shown to highlight the interconnectivity of the two chemosensory pathways at the level of the amygdala (gray square) and their convergence on outputs to the hypothalamus and bed nucleus of the stria terminalis. Note that most of the arrows are double ended – the influences go both ways. Additional links from the hypothalamus to the main olfactory bulb and accessory olfactory bulb and vice versa could be added.

Other abbreviations: ACo, anterior cortical nucleus; AON, anterior olfactory nucleus; BAOT, bed nucleus of the accessory olfactory tract; Me, medial nucleus; NLOT, nucleus of the lateral olfactory tract; OT, olfactory tubercle; PIR, piriform cortex; PMCo, posterior medial cortical nucleus; PLCo, posterior lateral cortical nucleus. Figure and caption after Brennan and Kendrick (2006).

In addition to the main olfactory system's indirect influence on GnRH neurons via the medial amygdala, genetic transneuronal tracers have revealed that in mice, the GnRH neurons receive pheromone and common odor inputs from the main olfactory system (including direct synapses from the olfactory cortex, olfactory amygdala, and directly and indirectly from the piriform cortex) (Boehm *et al.* 2005). Other experiments with different techniques confirm direct links from the olfactory bulb (Yoon *et al.* 2005). Both studies showed extensive feedback loops by which GnRH neurons could influence both odor and pheromone processing, as GnRH neurons innervate the olfactory cortex for example. The ~800 GnRH neurons communicate with ~50,000 neurons in 53 functionally diverse brain areas, some with sexually dimorphic connections (Section 9.5.3), helping to explain some of the far-reaching effects of pheromones and other olfactory inputs.

One further set of pheromone primer effects acts via hypothalamic neurons releasing kisspeptins, neuropeptides that have a crucial role in the initiation and maintenance of mammalian fertility. Kisspeptins stimulate GnRH neurons into releasing gonadotropin, activating the mammalian reproductive axis (Clarke 2011; Colledge 2009). Since GnRH neurons do not possess the relevant sex steroid receptors, a consensus is emerging that sex steroid feedback regulation of GnRH neurons is predominantly exerted via kisspeptin neurons (Clarke 2011). These neurons are at least "one step back" from the GnRH neurons, allowing for steroid feedback, season, stress, immune status, nutritional status, etc., to be integrated by the GnRH neurons and converted to a singular output of the brain that drives the reproductive system (Clarke 2011). Kisspeptin deficient mice or humans show delayed or impaired sexual maturation (puberty).

Kisspeptin also seems to be involved in the fine neuroendocrine control of the events leading to ovulation. In goats, it has been suggested that for the "male effect," male odors activate one population of kisspeptin neurons located in the arcuate nucleus, which generates the GnRH pulse generator (Murata *et al.* 2011; see also Hawken & Martin 2012). In mice, females exposed to male urinary odors show an activation of the other main population of kisspeptin neurons localized in the anteroventral periventricular (AvPv) nucleus (Bakker *et al.* 2010). The sexual dimorphism of these kisspeptin neurons may underlie the sex differences in response to male and female odors, via differences in the relay of olfactory signals to the GnRH neurons.

9.5 Pheromones, sex, and brain circuits

Sexual differentiation during development leads to profoundly different behavioral and physiological responses to pheromones. Males and females may produce different pheromones and have different chemosensory receptors, as found in moths and *Drosophila* CHCs. Alternatively, both sexes may have the same receptor but have sexually dimorphic brain circuits that treat the information differently, leading to sex-specific behavior, as is the case for *Drosophila* for another of its sex pheromones, *cis*-vaccenyl acetate (cVA). In mammals, there are examples of both scenarios but we know less about the details.

9.5.1 Moths

Moths have an obvious external sexual dimorphism as males typically have much larger antennae than females, carrying tens of thousands of sensillae containing pheromone-sensitive olfactory sensory neurons (Box 9.2) (Chapter 3) (Figure 9.1) (Hansson & Stensmyr 2011; Martin *et al.* 2011a). These pheromone-sensitive neurons are exclusive to males in most species and occur in large numbers (up to 40,000 or so), each expressing an OR tuned to one of the female's

pheromone components (Figure 9.4) (Section 9.1). The sexual dimorphism of the antennae is reflected in sexually dimorphic glomeruli in the antennal lobes: in males, the large numbers of OSNs with ORs tuned to pheromone components connect to an MGC of enlarged glomeruli, one for each component, in his antennal lobes (Section 9.1). In moth species such as *Manduca sexta*, the female instead has large female glomeruli fed by OSNs with receptors tuned to plant volatiles. The male also has OSNs with ORs tuned to the characteristic molecules from the pheromones of sympatric species – brain circuits stop his flight response if these are detected (Figure 9.4) (Section 9.1).

Each adult antenna develops from an imaginal disk on the larva. Experimentally, in *M. sexta* these can be surgically transplanted at the larval stage to give antennal male and female gynandromorphs (see in Kalberer *et al.* 2010). The male donor antennal disks develop male OSNs (in male-type sensilla) and induce the formation of male-specific MGC glomeruli in the female antennal lobe. These female gynandromorphs, with the induced male-specific MGC, fly upwind to find a pheromone source just as a male would. Male gynandromorphs with female antennae have large female glomeruli with projection neurons sensitive to the same plant volatiles as females and such moths are stimulated, like females, to fly upwind to plant volatiles (Kalberer *et al.* 2010).

A chemical signal that remained bound to the receptor, stimulating it for long periods, would be no use to a male moth needing to respond to the contact and loss of a pheromone plume over milliseconds (Chapter 10). As in other signaling systems, an "OFF" is needed to recognize the "ON" (Kaissling 2009b). In the male moth this is achieved by highly specific pheromone-degrading enzymes in the sensillum lymph (Figure 9.1), which rapidly inactivate the pheromone at the millisecond rates needed (Ishida & Leal 2005).

9.5.2 Drosophila

The courtship behavior of *Drosophila* tends to follow a stereotyped sequence that allows males and females to confirm the correct species, sex, and mating status

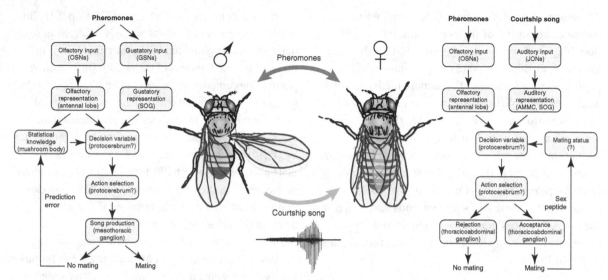

Figure 9.10 Mating decisions in *Drosophila* involve a complex duet of information exchanged between the male (left) and female (right). Parentheses indicate the relevant neurons or regions. OSNs, olfactory sensory neurons; GSNs, gustatory sensory neurons; JONs, Johnston's organ neurons; AMMC, antennal mechanosensory and motor center; SOG, suboesophageal ganglion. The decision variable reflects the likelihood of mating, constructed from sensory representations, acquired knowledge, and, for the female, current mating status. The decision variable guides a binary choice: for the male, to sing or not; for the female, to accept or reject the male. Figure and caption after Dickson (2008).

of the potential mate (Dickson 2008; Ferveur 2007, 2010). For both sexes, the courtship follows a series of conditional steps that integrate visual, auditory, mechanosensory, and chemosensory signals from volatile and non-volatile pheromones (Figure 9.10).

Visual and pheromone cues orient the male toward the female. After following her, he approaches and taps her with his front legs (stimulating male-specific gustatory receptors on his tarsi). If the female is from his species, he extends a wing and vibrates it in a species-specific courtship song. The female meanwhile has slowed her locomotion, turned toward him, sensed his CHC pheromones, and, if his song confirms he is of the correct species, presented her abdomen and protruded her genitalia. He licks her genitalia, gaining further chemical information. If accepted, the male will mount and copulate. Females reject males of other species as they offer the wrong song and pheromones. On largely chemosensory grounds, males reject already mated conspecific females, females from the wrong species, and males (Chapter 3).

Many of the receptors and brain circuits involved in the chemical and song courtship duet between males and females of the fruit fly, *Drosophila melanogaster*, have been genetically dissected, building on pioneering gynandromorph studies (which engineered flies each with a mosaic of male and female characteristics) (Billeter *et al.* 2006; Dahanukar & Ray 2011; Dickson 2008; Savarit *et al.* 1999). The key pheromones explored so far are the male-produced lipid pheromone 11-*cis*-vaccenyl acetate (cVA) and the CHCs produced by both sexes, and important for distinguishing species and the sexes (Chapter 3) (Billeter *et al.* 2009; Shirangi *et al.* 2009). cVA is passed to the female by the male in his ejaculate along with sperm and sex peptide (see below) when they mate.

Sexually dimorphic brain circuits produce different responses in males and females to the volatile male pheromone cVA: males do not sing to a female giving off cVA (which indicates she has already mated); cVA stimulates female mating behavior (and, with food odor is also an aggregation pheromone) (Dickson

2008; Ferveur 2010). Both sexes detect cVA with the same narrowly tuned receptor (OR67d) and the same glomerulus, DA1, is the target of these OSNs. The difference in male and female behavior comes from sexually dimorphic circuits that male and female projection neurons from DA1 make to higher parts of the fly brain (Dickson 2008; Ruta *et al.* 2010). The DA1 glomerulus is one of three sexually dimorphic glomeruli (Dickson 2008). The others are VL2a (which intriguingly receives a required input from a sex-specific receptor that is stimulated by plant odor molecules (Grosjean *et al.* 2011), and VA1v, which receives inputs from OSNs expressing the receptor OR47b, likely sensitive to CHCs produced by both sexes.

The sexually dimorphic brain circuits seem to be hard-wired during development (Dickson 2008; Ferveur 2010; Manoli *et al.* 2006). Sex in flies is primarily determined by the sex-specific splicing of two genes, *fruitless* (*fru*) and *doublesex* (*dsx*), working together. *Dsx* controls the initial differences in nerve number between the sexes, whereas *fru* controls later programmed cell death or arborization (such as that of the sexually dimorphic projection neurons from DA1, above) (Dickson 2008). There is little overt dimorphism in the central nervous system, but numerous fine sex differences have been reported that depend on either *fru*, *dsx*, or both. Neurons that express the sex-specific isoforms of the *fruitless* transcription factor ("Fru^M neurons") form the male circuits. There are ~2,000 Fru^M neurons in males, including sensory, central, and motor neurons (Dickson 2008). These include the sensory neurons and inter-neurons that detect and process pheromones (for example DA1's projection neurons), motor neurons that regulate the wing vibration song, and central neurons that contribute to the intervening neural processing and integration of courtship inputs. If the synaptic activity of all the Fru^M neurons is blocked, all aspects of male courtship including song are suppressed (Dickson 2008).

The integration of courtship signal inputs in the male seems to be carried out by one cluster of ~20 Fru^M neurons, named P1, in the lateral protocerebrum (lateral horn), a higher brain center that receives sensory inputs from the olfactory, gustatory, visual, and auditory systems (Figure 9.10) (Benton 2011; Dickson 2008). P1 neurons, only found in males, are activated by gustatory cues from foreleg contact with female cuticle extracts and inhibited by females carrying cVA (Kohatsu *et al.* 2011). These neurons may perhaps be performing a similar role to some neurons in the mammalian amygdala (Section 9.4).

The female equivalent of the male's P1 neurons for integrating the sensory inputs from the olfactory and auditory system and internal information from her reproductive tract, for her mating decision, have not been identified and the female mating decision could be based on quite different circuits (Figure 9.10) (Dickson 2008). Silencing synaptic transmission of the *fru* neurons inhibits female receptivity, which suggests that these neurons are involved in some way.

Cuticular hydrocarbons also have roles in mating status, sex, and species recognition (see Chapter 3 for CHCs in species isolation) (Ferveur 2010; Ferveur & Cobb 2010). Male-specific gustatory receptors on a male's forelegs and proboscis (labellum) detect male-specific cuticular molecules, leading him to reject males and also already mated females, which carry male CHCs transferred at mating (Everaerts *et al.* 2010; Inoshita *et al.* 2011; Montell 2009; Wang *et al.* 2011). These include 7-T and a compound (CH503) that remains on the female cuticle for at least ten days (Yew *et al.* 2009).

By contrast, on a virgin female fly, he detects the female-produced hydrocarbon 7,11-heptacosadiene, and perhaps other female hydrocarbons, via his Fru^M gustatory neurons in the foreleg, characterized by so-called "pickpocket" ion channels (Lu *et al.* 2012; Thistle *et al.* 2012; Toda *et al.* 2012). Fru^M neurons appear in the suboesophageal ganglion, the target of gustatory chemosensory neurons (Manoli *et al.* 2006). Later in the courtship, two female pentacosene isomers (7-P and 9-P) are detected by gustatory receptors during licking by the male (Ferveur & Cobb 2010). Females detect the male's CHC 7-T by an olfactory receptor(s), stimulating female mating (Ferveur & Cobb 2010).

Courtship usually occurs on a crowded fruit, not in isolation. Aggressive interactions between males are part of adult *Drosophila* life. The male CHC 7-T stimulates aggression between males, with cVA having an additive effect (Wang *et al.* 2011).

As well as the underlying "hard-wired" neural circuits, males also learn from experience to better discriminate receptive and non-receptive partners, such as receptive virgins from recently mated females (Dickson 2008; Griffith & Ejima 2009). A specific class of Fru^M neurons in the mushroom body may be involved in courtship learning (Dickson 2008; Manoli *et al.* 2006).

Once she has mated, other changes occur in the female's behavior and physiology, driven by sex peptides in the seminal fluid transferred by the mating male (Chapter 3) (Avila *et al.* 2011; Wolfner 2009). She then rejects other mates and starts to lay eggs. The changes come from activation of the sex peptide receptor on chemosensory neurons in her uterus (Figure 9.11) (Häsemeyer *et al.* 2009; Yang *et al.* 2009). These and other neurons expressing the female-

specific isoforms of *doublesex* (*dsx*) are required to induce the post-mating responses, from sensing the sex peptide to passing the signal to higher order circuits (Figure 9.11) (Rezával *et al.* 2012). The sex peptides are allohormone pheromones, bypassing the usual external sensory organs (Chapter 1).

9.5.3 Mammals

Male and female mammals seem to have the same olfactory receptors as each other, but male and female sexual behavior in response to pheromones is very different so, as in the *Drosophila* cVA-circuit case, there are likely to be sex-specific neural circuits. The best example so far (with ligand, receptor, possible circuit(s), and behavior explored) is the female lordosis response to the male mouse pheromone, ESP1 (exocrine gland-secreting peptide 1) (Haga *et al.* 2010). ESP1 is a 7-kDa peptide released into male tear fluids. Both males and females have the same narrowly

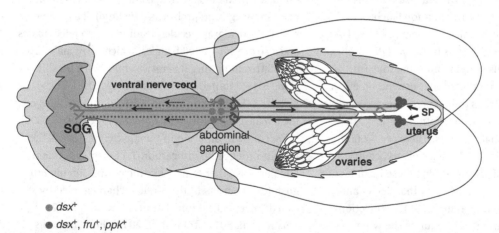

● *dsx*⁺

● *dsx*⁺, *fru*⁺, *ppk*⁺

Figure 9.11 A sex peptide (SP) in the male *Drosophila melanogaster*'s ejaculate changes the female's behaviors so that after mating she rejects other males and starts to lay eggs (Chapter 3). Female-specific neuronal circuits underlie the changes. Six neurons expressing the sex peptide receptor innervate the uterus (Häsemeyer *et al.* 2009; Rezával *et al.* 2012; Yang *et al.* 2009). The SP-sensor neurons and/or their efferent synapses are active in the pre-mating state but inhibited by the presence of SP after mating.

The SP-sensor neurons (dark neurons, *dsx*⁺, *fru*⁺, *pkk*⁺) project to two target regions in the central nervous system: the abdominal ganglion, which contains circuits for egg laying and possibly (dotted) to the suboesophageal ganglion (SOG), thought to contain auditory circuits tuned to the male courtship song. Ascending neurons from the abdominal ganglion target the brain, local inter-neurons and descending neurons innervate the genital tract (light gray neurons = *dsx*⁺ expression)(Rezával *et al.* 2012). See text for more detail. Diagram and caption adapted from Kubli and Bopp (2012) and Clyne and Miesenböck (2009).

specific vomeronasal receptor (V2Rp5) for ESP1 in their VNO and activation of the receptors and AOB neurons were the same in both sexes. However, different AOB targets in the higher vomeronasal centers were stimulated in males and females by ESP1, leading to transmission of ESP1 stimulation to the hypothalamus along a sex-dimorphic pathway (Figure 9.12) (Haga *et al.* 2010).

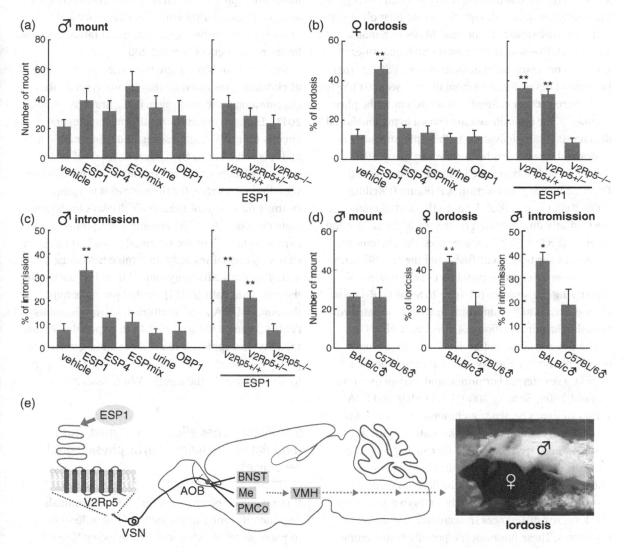

Figure 9.12 Female mice respond to the male peptide pheromone ESP1 (exocrine gland-secreting peptide 1) with sexually receptive lordosis behavior. ESP1 specifically stimulates the vomeronasal receptor V2Rp5 on the microvillar surface of a subset of vomerosensory neurons (VSNs) in the vomeronasal organ and elicits electrical signals. The signals are then transmitted by the vomeronasal axons to target glomeruli in the caudal accessory olfactory bulb (AOB) and subsequently to a spatially restricted subpopulation of AOB neurons. The ESP1 information is further integrated in higher brain areas including the bed nucleus of the stria terminalis (BNST), medial amygdaloid nucleus (Me), and posteriomedial cortical amygdaloid nucleus (PMCo), leading to female-specific activation in the ventromedial hypothalamus (VMH) (see also Figure 9.9). Finally, sexual receptive behavior is enhanced in female mice by neural circuitry that remains to be elucidated (dashed arrows). Figure and caption adapted from Haga *et al.* (2010).

The sex-specific stimulation of the hypothalamus by ESP1 results in female sex receptive behavior upon male mounting (lordosis), allowing successful copulation. Females engineered to have no ESP1 receptor in the VNO had no brain activation from ESP1 and rarely gave the lordosis response though the rest of their olfactory behavior was apparently normal. Males of a mouse strain that, like wild mice, secretes high quantities of ESP1 did not show brain activation from ESP1, perhaps because of desensitization from the self-secreted ESP1.

Numerous other examples of attraction to the pheromones of the opposite sex are known in mammals and differential brain activation to these pheromones in males and females has been shown in many other species in addition to mice including hamsters and ferrets, *Mustela furo* (Section 9.1) (Baum & Kelliher 2009; Keller *et al.* 2009). Primer effects of pheromones are similarly differentiated by the sex of the sender and receiver (Section 9.7). In some cases, the pheromones involved have been identified – and the sex differences in response form an essential part of the process of identifying pheromones (Chapter 2). Some GnRH neural circuits are sexually dimorphic in areas associated with sexual behaviors (Section 9.4) (Boehm *et al.* 2005).

The development and activation of the sexually dimorphic neural circuits above are profoundly influenced by sex steroid hormones produced by the gonads (Arnold 2009; Simerly 2002; Wu & Shah 2011). A subset of genes on the sex chromosomes, including a key gene *Sry* found only on the male-specific Y chromosome, are responsible for causing sex differences in development and adult phenotype. The presence of *Sry* leads to the development of testes rather than ovaries in the embryo and sets up life-long sex differences in secretion of gonadal hormones. These hormones, especially testosterone and estrogen, tightly regulate the development and function of the neural pathways that mediate sexually dimorphic behaviors. The developmental or "organizational" neonatal effect of sex steroids results in long-lasting changes in the nervous system and behavioral displays, whereas the subsequent "activational" effect of these hormones in sexually mature animals leads to transient alterations in neural function and behavior (Section 9.6.2) (Arnold 2009; Wu & Shah 2011). The sex steroid receptors are ligand-activated transcription factors that can directly modulate expression of target genes. However, some aspects of sexual differentiation due to *Sry* and other sex chromosome genes are direct rather than via hormonal influences (Arnold 2009).

On the day of birth, there is a male-specific surge of circulating testosterone that is converted by the enzyme aromatase to estrogen in the brain (Wu & Shah 2011). Estrogen masculinizes the neonatal neural circuits for mating, aggression, and urine marking, and it is sufficient to elicit baseline levels of male patterns of these behaviors in the adult. Testosterone, signaling via its receptor the androgen receptor, controls the extent of these male displays in the adult male (Section 9.6.2). The androgen receptor is expressed in each of the neuronal pools that has been shown to control sexually dimorphic behaviors, including the medial amygdala (MeA), bed nucleus of the stria terminalis (BNST), medial pre-optic hypothalamus (MPOA), and ventromedial hypothalamus (VMH) (Figure 9.9) (Section 9.4). Its expression is also sexually dimorphic: there are more androgen receptor positive neurons in the male BNST and POA compared to these centers in the female (Wu & Shah 2011).

9.6 Pheromones elicit stereotyped, but modulated, behavior and/or physiological responses

It is a common observation that different animals respond differently to the same stimulus. Responses to pheromones are characterized by being "innate" (see discussion in Chapter 1) but the responses can be conditional and vary according to context, time of day, and many other factors including the receiver's genetics, age, sex, hormonal state, dominance status, and experience. The same animal may also respond differently on different occasions. Innate does not mean unconditional or invariant.

Responses to pheromone signals may depend on the context. For example, honeybee alarm pheromone causes bees to attack the intruder if the pheromone is released near the nest but to flee if detected when the bee is far from the nest (Chapter 8).

With some species and stimuli, the changes in responsiveness to pheromones occur in the central nervous system processing. In others, the changes occur in the peripheral sensory system. Some changes may be developmental, having a lasting effect over a long time scale, whereas others may be temporary. Hormones or neuropeptides mediate many of the effects on responsiveness.

9.6.1 Developmental effects

Normal responses to pheromones may not develop unless species-specific conditions are met, which usually occur as a matter of course in normal development (Section 1.2). The behavioral response of young worker bees to queen mandibular pheromone depends on exposure to the pheromone soon after pupal emergence, via an effect on expression of a dopamine receptor gene in the OSNs (Vergoz et al. 2009) (see also Section 9.7.3).

Response may depend on age or developmental stage. For example, the response of fish to alarm cues grows in amplitude and specificity with developmental stage (size) rather than age itself (Døving & Lastein 2009) (see Johnson & Li 2010 for other factors affecting the response of fish to olfactory stimuli). Changes in hormone levels underlie the age-related changes in responses of moth males and honeybee workers to pheromones (see below).

Apart from sex differences (Section 9.5), some of the most profound developmental effects on responses are the changes leading to different castes in social insects (Section 9.7.3). In leaf-cutting ants, queens and different worker types develop different antennal lobes in the pupal stage, with major workers having a macroglomerulus (a feature reflected in greater sensitivity than small workers to releaser effects of its trail pheromone) (Kleineidam & Rossler 2009; Kuebler

et al. 2010). Worker honeybee brains have larger antennal lobes and mushroom bodies than queens (Sandoz 2012; Sandoz et al. 2007).

On a shorter time scale, prior exposure for some days to the female pheromone(s) in the pre-molt urine of the shore crab, Carcinus maenas, primes his later sexual behavioral responses such as cradling when he is later exposed to the female pre-molt urine (Ekerholm & Hallberg 2005).

9.6.2 Hormonal and other effects on central or peripheral responses to pheromones

Hormones help to orchestrate appropriate responses to pheromone signals, according to social status, reproductive maturity, sex, and environmental variables such as day length or temperature (Haupt et al. 2010). Hormones may influence either the olfactory sensory neurons (peripheral effects) or the neurons in the central nervous system involved in processing olfactory information.

At the periphery, the male Manduca sexta moth's olfactory sensory neuron sensitivity to sex pheromone is modulated by hemolymph levels of the hormone octopamine, following a circadian rhythm (Flecke & Stengl 2009). Hormones affect peripheral olfactory responses in males of the southeast Asian cyprinid fish Puntius spp. – higher blood levels of androgen increase the electro-olfactogram (EOG, Chapter 2) response to a putative female sex pheromone (15-keto-prostaglandin-$F_{2\alpha}$), giving a ten times increase in sensitivity specifically to this compound (Cardwell et al. 1995).

Hormones also mediate changes in the central nervous system in insects, leading to changes in responsiveness (Haupt et al. 2010). Freshly eclosed adult males of the moth Agrotis ipsilon do not respond to female pheromone. Full responsiveness develops over days, due to juvenile hormone-sensitive changes to the central neurons in the male antennal lobe, the primary olfactory center (review Anton et al. 2007). Older males do respond, as a result of activation of the antennal lobes in the brain by juvenile hormone (JH) produced by the corpora allata (CA) secretory glands.

In the rest of their adult life, continued male response to female sex pheromone depends on juvenile hormone released by the CA: mature males with the CA removed do not respond to female pheromone, but they regain their response if injected with JH (reminiscent of the responses in the male hamster brain dependent on testosterone, below). Together with JH, octopamine (a biogenic amine neuromodulator) is also important for the male response (Jarriault *et al.* 2009).

The response of the male moth to female sex pheromone is shut down in the antennal lobe for 24 hours after he has mated, a period he needs to replenish his accessory gland proteins (Anton *et al.* 2007; Barrozo *et al.* 2010). His antennae are still responsive to pheromone during this period. In another moth, *Manduca sexta*, a nocturnal rise of another neuromodulator, serotonin (5-hydroxytryptamine, 5HT), in the antennal lobe leads to his night-time response to female pheromone (Kloppenburg & Mercer 2008).

As in moths, responses to pheromones in mammals may be conditional on the hormonal state of the receiving animal (Section 9.4). For example, in male hamsters, brain circuits involved in responses to female pheromones are not active unless testosterone levels in the blood are above a threshold value to give adequate testosterone on the BNST and the MPOA in the hypothalamus (Figure 9.9) (Petrulis 2009; Swann *et al.* 2009). This mechanism provides an internal monitor of readiness to mate because only sexually mature, well-fed males produce sufficiently high testosterone levels. Castrated hamster males show no mating interest in females. Males need both testosterone from the testis and chemosensory input from the olfactory bulb (Swann *et al.* 2009). Injection with testosterone restores mating in castrated, but olfactory intact, males. Male mice only give their ultrasonic courtship song in response to female pheromones if there are sufficient androgen levels in their MPOA (Nyby 2009). In these and most other mammal species, steroid hormones produced in the testis are a prerequisite for expression of mating behavior (Baum & Kelliher 2009; Swann *et al.* 2009; Wu & Shah 2011). Gonadal steroids reflect social status, nutritional status, state of maturity, and stress levels, so they represent an integrated signal from the internal milieu to the central nervous system. As well as stimulation of steroid receptors on amygdala neurons, feedback inputs to the amygdala from GnRH neurons from the hypothalamus may add to the integration (Section 9.4) (Boehm *et al.* 2005).

Honeybee workers change their behavior and responses to pheromones as they age and change role in the hive, from nursing to foraging (Section 9.7.3). Newly emerged workers show little behavioral or electrophysiological response to queen-produced pheromones and alarm pheromones, but strong responses are apparent by the time workers are five to ten days old. By contrast, foragers are not attracted to queen mandibular pheromone though their electroantennogram responses are the same as those of young bees (Pham-Delègue *et al.* 1993; for peripheral molecular changes see also Vergoz *et al.* 2009). These changes in responses to pheromones are likely to be related to the hormonal changes that control the transition between the roles: juvenile hormone accelerates the maturation of worker response to alarm pheromone and octopamine makes workers more sensitive to the foraging-inducing effects of brood pheromone (Box 6.3 and Section 9.7.3) (see Grozinger & Robinson 2007).

Pheromones can have both short- and long-term effects (Chapter 1). The main component of honeybee alarm pheromone, isopentyl acetate (IPA), elicits a quick defensive response from honeybees (Chapter 8) and also causes a significant increase in colony response to IPA over a period of hours (Alaux & Robinson 2007). The short exposure to IPA also induced gene expression in the antennal lobes, perhaps leading to some of the lasting changes in behavioral response to the pheromone (Alaux & Robinson 2007; Alaux *et al.* 2009b).

9.6.3 Experience and learning

For some chemosensory systems, the responses of animals change with experience. For example, after laying an egg, the female apple maggot fly, *Rhagolitis pomonella*, marks the fruit with a host-marking pheromone from her ovipositor, which deters further egg laying in the fruit by herself or other females (Chapter 4) (Roitberg & Prokopy 1981). However, females seem to require experience of the host-marking pheromone before they can discriminate between marked and unmarked fruit. Contact with her own chemical mark is sufficient, so normally in the wild she would have this experience while criss-crossing her trail as she marks her first fruit. *Drosophila* males' ability to discriminate receptive from unreceptive females improves with experience (Section 9.5.2) (Griffith & Ejima 2009).

Experience facilitates the responses of both male and female mammals to pheromones (Chapter 1). Male mammals such as rats and mice may need sexual experience before they can distinguish estrus from non-estrus female odors (Swaney & Keverne 2011). In the mouse and hamster, for virgin males the VNO is essential for response to female pheromones, stimulating investigation and mounting (Meredith 1998). However, sexually experienced males no longer need the VNO input to stimulate copulation as they have learned other odor cues associated with females, detected by the main olfactory system (MOS) (Hurst 2009). One of the genes in mice that seems to be involved in learning female odor cues from sexual experience is *Peg3*, a paternally imprinted gene strongly expressed in areas of the hypothalamus involved in olfaction and control of male sexual behavior (Swaney & Keverne 2011; Swaney *et al.* 2007). Maternal behavior elicited by pheromones in a number of mammals, including sheep and mice, has elements of learning (Section 9.8). More specific cues, usually olfactory, are needed by animals giving birth for the first time, whereas more experienced (multiparous) females appear to have learned a wider range of associations (Lévy & Keller 2009).

Pheromones can prompt associative learning of other odors in the environment. For example, rabbit mammary pheromone (Chapter 1) facilitates the pups' learning of their mother's signature mixture odors

and these will then elicit suckling too (Chapter 1) (Coureaud *et al.* 2010; Schaal *et al.* 2009). In mice, contact with the protein pheromone darcin stimulates the female mouse to learn the male's volatile individual signature mixture and the location of the scent mark (Section 9.9) (Roberts *et al.* 2012; Roberts *et al.* 2010). The VNO is vital for the male hamster's first sexual experiences to be successful, even if this experience allows later substitution by learned odors detected through the MOE (Johnston 1998; Meredith 1998). Experience seems to sensitize the MPOA in the hamster male to chemosensory input and to re-route main olfactory lobe input so that it can substitute for vomeronasal input in driving the MPOA (Figure 9.9). In these examples, the pheromone-prompted learning has the advantage of flexibly tuning behavior.

9.7 Pheromone primer effects

Pheromone primer effects change the physiology and behavior of animals via modulation of hormone signaling and gene expression. These changes may be temporary, such as the primer effects on mammal reproduction (below) and on male sperm production in goldfish (Chapter 1). Alternatively, the changes can be long-lasting developmental changes such as the metamorphosis of marine plankton larvae to their sessile form and phase changes in locusts (Chapter 4). I discuss two other examples of major developmental changes affected by pheromones here: the formation of a resting stage by nematode larvae when conditions get tough, and the social organization and castes of social insects.

9.7.1 Mammals

Pheromone primer effects are important for co-ordinating reproduction in many types of organism (Chapters 1, 3, and 6). They have been particularly well studied in mammals, notably mice and domesticated livestock (Section 9.4). The reported primer effects in laboratory studies of rodents include puberty acceleration of females by male pheromones (the

Vandenbergh effect), estrus induction (the Whitten effect), estrus suppression by pheromones from other females in the absence of males (the Lee–Boot effect), and luteinizing hormone surges in males in response to female odors (Doty 2010; Halpern & Martinez-Marcos 2003). Many primer effects in rodents act in large part via the vomeronasal system (Section 9.4). However, in sheep, the stimuli for the primer male or ram effects act via the main olfactory system (see Sections 9.4 and 9.9).

A number of the primer effects, including pregnancy block (the Bruce effect) in female mice and the male effect in sheep, involve an interaction with the learning of signature mixtures of individual conspecifics. I discuss these in Section 9.9.

Primer effects can be harder to study than immediate behavioral responses (Chapter 2). Studies on the effect of the pheromones of dominant male mice in accelerating puberty in young females (the Vandenbergh effect) illustrate some of the challenges. Different laboratories, using different mouse strains, have proposed different molecules as the active pheromone(s). Among the proposed non-polar molecules in male urine are dehydro-*exo*-brevicomin 2-(sec-butyl)-dihydrothiazole (thiazole), 6-hydroxy-6-methyl-3-heptanone (from bladder urine), and α-farnesene and β-farnesene (secreted into urine from the preputial gland). All four are reported to be active in puberty acceleration and estrus induction, in the absence of protein (Ma *et al.* 1999; Novotny *et al.* 1999a,b). By contrast, other researchers have suggested that the primer effect is instead due to the major urinary proteins themselves (Mucignat-Caretta *et al.* 1995). However, in a thorough set of experiments, using one mouse strain throughout, which tested all of the previously proposed molecules, volatile and protein, Flanagan *et al.* (2011) found that none of them, separately or in combination, stimulated uterus growth (the Vandenbergh effect) as much as whole male urine did. Unknown hydrophilic and not particularly volatile molecules appeared to be the best candidates for the effect.

It has been suggested that group-living female mammals synchronize their estrous cycles in response to female pheromones. However, reports of synchrony in rats, other rodents, and non-human primates may be due to statistical artifacts and asynchrony may be common (Doty 2010, pp. 120 ff.; Setchell *et al.* 2011; Yang & Schank 2006). No estrous synchrony was found in a free-ranging population of mandrills, *Mandrillus sphinx*, a primate (Setchell *et al.* 2011). Human menstrual synchrony is similarly in doubt (Chapter 13).

9.7.2 Nematode dauer resting stage

The first multicellular organism to have its whole genome sequenced, the nematode worm *Caenorhabditis elegans*, is providing a model system for investigating chemosensory behavior, including responses to pheromones, using genetic mutants and selective destruction of identified olfactory sensory neurons with a laser (Bargmann 2006a; Edison 2009; Hart & Chao 2010). In the presence of ample food and low population density, the nematode goes from egg to adult in about 3.5 days, passing through four larval stages (L1 to L4) (Figure 9.13). At high population densities, high concentrations of a pheromone cause young nematode L2 larvae to enter the dauer stage, a long-lived stage specialized for dispersal and survival under harsh conditions (Edison 2009). When conditions improve (with ample food and low pheromone), the dauer larvae resume normal development as L4 larvae.

The multicomponent dauer pheromone is a synergistic mixture of five related ascaroside molecules (Chapter 1) (Figure 1.6) (Pungaliya *et al.* 2009; Srinivasan *et al.* 2008, 2012). At concentrations about 10,000 times lower than the concentrations that induce dauer response in larvae, a synergistic combination of a subset of these molecules forms the *C. elegans* sex pheromone (Chapter 1) (Figure 1.6).

Response to the dauer pheromone is dependent on the major chemosensory organ in *C. elegans*, the bilaterally symmetrical amphid sensillae, located in the tip of the anterior of the animal (Bargmann

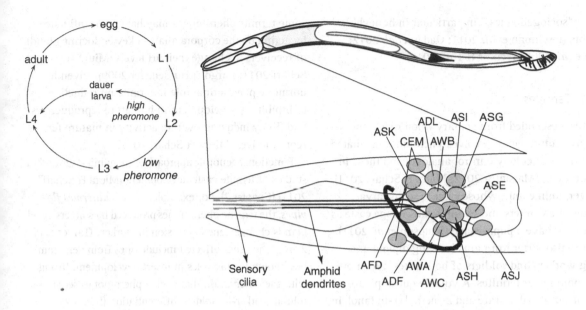

Figure 9.13 (a) The nematode worm, *Caenorhabditis elegans*, passes into an alternative resting stage, the dauer larva, if the conditions are poor, signaled by high concentrations of dauer pheromone released by conspecifics. When pheromone levels drop or food becomes abundant, the dauer larva comes out of diapause and rejoins the normal cycle at the L4 stage.

(b) (Top) A whole worm with the pores indicated. (Bottom) Diagram of the position of the sensory cells (amphid pores) and some of the amphid chemosensory neurons (ADF, ASI, ASK, and ASG) that control dauer formation, viewed from the left side of the animal.

Figure (a) after Thomas (1993); (b) by J Srinivasan, with information from Ortiz *et al.* (2006). See Chapters 1 and 2 for more background on the multicomponent dauer pheromone.

2006a). Each of these sensilla contains just 12 chemosensory neurons. Unlike mammalian OSNs, which express only one OR, each nematode chemosensory neuron expresses many different chemosensory receptor proteins. For example, one of the key sensory neurons involved in dauer responses, ASI, expresses nine different chemosensory receptors. Two of these chemoreceptors are sensitive to dauer pheromone component ascr#5 and mutations make the worms insensitive to the pheromone (Section 9.2.3) (McGrath *et al.* 2011). Though much of the detail remains to be worked out, it seems that in the absence of dauer pheromone, the chemosensory neurons secrete peptides (from the transforming growth factor-beta (TGF-β) family, insulin peptide families, and neuropeptide families), which act as

neuroendocrine signals to prevent dauer formation (Bargmann 2006a; Hu 2007).

9.7.3 Social insects

Pheromones mediate key aspects of the biology of social insects, from the morphological and physiological differentiation into reproductives and non-reproducing workers to the dynamic allocation of workers to tasks to match colony needs (Chapter 6). Early developmental switches caused by rearing an individual as a worker or as a reproductive lead to life-long differences in response to pheromones in social insects of different castes (pheromone production is similarly affected, Boxes 6.1, 6.2, and 6.3). Many of these effects can now be understood at the molecular level of "sociogenomics," in particular in honeybees and ants (Fischman *et al.* 2011; Gadau *et al.* 2012; Smith *et al.* 2008; Weiner & Toth 2012).

9.7.3.1 Termites

Termites, descended from solitary wood roach ancestors, have convergently evolved pheromone-mediated mechanisms of colony control analogous to those in ants (Box 6.2) (Matsuura 2012; Miura & Scharf 2011). However, unlike ants, whose caste is set as larvae, most termite workers and nymphs can change caste as they do not have a pupal stage (Simpson *et al.* 2011). Termites also differ from ants in being diploid and having workers and soldiers of both sexes. We know most about lower termites. A volatile queen pheromone, *n*-butyl-*n*-butyrate and 2-methyl-1-butanol, in the lower termite, *Reticulitermes speratus*, inhibits differentiation of female nymphs into queens (Box 6.2) (Matsuura 2012).

All termite eggs can potentially develop into workers, soldiers, or reproductives (the three morphologically differentiated castes). Control of caste development is, with a few exceptions, largely environmental, including the effect of pheromones, rather than genetic (Glastad *et al.* 2013; Vargo & Husseneder 2009) (for possible exceptions in termites and ants, see also Schwander *et al.* 2010). Castes differentiate based on gene expression responses to intrinsic and extrinsic factors. The environmental (extrinsic) factors include nestmates (soldiers and reproductives), pheromones, and physical conditions (temperature, etc.). Interacting with environmental factors are intrinsic ones, which include instar, nutritional status, and hormonal interactions including juvenile hormone (JH) titer in particular, allostatins (neuropeptide hormones that reduce JH), ecdysteroid hormones, and insulin signaling (Miura & Scharf 2011; Vargo & Husseneder 2009). Juvenile hormone itself was once thought to be a possible pheromone with primer effects but there is no direct evidence that JH is transferred among nestmates (Miura & Scharf 2011). However,

some termite pheromones may have their influence by acting on the corpora allata (a key endocrine gland) of receivers to change their JH levels (Miura & Scharf 2011; Vargo & Husseneder 2009). Juvenile hormone plays a dual role in termite reproduction, by (i) inhibiting development in immature reproductives and (ii) by inducing ovarian activity in mature female reproductives (Miura & Scharf 2011).

Functional genomic approaches are gaining ground in studies of termite caste development (Miura & Scharf 2011) (Chapter 2). For example, in *Reticulitermes flavipes*, the terpene pheromones produced by soldiers' head glands change gene expression in workers (Tarver *et al.* 2010). The genes affected include ones from gene families known to play roles in insect development, fitting with the observation that soldier pheromones seem to inhibit worker-to-soldier differentiation (Box 6.2).

The newly found termite queen pheromones in termites (Box 6.2) offer the possibility of understanding more about the mode(s) of action of pheromones in colony organization and caste development (Matsuura 2012; Miura & Scharf 2011). In honeybees we already have a much more detailed understanding about the action of pheromone primer effects at the molecular level (below).

9.7.3.2 Ants

Pheromone primer effects influence, directly or indirectly, many physiological and developmental pathways in ants (Vargo 1998). For example queen pheromones in the fire ant, *Solenopsis invicta*, including venom gland piperidine alkaloids and pyranones, affect almost every aspect of colony life though how many different pheromones are involved in these varied effects is not known (Eliyahu *et al.* 2011; Vargo 1998). Queen pheromones inhibit virgin daughter queens in the nest from losing their wings (dealating) and starting to lay male eggs. These inhibitory effects seem to act by suppression of JH secretion (and can be seen at the level of gene expression, Wurm *et al.* 2010). *Solenopsis invicta* queen pheromones also inhibit the production of new reproductives (queens and sexual males) through an indirect effect on caste determination (by causing workers to feed female larvae less, resulting in their development as workers, and killing male larvae; Vargo 1998). In some ant species with many queens in each colony, there may be an inverse relationship between queen number and individual fecundity. In the fire ant this effect is caused by a mutual inhibition mediated by pheromones (Vargo 1992). In *Lasius niger*, rather than inhibiting each other, multiple founding queens may be selfishly holding back from egg production so as to conserve resources for the later queen-elimination phase (Figure 6.2) (Holman *et al.* 2010a).

9.7.3.3 Honeybee queens and workers

Queens have a different morphology, physiology, and behavior from workers. Queens also live about ten times as long. Allohormone pheromones drive the fundamental developmental switch to develop as worker or queen (Chapter 6). Colony pheromones then co-ordinate and allocate the activities of the workers inside and outside the hive, meeting the changing needs of the colony (Boxes 6.1 and 6.3) (Alaux *et al.* 2010; Grozinger 2013; Le Conte & Hefetz 2008).

Female honeybee larvae have the potential to develop into either workers or queens. The dimorphism is not due to genetic differences. Instead, a larva's path depends on its environment at critical periods of caste determination. For the first three days, all larvae are fed with royal jelly, produced by nurse worker bees from their hypopharyngeal and mandibular-gland secretions (Miklos & Maleszka 2011; Winston 1987). After three days, larvae destined to become workers are fed nectar plus pollen, whereas larvae to become queens are fed copious quantities of royal jelly for the rest of their larval stages. The worker–queen switch is irreversible.

A key signaling molecule in royal jelly, which induces the differentiation into a queen, has been identified as royalactin, a 57-kDa glycoprotein

allohormone pheromone (Kamakura 2011). Honeybee larvae fed purified royalactin showed the queen characteristics of increased body size, larger ovaries, and shortened development time. Surprisingly, royalactin had the same effects when fed to *Drosophila* larvae, which allowed Kamakura to use them to ingeniously explore royalactin's mode of action. RNAi techniques revealed that royalactin stimulates the epidermal growth factor receptor (EGFR) on cells in the fat body, a tissue analogous to vertebrate liver and adipose tissues. The EGFR activation of fat body cells by royalactin causes a signaling cascade, activating a kinase responsible for stimulating an increase in body size, another kinase involved in decreased development time (by stimulating synthesis of the hormone 20-hydroxyecdysone), and increasing the titer of JH, an essential hormone for ovary development (Kamakura 2011). *Egfr* RNAi knockout in the honeybee fat body confirmed royalactin works via EGFR signaling in the way revealed in *Drosophila*. Kamakura established that the action of royalactin was not a nutritive effect: feeding with another protein, casein, to the same energy value did not have the effect. Royalactin is thus an allohormone pheromone (Chapter 1), acting directly on tissues, not via chemosensory neurons.

Caste determination also involves epigenetic regulation, ultimately triggered by royal jelly (Weiner & Toth 2012). Epigenetic DNA methylation (present in the honeybee) may serve as a developmental switch, silencing crucial downstream genes involved in caste formation. Worker development is associated with increased DNA methylation (Lyko *et al.* 2010; Miklos & Maleszka 2011) (though queen–worker methylation differences were not found by Herb *et al.* 2012). Other molecules in royal jelly may yet be found to have a role in caste determination – for example (*E*)-10-hydroxy-2-decenoic acid, a major fatty acid constituent of royal jelly from worker mandibular glands, may influence the pattern of methylation (Spannhoff *et al.* 2011). DNA methylation is also found in termites and may play a similar role (Glastad *et al.* 2013).

Maintenance of control of worker reproduction

Workers' ovaries are small as a result of the developmental pathway triggered by royal jelly but continued inactivation of ovaries in workers is maintained by the combined effects of queen mandibular pheromone (QMP) (Hoover et al. 2003) and brood pheromone (BP) (in particular the components including ethyl palmitate and methyl linolenate), emitted by larvae (Chapter 6) (Boxes 6.1 and 6.3). These two pheromones inhibit JH synthesis by the corpora allata but, unlike other insects, no direct effects of JH on ovary development have been found (Alaux et al. 2010; Le Conte & Hefetz 2008). In the honeybee, unlike other insects, JH and vitellogenin (the major yolk protein precursor) are inversely related in a complex regulatory loop. Vitellogenin may have co-ordinating effects on social organization (Alaux et al. 2010; Le Conte & Hefetz 2008).

Queen mandibular pheromone causes changes in expression of some 2,000 genes in the brain of workers compared with queens (Grozinger et al. 2003, 2007b; Zayed & Robinson 2012) (see below for behavioral effects). Evidence of which genes are involved in maintaining worker ovary inactivation could come from comparisons of gene expression in wild type and mutant "anarchic" bees, which activate their ovaries despite the presence of QMP and BP (Thompson et al. 2006). Queens of anarchistic honeybee colonies still produce normal QMP so the mutants may differ in the transmission of, reception of, or sensitivity to queen pheromones (Hoover et al. 2005).

Colony dynamics and worker task allocation

A complex of interlocking feedbacks, including a large number of pheromones from the queen, brood, and foraging workers, ensures that an optimum effort goes into nursing new larvae and bringing in the right proportions of pollen and nectar food, reflecting the changing needs of the colony (Alaux et al. 2010; Grozinger 2013; Le Conte & Hefetz 2008; Page et al. 2012). The response by individual bees to social and pheromone cues leads to co-ordination without any central direction or blueprint, an idea explored in Chapter 7. There is likely to be some colony-level selection, reflected in the number of reproductives produced (Bourke 2011b, pp. 57 ff.).

Honeybees show an age-related polytheism or division of labor (Alaux et al. 2010; Le Conte & Hefetz 2008). For the first two to three weeks of adult life, young workers tend to perform in-nest tasks such as feeding and caring for brood ("nursing") and comb building. Older bees switch to outside tasks such as foraging for pollen and nectar for the rest of their five- to seven-week life. The timing of bees' transition from nursing to foraging is responsive to the needs of the colony, largely communicated by pheromones. Queen mandibular pheromone delays the onset of foraging. A volatile pheromone produced by young larvae, (E)-β-ocimene, encourages workers to forage earlier in life (Maisonnasse et al. 2010). Brood pheromone (BP) (Box 6.4) has dose-dependent effects: at low concentrations BP stimulates behavioral maturation (to foraging), high doses (from large numbers of larvae) delay the transition (Alaux et al. 2010). Ethyl oleate, also produced by the queen and by the brood, has the additional effect of stimulating the growth of the hypopharyngeal gland used by nurse bees to feed larvae (Alaux et al. 2010). Thus if a colony has a shortage of foragers, lower ethyl oleate levels cause nurse workers to change to foraging up to two weeks earlier than they normally would. Brood pheromone has a different effect on foragers: it increases their pollen gathering (see gene expression below) (Chapter 12) (Pankiw 2004). Perception of ethyl oleate appears to be by olfaction, at close range or during contact (Muenz et al. 2012).

The QMP and BP influence the reversible transition from nurse to forager by orchestrating large-scale changes in brain gene expression, particularly affecting genes encoding transcription factors involved in regulating other genes (many of which have cis-regulatory elements in their promoter regions) (Alaux et al. 2009a; Smith et al. 2008; Zayed & Robinson 2012). Comparing the QMP-regulated genes

(Grozinger *et al.* 2003) with lists of genes differentially expressed in nurse and foraging bees' brains (Whitfield *et al.* 2003) showed that QMP consistently activated "nursing genes" and repressed "foraging genes," suggesting that QMP may delay behavioral maturation by regulating gene expression (Grozinger *et al.* 2003). The changes in gene expression may be mediated by reversible DNA methylation patterns that are different in forager and nurse bees (Herb *et al.* 2012; Zayed & Robinson 2012). The effects of BP are age dependent. In nurse bees, like QMP, BP similarly tended to upregulate "nursing genes" and repress "foraging genes" (Alaux *et al.* 2009a). This pattern was reversed in forager bees. The size of worker bees' ovaries may also affect the timing of their transition to foraging (Wang *et al.* 2012).

Pheromone effects on the transition may be mediated by influencing rates of JH synthesis, perhaps via the action on gene expression (Le Conte & Hefetz 2008; Zayed & Robinson 2012). For example, a major component of QMP, 9-oxo-2-decenoic acid (9-ODA), inhibits the rate of JH biosynthesis in workers leading to a JH-based delay in the transition (Le Conte & Hefetz 2008; Slessor *et al.* 2005). Brood pheromone also seems to act on JH levels (Le Conte & Hefetz 2008).

Co-option of genes from solitary species

The similarity of responses of *Drosophila* and honeybee larvae to royalactin (above), despite a divergence between flies and Hymenoptera of some 300 million years, is an example of the way pre-existing mechanisms have been repurposed in social insects for social behavior (Tóth & Robinson 2007). Other examples come from the pheromone-regulated genes influencing honeybee worker behavior (Tóth & Robinson 2007; Zayed & Robinson 2012). The honeybee version of the *foraging* gene, first identified in *Drosophila*, is one of the genes upregulated in the brains of forager bees (Whitfield *et al.* 2003). In *Drosophila*, a solitary insect, the *foraging* gene influences activity; in honeybees, the transition to foraging. Another *Drosophila* gene, *malvolio*, is involved in responsiveness to sucrose. The honeybee version of the *malvolio* gene is upregulated

in foragers (Ben-Shahar *et al.* 2004). Brood pheromone causes an upregulation of *malvolio* in foraging bees but not in younger ones (Alaux *et al.* 2009a). Variation in the pace of the transition to foraging also has a genotypic component, which can be seen in different honeybee subspecies or artificially selected strains or different patrilines (Tóth & Robinson 2007; Zayed & Robinson 2012). Selection is likely to be acting on many overlapping networks of genes, a complexity hinted at by the number of different pheromones and social influences that affect the transition.

9.8 Learning of signature mixtures

Many discriminatory behaviors, such as a parent feeding only its own offspring or a guard bee letting another bee into the nest, are based on learned olfactory cues (signature mixtures) (see recognition mechanisms, Section 1.7). In vertebrates such as mammals, these may be forms of imprinting (biologically relevant learning during a sensitive period defined by a particular developmental stage or physiological state, Hudson 1993). Learning of important odor cues can happen early in life or when adult, as appropriate. One example of early learning is the way mouse pups learn the signature mixture of their mother as the cue for suckling (in contrast to rabbit pups, which use a species-wide pheromone, Chapter 1) (Logan *et al.* 2012).

The subtlety of signature mixtures probably depends on olfactory systems with glomeruli and combinatorial processing. Signature mixtures, as they involve learning, may be processed in different glomeruli from pheromones, though this is still not fully understood (Wyatt 2010).

We do not know much about the detailed neurobiology of learning of signature mixtures in insects and other invertebrates. Individual recognition in lobsters and spiny lobster responses to conspecific urine signals are mediated by olfactory (= "aesthetasc") pathways though currently not enough is known about olfactory processing in lobsters to know if pheromones are processed differently (Aggio & Derby 2011; Horner *et al.* 2008; Johnson & Atema 2005; see also Schmidt & Mellon 2011). In social insects, a colony's shared labels are constantly changing and thus the learned signature mixture has to be constantly reinforced and fine tuned (Bos & d'Ettorre 2012; van Zweden & d'Ettorre 2010). In ants, one study has suggested recognition of colony members of *Camponotus japonicus* occurs at the level of sensilla on the antennae, which respond only to non-nestmates (peripheral filtering) (Ozaki *et al.* 2005) though this seems not to be the case in *Camponotus floridanus* (Zube *et al.* 2008). The length of time taken to reform nestmate-recognition templates, more than two hours, suggests that this is a process occurring in the central nervous system rather than chemosensory neurons (Leonhardt *et al.* 2007). Antennal recordings, again in *C. floridanus*, do not support the idea of peripheral filtering as nestmates did stimulate the olfactory system (Brandstaetter & Kleineidam 2011; Brandstaetter *et al.* 2011). Calcium imaging of the antennal lobe found variable activity from repeated stimulation with the same colony odor but it is early days for imaging of nestmate-recognition cues. It seems likely to me that template formation (identity learning) in social insects will occur in the mushroom bodies as these are involved in insect memory (Gronenberg & Riveros 2009; Kleineidam & Rossler 2009) though pheromones may also be processed there too (Yamagata *et al.* 2005). Bos and d'Ettorre (2012) discuss our understanding of templates in ants, including a possible role for habituation.

Two mammalian odor recognition systems described in Chapter 6 have provided detailed model systems for investigating the mechanisms of olfactory memory of signature mixtures in vertebrates (next sections) (reviewed by Brennan & Kendrick 2006; Lévy & Keller 2009; Sanchez-Andrade & Kendrick 2009). The first is the way a mother sheep (ewe) learns the odor of her lamb, via the main olfactory system; the second, via the vomeronasal system, is the female mouse's learning of her mate's odor, which is the basis of pregnancy block (the Bruce effect) (Section 9.9.3). Both model systems require simultaneous appropriate odor stimulation and norepinephrine (noradrenaline) release in the olfactory bulbs by nerves stimulated by birth or mating, respectively. One exposure to the odor is enough for learning. In both systems, the sensitive period is characterized by changes in the animal's hormonal state that facilitate the learning (Hudson 1993).

Information about individual identity from the VNO and MOE is integrated via their combined inputs to the medial amygdala in a variety of mammals (Brennan & Kendrick 2006; Lévy & Keller 2009; Sanchez-Andrade & Kendrick 2009). The same molecules, for example MHC peptides (Box 3.1), can be processed in parallel by both systems, giving different kinds of information (Section 9.4).

9.8.1 Maternal memory of offspring by sheep

In sheep, an enduring bond between a ewe and her lambs is established very rapidly, usually within two hours of giving birth (parturition) (Lévy & Keller 2008, 2009; Sanchez-Andrade & Kendrick 2009). Olfactory cues perceived by the main olfactory system, not the VNO-accessory olfactory system, are key to the process. Stretching of the vaginocervical area while giving birth sends nerve signals to the brain that trigger a cascade of neurobiological and hormonal mechanisms resulting in three changes in behavior: first, release of oxytocin, a peptide hormone, mainly in the paraventricular nucleus of the hypothalamus, leads to maternal behavior toward lambs in general; second, the ewe's response to amniotic fluid changes, making it very attractive (whereas before it was repulsive); and, third, norepinephrine and other neurotransmitters released in the main olfactory bulb cause her to learn the individual odor of her lamb while sniffing and licking the amniotic fluid. Afterwards she will only suckle her own lambs, recognized by smell (see also Chapter 6). The sensitive period for learning lasts for between four and twelve hours after giving birth. In this early period, but not later, an orphan lamb will be accepted and adopted, particularly if coated with amniotic fluid (a method traditionally used by farmers). Sufficient levels of estrogen are also important for all the maternal behavior described above (Sanchez-Andrade & Kendrick 2009).

The formation of memory has been studied at the level of mitral cells in the mother's olfactory bulb (Figure 9.14) (Lévy & Keller 2009; Sanchez-Andrade & Kendrick 2009). Before giving birth the majority of these mitral cells respond to food odors. After giving birth there is an increase in the proportion of mitral cells that respond selectively to lamb odors. These shifts are reflected in concurrent changes in the release of the neurotransmitters γ-aminobutyric acid (GABA) and glutamate within the main OB, the release of norepinephrine by the noradrenergic afferent system (relaying the vaginocervical stimulation of giving birth), and the release of oxytocin (Lévy & Keller 2009). All these neurotransmitters are important for olfactory learning associated with individual lamb recognition. After learning her lamb's odors, only mitral cells activated by the odor of the familiar lamb are stimulated, other mitral cells are inhibited. The cortical and medial amygdala, which receive olfactory information from the olfactory bulb, also play a part in memory formation (Lévy & Keller 2009).

Maternal experience affects the ewe's response to stimuli. Ewes giving birth for the first time need exposure to amniotic fluid on the lamb to develop maternal care, whereas experienced mothers do not reject lambs that have been washed (Lévy & Keller 2008). Maternal experience both increases the sensitivity of the brain to the effects of oxytocin in all future births and also facilitates the learning of their offspring's individual odor.

In humans, close olfactory contact between mother and baby in the first hours after birth is important for each to learn the odor signature of the other (Porter & Winberg 1999). Olfactory recognition of the baby and responses to infant odors can be related to the hormone levels of the mother (Chapter 13) (Fleming *et al.* 1997). In parallel, in the process of being born, babies

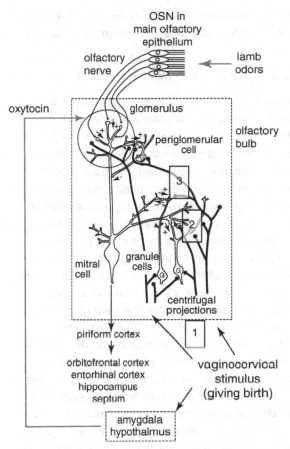

Figure 9.14 Mother sheep learn the odor of their newly born lambs. The changes in brain neurochemistry stimulated by giving birth that facilitate the mother's changes in behavior and odor learning have been mapped. The olfactory memory underlying selective recognition of lamb odors involves plastic changes occurring within the olfactory bulb as well as the cortical and medial amygdala (see Figure 9.9).Within the olfactory bulb, memory formation is associated with a strengthening of the mitral to granule and periglomerular cell synapses to bias the network to respond to that lamb's odors. Vaginocervical stimulation leads to enhanced release of norepinephrine from the centrifugal projections (1), reducing GABA release onto mitral cells (2). Olfactory sensory neuron activation stimulates glutamate release inducing nitric oxide/cGMP release (3). These will potentiate the release of glutamate from mitral cells and subsequent GABA release from granule and periglomerular cells. (OSN, olfactory sensory neuron). Figure and caption after Kendrick *et al.* (1997) and Sanchez-Andrade & Kendrick (2009) (caption).

seem to be primed for olfactory learning of their mother's odor (Mizuno 2011; Romantshik *et al.* 2007).

9.8.2 Mate choice to avoid inbreeding

In young mammals, the negative imprinting of familial odors, notably those associated with the MHC, affects their later mate choice (see Chapters 3 and 6) (reviewed by Brennan & Kendrick 2006). When adult, mice choose mates with a different MHC haplotype from their siblings, learned when growing up. The main olfactory system is sufficient for mate choice by olfactory cues, which include a complex signature mixture consisting of volatile molecules associated with the MHC (the odor type) and MHC peptides (Restrepo *et al.* 2006). The processing of the volatile odor types is combinatorial, demonstrated by mapping c-*fos* activation of glomeruli across the olfactory bulb by urine odors from different strains (Schaefer *et al.* 2002). The MOE also detects some MHC peptides (Spehr *et al.* 2006a). These gain access to the MOE during the close contact of mouse social interactions. Peptide recognition in the MOE probably occurs in a combinatorial manner too (Spehr *et al.* 2006a). In mice MUP haplotype may also be important in mate choice (Hurst 2009; Hurst & Beynon 2013). However, many mammal species have only a single MUP iso-form so this would not be useful for identity (Logan *et al.* 2008).

9.8.3 The Coolidge effect

The renewed stimulation of sexual behavior in a sexually satiated animal by a new unfamiliar mate, shown in a wide range of animals from crickets to chickens, is called the Coolidge effect (Chapter 3). For hamster males, it seems that the main olfactory system is most important for distinguishing the new individual from familiar females (Johnston & Rasmussen 1984) and that processing by the entorhinal cortex is involved (Petrulis 2009). Other experiments have suggested the VNO may be involved instead (Steel & Keverne 1985).

9.8.4 Monogamous partner recognition in prairie voles

Unlike most rodents, many prairie voles, *Microtus ochrogaster*, are socially monogamous and form long-lasting pair bonds, mediated in part by memory of partner signature mixtures detected via the main olfactory system, which sends its signals to the medial amygdala. Both parents share in rearing the young (Chapter 6). Mating facilitates the formation of partner preferences but longer co-habiting without mating can also lead to a partner preference (McGraw & Young 2010). Within each animal's brain, the neurotransmitters oxytocin, vasopressin, and dopamine act on specific neuron circuitry to facilitate social attachment in a sex-specific way. In the female, oxytocin plays a crucial role in formation of partner preference. In males, arginine vasopressin is most important though oxytocin may also have a role. The pair bond and an olfactory memory for the partner's signature mixture are formed by the simultaneous stimulation of dopamine receptors and in females, oxytocin, or in males, vasopressin receptors, at mating (McGraw & Young 2010).

9.9 Interactions between signature mixtures and pheromones

In mammals, behavioral responses to signature mixtures can affect exposure or response to pheromones and vice versa. These effects illustrate why it can be useful to separate the ideas. I discuss four different kinds of interaction here: first, a pheromone prompts a female mouse to learn a male signature mixture; second, young female prairie voles avoid picking up pheromone from familiar males; third, the female mouse's memory of the signature mixture of the mating male prevents his pheromone blocking pregnancy (the Bruce effect); and fourth, a novel male sheep with an unfamiliar signature mixture allows a male pheromone to act on the female. See Chapter 1 for detailed discussion of the terms "signature mixture" and "pheromone".

9.9.1 Male mouse pheromone darcin prompts female memory

A dominant male mouse marks his territory extensively with urine marks containing high concentrations of a variety of MUPs (Chapter 5) (Hurst & Beynon 2013). Most MUPs are highly polymorphic but one MUP, Mup20 (named darcin), is the same in all male mice and is a pheromone (Roberts *et al.* 2010).

A female visiting the territory is attracted by volatiles coming from the urine marks. When she contacts and sniffs a urine mark, darcin reaches her VNO, induces her to spend more time sniffing the urine, and stimulates her to learn the location of the mark and the marking male's volatile individual signature mixture (Roberts *et al.* 2010, 2012). She is then attracted to this male, recognizes him by his airborne signature mixture, and mates with him. She is uniquely attracted to him, not other males at this point. Mup20 is probably the MUP that elicited aggression in male mice in earlier experiments by Chamero *et al.* (2007). Presumably, other territorial males will treat Mup20 as an indication of a potential challenger. Darcin binds and releases the male-specific small molecule mouse pheromone, 2-*sec*-butyl 4,5 dihydrothiazole (thiazole).

9.9.2 Recognition of family, avoiding pheromone contact

In the co-operatively breeding prairie vole, *Microtus ochrogaster*, subordinate female prairie voles remain pre-pubescent and help their mother (Carter & Roberts 1997). Young females delay puberty so long as they are exposed to only familiar males (father or male sibs), recognized by their signature mixture odors. Because familiar males do not evoke mutual nasogenital grooming, the male's urinary pheromone does not get into the VNO of the female. However, if urine is placed on the vole's nose by the experimenter, it is effective even from a sib male (Figure 9.15). There do not appear to be any inhibitory pheromones in the urine of their mother, as juvenile females became pregnant if kept with a strange male, whether their mother was present or not (Wolff *et al.* 2001).

9.9.3 Memory and pregnancy block (the Bruce effect)

In the laboratory, exposing a recently mated female mouse to the urinary chemosignals of an unfamiliar male mouse initiates a neuroendocrine reflex that results in pregnancy failure and a return to estrus (the Bruce effect) (Brennan 2009; Brennan & Kendrick 2006). A pheromone in the strange male's urine is sensed by her VNO and activates an excitatory pathway, via the accessory olfactory bulb and medial amygdala, to the hypothalamus. This increases dopamine release, which in turn inhibits prolactin production by the anterior pituitary. Without prolactin, progesterone production by the corpora lutea declines, implantation of the developing embryos fails, and the female returns to estrus.

All male mice produce the pheromone but pregnancy block does not happen when the female meets the male she recently mated. Why not? The reason is that when she mated, vaginocervical signals prompted the individual signature mixture of her male mate to be learned in her accessory olfactory lobe. This memory selectively prevents his pheromones from eliciting the pregnancy block. Thus there are two distinct kinds of chemical information, a male testosterone-dependent pheromone (the same for all males), as yet unidentified (though it is of low molecular weight; Peele *et al.* 2003), and the male's individual chemical profile including his urinary odor type and peptides related to the MHC. Males that differ only at one MHC locus (Chapter 3) can cause pregnancy failure (Box 3.1) (Yamazaki *et al.* 1983). Major histocompatibility complex peptides from an unfamiliar male of a different haplotype can induce pregnancy block and can be detected directly by the V2R receptor expressing zone of the vomeronasal

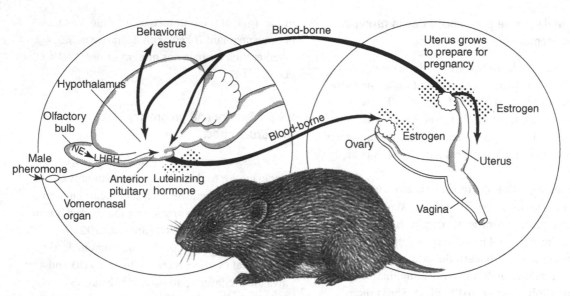

Figure 9.15 The hormonal cascade that triggers estrus in the female prairie vole, *Microtus ochrogaster*, begins when she sniffs an unfamiliar male (she does not sniff her father or sibs – see text for discussion).The vomeronasal organ picks up pheromones, stimulating the olfactory bulb. Norepinephrine (NE) (noradrenaline) and luteinizing hormone-releasing hormone (LHRH = GnRH) are secreted and start the production of luteinizing hormone. Luteinizing hormone reaches the ovaries via the bloodstream and stimulates them to produce estrogen. Estrogen is then carried to the hypothalamus, where it induces estrus (see Section 9.4.1). Figure adapted from Carter & Getz (1993). Original artist Patricia Wynne.

epithelium (Leinders-Zufall *et al.* 2004). However, MHC peptide ligands have yet to be identified in mouse urine so there may be other classes of individuality chemosignals (Brennan 2009). Major urinary proteins can carry many different kinds of odor molecules so they may have a role in carrying the odor molecules of individual difference. It is possible that the female recognizes whatever is different from the signature mixture she has remembered of the first male.

The female's long-term memory of her mate's signature mixture is formed when the mating male's individual odors stimulate a variety of VSNs in the VNO to send signals to a subpopulation of mitral cells and associated granule cells in the AOB glomeruli, the first relay in the vomeronasal pathway (Brennan 2009). The conjunction of the mate's odors and higher levels of norepinephrine (noradrenalin) after mating are thought to prime these inhibitory synapses so that if the same individual odor is smelled again, transmission to the

hypothalamus of the pregnancy-blocking signal from male urinary pheromone is stopped. For the memory to be formed, mating and then exposure to the mate's odor during the sensitive period (3 to 4.5 hours after mating) are required.

Contact with the odor source seems to be needed for pregnancy block (Brennan 2009) but the VNO can also detect volatiles so the need for contact may reflect the way molecules enter the VNO. The male pheromone responsible for the Whitten effect (Section 9.7.1) is different from the one(s) involved in the Bruce effect as males from some mouse inbred strains induce the Bruce effect but not the Whitten effect (Doty 2010).

Currently there is no evidence that the Bruce effect occurs in mice or other rodents under natural conditions (Wolff 2003). One contributory explanation for this might be that in natural conditions the female may control exposure to male odors herself and thus control whether the Bruce effect occurs (Becker &

Hurst 2009). Nonetheless, the Bruce effect has become a very valuable model system for studying olfactory learning and the brain (Brennan 2009).

While the Bruce effect works through the VNO–AOB, there may also be inputs to maintain pregnancy via the MOE–MOB (Brennan 2009; Sanchez-Andrade & Kendrick 2009), again emphasizing the integration of the two systems in higher levels of the brain. Four days or so after mating, the drop in sensitivity to the Bruce effect might be caused by the effect of the post-mating rise in dopamine in the main olfactory bulb (Serguera *et al.* 2008). This rise stimulates dopaminergic neurons that modulate inputs from olfactory sensory neurons to mitral cells in the MOB, making the MOB less sensitive to social odors.

9.9.4 The male effect in sheep and goats

For species such as sheep and goats that breed only at certain times of year, there may be suppression of estrus for most of the year, often controlled by day length, broken by contact with male odors in the breeding season (the "male effect," Delgadillo *et al.* 2009; Hawken & Martin 2012) (Chapter 12) (Section 9.4.1). In sheep, novel males have a much greater effect. This leads me to speculate that the females detect the signature mixture of an unfamiliar male and that his pheromone, the same in all males, is "allowed" to act, in a way analogous to the pheromone of the strange male in the Bruce effect of mice (above). In sheep, the memories of the familiar male occur in the olfactory bulb rather than accessory bulb.

Summary

The olfactory systems found in most multicellular organisms appear to have the same design features, ones that give olfaction a remarkable flexibility for signaling to evolve. These features include olfactory sensory neurons that each express just one or two receptor protein types, and the linking of all sensory neurons with a particular receptor to the same few glomeruli for integration and onward projection of signals to the rest of the brain. Receptor families in different animal taxa have evolved independently from a wide range of different receptor proteins already on the surface membrane. In insects, some CHC pheromones are detected by the gustatory system.

In terrestrial vertebrates both the VNO and the MOE are stimulated by pheromones and signature mixtures. Inputs from the MOE and VNO are integrated at higher brain levels including the amygdala. In social insect colonies, primer effects of pheromones help match caste development of colony members, and their behaviors, to colony needs. Signature mixtures are detected and processed via the olfactory system in both vertebrates and invertebrates. The processing of signature mixtures and pheromones can influence each other.

The recent rapid progress in understanding chemosensory receptors and the organization of chemosensory systems will continue as techniques in molecular biology, behavioral analysis, and neuroanatomy are combined with opportunities made possible by the growing genomics revolution.

Further reading

The Nobel lectures of Linda Buck and Richard Axel offer clear, freely available overviews of the combinatorial logic of olfaction (www.nobelprize.org) (Axel 2005; Buck 2005). Good review chapters on olfaction in *C. elegans*, insects, and mammals in Menini (2010) are currently freely available online. Chapters in Allison and Cardé (2014) cover processing of pheromones by moths and how this evolved. Developments in our understanding of receptors and neural circuits are moving so rapidly that the best sources are recent reviews, for example, in *Annual Reviews*, *Trends*, and *Current Opinion* series.

Chapters in Touhara (2013) cover modern techniques used in pheromone research (on a wide range of invertebrates and vertebrates), including electrophysiology, molecular biology, and bioassays.

Chapters in Wolff and Sherman (2007) cover rodent biology including the ecological background to reproductive pheromones. Chapters in Choleris *et al.* (2013) review the roles of oxytocin and other peptides. For a skeptical view on primer effects in mammals see Doty (2010) (though in my view not all his criticisms are justified) (Chapter 1).

The way our own brain creates flavor from smell and taste is explored by Shepherd (2012).

Be aware when reading the past and current literature that the term "pheromone" is often used ambiguously and may be used in contexts where "signature mixture" or "olfactory cues" might be more accurate or helpful.

For molecular structures see sites such as www.chemspider.com, which allows you to search by common name and shows synonyms as well as the systematic names. See also the Appendix for a short guide to the terminology (available for free download from the website associated with this book). Many pheromones are illustrated on Pherobase www.pherobase.com (El-Sayed 2013).

10 | Finding the source: pheromones and orientation behavior

Finding the source of a pheromone plume in a turbulent flow is a greater challenge than finding an animal producing sound or visual signals. Nonetheless, animals responding to pheromones offer some of the most spectacular examples of remote responses to stimuli. These include male moths attracted to females over hundreds of meters and perhaps even farther. Over evolutionary time, receiving organisms have been selected to search efficiently and to find a source as quickly as possible: many odor sources (whether food or mates) do not last long – the odor source moves or another animal will get there first (Cardé & Willis 2008).

Searching animals exploit the invisible odor "landscape" created by high and low concentrations of countless molecules in overlapping plumes released by other organisms (Atema 2012; Hay 2009; Nevitt 2008). These chemical plumes occur in a wide range of spatial scales and durations. For example, in the sea, they range from the pheromone plume released by a single planktonic copepod, a few millimeters long, to the dimethyl sulfide odor plumes from a plankton "hotspot," kilometers wide. Odor signals have a significant but limited life before the molecules are dispersed or broken down.

How do animals distinguish chemical signals from this chaotic chemical world? Signals may be identified by characteristic fine-scale changes in concentration over time, or by chemical specificity, or both (Atema 1995, 2012; Vickers 2000). Chemical specificity may come from unique compounds, but more often by unique mixtures, including common molecules in specific combinations (Chapters 1 and 3). Animals may evolve to respond to different aspects of these potential signals. Each animal species extracts its unique information from the chemical world and uniquely contributes to it (Atema 1995).

The ability of animals to use pheromones for low-cost, long-distance communication may have important evolutionary implications by allowing animals at low densities to find mates (Chapter 4). For example, in moths, it may facilitate speciation associated with larval specialization on rare plant species (Chapter 3) (Cardé & Mafra-Neto 1997). Similarly, like other deep-sea fish, hatchet fish, *Argyropelecus hemigymnus*, live at extremely low densities and males have the biggest nasal organ relative to body size of any vertebrate, implying that olfaction is very important in mate location (Jumper & Baird 1991).

In this chapter I start by describing some of the mechanisms that animals use to find odor sources before discussing the different kinds of odor stimulus, from diffusion gradients to odor plumes, and the ways that animals have evolved to respond. We know most about the mechanisms used by invertebrates but the principles of olfactory orientation seem to apply to all animals.

10.1 Investigating orientation behavior mechanisms

One of the most successful approaches to understanding orientation behavior has been to explore the precise stimuli that animals use. The different kinds of responses to stimuli are outlined in Box 10.1. Teleological terms, such as "attraction" describe the endpoint or provide a handy metaphor for the "function" or ultimate explanation of a behavior (Kennedy 1986, 1992). However, such terms do not take us nearer the mechanisms. For example, an aggregation of animals near a pheromone source could have been attracted from a distance or instead formed by arresting (stopping) animals that came close to the source by chance. For example there are different volatile attractant and contact arrestant pheromones in the common bedbug, *Cimex lectularius* (Siljander *et al.* 2007, 2008).

At first sight, many animals behave as if they are aiming for a "goal" or target but, without other evidence, it is more accurate to describe their movements as responses to the stimuli received at or up to that point (Kennedy 1986). Most orientation behavior, chemo-orientation in particular, is based on immediate responses to stimuli. However, behaviors initiated and modulated but not steered by the odor stimulus are just as important as external stimuli. These "self-steered" behaviors are internally set (idiothetic) behavior patterns triggered by odor contact (Kennedy 1986).

Box 10.1 The basic mechanisms used by animals for orientation toward or away from a stimulus

Indirect guiding: no directional bias

Kinesis: The animal's movement is affected by the intensity of the stimulus but the direction of turns and movements is not related to gradients in the stimulus concentration.

Orthokinesis: The animal moves faster or slower depending on concentration.

Klinokinesis: The animal turns more or less, makes more or larger, or fewer or smaller turns, depending on concentration.

Direct guiding: directed movement toward a stimulus source

Taxis: The direction of turns and movement is related to the pattern of concentration. The different types of taxis reflect the different ways that animals detect the gradient or pattern of concentration, from single or paired sensors.

Klinotaxis: Directed turns based on successive samples of concentration in different places (one sensor moved from side to side, for "transverse klinotaxis"; one sensor, sampling successively along the animal's path, "longitudinal klinotaxis").

Tropotaxis: Directed turns based on simultaneous comparison of stimulus intensity on two points of the body, or for example the simultaneous inputs from two antennae. This is possible only where the chemical gradient is steep enough to stimulate the two sensors with different intensities.

Teleotaxis: Directed turns based on input from many receptors in a "raster," such as vertebrate or insect eyes (and so far only demonstrated for responses to light). An extreme form of tropotaxis.

Other prefixes can be added: chemo- (in response to chemical stimuli), osmo- (response to molecules in gas or water phase), anemo- (oriented to the wind), rheo- (oriented to water current flow), meno- (orientation at an angle to the direction of the stimulus field). For example, osmotropotaxis is the ability to detect and move up gradients of molecular concentration by simultaneous stimulation of a pair of sensors (as in the honeybee opposite). Designing experiments to distinguish these possible responses can be hard.

Box 10.1 **(cont.)**

(a) klinokinesis

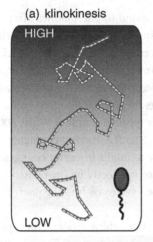

(b) klinotaxis

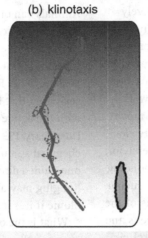

(c) tropotaxis

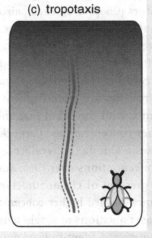

Illustration of the orientation strategy employed by bacteria (a), fruit fly larvae (b) and walking bees (c) in a chemical gradient. The trajectory of the body and chemosensors is represented as a plain and a dashed curve, respectively. Bacteria implement a biased random walk where runs are elongated in the direction of the gradient (klinokinesis) (Berg 2004). Fruit fly larvae are capable of directing their turns toward the gradient (Gomez-Marin *et al.* 2011; Louis *et al.* 2008). This mechanism involves active sampling through lateral head movements (klinotaxis) (Gomez-Marin *et al.* 2011). To orient in an odor gradient, walking bees compare the inputs from their left and right antennae and veer toward the side of highest stimulation (tropotaxis) (Martin 1965). Caption and figure from Gomez-Marin and Louis (2012).

Classification largely based on the scheme of Fraenkel and Gunn (1940) since modified in many ways; see Dusenbery (1992, p. 414), Gomez-Marin and Louis (2012), and Webster and Weissburg (2009).

Movement toward or away from a higher concentration can occur by *kinesis* (indirect) or by *taxis* (direct guidance) (Box 10.1) (Dusenbery 1992; Webster & Weissburg 2009). Depending on the circumstances and their size, many, perhaps most, organisms are capable of using more than one of these strategies (Dusenbery 1992, p. 431).

Kineses are indirectly guided responses, in which the movements of animals are not directed by concentration gradients. Instead, the instantaneous effect of concentration may be to change, for example, the length of path before the next turn. Kinesis movements may be far from random since pheromones may trigger self-steered (idiothetic) turning patterns (Kennedy 1986). Kineses are used by bacteria, sperm, and a variety of small animals, for example by planktonic animals, including fish larvae, to find and stay in food patches (Dusenbery 1992, 2009).

Taxis is orientation with respect to the odor concentration gradient, leading to directly guided behavioral responses toward or away from the higher concentration. These include *klinotaxis*, based on successive sampling with one sensor, and *tropotaxis* (stereo, comparing simultaneous inputs from two widely separated sensors) (Box 10.1). Klinotaxis seems to be limited to smaller organisms. Anemotaxis and rheotaxis are orientations to currents in air and water respectively and many animals use these cues, not concentration gradients, to orient to odor plumes (Section 10.3.3).

The nematode *Caenorhabditis elegans* uses both klinokinesis and klinotaxis (Iino & Yoshida 2009; Lockery 2011). Although it has pairs of left–right chemoreceptors (Chapter 9), each pair effectively acts as one sensor. Straight "runs" are punctuated by "pirouettes" about every 30 seconds, when the animal briefly reverses, makes a deep body bend, and goes forward in this new direction. Klinokinesis occurs as pirouettes are more frequent at lower concentrations, with the effect that animals move toward higher concentrations. *Caenorhabditis elegans* also uses a klinotaxis "weathervane" behavior, sampling concentrations over time, and responding to a spatial gradient of chemoattractant, producing a curving track toward higher concentrations of the chemical. Laser ablations of single sensory cells and neurons have helped identify candidate neural circuits for both behaviors (Iino & Yoshida 2009; Lockery 2011).

Drosophila melanogaster larvae use klinotaxis. At intervals as it crawls forward, a *D. melanogaster* larva casts its head from side to side to actively sample left and right (Box 10.1) (Gomez-Marin & Louis 2012). A larva genetically engineered to have just a single olfactory neuron on one side of the head can still successfully orient by klinotaxis (Louis *et al.* 2008). However, larvae orient more accurately if, as normal, they have bilateral inputs (though this benefit seems to be the pooling of signals to improve signal detection from noise rather than the stereo benefit of tropotaxis).

A first example of tracking odor sources by comparing bilateral arrival time of the odor (exploiting the analogous mechanism that allows us to locate a sound by the arrival time difference to our ears), has been demonstrated in sharks (Gardiner & Atema 2010). The wide separation between the nostrils (nares) of the hammer head shark may have evolved to facilitate this.

Crabs use simultaneous sampling from chemosensors on their right and left side limbs, allowing them to track an odor plume by tropotaxis combined with current flow (Section 10.3.3.2) (Weissburg 2011).

10.2 Ranging behavior: search strategies for finding odor plumes, trails, or gradients

Animals can only respond to pheromones after entering the active space, defined as the zone where the pheromone concentration is at or above their threshold for detecting it (Bossert & Wilson 1963; Wilson 1970). Ranging behavior is the initial stage of searching that brings the animal into first contact with the active space (plume) (Cardé & Willis 2008; Dusenbery 1992, pp. 385 ff.). Having contacted the chemical stimuli, the animal's behavior changes to one guided directly or indirectly by the odors. Ranging may also be used to re-find the pheromone plume if it is lost.

What is the optimal searching strategy for rapid or energetically efficient plume contact? There is surprisingly little experimental data on the search patterns animals use in practice and most work is based on models (Cardé & Willis 2008; Dusenbery 1992, pp. 385 ff.). Much depends on the model's assumptions, including the size of the odor source, flow direction constancy, and the sensory and cognitive abilities of the organism. Should the animal search across the current, upcurrent, or downcurrent? Other possible strategies are random walks (or a subclass of these, Lévy walks), which have trajectories and turn frequencies that are not oriented by the flow direction (James *et al.* 2011; Reynolds 2010). In recent years, various kinds of Lévy walks have been proposed as the optimal mechanisms for many searching organisms, but this is much debated not least because foraging by real animals may not fit the models' assumptions (James *et al.* 2011).

A number of models suggest searching crosswind when the wind direction is steady, but in shifting winds to fly parallel to the average wind direction (see references in Cardé & Willis 2008; Cardé *et al.* 2012). *Drosophila* fruit flies show this behavior in a windtunnel (Zanen *et al.* 1994). Albatross fly a crosswind path in the constant open-ocean winds, turning upwind when they detect food odors (Section 10.3.3) (Nevitt *et al.* 2008). Crosswind flight is also the most energetically efficient mode of flight for albatross. However, there are many examples of insects seeming to make ranging flights in random directions in relation to wind direction and some studies show downwind searching flights (Cardé & Willis 2008; Cardé *et al.* 2012), for example in moths tagged with transponders (Reynolds *et al.* 2007).

10.3 Finding the source: orientation to pheromones

All life lives in a fluid medium. Although air and water differ in characteristics such as density and viscosity, the same laws apply to both media (Denny 1993; Dusenbery 1992; Koehl 2006; Webster & Weissburg 2009). Within either air or water, animals may respond to three main kinds of odor stimulus. The first is short-range diffusion of molecules from a point source, such as an injured ant releasing alarm pheromone; the second is a trail of pheromone laid on the substrate behind a moving animal; the third is a plume of molecules carried away from the odor source by turbulent wind or water currents, in which diffusion plays almost no role (Box 10.2). The rest of this chapter is organized around the characteristics of these three types of odor stimulus and how animals have independently evolved orientation mechanisms to cope with the different challenges these stimuli offer.

10.3.1 Diffusion very close to the source: orientation of animals to diffusing signals

A limited number of pheromone signals, particularly for animals the size of ants and smaller, act within a short range and reach the receiver by diffusion alone (Dusenbery 2009; Webster & Weissburg 2009). This is a special case: the pheromones of most animals much more than a centimeter long are carried by currents of air or water (see Section 10.3.3).

From an instantaneous release of pheromone near the ground, the active space (which is the signal) expands as a hemispherical cloud of pheromone (Bossert & Wilson 1963; Dusenbery 1992). The radius of the active space grows to a maximum, and then starts to contract and fades away to below the behavioral threshold as outwards diffusion continues. In models of the diffusion of pheromone molecules the active space is related first to the *quantity* of pheromone (Q); second, to the sensitivity of the receiver, its *threshold* (K), and; third, to the *diffusion coefficient* (D) of the pheromone (which is correlated with its molecular weight) (Bossert & Wilson 1963). In response to the selection pressures on the signaling system (Chapter 1), over evolutionary time each of these values can be changed to give an active space that is larger or smaller, quicker or slower, to expand to its maximum, faster or slower to fade. The characteristics of a pheromone system can be expressed as the Q:K ratio.

For an instantaneous pheromone signal, the maximum detection distance depends only on the Q:K ratio. An increase in the active space can be achieved by either increasing the amount of pheromone released (Q) or lowering K (more sensitive reception), or both. Increasing the sensitivity is more likely as increases of orders of magnitude in sensitivity of chemoreceptors are possible, whereas producing more pheromone to have the equivalent effect would require a big increase in gland size (Hölldobler & Wilson 1990, p. 244). Aquatic organisms at the small scale of rotifers may find this particularly limiting (Box 10.3) (Dusenbery & Snell 1995).

For continuously emitted signals, maximum range and duration are directly proportional to the Q:K ratio and inversely proportional to the diffusion coefficient (D) of the pheromone. For example, a shorter fade-out time, making the signal more sharply pinpointed in space and time, can be achieved by lowering the quantity of pheromone (Q) or raising the threshold, K (making the receiver less sensitive), or both (Hölldobler & Wilson 1990, p.244).

Box 10.2 Describing currents and flow

Explaining chemical communication at a distance, with molecules carried in air or water between signaler and receiver, requires an understanding of fluid properties and flow: the fluid environment determines the spatial and temporal patterns of the arrival of odor molecules to the receiver's sensory system (Weissburg 2000, 2011).

The same fluid laws apply to both air and water. The transmission and delivery of chemical cues are controlled by flow physics, and the mode of transport is dictated by a combination of Reynolds number (a product of speed × size, divided by the viscosity of the medium):

$$Re = \frac{\rho UL}{\mu} = \frac{UL}{\nu}$$

and Péclet number:

$$Pe = \frac{UL}{\Gamma}$$

where U is the characteristic velocity; L is the characteristic length; ρ is the fluid density; μ and ν are the dynamic and kinematic viscosities, respectively; and Γ is the molecular diffusivity (Vogel 1994; Webster & Weissburg 2009; Weissburg 2000). Although air is less viscous than water, it is also less dense and has a higher kinematic viscosity (e.g., at 20°C, $\nu_{air} = 15 \times 10^{-6}$ m^2/s and $\nu_{water} = 1 \times 10^{-6}$ m^2/s (Vogel 1994, p. 23)). Incidentally, as Re is a dimensionless parameter you can explore the flow effects in air of a structure such as a wing or antenna by using the greater density of water and appropriately scaling down the size of the model or decreasing the fluid speed (Vogel 1994).

If there is no flow then molecular diffusion dominates (Section 10.3.1). With an Re of less than about 100, viscosity dominates, and the character of the flow is orderly and predictable *laminar-advective* transport. Particles move along streamlines that remain parallel to each other, a characteristic referred to as laminar flow (such as experienced by plankton, Box 10.3) (Webster & Weissburg 2009; Weissburg 2000). Larger values of $Re > 10^4$–10^6 generally indicate turbulent conditions in which the dominant flow is *turbulent-advective* transport (Figure 10.4) with chaotic patterns of filaments of air or water containing odor molecules.

The value of the Péclet number is a ratio of flow-driven transport to molecular diffusion – so at low Pe, diffusion dominates molecular movement; at high Pe, advection (bulk flow) dominates.

The boundary layer close to a surface affects flow speeds and patterns. Boundary layers and Re are also important for the design of sense organs (Boxes 9.2 and 9.3) (Koehl 2006, 2011).

10.3.1.1 Air vs. water

As the maximum range of a diffusing chemical signal is independent of the diffusion rate (D), the range of signals is approximately the same in air and water. However, in water the signal takes far longer to travel and then fade; as the time scale is inversely proportional to D, the kinetics affecting rise-time and fade-out will be slowed 10,000 times in water as typical diffusivity values (in m^2/s) for small molecules are 2×10^{5} in air, but 2×10^{9} in water (Dusenbery 1992, p. 62; Koehl 2006). For the same substance to give the same times to maximum radius and fade-out in water, the Q:K ratio would need to be about a million times greater in water than air (Wilson 1970). That is, the amount of pheromone released (Q) would have to be a million times greater, or the sensitivity threshold lowered by the same amount. Wilson suggests that by using highly soluble polar molecules (for example, anthopleurine) (Chapter 8), such increases in Q might be possible in water.

Animal size will affect communication strategies. Smaller animals, 1 to 2 mm in diameter, live in quite a different fluid world (Box 10.3) in which laminar flow and diffusion dominate (Dusenbery 2009). However, larger animal species rely instead on natural currents, not diffusion, to carry their messages to the receiving animal (Box 10.2) (Section 10.3.3). Animals that live within the boundary layer (see Section 10.3.3) may be able to exploit natural currents if they can release their pheromone into the faster flowing fluids above, by lifting their pheromone gland up off a leaf (in air) or squirting pheromone out with some force (in water) (Conner & Best 1988; Vogel 1994). It is energetically too expensive to propel water or air very far but some animals briefly create their own current for short-range signaling. For example, male lobsters jet their urine signal out of their dens toward females (Chapter 3) and male blue crabs use their legs as paddles to waft water containing pheromone toward females (Kamio et al. 2008). Similarly, when male moths reach a female, they direct their pheromones toward her by vigorously fanning their wings (Figure 1.5).

10.3.1.2 Ant alarm pheromones: a case study of a diffusing signal

The active space of the alarm signal of the harvester ant, *Pogonomyrmex badius*, would have a maximum radius of about 6 cm, given its Q:K and D (Bossert & Wilson 1963, p. 208; Hölldobler & Wilson 2009). This would be reached in 13 seconds and the active space signal would shrink to zero in about 30 seconds (ideal for an alarm pheromone, with a quick rise time, short range, and rapid fade).

The alarm pheromone of *P. badius* has only one component (4-methyl-3-heptanone), but many social insects have multicomponent alarm pheromones (Chapter 8). The various components may simultaneously alert, attract, and evoke aggression (Hölldobler & Wilson 1990, p. 261; 2009, p. 209). Is there one active space or several (one for each component)? The African weaver ant, *Oecophylla longinoda*, has four major alarm pheromone components, which diffuse from the source to create overlapping independent active spaces expanding at different rates and eliciting different behaviors (Figure 10.1) (Bradshaw et al. 1975, 1979). The work so far on ant alarm pheromones suggests that in the nest and close to the substrate outside it, ants are working within the boundary layer of slow air next to the surface and diffusion dominates so that each pheromone component has its own active space. I wonder if this could be reinvestigated now that we can follow the responses of individual ants and compare this with models of their behavior and the stimulus they will be receiving (e.g., Sumpter et al. 2012).

10.3.2 Following trails laid on a substrate

Trails consist of odor molecules left on the substrate by the signaling animal as it moves along (Chapter 7). They are important for a wide range of animals from snails to vertebrates, including hamster and salamander males following female pheromone trails (Chapter 3). However, the mechanisms used in trail following have only been studied in a few species.

Ants use pheromones to mark trails back to the nest from good food sources (Chapter 7). An indication of the sensitivity of ants is that 4 fg/cm of its trail pheromone is sufficient to elicit trail following in half

the leafcutter ants, *Atta vollenweideri*, tested (one milligram of the pheromone could lead a worker 60 times round the world) (Hölldobler & Wilson 2009, p. 206; Kleineidam *et al.* 2007) (see also Chapter 9 for caste-specific neural specializations for trail following).

Ant trail pheromones are volatile low molecular weight compounds so the signal is detectable only

---

Box 10.3 Pheromone signals in the plankton: 3D-trails in water

Pelagic copepods are small (0.1 to 10 mm) aquatic crustaceans that swim in the plankton, often separated by thousands of body lengths (Kiørboe 2011; Yen & Lasley 2011). In order to increase the chance of finding mates and prey, they aggregate, often using chemokinesis to remain within richer patches (see Weissburg *et al.* 1998).

Planktonic animals of this size live at a scale with low (<10) Reynolds numbers (*Re*) (Box 10.2), dominated by high viscous forces and laminar flow, likened to swimming in treacle (Denny 1993; Dusenbery 2009; Vogel 1994). In these conditions, pheromone trails left by moving female copepods persist as trails for seconds or minutes, like vapor trails behind airplanes. The trail that a female copepod leaves increases her effective size to 130 times her body length, thus also increasing the area encountered by the searching male (see figure opposite). Male copepods use chemotaxis to exploit these unique 3D pheromone trails, detecting trails up to 10 to 20 seconds old and pursuing females up to 100 body lengths away along the trail (Bagøien & Kiørboe 2005; Doall *et al.* 1998; Weissburg *et al.* 1998). The males may use receptors on their paired antennae for tropotaxis as well as exploiting sequential klinotaxis to follow the chemical trail. The ability to use just chemical cues to follow a trail in 3D, as ants do in 2D, is most unusual (Yen & Lasley 2011; Yen *et al.* 1998). Initial tracking is mediated by the diffusible pheromone while the final leap of the male to contact the female is mediated by the hydrodynamic wake generated by the female (Yen *et al.* 1998). Males also seem able to detect and reverse if they find they are going in the wrong direction along the trail. In some other species of copepod, the female hovers on the spot and creates a near-spherical cloud of pheromone that males zigzag through to find her (Bagøien & Kiørboe 2005).

Modeling of diffusion ranges and costs of locomotion suggests that mate location using diffusable pheromones in water is worthwhile only above a critical size of animal, larger than 0.2 to 5 mm or more (Dusenbery 2009; Dusenbery & Snell 1995). Thus planktonic copepods, which are within this size range, do use pheromones but most rotifers, which are smaller, do not.

Box 10.3 (cont.)

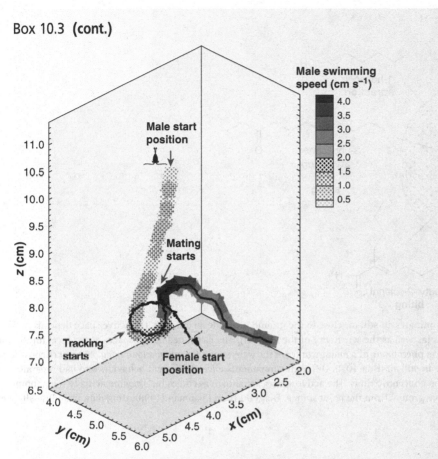

The mate-tracking trajectory of the male copepod, *Temora longicornis*, which has a body length of approximately 2 mm, is shown via the thick line, with the color indicating swimming speed. The trajectory of the female, which is coincident with her pheromone trail, is illustrated by the thin black line. Male swimming speed increases after contacting the female trail, and his path accurately follows the female trajectory. Data courtesy of Jeannette Yen. Figure and caption adapted from Webster and Weissburg (2009).

briefly before evaporating. A fire ant, *Solenopsis saevissima*, trail on glass lasts just a few minutes but other ant species may have trails lasting days. The short life of most trail pheromones matches foraging effort to the changing value of the resources at the trail end: when the ants no longer find food at the end of the trail they no longer mark the trail on their return and thus the trail fades away (Chapter 7). If the molecules were too persistent, trails would lead to resources long gone.

Ants following a trail respond to the airborne molecules of pheromone diffusing from the marks rather than with a contact response to the trail itself: fire ants can be drawn along in response to molecules diffusing from a droplet of pheromone held on a glass rod above them (Wilson 1962). Over the trail there is a semi-ellipsoidal active space (Bossert & Wilson 1963). As ants travel through this "vapor tunnel," they sweep their antennae from side to side (Hölldobler & Wilson 1990, p. 268). The ant uses tropotaxis, comparing

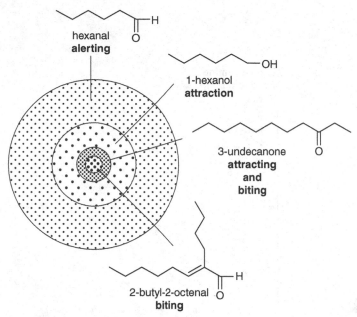

hexanal
alerting

1-hexanol
attraction

3-undecanone
**attracting
and
biting**

2-butyl-2-octenal
biting

Figure 10.1 Where diffusion dominates, in still air close to the ground or inside an ant nest, the active space depends on the diffusion speeds of molecules (as well as the sensitivity of the receiver). The calculated active spaces at ant level for the four major components of the alarm pheromone of a major worker of the weaver ant, *Oecophylla longinoda*, 20 s after deposition at a central point of a flat surface in still air (Box 10.2). Different components elicited different behaviors and had separate active spaces represented here by the concentric rings. The active spaces are shown as circles but they are actually overlapping hemispheres that spread above ground from the point source. Diagram from Chapman (1998) using data from Bradshaw *et al.* (1975).

stimulus inputs from left and right antennae (Calenbuhr & Deneubourg 1992; Calenbuhr *et al.* 1992; Hangartner 1967). However, when one antenna is removed the ant can still follow the trail, presumably by klinotaxis, although it does overshoot more (Hangartner 1967).

Both tropotaxis and klinotaxis are implicated in different species of termites, which have evolved trail laying and trail following behavior independently of ants (Leuthold 1975, cited in Calenbuhr *et al.* 1992). Termites respond to trails of sternal gland marks by olfaction, like the ants above, but termite responses to labial gland pheromone involve contact as does trail following by tent caterpillars (Chapter 7) (Bordereau & Pasteels 2011; Kaib 1999; Roessingh *et al.* 1988).

Trail following is important for snakes and some lizards for finding mates and food (Mason & Parker 2010). Snakes following a trail show a characteristic rapid tongue-flicking, touching the ground about once a second, picking up molecules to deliver to the paired VNO openings in the mouth roof. All snakes have forked tongues, as do lizard species that follow scent trails to hunt prey (Schwenk 1994). It has been suggested that, as with ants, snakes perhaps use tropotaxis to follow the trail, comparing the left and right inputs (Schwenk 1994). Ford (1986) suggested this explained the behavior of male garter snakes, *Thamnophis sirtalis parietalis*, following a female's trail. However, eliminating the VNO input on one side in rattlesnakes did not affect trailing behavior (Parker *et al.* 2008), so perhaps the forked tongue's spread may be to sample a wider area.

Whether or not it is using tropotaxis, if both tongue tips leave the trail, the snake stops and swings its head from side to side, in behavior more like klinotaxis, tongue-flicking until contact is made with the trail again (Ford 1986). This gives a zigzag motion (though with the head rather than the whole body), a movement pattern seen in swimming and flying animals (Section 10.3.3). Humans show a similar zigzag path when blindfolded and asked to use smell to follow, on their hands and knees, a trail of chocolate odor on grass (Porter et al. 2007). Human nostrils provide spatially distinct information and people were more accurate and faster in following the trail when sniffing from two separate nostrils rather than just one. Rats also smell in stereo and can use simultaneous comparison of inputs from their left and right nostril to identify the direction of an odor source (Rajan et al. 2006); brain circuits potentially involved in comparing the inputs have been found in the anterior olfactory nucleus of the olfactory cortex (Kikuta et al. 2010).

Dogs also zigzag along a trail but, unlike ourselves, they seem to be able to work out the direction of the trail (below).

10.3.2.1 Trail orientation: which way to go?

Most animal species cannot detect the direction of a trail. Tropotaxis does not provide any information on which direction the trail goes, as the longitudinal concentration gradient would be too small to detect. There is no evidence that ant trails are polarized or that ants can determine which way is home from the pheromone trail alone (Hölldobler & Wilson 1990, p. 269 ff.; Jackson et al. 2004). Many ant species use celestial or visual landmarks for return to the nest and do not lay trails outbound (Hölldobler & Wilson 2009). However, the almost blind Pharaoh's ants, Monomorium pharaonis, can use the branching angles of their foraging networks for orientation: in experiments, nest-bound foragers turned round if trail branching angles indicated they were walking away from the nest (Figure 10.2)

(Jackson et al. 2004). The pheromone trail itself had no polarity.

Some animals can apparently detect trail polarity. Trigona stingless bees lay a trail of pheromone "spots" (Chapter 7) and the stingless bees can follow the polarity of the trails in the direction of the food source (Nieh et al. 2004). Some snails and slugs have the ability to follow conspecific trails in the direction they were laid down but the mechanism is not known (e.g., Clifford et al. 2003; Davis 2007; Wells & Buckley 1972). Planktonic copepod males can follow females in the correct direction (see Box 10.3).

Snakes seem to be able to determine trail direction. Male garter snakes, Thamnophis sirtalis parietalis, can correctly follow the direction of a female's trail but only if she has moved on a rough surface, not a smooth one (Ford & Low 1984). The males seem to be able to detect which side of the pegs the female pushed against while moving forward and choose to go in that direction. Although males can follow female trails left on a smooth paper surface, they are just as likely to go in either direction.

The ability to detect the direction of track would also be very useful to predators such as wolves and dogs. Müller-Schwarze (2006, p. 414 ff.) describes ingenious experiments on dog tracking behavior from the 1920s and 1930s, which allowed the separation of human odors and the odors of crushed plants and shoe leather left by the footprints. In more recent experiments, four trained German shepherd dogs always followed a person's track in the correct direction (Figure 10.3) (Thesen et al. 1993). The tracking behavior was divided into three phases. During the searching phase (equivalent to ranging) they moved quickly, sniffing ten to 20 times per breath. Once the dogs had found the trail, in the deciding phase, they sniffed for a longer period and slowed down, sniffing at two to five footprints. In the tracking phase, once the direction had been established, the animals moved quickly. If odors are their cue, remarkably the dogs are detecting a difference in the concentration of scent in the air just above two consecutive footprints, made one second apart, up to 20 minutes earlier (Thesen et al. 1993). Hepper and Wells

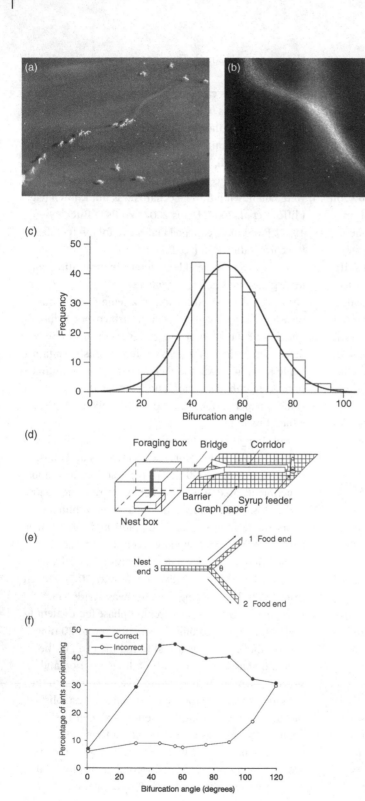

Figure 10.2 How the Pharaoh's ant, *Monomorium pharaonsis*, uses trail angles at bifurcations to decide which direction to follow. (a) Trails on a smoked glass surface. Part of a network showing bifurcations to smaller trails (scale bar, 1 cm). (b) Close-up of a single bifurcation (scale bar, 0.5 cm). (c) Analysis of bifurcation angles from four separate trail networks: mean angle 53.4°, s.d. ± 14.8°, n = 321; mean distance between bifurcations = 2.9 cm, s.d. ± 2.3 cm, n = 485. Solid line, normal distribution curve. (d) Experimental set-up used to form straight trails and assembly of trail bifurcations. Foraging ant traffic was constrained to produce straight pheromone trails on paper, which were cut into sections (e) and reassembled to form the requisite trail bifurcations with variable angle θ = 0–120°. Start locations were designated as 3, the nest direction (left) and 1 or 2, the food directions (right). Arrows show that the original direction to food on the trail was preserved when bifurcations were prepared. (f) Outcomes of individual trail following tests using straight (0°) and bifurcating (30°–120°) trail sections. Percentage of foragers making correct and incorrect reorientations on meeting a bifurcation (correct for fed workers would be nestward, for unfed workers would be toward food). For each point n = 200, except at 0° where n = 100. At 120° the angle gives no information on trail polarity so the ants make equal numbers of correct and incorrect choices. Figure from Jackson *et al.* (2004).

(a)

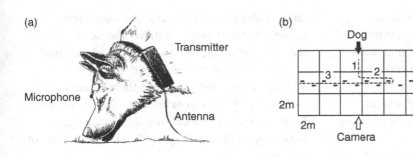

(b)

Figure 10.3 Dog behavior while tracking footprints. (a) The position of the microphone and transmitter while the dog was tracking. (b) How the footprints were placed in the 2 × 2 m grid, and one example of a dog's path during the searching phase (1), the deciding phase (2), and the tracking phase (3). Only 18 of the 40 squares are shown. Figure from Thesen *et al.* (1993).

(a)

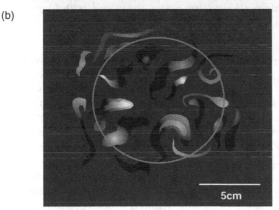

(b)

Figure 10.4 Two perspectives on an instantaneous view of smoke visualizations of wind-borne odor plumes. Plumes of dye in water look the same. (a) A single field from a video recording of a TiCl$_4$ smoke plume in a laboratory wind tunnel viewed from above. The time-averaged plume boundaries are represented by lines running from the upwind to the downwind side of the image (i.e., from left to right). In this image, the wind is blowing at 75 cm/s, and the source is a 0.7 cm filter paper disk to which a drop of TiCl$_4$ has been applied. Since this is the same source used in insect odor-tracking experiments, according to the laws of fluid dynamics, the structure of the plumes should be similar to the odor plume experienced by the insects. (b) A schematic approximation illustrating the tracker's eye view of the smoke plume visualization depicted in (a) above. The circle represents an extension of the time-averaged plume boundaries in (a) onto this cross section. White dashed line is approximately 1 m downwind from the source. (a) Reproduced from Webb *et al.* (2004). Whole image and caption from Willis (2008a).

(2005) also found that dogs needed about five footsteps to establish the direction of the trail.

10.3.3 Orienting to an odor plume in air or water

For most animals larger than 1 mm, pheromone communication probably occurs outside the smooth boundary layers of still air or water in the millimeters next to the surface, and quite beyond the physical scale at which diffusion dominates (Box 10.2) (Denny 1993; Koehl 2006; Vogel 1994; Webster & Weissburg 2009). Odorant molecules are carried great distances by "advection," the bulk flow of air or water across a habitat. Far from being a smooth concentration gradient from the source, odor plumes are turbulent and unpredictable (chaotic) (Figure 10.4) (Atema 2012; Koehl 2006; Webster & Weissburg 2009). Gradients along plumes are too small or noisy to detect so animals cannot evaluate progress toward an odor source by the increase or decrease in odor concentration. The chaotic nature of the signal means it is very difficult to compute the average concentration at any one point, even if time allowed. It takes too much sampling (Webster & Weissburg 2001). Nevertheless, despite the formidable difficulties of following the track of a fluctuating, meandering pheromone plume, animals can find the odor source.

In most cases, animals require information from multiple sensory systems to track a plume successfully, allowing the animal to head upcurrent while in contact with the odor plume (Willis 2008a,b). Superficially very similar behavior is shown by animals following an odor plume, whether in air or water, or flying, swimming, or walking: insects, fish, birds, and mammals all follow a zigzag path upcurrent (Figure 10.5)(Willis 2008a,b). The orientation behavior can be to any odor and is not specific to pheromones: the orientation behavior of male moths to pheromone and female moths to plant odors is almost indistinguishable – and indeed gynadromorphs behave with the other sex's searching behavior if given transplanted antennae of the other sex at the pupal stage (Chapter 9) (Kalberer *et al.* 2010). However, these zigzag tracks, common to animals from insects to fish, could result from different behavioral solutions, evolved independently, to the challenges of locating the source of a plume.

10.3.3.1 Dynamics of plume structure

Turbulent odor plumes form as air or water currents disperse odor molecules from their source. Odor plumes are of course normally invisible but swirling smoke clouds from a chimney provide a good visual analogy of the important features (Atema 2012; Murlis *et al.* 1992). The smoke forms a meandering cloud that snakes downwind. If you get closer, you can see the fine-scale structure within the clouds, with filaments

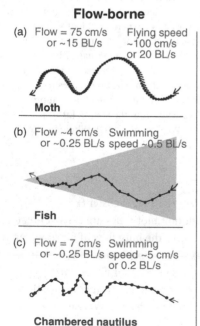

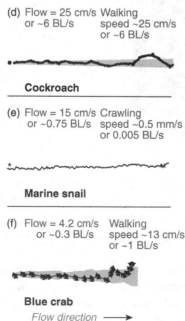

Figure 10.5 Paths of swimming or flying plume trackers contrasted with substrate-based ones (all with speed converted to body lengths, BL). (a) male moth tracking female pheromone (Webb *et al.* 2004). (b) Antarctic fish tracking a water-borne plume of prey extract (Montgomery *et al.* 1999). (c) Nautilus tracking a water-borne plume of prey extract (Basil *et al.* 2000). (d) Male cockroach tracking female pheromone (Willis & Avondet 2005). (e) Predatory marine snail tracking prey odor (Ferner & Weissburg 2005). (f) Path of a blue crab tracking prey odor (see also Figure 10.8)(Zimmer-Faust *et al.* 1995). In all images the fluid flows from left to right. Shading = time-averaged plume, BL/s = body lengths/second. See also the track of a flying bird on a scale of kilometers, Figure 10.9. Figure and caption from Willis (2008b).

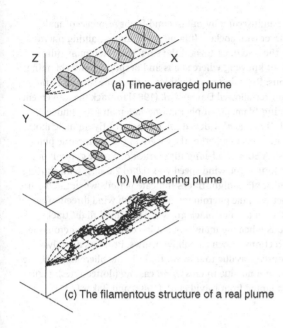

(a) Time-averaged plume

(b) Meandering plume

(c) The filamentous structure of a real plume

Figure 10.6 Models of odor plumes have changed in recent years, reflecting increasing understanding of actual plumes. The first models proposed (a) a time-averaged Gaussian plume. A more accurate model (b) includes the meandering caused by changing wind direction in a plume model with concentration in each disk distributed normally about the meandering center line. The structure of a real plume (c) in air or water has a filamentous structure of odor-filled filaments separated by clean medium. Figure from Murlis *et al.* (1992).

(patches) of high concentration interspersed with cleaner air.

As a cloud of odor molecules moves from the source, turbulence tears apart the cloud into elongated odor-containing filaments, each only a few millimeters wide, separated by "clean" water or air containing few or no odorant molecules (Figure 10.4, Figure 10.6) (Atema 2012; Koehl 2006; Webster & Weissburg 2009). The turbulent effects are greater than diffusion (which is comparatively slow, about 10^{-9} m^2/s in water, see above) and an important consequence is that a plume is far from a uniform cloud of pheromone drifting downwind, rather it is composed of filaments that maintain the concentration and chemical composition they start off with from the calling female moth, for example – which allows male moths to respond only to filaments containing the multicomponent blend from their species (Chapter 3) (Section 10.3.3.3) (Linn *et al.* 1987). Thus the pheromone concentrations within the filaments will be above the response threshold much further downstream than a diffusion model would predict – but in a spreading plume, far

downstream, the odor filaments may be widely spaced.

Two characteristics of odor plumes combine to give the intermittent signal that presents so many challenges to a receiver downstream. The first is the fine filament structure and small-scale eddies. The second is the changing current direction, on a scale of seconds or minutes, which leads to the plume wandering across the landscape (made visible in field experiments with neutrally buoyant balloons or soap bubbles) (Figure 10.7). The result is that an animal (or detector) at a point downstream receives a highly intermittent signal with two time scales of variation: of minutes or seconds with signal, and as much as 80% of the time without, as the plume meanders over the detector and away again; and on a finer scale of milliseconds while the plume is over the detector, as individual odor filaments in the plume touch the detector (Koehl 2006). Because of water's greater viscosity, relevant scales of turbulence for plumes in water are orders of magnitude smaller than in air (Box 10.2) (Koehl 2006; Webster & Weissburg 2009). Incidentally, the levels of turbulence in air or water make it unlikely that releasing pheromone in pulses, as

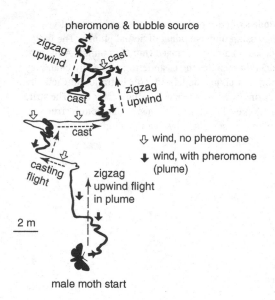

pheromone & bubble source

zigzag upwind

cast

cast

zigzag upwind

cast

casting flight

zigzag upwind flight in plume

⇩ wind, no pheromone

⬇ wind, with pheromone (plume)

2 m

male moth start

Figure 10.7 The millisecond by millisecond flight response of male moths to contact with pockets (filaments) of air containing phero-mone allows them to track toward the source of pheromone (the female) without knowing where she is and despite rapidly changing wind directions. Releasing soap bubbles at the same spot (*) as the pheromone source allowed David *et al.* (1983) to track the movement of air containing filaments of pheromone-laden air (the "plume") across a field. David *et al.* videoed the response of flying male moths and could match this to whether they were in the pheromone plume (with bubbles). A 50-second-long flight track of a male gypsy moth is shown in the figure. The wind speed was 0.8 to 2.0 m/s, with shifting directions. Thick black flight tracks indicate periods when the moth was in contact with the pheromone plume, with wind directions at these times shown as thick black arrows. Thin-lined flight tracks indicate periods when the male was not in contact with pheromone, with wind directions shown as hollow arrows. Progress directly toward the source was due to upwind flight in the pheromone plume (dashed lines) and not due to crosswind casting (dotted lines). Figure and caption adapted from David *et al.* (1983) and Baker (2011).

one moth species, *Utetheisa ornatrix*, does (Conner *et al.* 1980), could be detected any distance downwind (even if, theoretically, pulses with a higher starting concentration might be detectable further away, Dusenbery 1989).

The chemosensory cells and central processing cir-cuitry of orienting animals have evolved to respond on a millisecond time scale to instantaneous changes in concentration as they contact the fine filament struc-ture, rather than (with few exceptions) time averaging. Detecting the fine structure of odor plumes at the spatial and temporal resolution relevant to animals has been a major challenge to scientists. Animals are many orders of magnitude more sensitive than the most sensitive instruments. One way to reveal the fine structure of plumes is to add a tracer gas such as propylene that can be detected by physical or chemical instruments or a dye that fluoresces in a laser beam (Figure 10.8) (Koehl 2006; Vickers 2006; Webster & Weissburg 2009). Electrical recordings from insect antennae in the field confirm that the insect does receive the kind of intermittent signal predicted by tracer studies (Chapter 2) (e.g., van der Pers & Minks 1997).

10.3.3.2 Animals moving in contact with the substrate

Walking and crawling animals in many habitats (on land as well as in water) orient to odor plumes (Weissburg 2011; Willis 2008a,b). As the molecules move downstream in the flow, going upstream will bring you toward the source. Animals respond to instantaneous contact with odor filaments by odor-gated rheotaxis (rheo = current) in water and anemo-taxis (anemo = wind) on land. Animals in contact with the substrate can detect the flow direction by deflec-tion of their mechanoreceptors by the current. Flying insects and fish also use these odor-gated mechanisms, though detecting the flow direction is more challeng-ing (Section 10.3.3.3).

In walking decapod Crustacea, hydrodynamic mechanoreceptor and chemosensory inputs from its antennules are integrated by inter-neurons in the brain, perhaps allowing them to detect the boundaries of odor plumes (Mellon 2012; Schmidt & Mellon 2011). The detection of the odorant molecules by crustaceans is facilitated by the flicking of their antennae to "sniff" (Boxes 9.2 and 9.3), speeding up contact with their

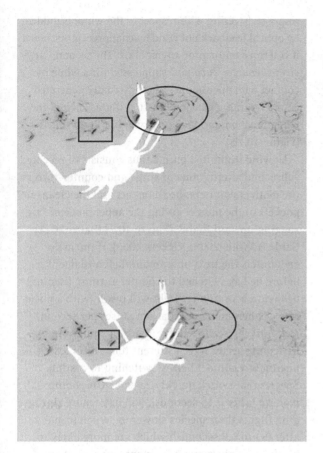

Figure 10.8 Real-time visualization of a blue crab, *Callinectes sapidus*, navigating to a turbulent plume. Panels show two successive frames illuminated by a thin laser sheet approximately 3 cm above the bed, separated by 200 ms. Contact with small odor filaments at the antennules (boxed region) guides upstream progress and turning is mediated by the spatial asymmetry, e.g., more filaments arriving at the right legs (circled region) compared to left legs. This animal surged upstream and to its right (arrow) subsequent to this encounter. Caption and figure from Weissburg (2011).

chemoreceptors (Koehl 2006, 2011; Reidenbach & Koehl 2011).

Some species can use a chemotactic response to odor, via klinotaxis or tropotaxis, as well as rheotaxis or anemotaxis triggered by the presence of odor. For example, walking male cockroaches use positive anemotaxis to walk upwind to a pheromone source but also seem to guide many of their turns back into the plume by bilateral comparisons (tropotaxis) of the inputs from their two antennae, as do lobsters (Weissburg 2011; Willis 2008a,b). Blue crabs, *Callinectes sapidus*, in estuarine creeks use a combination of odor-triggered rheotaxis and tropotaxis to orient up odor plumes (Figure 10.8) (Weissburg 2011). The crabs keep within the plume by chemotactic

responses, turning back if the chemoreceptors on some legs hit clean water (Page *et al.* 2011). For both cockroaches and crabs, tropotaxis may work because of their slow speed of movement relative to the changing filaments, whereas for moths moving at speed such comparisons may not work (Weissburg 2011; Willis 2008a,b). It may also be a matter of size – crabs and cockroaches are large enough relative to their plume structures to detect useful differences between sensors in and out of the plume (Weissburg 2011).

It is possible that a few slow-moving animals, such as large marine snails travelling at just millimeters per second, may be able to use a different orientation mechanism, klinotaxis by time averaging over a longer period and comparing successive sampling to determine in which direction to move (Figure 10.5) (Ferner & Weissburg 2005; Weissburg 2011).

10.3.3.3 Swimming and flying animals

Like walking animals, flying or swimming animals orienting up to an odor plume need to combine olfaction with another sensory input to guide them upcurrent. Unlike walking animals, they cannot use simple deflection of mechanoreceptors to know if they are making progress upcurrent – their entire measurable world is drifting with them. Imagine swimming in a river: with your eyes shut you cannot tell if you are making any progress against the current; only seeing landmarks on the bank shows you. This is the mechanism used by moths: seeing the movement of the ground below tells them how much they are progressing upwind (optomotor anemotaxis, see below). Fish use their lateral line system to direct movement upcurrent for rheotaxis. These mechanisms are explored below.

The best studied responses of flying animals to pheromones are undoubtedly those of male moths responding to female pheromones (Baker 2011; Cardé & Willis 2008). The turbulent, filamentous nature of odor plumes plays a crucial role in upwind orientation. Most moths cannot fly upwind without the intermittent signal from the pheromone filaments in a turbulent plume. (*Drosophila* may be different, as experiments suggest they continue to fly upwind to food-related odors in a homogeneous cloud (Duistermars *et al.* 2009; Frye 2010; Gaudry *et al.* 2012)).

Upwind pheromone-modulated orientation by moths is the product of two parallel control systems working together (Baker 2011; Cardé & Willis 2008; Vickers 2006). The first, "*optomotor anemotaxis*," activated on contact with pheromone, steers the moth to fly more directly upwind, resulting in a brief "surge" upwind (Figure 10.7). This upwind anemotactic response (steering with respect to the wind) is guided by optical feedback not wind mechanoreceptors, hence it is termed optomotor anemotaxis. The second, "*self-steered counter-turning*," suppressed for a while by contact with pheromone, is an internally generated counter-turning causing "casting flights" across the wind direction, producing zigzagging flight tracks (Figure 10.7).

Upwind flight in a plume thus consists of repeated subsecond alternations of surge and counter-turn as the moth encounters the filaments and the clean-air pockets of the plume, giving the appearance of "zigzagging upwind flight" (Figure 10.7) (Baker 2011; Cardé & Willis 2008; Vickers 2006). If the male encounters the next pheromone-laden filament before he has switched to counter-turning (casting) behavior, a straighter flight will result (with a pulse rate of pheromone puffs at ten pulses per second, *Heliothis virescens* will fly almost a straight track). Different species have different times after losing pheromone contact before switching to casting. Species that switch quickly, such as *Grapholita molesta* (after 0.15 seconds), will have more zigzagging flights than species slower to switch (usually after 0.3 to 0.6 seconds), which are more likely to hit the next pheromone filament before switching to casting. If the moth loses the plume altogether the casting behavior becomes more marked and covers a wider left-right crosswind "casting flight" (Figure 10.7). The casting behavior increases the chance of regaining contact with the shifting plume.

Unlike the walking crabs, the counter-turning in moths is generated internally and is not a tropotactic response to concentration. The casting turns when pheromone is lost are self-steered, not chemotaxis. Male moths are not comparing inputs from their paired antennae: with only one antenna they can still fly up the center of an odor plume just as well (Vickers & Baker 1991). (Again, *Drosophila* may be different, as they seem able to compare odor inputs between the antennae, though how they do this is unclear (Duistermars *et al.* 2009; Frye 2010; Gaudry *et al.* 2012)).

Male moths make subsecond orientation responses based on both the fine temporal structure and the chemical content of filaments in pheromone plumes (Baker 2011; Cardé & Willis 2008; Vickers 2006). The fast response may be facilitated by the fast "ON"(millisecond or submillisecond) ionotropic transduction mechanism used in insect olfactory sensory neurons (Silbering & Benton 2010) (see Chapter 9 for this and the fast "OFF" to signals). Moth species have different multicomponent pheromones (Chapter 3) (de Bruyne & Baker 2008). Male moths can fly up a plume made up of alternate puffs of their own species' sex pheromone and puffs of the pheromone of a sympatric species, even if the puffs are separated by just 1 mm and by at most 0.001 seconds (Baker *et al.* 1998). This is achieved by the male having the sensory cells for key components of his own species' pheromone and sensory cells for antagonist components (characteristic of sympatric species, Chapter 3) in the same sensory hairs on the antenna (Figure 9.4): the male will only make a surge upwind to a filament containing the right compounds (and no antagonists). This is how male moths find their female despite the complex chemical background: even though plumes from females of different species will be intermingling at a coarse scale, at a fine scale the filaments remain separate (Lelito *et al.* 2008; Liu & Haynes 1992).

Fish and birds

Throughout the rest of this book there are numerous examples of vertebrates that find mates or food by orienting up odor plumes. However, there are relatively few detailed studies of the mechanisms that swimming and flying vertebrates may be using (DeBose & Nevitt 2008). It does seem that, like insects, fish and birds track plumes by using olfaction in combination with other sensory cues to gauge flow direction.

Most examples relate to orientation to food. One exception is the response of adult sea lamprey, *Petromyzon marinus*, an anadromous fish that relies on larval pheromones, a mixture of sulfated steroids, to locate spawning streams (see Figure 12.3) (Sorensen & Hoye 2007; Vrieze *et al.* 2011). Until they detect the mouth of a stream with larvae, lampreys pursue relatively straight bearings parallel to the shoreline while also changing their depth frequently. In contrast, when within the stream plume containing the pheromone, lampreys make large turns, enter and swim up the stream.

Studies of the mechanisms in other fish have confirmed that they can use a combination of olfaction and odor-gated rheotaxis to locate the source of an odor plume, with cross-stream casting behaviors used to search for the odor source or to stay within the plume and to head upcurrent when the plume is detected (Figure 10.5) (DeBose & Nevitt 2008). Fish can use visual feedback if available. The rheotaxis is largely guided by inputs from the lateral line, in particular its "superficial neuromasts system," the displacement-sensitive hairs in structures superficially embedded in the skin (Montgomery *et al.* 2000). These can detect large-scale turbulence (much larger than the animal) allowing animals to orient to the mean flow. For example, the nocturnal banded kokopu fish, *Galaxias fasciatus*, uses its superficial lateral line system to guide its rheotaxis toward an odor source (Baker *et al.* 2002).

Sharks also use the superficial lateral line information for rheotaxis-guided orientation upcurrent to find an odor source (Gardiner & Atema 2007). Sharks, and other fish such as nocturnal catfish, can use the other lateral line system of canal neuromasts to detect fine-scale turbulence. Combined with olfactory information, sharks may be able to use "eddy chemotaxis" to track the fine-scale turbulence of the plume toward the source. Sharks can also turn toward odors by responding to the time delay of arrival at their widely spaced bilateral nasal sensors (Gardiner & Atema 2010).

Some fish, such as catfish, can use tropotaxis to find an odor source in still water but such bilateral comparison does not seem to have a role in orientation in the banded kokopu (above) (Baker *et al.* 2002).

Olfaction is important for the orientation behavior of some birds and may be used in combination with visual cues or geomagnetic information, depending on species and the spatial scale of the behavior (DeBose & Nevitt 2008). Many species of procellariiforms, such as albatross, perform a characteristic zigzag upwind search in response to odor cues. High-precision global positioning system (GPS) allowed Nevitt *et al.* (2008) to follow the tracks of wandering albatrosses, *Diomedea*

exulans, over the open ocean (Figure 10.9). The birds flew across-wind until turning upwind with a zigzagging flight to end with a food meal. The zigzag approaches started, on average, about 2.5 km downwind of the prey capture site, with maximum detection distances observed at 5 km. Almost half the food gathered by the albatross used this olfactory search. The feedback input for determining wind direction is not known but it is likely to be visual, as for a moth.

One of the most exciting and challenging future research areas in animal orientation behavior will be to investigate vertebrate orientation mechanisms in more detail.

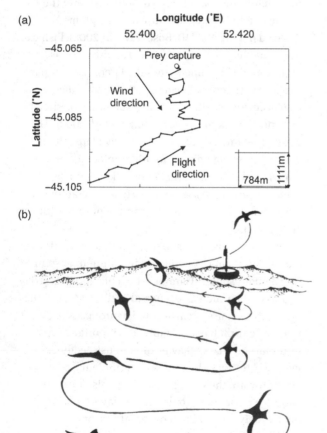

(a)

Figure 10.9 (a) Global positioning system (GPS) track of a wandering albatross, *Diomedea exulans*, also equipped with a stomach temperature recorder to monitor feeding events. The albatross had been followed over many kilometers flying across wind before the track section shown here. It turns upwind presumably when it crosses the odor plume and switches to a zigzag upwind flight before finding the food and feeding (open circle). Arrows refer to flight and wind direction as indicated. The horizontal and vertical distances represented by the smallest grid rectangle are shown in the lower right corner. Figure and caption from DeBose and Nevitt (2008). (b) Artist's impression of an albatross approaching a food source. From Hutchison and Wenzel (1980).

Summary

The olfactory world consists of chaotic overlapping and intermittent chemical signals yet animals have evolved exquisite ways of extracting information that will lead them to mates, food, and other resources. Ranging behavior is the initial stage of searching and brings the animal into first contact with the active space, in which it can detect the signal. Mechanisms for orienting to the olfactory sign include taxis (guided by gradients) and kinesis (not directly guided). In air or water, animals may respond to three very different kinds of odor stimulus: short range diffusion, a trail, or a plume. The orientation of most animals occurs in turbulent plumes. Walking animals can detect currents and orient up them. Flying and swimming animals need visual or other feedback to orient upstream. Zigzag tracks of animals working up an odor plume are very similar in appearance across the animal kingdom.

Further reading

Atema (2012), Koehl (2006), Webster and Weissburg (2009), and Willis (2008a,b) discuss the characteristics of chemical signals carried in water or air and how animals orient to these signals. Chapters in Allison and Cardé (2014) cover moth orientation behavior. Vogel (1994) and Denny (1993) discuss the flow of fluids in more detail and how this influences chemical signals. Dusenbery (1992) provides a very good account of chemical signals, ranging, and searching behavior. Kennedy (1992) has interesting views on the design and interpretation of experiments on orientation behavior. Schöne (1984) and Bell (1991) review orientation and searching behavior in general. For chemical signals and life at micro-scale, see Dusenbery (2009).

For molecular structures see sites such as www.chemspider.com, which allows you to search by common name and shows synonyms as well as the systematic names. See also the Appendix for a short guide to the terminology (available for freedownload from the website associated with this book). Many pheromones are illustrated on Pherobase www.pherobase.com (El-Sayed 2013).

11 Breaking the code: illicit signalers and receivers of semiochemicals

Any pheromone or chemical recognition system described in the other 12 chapters in this book could potentially be exploited by other organisms, whether conspecifics or predators and parasites. Exploitation can take many forms, from parasitoid flies eavesdropping the alarm pheromones of fighting ants (Mathis & Philpott 2012), to the production of insect sex pheromone molecules by orchids to deceive male bees and wasps into becoming inadvertent pollinators (Gaskett 2011). However, not all relationships involve exploitation: mutualistic relationships, such as those between sea anemones and anemonefish, can also involve chemical cues. Inter-specific semiochemicals (allelochemicals) in mutualistic relationships are called synomones; those used in active deception, allomones, and when eavesdropped, kairomones (see Table 11.1 and Table 1.1 for more details).

Given the subtleties revealed in other chapters about chemical communication *within* species, we should expect that inter-specific interactions will be no less extraordinary. Many of the examples are of insect interactions, particularly those between ants and their guests and parasites. This is largely because we know most about chemical communication in insects. Investigation of the chemical ecology of other animals in similar detail is likely to reveal a near ubiquitous role of chemical cues in other inter-specific relationships. Chemical detection and interaction of inter-specific partners and enemies may turn out to be the rule rather than the exception.

Herbivore-induced plant volatiles can attract natural enemies of the herbivores, such as parasitoids and predators, a phenomenon called indirect plant defense (for reviews of these tritrophic systems see e.g., Clavijo McCormick *et al.* 2012; Dicke & Baldwin 2010).

The design features of olfactory systems, described in Chapter 9, make every animal a potential eavesdropper and the evolution of inter-specific eavesdropping of olfactory cues or "code breaking" both possible and likely. These features include a basic broad sensitivity to any odor, including novel ones, and the ability to learn or evolve responses to any odor that provides selective advantage. Some specialist eavesdroppers may be as, or more, sensitive to the pheromone as the legitimate receiver. Production of chemicals to mimic those used by prey species is facilitated by the basic biochemical machinery that all animals share by common descent and which can evolve to produce new end products.

Like other interactions between organisms, eavesdropping and deception involving chemical communication can create a co-evolutionary arms race, with strong selection pressures leading to changes, for example, in the pheromone blends and the behavior of the exploited organism, followed by counter-evolution by the exploiter.

11.1 Eavesdropping

Any broadcast signal can be intercepted by illicit, eavesdropping receivers.

11.1.1 Predators and parasitoids

Some of the first examples of chemical eavesdropping were discovered by accident, when traps baited with bark beetle pheromones attracted predators and parasitoid wasps as well as the intended bark beetles. The predators use the bark beetle pheromones as kairomones to find their prey (Table 1.1). A kairomone is a chemical emitted by an organism, which attracts

Table 11.1 **Classification of semiochemicals by cost and benefit to emitter and receiver. After Alcock (1982).**

Semiochemical [a]	Effect on emitter	Effect on receiver	Communication
Allomones	+	–	Deceit, propaganda
Kairomones [b]	–	+	Eavesdropping
Synomones	+	+	Mutualism

[a] Definitions of allomones, kairomones, and synomones are given in Chapter 1.
[b] A chemical used intra-specifically as a pheromone is termed a kairomone when eavesdropped by a predator.

exploiters of another species. Any kind of pheromone can be eavesdropped. Surprisingly, there are some rare examples of semiochemicals being eavesdropped by visual cues. The protein-rich urine marks of voles are conspicuous in the ultraviolet part of the spectrum visible to birds: the kestrel, *Falco tinnunculus*, and the rough-legged buzzard, *Buteo lagopus*, may be able to use the vole scent marks as a cue when searching for profitable hunting and breeding areas (Koivula & Viitala 1999; Viitala *et al.* 1995; Zampiga *et al.* 2006).

11.1.1.1 Sex pheromones

Many predators and parasitoids use their victims' sex pheromones to locate them (Zuk & Kolluru 1998). For example, even we can smell the pheromone coming from leks of Mediterranean fruit-fly males, *Ceratitis capitata*, and yellowjacket wasps, *Vespula germanica*, use this odor to locate them (Hendrichs *et al.* 1994; Hendrichs & Hendrichs 1998).

Parasitoid wasps that parasitize eggs of other insects use a variety of host semiochemicals including aggregation and sex pheromones of their hosts to locate the small and inconspicuous eggs (Colazza *et al.* 2010; Fatouros *et al.* 2008; Hofstetter *et al.* 2012). For example, some of the sex pheromone released by a calling female cabbage moth at night is adsorbed on the plant and when the pheromone is slowly desorbed during the day this provides a "bridge-in-time" for daytime-searching parasitoids to locate the eggs she laid (Noldus *et al.* 1991).

The parasitoid wasp, *Telenomus euproctidis*, has a more direct way of finding the eggs. Having located a virgin moth female, *Euproctis taiwana*, by her pheromone ((Z)-16-methyl-9-heptadecenyl isobutyrate), the parasitoid hitches a ride (phoresy) on the moth until the moth female oviposits. The parasitoid then jumps off and parasitizes the newly laid eggs (Arakaki *et al.* 1996). *Telenomus euproctidis* also demonstrates that eavesdropping evolves in response to local opportunities: populations of the wasp in different parts of Japan use different, allopatric species of moth as hosts and respond only to the sex pheromones of the tussock moth species common at that locality (Arakaki *et al.* 1997, 2011).

Egg parasitoid wasps such as *Trichogramma brassicae* also hitch a ride with their host, the large cabbage white butterfly, *Pieris brassicae*, but instead find her after she mates, by the anti-aphrodisiac pheromone, benzyl cyanide, transferred by the male butterfly at mating (Chapter 3) (see Huigens *et al.* 2011). Another *Trichogramma* species, *T. evanescens*, which is a generalist parasitoid, can learn the anti-aphrodisiac when it parasitizes freshly laid eggs, allowing her to then follow that butterfly species (Huigens *et al.* 2009).

One evolutionary response to parasitoid pressure is a change in calling behavior. Males of the spined soldier bug, *Podisus maculiventris*, release pheromones to attract females, but unfortunately these pheromones also attract specialist tachinid parasitoid flies that lay eggs on the male (Aldrich 1995). Unlike most other bugs, *P. maculiventris* males have pheromone glands that can be closed (perhaps evolved in response to parasite pressure). Some *P. maculiventris* males, attracted to the pheromone released by calling males, pursue a "satellite male" strategy, staying "silent" themselves and attempting to intercept females responding to the calling male (Chapter 3) (Aldrich *et al.* 1984). Silent satellite males may get fewer matings but they may be parasitized less.

11.1.1.2 Territorial marking pheromones

East African klipspringer antelopes, *Oreotragus oreotragus*, mark their territories with a resinous secretion from their antorbital gland, smeared onto twigs (Chapter 5), and *Ixodes neitzi* ticks aggregate on these territorial marks (Rechav *et al.* 1978). The ticks respond to phenolics, leached out of the scent marks by rain, by climbing up the twig. For parasites, territorial marks are an ideal place to wait, as the territory owner will return repeatedly to renew the mark and conspecifics will also approach the marks to sniff them. Responses by North American *Ixodes scapularis* ticks to pheromone secretions of their deer hosts are the likely cause of clustering of ticks along deer trails (Carroll *et al.* 1996). See Allan (2010) for further discussion.

11.1.1.3 Ant alarm pheromones attract parasitoids and predators

Phorid flies, which lay their eggs in the head of ants (which eventually falls off, hence the common name for these phorids of "ant-decapitating flies") are attracted by the pheromones of their host ants (Mathis & Philpott 2012). In Central America, female phorid flies, *Apocephalus paraponerae*, are attracted to the mandibular pheromones, 4-methyl-3-heptanone and 4-methyl-3-heptanol, released by fighting or injured individuals of their host, the giant tropical ant, *Paraponera clavata* (Feener *et al.* 1996). Male flies are also attracted by these compounds as they are assured of finding females on the ant battleground. The phorid fly, *Pseudacteon tricuspis*, introduced to the USA as a possible biological control of fire ants, *Solenopsis* spp.,

is similarly attracted to the ants' alarm and venom gland constituents (Sharma *et al.* 2011).

In Australia, a spider, *Habronestes bradleyi*, which is a specialist predator of the meat ant, *Iridomyrmex purpureus*, is attracted to territorial battles by the ants' alarm pheromone 6-methyl-5-hepten-2-one (Allan *et al.* 1996).

11.1.1.4 Bark beetles, aggregation pheromones, and predators

Host tree volatiles and aggregation pheromones released by the first bark beetles arriving on a suitable tree attract conspecifics, leading to a mass attack of the tree (Chapter 4). These pheromones are eavesdropped both by other bark beetle species, which may avoid competitor species, and by a host of natural enemies (Raffa 2001; Raffa *et al.* 2007). Eavesdropping responses by predatory beetles and parasitoid wasps and flies can be dramatic: for example, traps baited with *Dendroctonus brevicomis* aggregation pheromone caught 86,000 predatory beetles, *Temnochila chlorodia*, along with 600,000 of the bark beetle (Wood 1982). Some of these predatory species may be more sensitive to the pheromones of their prey than the bark beetles themselves.

This intense predation pressure can select for pheromone blend changes for partial escape from bark beetle predators in a locality (Raffa 2001; Raffa *et al.* 2007). In Wisconsin, the pine engraver bark beetle, *Ips pini*, attracts two main predatory beetles with its pheromone, ipsdienol: *Thanasimus dubius* and *Platysoma cylindrica*. Ipsdienol comes in two stereoisomers (*S*)-(+) and (*R*)-(–) (see Appendix), and the predators respond to different ratios: *T. dubius* to (*S*)-(+) stereoisomer and *P. cylindrica* to the (*R*)-(–) form. Wisconsin *Ips pini* produce and are attracted to ratios midway between the two predators' preferences (Figure 11.1). The predators do not respond to an additional component in the *Ips pini* aggregation pheromone blend, lanierone (which acts synergistically for *I. pini*, see Chapter 1). *Ips pini* responding to the blend will be able to respond to lower

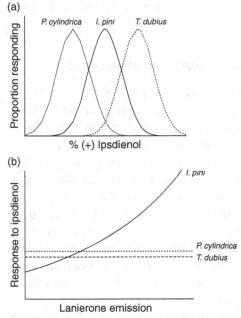

(a)

(b)

Figure 11.1 (a) Predator pressures have selected the bark beetle, *Ips pini*, in Wisconsin to use a proportion of (*S*)-(+)-ipsdienol which "hides" between the response profiles of the two key predatory beetles *Platysoma cylindrica* and *Thanasimus dubius*. (b) In Wisconsin *Ips pini* have an additional, synergistic pheromone component, lanierone, which further differentiates attraction to the pheromone between the bark beetle and its predators, which do not respond to lanierone. Low concentrations of ipsdienol combined with lanierone attract *I. pini* but do not attract the predators. Figure and caption Raffa (2001). Data on responses in Raffa *et al.* (2007).

Table 11.2 **Relative responses of local *Ips pini* and predators to local vs. distant populations of *I. pini*. Data from a reciprocal exchange experiment in Raffa and Dahlsten (1995). Values indicate ratio of insects caught in response to *I. pini* from different regions (about 1,500 miles apart). Within each cell the relative response to the preferred source is indicated. Data for *I. pini*, its most abundant predatory beetle in each region, and a parasitoid wasp in California are shown. From Raffa (2001).**

Responding insect	Preferred source of *Ips pini*	Ratio of preferred to less preferred source
California *Ips pini*	California	11.0
Wisconsin *Ips pini*	Wisconsin	5.3
California predator	Wisconsin	2.6
Wisconsin predator	California	1.8
California parasitoid wasp	California	4.9

concentrations of ipsdienol than the predators, giving greater escape from eavesdropping. A similar pattern occurs in California, where *I. pini* produce (*R*)-(−)-ipsdienol, and the major predators prefer blends with a higher (*S*)-(+) ratio (see Raffa *et al.* 2007).

The proportions of the (*S*)-(+) and (*R*)-(−) stereo-isomers of ipsdienol used as the aggregation pheromone by *I. pini* always differ from the preferences of the main local predatory beetles in a cyclical game of evolutionary hide-and-seek in the space defined by possible blends of (*S*)-(+) and (*R*)-(−) ipsdienol (Table 11.2). The difficulties involved in tracking

prey are further compounded by spatial and temporal variation in prey signaling on a local scale of kilometers.

It seems that *I. pini* can keep "ahead" of the predators locally in part because the predators have alternative bark beetle species with different pheromones to follow, if *I. pini* disappear from their chemosensory radar. By contrast, specialist parasitoid wasps such as *Tomicobia tibialis*, without alternative hosts, closely track the local *I. pini* blend, including lanierone and the appropriate ratio of ipsdienol isomers (Raffa *et al.* 2007). Specialist parasitoids in California similarly tracked the local *I. pini* pheromone blend (Table 11.2).

By definition, aggregation pheromones attract conspecifics of both sexes (Chapter 4), which distinguishes them from sex pheromones that only attract the opposite sex. However, many aggregation pheromones may really be intra-specific eavesdropping of sexual communication directed at the opposite sex but also responded to by the same sex looking for mates, as for the spiny soldier bug (Chapters 3 and 4) (Section 11.1.1).

11.1.2 Prey responses to predator semiochemicals

Many species show escape or avoidance responses to the odors, kairomones, of their predators. In streams, mayfly nymphs change their behavior in response to odor cues from both fish and invertebrate predators such as stonefly nymphs, by reducing activity, drifting downstream (and changing the time of day of drifting to avoid peak predation times) (Boone *et al.* 2008). The freshwater gammarid, *Gammarus roeselii*, avoids the odors of predatory crustaceans and many fish (Hesselschwerdt *et al.* 2009). Vertebrates, from fish to mammals, also typically avoid the odors of their predators (Ferrero et al. 2011; Müller-Schwarze 2006). For example, many snakes show characteristic defensive behaviors to the odors of snake-eating snakes (Mason & Parker 2010). Rats show an innate fear response to cat odor (Chapter 9). The protozoan parasite, *Toxoplasma gondii*, manipulates the behavior of rats it infects, changing the fear response to cat odor into attraction, making the rats more likely to be eaten by *T. gondii*'s primary host, the cat (House *et al.* 2011; Vyas 2013).

Any cues that give early warning of hunting predators will have survival value for prey but few responses are to odors with a demonstrated role in the predator's own intra-specific communication. However, two examples where this is the case come from social insects: Japanese honeybees respond to the giant hornet marking pheromone when it marks the bee nest for attack (see Chapter 8) and the stingless bee, *Trigona angustula*, reacts to the pheromones of the robber bee, *Lestrimelliata limao* (Section 11.3.4).

11.1.3 Egg dumping: parasitism or mutualism?

In some insect species with parental guarding of eggs, the odors of recently deposited eggs are exploited by other females as a cue to add their own eggs to the egg mass (Tallamy 2005). This leads to alloparental care in which parents look after eggs and young that are not their own. An intra-specific example is egg dumping by females of the aubergine lace bug, *Gargaphia solani* (Monaco *et al.* 1998; Tallamy 2005). In this species, the female stands over her eggs and defends these and first-instar nymphs from predators (see Chapter 8). The odor of newly laid eggs attracts other females to approach and add their eggs, which the mother allows.

In fish there are numerous instances of inter-specific egg dumping leading to alloparental care (Wisenden 1999). One well studied association is between the green sunfish, *Lepomis cyanellus*, and the redfin shiner, *Notropis umbratili*, which deposits its eggs in the nests of male sunfish. Chemicals are important proximate cues. Hunter and Hasler (1965) were able to attract male and female shiner fish with milt (semen) and ovarian fluids of the sunfish host. If predators are around, sunfish nests with shiner nest associates produced four times as many sunfish fry as those without (Johnston 1994). Other experiments show that the benefit is mutual, as shiner nest associates benefit from parental care by the hosts.

In these fish and insect examples, the egg dumping may be a mutualism benefiting both parties (see next section). A common feature of both the *Gargaphia* insect and the fish systems is that the young are precocial (independent from birth and able to feed themselves) so looking after additional eggs or young is not costly, unlike hosting a young cuckoo (Tallamy 2005). Instead, in these species egg dumping could be thought of as a mutualism. The benefits include the way additional eggs create a protective buffer zone around the *Gargaphia* female's own eggs.

For *Gargaphia* and the fish examples, once the eggs hatch, the dilution effect will reduce the individual risk of loss to predators for the guarder and the dumper alike (Hamilton 1971).

11.2 Chemical communication in mutualisms

Just as interactions within ant colonies are largely mediated by pheromones, mutualisms between ants and their myrmecophilous (ant loving) "domesticated animals" rely heavily on chemical cues. Chemical cues are also important in the symbiosis between anemones and anemonefish.

11.2.1 Ants and aphids, leafhoppers, and scale insects

The mutualistic association between ants and honeydew-producing homopterans (plant-sucking bugs: aphids, leafhoppers, and scale insects) is well known: the ants gain a rich sugar source and the homopterans benefit by defense against predators and parasitoids (Lin 2006; Stadler & Dixon 2005). The close associations that have evolved between some ant species and homopterans may have led to changes in the behavior of both partners. Whereas aphids of non-ant-associated species tend to drop from the leaf (to escape the predator) when they detect the alarm pheromone released by an aphid under attack (Chapter 8), aphids of ant-associated species remain feeding and rely on their ant guards to protect them (Nault *et al.* 1976). The suppression of the dropping response is even stronger if ants are present. For their part, the ants are attracted by aphid alarm pheromone, attacking any predator menacing their "cows."

11.2.2 Ants and lycaenid butterfly caterpillars

Chemical communication is the key to the close relationship between ants and caterpillars, found in an estimated 75% of the 6,000 or so species of lycaenid

butterflies (Fiedler 2012; Pierce *et al.* 2002). These associations can be mutualistic (the majority) or parasitic (Section 11.4). The associations range from loose facultative interactions in which larvae are only occasionally tended by several species of ants (about 45% of associations), to complex obligate associations in which larvae are always tended by ants, often by only a single species (30%) (Pierce *et al.* 2002). As well as chemical signals, the caterpillars also use acoustic signals (see Section 11.4.4) (Barbero *et al.* 2009).

In the mutualisms, both species benefit: the ants guard the caterpillars from parasitoids and predators in return for sugar and/or amino acid secretion from the caterpillars' specialized dorsal nectary organ. Caterpillars maintain a "standing guard" of attending ants and if there is greater danger they strategically release more secretory reward, attracting more guards (Pierce *et al.* 2002).

There are suggestions that lycaenid larvae suppress ant aggression in part by mimicking aspects of the hydrocarbons of ant brood (Section 11.4.4) (Akino 2008; Nash & Boomsma 2008). Attendant ants lick and antennate lycaenid larvae much as they do their own brood.

Many lycaenid caterpillars can signal distress to their ant guards, from eversible tentacle organs that release compounds that seem to be ant alert or alarm pheromone mimics, and stimulate the ants to activity (Pierce *et al.* 2002).

11.2.3 Sea anemones and anemonefish

Coral reef anemonefishes are well known for their ability to live unharmed among the stinging tentacles of sea anemones (Mebs 2009). The 26 species of anemonefishes in the genera *Amphiprion* and *Premna* are obligate symbionts of ten species of sea anemones. Some anemonefish species are associated with only one anemone species, recognized by chemical cues (Arvedlund *et al.* 1999), but a few can live in many anemone species. The anemonefish gains protection and in return vigorously defends the anemone from predators such as butterflyfish.

Box 11.1 Using disguise to evade detection

Avoiding detection is a problem shared by animals attempting to live in a social insect colony, or inside another organism as a parasite (such as a malaria trypanosome or tapeworm), or live on a sea anemone without being stung.

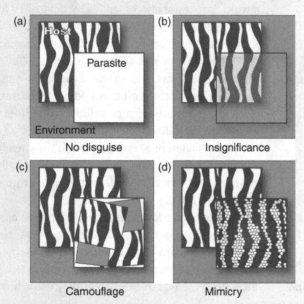

With no disguise: (a) a parasite's own molecular coat makes it stand out and easy to detect by a potential host. There are three main ways of disguise to evade detection: (b) insignificance, offering few or no cues detectable by the host (a "stealth" approach, visually equivalent to transparency); (c) chemical camouflage ("stealing a uniform" by picking up the surface recognition molecules from the host or its nest); and (d) chemical mimicry (synthesizing a "counterfeit uniform"). Such molecules may be produced by different pathways from the host, and may not match all aspects of the cues. Figure and adapted text from Nash and Boomsma (2008).

The anemone and its zooanthellae (endosymbiotic dinoflagellate algae living in its cells) also gain nutrients excreted by the anemonefish including ammonia, sulfur, and phosphorus (Cleveland *et al.* 2011; Porat & Chadwick-Furman 2004).

A protective mucous coat allows the anemonefish to contact the tentacles of their host anemone without being recognized as "non-self," and being stung by the anemone's nematocysts (Elliott *et al.* 1994). However, whether the immunity comes from molecular camouflage, mimicry, or insignificance (not having the sting trigger-compounds in their

mucus) is still not clear (Mebs 2009) (see Box 11.1 for definition of terms). Some anemonefish species appear to use camouflage, stealing their identity from their host, like some ant guests, by picking up anemone substances during the tentative "acclimation" behavior when they are first introduced to the anemone. However, some anemonefish species can join their anemone without acclimation behavior. Sea anemones also have symbioses with crustaceans, mediated by chemical cues and perhaps involving acclimation (Guo *et al.* 1996; Mebs 2009).

11.3 Deception by aggressive chemical mimicry

Illicit signalers can exploit the strong selection pressures on males to respond to sex pheromones by mimicking these for gain (Vereecken & McNeil 2010). This could be seen as extreme sensory exploitation (Chapter 1). Why do male moths still respond to the lures of bolas spiders and why do ants feed cuckoo caterpillars (Section 11.4)? If the deceiver is at relatively low frequency, then the cost of not treating the signal as genuine is too high as most signals are genuine and legitimate use of the signal increases reproductive success or survival. As in other cases of mimicry, as long as bolas spiders are comparatively rare compared with genuine female moths, male moths must respond to female moth pheromone even if sometimes it is counterfeit and leads to a sticky end.

11.3.1 Bolas spiders and moths

Adult female bolas spiders lure male moths by mimicking the sex pheromones emitted by female moths of those species (Gemeno *et al.* 2000; Haynes *et al.* 2002). The bolas spider's web is reduced to a sticky ball on a thread, the bolas, held by one of her forelegs (Figure 11.2). When a male moth is lured by the pheromone, she detects his wing vibrations, draws back her foreleg, and swings the bolas at the moth. If the sticky bolas makes contact, the moth rarely escapes.

At different times of night the bolas spider, *Mastophora hutchinsoni*, in Kentucky USA, attracts males of two noctuid moth species, *Lacinipolia renigera* and *Tetanolita mynesalis*, which have completely different pheromone blends (Chapter 3), respectively (*Z*)-9-tetradecenyl acetate (*Z*9–14:Ac) and (*Z,E*)-9,12-tetradecenyl acetate (*ZE*-9,12–14:Ac) and a 2:1 blend of (3*Z*,9*Z*)-(6*S*,7*R*)-epoxy-heneicosadiene and (3*Z*,6*Z*,9*Z*)-heneicosatriene (Haynes *et al.* 2002). The spider's early evening blend is a compromise, not perfect for either species, but

Figure 11.2 Bolas spiders (*Mastophora yeargani* is shown here) aggressively mimic the sex pheromone blends of their moth prey. When an attracted male moth approaches within striking distance, the spider draws back the bolas, the sticky ball of glue suspended on a short thread, and swings it at the moth. Photograph by K. F. Haynes.

sufficiently close to attract *L. renigera* males, which are not deterred by *T. mynesalis* pheromone components. *Tetanolita mynesalis* flys later and over the evening the spider greatly decreases its production of the early-flying moth's main pheromone molecule (*Z*9–14:Ac), which would deter *T. mynesalis*, so as to better attract the late flying species.

Spiderlings and the small males of the spider instead synthesize other pheromone molecules to attract small pyschodid flies (moth flies) (Yeargan & Quate 1997).

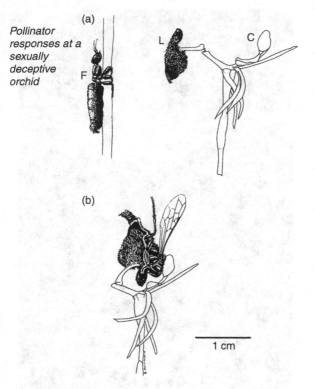

(a) Pollinator responses at a sexually deceptive orchid

(a)

L   C

F

(b)

1 cm

Figure 11.3 (a) The flower of the Australian orchid, *Drakaea glyptodon*, (top right) mimics the look and pheromone of wingless female thynnine wasps, *Zaspilothynnus trilobatus* (top left): C, column (the fused stamen and style bearing the pollinia); F, female wasp; L, the labellum, twisted to form a landing platform and mimicking the female). (b) When a male wasp is deceived by the orchid flower and attempts to fly off with the "female", he is catapulted into the pollinia (the orchid's pollen packet), which glues to his body. If he is fooled by another *D. glyptodont* flower he will deliver the pollen. Drawings by G. B. Duckworth in Peakall (1990). In another Australian thynnine wasp–orchid pair, the *Chiloglottis trapeziformisa* orchid flower produces ten times more authentic pheromone than a real female *Neozeleboria cryptoides* wasp (Schiestl *et al.* 2003). In a further Australian species, the male wasp may even ejaculate on the orchid flower (Gaskett *et al.* 2008).

11.3.2 Pollination by sexual deception

Instead of producing nectar, many orchids use aggressive mimicry of the female sex pheromone of local solitary bee and wasp species to lure males to attempt to mate with the flower, thus ensuring that they pick up the pollen packet and inadvertently act as pollinators (Figure 11.3) (see reviews by Gaskett 2011; Vereecken & McNeil 2010). These associations have evolved multiply and independently on different continents. Each orchid species mimics the female pheromone of only one or a small range of pollinator species so that as the male bee or wasp continues his search for real females, he is likely to transfer the pollen to another flower of the same orchid species with high fidelity (Xu *et al.* 2012). Floral odor is the key factor for specific pollinator attraction though the deception is often completed by mimicry of contact pheromones and visual and tactile cues (Gaskett 2011; Xu *et al.* 2012). In turn, pollinator-mediated selection drives floral scent composition. For example, the multicomponent female hydrocarbon sex pheromone blend of the solitary bee, *Andrena nigroaenea*, is almost exactly mimicked by the European orchid, *Ophrys sphegodes* (Figure 11.4) (Schiestl *et al.* 1999). However, in the association between the solitary bee, *Colletes cunicularius*, and its specific orchid mimic, *Ophrys exaltata*, the orchids in a locality produce a volatile pheromone-mimic blend that is subtly different from local female bees, which paradoxically makes the orchids more attractive than a perfect mimic (Vereecken & Schiestl 2008). Male *C. cunicularius* bees are more attracted to females from neighboring populations (which have a slightly different pheromone blend) than their own females, for optimal outbreeding (Chapter 3), and the orchid seems to be exploiting this.

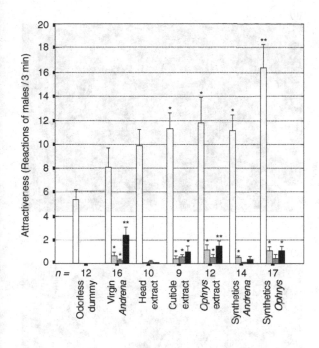

Figure 11.4 The flowers of the European orchid, *Ophrys sphegodes*, almost exactly chemically mimic the multicomponent female hydrocarbon pheromone of the solitary bee, *Andrena nigroaenea*. This was tested in the field by offering male bees odorless dummies (chemically extracted and dried *A. nigroaenea* females) scented with different samples and synthetic hydrocarbon blends reproducing the female bee or orchid flower odors. Attractiveness is given as means ± SEM of approaches to the dummy (white bar), pouncing on the dummy (light gray), alighting (dark gray) and copulation attempts (black). Asterisks indicate significant differences between the reaction types of the odorless dummy group and each of the other test groups. Figure from Schiestl *et al.* (1999).

Subtle changes in floral odor blend can be important in orchid speciation as different pollinator species will be attracted (Gaskett 2011; Xu *et al.* 2012). Selection on floral scent is likely to be divergent among closely related orchid species, which attract different pollinators with different floral scent compositions (Xu *et al.* 2012). In Australian *Chiloglottis* orchid species, 95% of closely related species pairs attracted different wasp pollinator species, by variations in pheromone blend either of single compounds or a variation of two compounds from a range of related "chiloglottones" (2,5-dialkylcyclohexan-1,3-diones) (Peakall *et al.* 2010; Xu *et al.* 2012). Speciation was always associated with pollinator switching and usually underpinned by chemical change in floral blend.

Orchids are not limited to mimicking sex pheromones. One orchid species mimics an alarm pheromone component, (Z)-11-eicosen-1-ol, of honeybees to attract hornets as pollinators (Brodmann *et al.* 2009) and another mimics aphid alarm pheromone to attract hoverflies (Stökl *et al.* 2011).

11.3.3 Blister beetle larvae dupe male host bees

Aggregations of parasitic blister beetle larvae, *Meloe franciscanus*, use visual and chemical mimicry of a female bee to attract males of their host species of solitary bee, *Habropoda pallida*, to get themselves transported to the nest of their hosts (Hafernik & Saul-Gershenz 2000; Saul-Gershenz & Millar 2006). When the male bee attempts to mate (pseudocopulate) with the bee-like aggregation, the blister beetle first instar (triungulin) larvae quickly attach to the male and then later transfer to a female bee when he mates or attempts to mate (Figure 11.5) (Hafernik & Saul-Gershenz 2000). The female bee carries the triungulins to her nest, where they dismount to feed and develop on the nest's pollen, nectar provisions, and possibly the bee egg. The beetle larvae complete their development in the nest and the gravid female beetles emerge and lay their eggs at the base of plants to complete the cycle.

Although the visual mimicry of a female bee by the triungulin groups is striking, their chemical mimicry of the female bee's volatile pheromone is crucial to attract

Figure 11.5 Flightless larvae of the blisterbeetle, *Meloe franciscanus*, aggregate to visually and chemically mimic a female of the solitary bee, *Habropoda pallida*. (a) Triungulin aggregation on a grass stem in the Mojave Desert. The aggregations are typically about 1 cm in diameter, the size of a female bee. Male bees are attracted by sight and smell to the triangulin aggregation. (b) A male *H. pallida* bee inspecting an aggregation of triungulins on the branch tip of the dune plant *A. stragalus lentiginosus* var. *borreganus*. If he makes contact, the larvae transfer to his back (c). (Saul-Gershenz & Millar 2006). Photographs copyright Leslie Saul-Gershenz (www.lsaul.com). Photographs (a) and (b) published with permission from PNAS.

the male bee. The triungulin larvae produce a subset, almost exclusively of (*Z*)-9- and (*Z*)-11-tricosenes and pentacosenes, of the female bee's full alkene hydrocarbon blend of (*Z*)-alkenes with double bonds in positions 9 to 15 (Saul-Gershenz & Millar 2006). The triungulin larvae move as one to find a perching position on the plant. Co-operation of the larvae is the key to success as only collectively can they produce enough pheromone to attract the male and form a bee-sized visual target. The group of 120 to >2,000 larvae are typically siblings from a single egg mass, so the degree of relatedness within an aggregation is high (Saul-Gershenz & Millar 2006).

11.3.4 Propaganda: mimicry of alarm pheromones

The powerful releaser effects of social insect alarm pheromones (Chapter 8) can be exploited by other species to devastating effect in warfare, colony take-over, and theft. For example, "slave-making" (dulotic) ants produce and release mimics of the alarm phero-mones of victim species to spread confusion and intra-colonial fighting during the takeover of nests (after which they carry back pupae from which workers will emerge to live and work in the slave-maker's nest) (Akino 2008; Morgan 2008). Workers of the "slave-maker" ant, *Formica subintegra*, have enormously enlarged Dufour's glands loaded with approximately 700 µg (almost 10% of the ant's body weight) of a mixture of decyl, dodecyl, and tetradecyl acetates (Regnier & Wilson 1971). The Dufour's gland secre-tions are sprayed at the defenders during raids. These high concentrations attract other *F. subintegra* work-ers but panic and disperse the defenders, which cannot help but respond to them as alarm pheromones. In the European slave-maker species, *Polyergus rufescens*, the queen's Dufour's gland contents decrease the aggressive response of the *Formica cunicularia* work-ers, acting as an appeasement pheromone or perhaps as a repellent (Akino 2008; Lenoir *et al.* 2001). To safely reach its host, the ant-guest caterpillar, *Maculinea rebeli*, which lives deep in *Myrmica schencki* ants' nests, the ichneumonid parasitoid wasp, *Ichneumon eumerus*, uses propaganda molecules to cause confusion and infighting among the ants (Thomas *et al.* 2002). The myrmecophilous staphylinid, *Pella funesta*, synthesizes the alarm pheromone, 6-methyl-5-hepten-2-one, of its ant host, *Lasius fuli-ginosus*, to enable its escape (Stoeffler *et al.* 2011).

Stingless bees in two genera, *Lestrimelitta* in South and Central America and *Cleptotrigona* in Africa, use propaganda pheromones as part of their cleptobiotic strategy to steal honey, pollen, and other resources from other stingless bee species, never visiting flowers themselves (Breed *et al.* 2012; Jarau 2009). During a raid on a *Trigona subterranea* nest, the *Lestrimelitta limao* robber bees release massive amounts of "prop-aganda" pheromone including citral, a mixture of the two stereoisomers, neral and geranial, *T. subterranea*'s own alarm pheromone. This causes the *T. subterranea* workers and queen to retreat far into the nest (Jarau 2009). The same chemicals recruit more robber bees to the nest.

However, another stingless bee, *Tetragonisca angustula*, resists raids by *L. limao*, in part because it uses benzaldehyde rather than citral for its alarm pheromone, so its defense is not disrupted by the rob-ber bees' propaganda, but also because of the defensive behavior of an unusual specialist caste of soldier-guards, which constantly hover in front of the nest (Grüter *et al.* 2012; van Zweden *et al.* 2011; Wittmann *et al.* 1990). These guards are alerted by the size, color, and the characteristic pheromones (citral and 6-methyl-5-hepten-2-one) of *L. limao* scouts. The guards clamp their mandibles on the wings or antenna of the *L. limao* scouts, bringing them to the ground. The guards recruit other guards with their alarm phero-mone benzaldehyde.

Plants can exploit aphid alarm pheromones for their own defense. Wild potatoes, *Solanum berthaultii*, produce sufficient aphid alarm pheromone, (*E*)-β-farnesene, in glandular leaf hairs to repel aphids (Gibson & Pickett 1983). (See Chapter 12 for potential exploitation of this for crop protection.)

11.4 Social parasites using disguise to escape detection by social insect hosts

Social insects have evolved into supreme users of chemical communication (Chapters 6, 7, and 8). Indeed complex sociality depends on it: only pheromone communication and chemical cues could enable colo-nies containing millions of workers to act coherently. However, this makes them vulnerable to various forms of chemical deception by intruders (Akino 2008;

Bagnères & Lorenzi 2010; Nash & Boomsma 2008). For any organism that can breach the defenses, social insect colonies offer rich resources of stored food and vulnerable brood, as well as a place protected from predators and environmental extremes of temperature and humidity.

Perhaps because their colonies are the largest and longest lived, termites and ants seem to have most "guest" species. These are the myrmecophilous or termitophilous species ("ant-" or "termite-loving" respectively – "loving" only in the sense of being found in close association with social insects, as the relationship may be parasitic or predatory). Myrmecophilous species include millipedes, mites, spiders, isopods, crickets, flies, butterflies, beetles, and even snakes (Bagnères & Lorenzi 2010; Hölldobler & Wilson 1990; Kronauer & Pierce 2011; Rettenmeyer *et al.* 2011). Animals can be accepted even if they look nothing like the hosts, so long as they smell right. Once accepted, the guests can solicit food from workers or even eat the host brood, while hidden in an olfactory cloak of respectability (Figure 11.6). It's as if we shared our homes with alligators, which we insisted on feeding ahead of our own children and which ate our children without us noticing (Wheeler in Hölldobler & Wilson 1990).

The most integrated social parasites may be social insect species that take over other social insect colonies. These occur in all the groups of social Hymenoptera (ants, bees, and wasps) and the social parasites often come from the same genus as their hosts (Bagnères & Lorenzi 2010; Buschinger 2009; van Zweden & d'Ettorre 2010). In the most extreme forms, the parasitic species has no workers and cannot survive without the aid of the host species.

Guest species of all kinds must not only match species-specific molecules but also the colony odors, which are the key to the discrimination mechanisms of social insects, enabling them to exclude conspecific non-nestmates and other intruders (Chapters 1 and 6). As so much of the gatekeeping in social insects

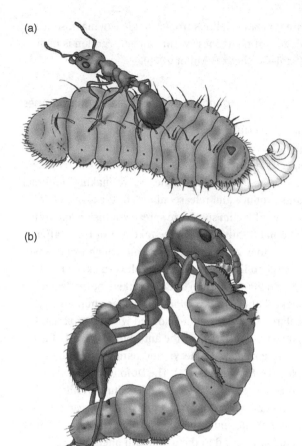

Figure 11.6 Exploitation of *Myrmica* ants by *Maculinea* caterpillar larvae once inside the host nest. (a) The predatory species, *Maculinea arion*, feeds on host ant larvae, and otherwise tries to avoid contact with worker ants. (b) The cuckoo species, *Maculinea alcon*, both feeds directly on ant brood, and is fed trophallactically by workers of its host. Larvae stay among the brood and have frequent contact with host workers. Figure and caption from Nash and Boomsma (2008).

involves CHC cues (Chapters 1 and 6), it not surprising that illicit entry often involves these.

Social parasites use three main ways of disguise: insignificance (visually equivalent to transparency), chemical camouflage ("stealing a uniform" by picking up the surface recognition molecules from the host or its nest), and chemical mimicry (synthesizing a "counterfeit uniform") (see Box 11.1) (Nash & Boomsma 2008). Social parasites may use any combination of these strategies.

11.4.1 Chemical insignificance

Chemical insignificance, lacking the chemical recognition cues that are noticed by the host, may be important for the initial entry into colonies by some social parasites but, generally, later acceptance depends on camouflage or mimicry as described below (Bagnères & Lorenzi 2010). For example the queen of the social wasp parasite *Polistes semenowi* carries a relatively small quantity of CHCs and lacks its own characteristic CHCs on its cuticle when it invades its host, the wasp *P. dominulus* (Bagnères & Lorenzi 2010). Some interspecific parasitic ant species such as *Acromyrmex insinuator* appear to use this strategy, with this species lacking the C_{29} to C_{45} hydrocarbons of its host but instead having long chain C_{43} to C_{45} CHCs (d'Ettorre & Lenoir 2010). However, lacking recognition molecules may not be enough for chemical "invisibility," as social insect recognition can also involve the *presence* of the right molecules (Nash & Boomsma 2008; van Zweden & d'Ettorre 2010) (Chapters 1 and 6).

11.4.2 Chemical camouflage: stealing the cuticular hydrocarbon "uniform"

Camouflage exploits the way that social insects themselves acquire their shared colony odors by mutual grooming, or in paper wasps from the nest, which allows exchange of the shimmering pool of hydrocarbons on their cuticular surface. This underlies the "gestalt model" of colony recognition (Chapter 6) (van Zweden & d'Ettorre 2010). Parasites often exhibit behaviors such as licking, grooming, or stroking their hosts, a behavior that gives many species of the social parasite genus *Formicoxenus* the common name "shampoo ant" (Bagnères & Lorenzi 2010). A myrmecophilous salticid spider, *Cosmophasis bitaeniata*, acquires the colony-specific CHCs by eating or

handling the larvae of its host, *Oecophylla smaragdina* (Elgar & Allan 2004).

11.4.3 Chemical mimicry by counterfeiting pheromones and colony recognition cues

An alternative strategy to acquiring molecules from the host is to synthesize them yourself, using your own biosynthetic pathways. Radiolabeling experiments show that the termitophilous beetle, *Trichopsenius frosti*, synthesizes its own CHCs to match those of its termite host, *Reticulitermes flavipes* (Howard *et al.* 1980). Different *Trichopsenius* species synthesize the CHCs that are most like the termite caste they spend most time with (queens for *T. frosti*, workers for the other *Trichopsenius* species). Different cuckoo bumblebee, *Bombus* (*Psithyrus*), species synthesize the species-specific alkene positional isomer molecules of their particular *Bombus* bumblebee host species before attempting to enter the host colony (Martin *et al.* 2010b). Many guest species use a combination of strategies: synthesizing some recognition CHCs (particularly the more general ones, which may vary less between species and colonies) and, once safe in the nest, acquiring the colony-specific profile by mutual grooming (Howard & Akre 1995).

The molecules that the chemical mimic species synthesize can reveal which molecules are important for the host's species or colony recognition, as typically a chemical mimic only produces a subset of the molecules found on the host (Nash & Boomsma 2008; van Zweden & d'Ettorre 2010).

Instead of matching the host odors, an unusual and completely different approach is taken by some *Polistes* cuckoo wasp species, which instead change the colony recognition odor to match their own by adding their parasite-specific hydrocarbons to the nest-paper (which wasps use as the reference for colony recognition; Chapter 6) (Bagnères & Lorenzi 2010; Lorenzi 2006). This sounds to me like the chemical equivalent of a burglar "changing the locks."

11.4.4 Co-evolutionary arms races between social parasite mimicry and hosts

Social parasites that use chemical mimicry tend to be those that have co-evolved over long periods with their hosts in an arms race as the hosts defend themselves against attack, which in turn selects for counter-strategies in the parasite. Like the bird hosts of the original cuckoo, social insect hosts discriminate against social parasites if they have a significant impact and can be detected, selecting for parasites that can better counterfeit the host hydrocarbon signature, as in the examples above (Kilner & Langmore 2011). Just as bird brood parasites select for diversity in the egg patterns of their cuckoo hosts, successful mimicry by parasites selects for diversity in the chemical profiles of social insect hosts. For example, populations of *Formica fusca* under more intense pressure from socially parasitic ant species have much greater diversity of CHCs, in particular the number of C_{25}-dimethylalkane isomers used for colony recognition, as well as greater egg and nestmate discrimination abilities lost in populations under less parasite pressure (Martin *et al.* 2011b). The socially parasitic bumblebee, *Bombus* (*Psithyrus*) *bohemicus*, in the United Kingdom may be tracking the chemical profiles of a number of related host species in an ongoing "arms race" (Martin *et al.* 2010b).

Patchy populations with little gene flow between them can lead to a geographic mosaic of co-evolutionary hotspots between the host and parasites (Thompson 2009). The caterpillars of the large blue, *Maculinea*, butterflies (Lycaenidae) are carried into the nest of their ant host and once there are fed like a cuckoo by the ants and also feed on the ant larvae (Figure 11.6) (Pierce *et al.* 2002; Thomas *et al.* 2005). Key to successful "adoption" of the caterpillar by the ants is how well its chemical mimicry of host hydrocarbons matches those of local ants. The Alcon blue, *Maculinea alcon*, can have a virulent effect on small colonies of its ant host, *Myrmica rubra*. This leads to selection for changes in host surface hydrocarbons in ant populations parasitized by the butterfly and then co-evolutionary changes in the caterpillar hydrocarbons to track these (Nash *et al.* 2008). The patchy distribution of host plants for the early, vegetarian, life of the caterpillar before adoption by the ants, and the slow gene flow between neighboring populations of the ant host, creates a geographic mosaic of isolated co-evolving populations, with a continuing arms race in chemical mimicry (Nash *et al.* 2008). While *Maculinea* caterpillars use chemical mimicry to gain entry to ant colonies they also mimic the distinctive sounds made by the ant queen, misleading the ants into treating them like royalty, and the ants will feed them in preference to their own larvae (Barbero *et al.* 2009).

Co-evolution between the hoverfly, *Microdon mutabilis*, and its ant host, *Formica lemani*, is even more local, at the level of a single ant population (Schönrogge *et al.* 2006). The fly lays its eggs near a *F. lemani* nest and relies on the ants picking up its eggs and carrying them into the nest. Egg survival drops rapidly with distance from the natal nest, from 100% if placed next to a nest within 100 m to 50% by 600 m. The reason for extreme local host specificity is that the female *Microdon* fly tends to lay her eggs at the nest she emerged from, so *Microdon* lays in the same nest, generation after generation. The fly larva can eat so much brood that the survival of the ant colony is threatened but, paradoxically, the way *Microdon* selectively feeds on host eggs and small ant larvae causes the surviving ant larvae to switch development into becoming queens, allowing persistence of the host ant genotype (Hovestadt *et al.* 2012). The chemical co-evolution is likely driven by host ant discrimination of the mimicked host surface hydrocarbons the female fly coats her eggs with – the mimicry only works for the local co-evolved population of ants.

Summary

As "code breaking" is a capability built into all olfactory systems, and chemical signals dominate the lives of most animal species, chemical eavesdropping and deception are likely to be common. Just as every animal has its parasites, every pheromone communication system has the potential to be exploited. Eavesdropping by conspecifics, predators, and prey, is perhaps the simplest exploitation of chemical signals. The effect of predator eavesdropping can lead to selection of different signals, as found in bark beetles. Chemical communication also plays a role in mutualisms, in anemonefish and their anemones and numerous examples of ants interacting with "farmed" symbionts such as aphids.

Aggressive chemical mimicry is used by orchids to dupe male Hymenoptera into carrying pollen from flower to flower and by bolas spiders to lure male moths. The powerful releaser effects of alarm pheromones on social insect colonies can be exploited by other species to devastating effect in warfare, colony take-over, and theft.

The first defense of social insect colonies is species and colony recognition of potential intruders. As this task relies heavily on pheromones and colony odor, deception by guests often involves displaying chemical cues and signals, either acquired from the host for camouflage or synthesized in chemical mimicry. Selection on the social insect hosts to change their cues and parasites to match them better can lead to co-evolutionary arms races.

Further reading

Vereecken and McNeil (2010) give a wide-ranging review of aggressive mimicry of pheromones in arthropods.

Nash and Boomsma (2008), Bagnères and Lorenzi (2010), and van Zweden and d'Ettorre (2010) explore the ways social parasites interact with their hosts and are especially strong on the role of surface hydrocarbons. The biology of guests and ant hosts is covered in Hölldobler and Wilson (1990, 2009). Kilner and Langmore (2011) compare the co-evolutionary arms races between bird cuckoos and social insect parasites and their respective hosts.

There is a special issue on ants and their social parasites in a 2012 issue of the open access journal *Psyche* edited by Lachaud, Lenoir, and Witte www.hindawi.com/journals/psyche/si/726548/.

For molecular structures see sites such as www.chemspider.com, which allows you to search by common name and shows synonyms as well as the systematic names. See also the Appendix for a short guide to the terminology (available for free download from the website associated with this book). Many pheromones are illustrated on Pherobase www.pherobase.com (El-Sayed 2013).

12

Using semiochemicals: applications of pheromones

The importance of smell in the natural behavior of animals has long been recognized and, long before it was known what semiochemicals were, people used them to manipulate the behavior of animals. For example, traditionally farmers have encouraged a mother sheep (ewe) to adopt a strange lamb if her own died at birth, by covering the strange lamb with the skin of her dead lamb (see Chapter 9).

The clear potential for applied uses of pheromones was an early encouragement for research. At the turn of the twentieth century and in its first few decades, the potential of synthetic chemical signals to control insect pests was anticipated both in North America and in Europe (Witzgall *et al.* 2010). There is now increasing use of an understanding of semiochemicals to affect the behavior of domesticated animals, from bees to sheep, as well as use as "greener" alternatives to pesticides, largely for the control of insect pests but also potentially for vertebrate pests. However, no matter how elegant the science, pheromones will be exploited only if they are commercially viable, though that can be strongly influenced by government policy (Chandler *et al.* 2011; Winston 1997). Human odors will be discussed in the next chapter (Chapter 13).

12.1 Semiochemicals used with beneficial and domestic animals

12.1.1 Manipulating the behavior of beneficial insects

Pheromones orchestrate or modulate every aspect of honeybee life (Chapter 6) and offer many opportunities for intervention (Grozinger 2013; Slessor *et al.* 2005). For example, swarming, which results in decreased honey yields, can be delayed by adding controlled-release sources of synthetic queen mandibular pheromone (QMP) to the hive (Winston & Slessor 1998). Worker bee Nasonov pheromone is routinely used to bait swarms into trap hives to help bee-keepers (Free 1987, p. 129).

Queen mandibular pheromone can be sprayed on fruit trees or other crops to increase bee visits: almost twice as many bees visited the sprayed areas as unsprayed areas. This leads to higher pollination rates (often a limiting factor) and increased yield of some fruits, for pears by about 6% and for cranberries 15% by weight, giving increases in grower income of US $1,000 to 4,000 per acre (Winston & Slessor 1998). Brood pheromone can be used to increase pollination effort by workers (Pankiw 2004) and this has for example increased pollination rates and yield in triploid watermelon (Guerra Sanz & Roldán Serrano 2008). However, worker bee Nasonov pheromone does not seem to show significant yield increases in field trials (Pettis *et al.* 1999).

Bumblebee, *Bombus terrestris*, pollination activity in greenhouses might be stimulated by artificial foraging recruitment pheromone (Molet *et al.* 2009).

12.1.2 Reproduction of farm animals

Semiochemicals can be used for the manipulation of reproductive behavior in domestic mammals, though many of the molecules involved have not been identified (Chemineau 2011; Rekwot *et al.* 2001). Modern farming practices, including isolation of the sexes, crowding, and artificial insemination (AI), may impede pheromone mechanisms evolved before domestication (Izard 1983). Using a greater understanding of olfactory stimuli including semiochemicals is part of a "clean, green, and ethical," hormone-free approach to improve both reproductive efficiency and animal welfare (Martin *et al.* 2004).

12.1.2.1 Priming effects

The biostimulatory effects of semiochemicals to bring forward, delay, or synchronize puberty and estrus in mammals (Chapters 6 and 9) have been widely used with domestic farm animals (Booth & Signoret 1992; Chemineau 2011; Rekwot *et al.* 2001; Ungerfeld 2007).

Priming effects and puberty

A brief daily exposure to a boar (adult male pig), or his odor advances puberty in juvenile female pigs by at least a month; the priming pheromone is probably 3α-androstenol in the boar's saliva (Booth & Signoret 1992). Exposure to the boar can counter the delaying effect on puberty of the confinement of young female pigs in intensive pig production systems (Izard 1983). Pheromones could be cheaper to administer, involve less labor on the farm, and cause fewer human health concerns than the hormone treatments currently used to cause earlier puberty in pigs and cows. Priming effects of males on puberty are also shown in a variety of ruminants including sheep, goats, and cattle (Ungerfeld 2007).

Priming "male effects" and ending seasonal anestrus

Female sheep and goats undergo seasonal anestrus during the winter and spring. In spring, introduction of an unfamiliar, sexually active male to females results in synchronized estrus some days later (novelty of the male is crucial to the effect, suggesting involvement of signature mixtures as well as a male pheromone(s)) (Delgadillo *et al.* 2009; Hawken & Martin 2012) (see Section 9.9.4). In sheep farming, this synchronized estrus gives a desirable clumping of lambing, which helps labor management and means that the lambs can be marketed as a uniform group (Izard 1983; Martin *et al.* 2004).

Priming effects and the postpartum interval

To ensure efficient reproduction in pigs and cattle, farmers must rebreed sows and cows as soon after they give birth as possible (Izard 1983). The introduction of a boar to a group of sows and their litters brings forward their next estrus. The effect is due to priming pheromones: the time to next estrus in pigs after pregnancy is reduced from 27 days to 10 days after weaning if they are sprayed with the commercial 5α-androstenone pheromone (Boarmate™) within two days of weaning (Hillyer 1976, cited in Booth & Signoret 1992). Ungerfeld (2007) reviews similar effects in ruminants.

12.1.2.2 Signaling

In many ungulates, females signal when they are coming into estrus: using pheromones and distinctive behavior, they solicit approaches by the male (Chapter 3). The farmer using artificial insemination for animal breeding also needs to choose the time with the best chance of fertilization. The time window may be narrow: in cows it is a 12- to 22-hour period on the day of estrus. Though dogs and rodents can be trained to recognize the estrus changes in odor of cow urine and other secretions, identification of the female cow sex pheromone is still elusive (Fischer-Tenhagen *et al.* 2011). A component of the female sex pheromone in the horse, *Equus caballus*, has been identified as *p*-cresol, found in the urine of estrous mares (Būda *et al.* 2012).

Under modern farm management, a pheromone signal (the salivary steroid pheromones 3α-androstenol and 5α-androstenone) from the boar that would normally elicit behavioral signs of estrus in female pigs (sows) may be missing as the sexes are separated. Most estrous sows will respond to pressure on the back, with a stereotyped "lordosis" or "standing reflex" response with the ears cocked, an immobile posture, and ventrally arched back, which would allow the male to mount and mate (Booth & Signoret 1992). About a third of sows in estrus do not react to pressure alone.

More than half of these will, however, respond to the pressure test if they are sprayed with synthetic 5α-androstenone (Boarmate™). Farmers thereby miss fewer estrous females that could have been inseminated.

12.1.3 Aquaculture

The use of pheromones to manipulate reproduction in aquaculture of crustaceans and fish has a promising future but there is little practical use as yet (Barki *et al.* 2011; Mylonas *et al.* 2010). One exception in fish is the demonstration that sex organ extracts and synthetic steroid and prostaglandin sex pheromones could attract aquacultured Chinese black sleeper fish, *Bostrichthys sinensis*, to artificial brood burrows and induce spawning (Hong *et al.* 2006). If the use of pheromones in aquaculture of either crustaceans or fish does become widespread then the effects of these on non-target organisms could become a concern (Lürling 2012; Olsén 2011; Zala & Penn 2004).

Many hatchery-reared fish released into the wild do not survive long because they do not associate predators with danger. Brown and Smith (1998) describe how naïve juvenile rainbow trout, *Oncorhynchus mykiss*, could be taught to associate predator odor (from the northern pike, *Esox lucius*) with the chemical alarm cue from trout skin extract (Box 8.1) before being released to stock rivers.

12.1.4 Semiochemicals, conservation, and animal welfare

A knowledge of semiochemical communication, largely gained from studies of domesticated and laboratory animals, can be important for the success of captive breeding programs for rarer species (Dehnhard 2011; Swaisgood & Schulte 2010). For example, primer effects of pheromones and familiarity with other odor cues can be manipulated to increase the success of captive breeding in the giant panda, *Ailuropoda melanoleuca* (see Swaisgood & Schulte 2010). An understanding of signature mixtures in kin recognition and

their role in mediating incest avoidance (Chapter 1) may help explain why a rare lowland gorilla female was not interested in males she had lived around since birth, but was successfully mated with a male from another zoo (Pfennig & Sherman 1995) (odor cues are also important in human mate choice, see Chapter 13). Chemical cues were manipulated to encourage females of the threatened pygmy loris, *Nycticebus pygmaeus*, to mate with particular males chosen for optimal outbreeding (see Swaisgood & Schulte 2010). The females were presented with artificial marks, which indicated that the selected males were territory holders marking a territory and over-marking competitor marks (Chapter 5). Females had a ten-fold preference for males made familiar in this way and an approximately two-fold preference for the over-marking male. Similar techniques were used to influence female choice in captive breeding of harvest mice, *Micromys minutus* (Roberts & Gosling 2004).

Pheromone cues for aggregation and social organization of animals, particularly in relation to Allee effects at low population levels (Chapter 4), could be an important tool in conservation efforts for the maintenance and reintroduction of rare species (Dehnhard 2011; Gascoigne *et al.* 2009; Swaisgood & Schulte 2010).

With captive animals we may act inadvertently in ways that cause greater stress because we do not understand their olfactory world. For example, partial cage cleaning, by replacing the sawdust substrate without completely cleaning the cage of odor marks, can increase aggression in caged groups of male laboratory mice because social odor cues are disrupted (Chapter 5) (Gray & Hurst 1995).

Like other mammals, cats and dogs mediate much of their behavior by chemical communication. Proprietary "feline facial pheromone" and "dog-appeasing pheromone" respectively have been used to manage spray-marking by male cats and anxiety in puppies, for example. However, full details of experiments to establish the identity of these products have not been published. A systematic review suggests that further work is needed before the clinical efficacy of the treatments can be established (Frank *et al.* 2010). Boarmate™ (Section 12.1.2) has been reported to reduce aggression among newly mixed young pigs (McGlone & Morrow 1988).

Pheromone trapping can provide a good way to monitor endangered species or their prey. For example the threatened scarab beetle, *Osmoderma eremite*, in European woodland can be monitored with its male pheromone (which also attracts females of its predator, the beetle *Elater ferrugineus*) (Larsson & Svensson 2009). In the USA, the red-cockaded woodpecker, *Picoides borealis*, depends on the wood cockroach, *Parcoblatta lata*, as prey. The effects of forest management efforts on the cockroach could be monitored with traps baited with its female pheromone (Eliyahu *et al.* 2012).

12.2 Pheromones in pest management

Currently the most successful applications of pheromones are for insect pest management, with significant cost and environmental benefits to the farmer, the consumer, and society. There are many successful schemes using pheromones, at least as or more effective than the conventional pesticides they have replaced, for the direct control of insect pests over millions of hectares (Baker 2011; Witzgall *et al.* 2010). Pheromones for many insect pests have been identified (El-Sayed 2014). A major strength of pheromones is their low toxicity, which makes them ideal as part of integrated pest management (IPM) schemes, which include biological control agents and other beneficial invertebrates such as bees and spiders (van Lenteren 2012; Witzgall *et al.* 2010). Pheromones fit neatly into the virtuous spiral, for example in greenhouse IPM, where the use of bumblebees for pollination or one biological control agent such as a predatory spider mite encourages (or requires) ending the use of conventional pesticides for other pests (Pilkington *et al.* 2010; van Lenteren 2012).

Although most uses have been in agriculture and forestry, manipulation of vector insects with

behavior-modifying chemicals looks increasingly possible in medical and veterinary entomology (Carey & Carlson 2011; Pickett *et al.* 2010). The main, and sometimes overlapping, ways of using pheromones to control pests are monitoring, mating disruption, "lure and kill" or mass trapping, and other manipulations of pest behavior. Some of these techniques have also been used to control vertebrate pests.

12.2.1 Monitoring

An important use of pheromones is for baiting traps for monitoring populations of insect pests of stored products (Phillips & Throne 2010; Trematerra 2012), crops, orchards, and forestry (Witzgall *et al.* 2010). Pheromone-based monitoring provides one of the most effective survey methods for detecting the presence and density of pest and/or invasive species. Thanks to the specificity of insect pheromones, almost all animals attracted to the trap will belong to that species. For example, the rapid spread of the larger grain borer beetle, *Prostephanus truncatus*, across Africa after accidental introduction from Central America in the 1980s, has been monitored with pheromone traps baited with the male aggregation pheromone (Chapter 4). The resurgence of the bed bug, *Cimex lectularius*, has made development of a monitoring trap a priority (Weeks *et al.* 2011). The aggregation pheromone looks most promising as a lure (Siljander *et al.* 2008).

Pheromone traps can be used in "supervised" or "guided" control programs to monitor the number of pests so that pesticides can be targeted only when and where they are actually needed (van Lenteren 2012). This can give improved pest control and significant reductions in pesticide use. Pheromones can also be used to monitor vertebrate populations (Müller-Schwarze 2006). For example, traps baited with mink scent glands were more effective than fish-baited traps for catching American mink, *Mustela vison*, and could help in wide-scale eradication of this predator (Roy *et al.* 2006).

The scent marks (Chapter 5) left by the animals themselves can also be used for density estimates. Visible scent marks such as the fecal piles of muskrats, scent mounds of beavers (*Castor canadensis*), carnivore droppings on conspicuous features or along trails, and urine marks in the snow, have all been used to estimate how many individuals are living in a given area (Müller-Schwarze 2006).

12.2.2 Mating disruption

Mating disruption is one of the most successful uses of insect pheromones covering around one million hectares worldwide, as illustrated by schemes against the codling moth, *Cydia pomonella*, in orchards, among many examples (Witzgall *et al.* 2010).

The aim of mating disruption is to prevent adult males and females finding each other, thereby stopping fertilization of eggs and thus the caterpillars that cause the damage. For most moths, which rely on pheromones for the sexes to find each other (Chapters 3 and 10), this can be achieved by flooding the air in the host crop with synthetic pheromone. Area-wide treatments that involve every farmer over a region are the most effective (Figure 12.1) (Witzgall *et al.* 2010).

A variety of slow-release formulations has been developed to release small quantities of volatile pheromone over the months of the insect pest season (Heuskin *et al.* 2011; Reddy & Guerrero 2010; Rodriguez-Saona & Stelinski 2009; Witzgall *et al.* 2010). These include polyethylene rope (sealed tubes containing pheromone that diffuses through the tube walls), hollow fibers, rubber septa or PVC beads impregnated with pheromone, microcapsules, and timed aerosol puffers. As a general guide, application rates of between 10 and 100 g per hectare per season are required to achieve communication disruption, giving aerial concentrations of at least 1 ng/m^3 (see Witzgall *et al.* 2010). A novel way of releasing moth pheromone is to apply it to sterile med flies, *Ceratitis capitata*, which then create "mobile mating disruption" (Suckling *et al.* 2011). As the sterile flies are already reared and released in their hundreds of millions, this could provide an effective alternative to spraying moth pheromone for mating disruption of the light brown apple moth, *Epiphyas postvittana*, in urban areas of southern California.

The potential mechanisms for mating disruption include: (1) desensitization (sensory adaptation of the olfactory sensory neurons (Chapter 9) or habituation in the central nervous system); (2) false-plume following (competition between natural and synthetic sources); (3) camouflage of natural plumes by ubiquitous high levels of synthetic pheromone; (4) imbalance in sensory input by massive release of a partial pheromone blend; and (5) the effects of pheromone antagonists and mimics (Cardé & Minks 1995; Rodriguez-Saona & Stelinski 2009). For moths, experiments to date suggest that the dominant disruption mechanism is false plume following, combined, in some species, with habituation when plume following brings males close to synthetic pheromone sources (Rodriguez-Saona & Stelinski 2009). Support for this also comes from experiments in large cages, each enclosing 12 apple trees, which also allowed general principles of attraction and competitive attraction to be confirmed (which turned out to have similar kinetics to substrate–enzyme interactions) (Miller *et al.* 2010b). Many trials of mating disruption of moth pests suggest that the full blend offers the most effective disruption, at the lowest dose (Minks & Cardé 1988) as the correct blend needs to hit the moth antenna as a unit to elicit the full behavioral response (Chapters 9 and 10). However, less precise or off-blends have been used successfully in some systems, with significant cost savings (Witzgall *et al.* 2010).

Another, complementary, way in which mating disruption reduces crop damage is by reducing or eliminating the need for pesticides, thereby keeping more natural enemies alive, so the few caterpillars present have a greater chance of being killed by predators or parasitoids. In addition, pollinating insects are unharmed, which increases fruit set.

These effects are demonstrated in an IPM scheme for tomatoes grown for export in Mexico (Trumble 1997).

Figure 12.1 Area-wide use of pheromones for mating disruption of the grapevine moths *Lobesia botrana* and *Eupoecilia ambiguella* in the Mezzocorona vineyards in Trento, northern Italy. Figure from Witzgall *et al.* (2010). Using data from Mauro Varner and Claudio Ioriatti. Photo by Mauro Varner.

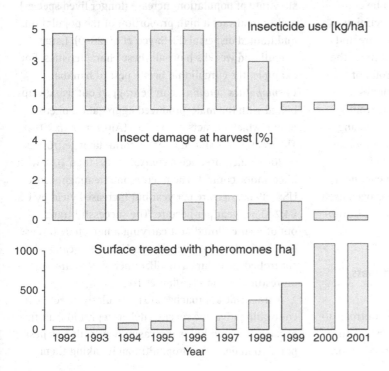

The tomato pin worm moth, *Keiferia lycopersicella*, had become a major pest as it developed resistance to a wide variety of pesticides. In an attempt to control the moth, 20 to 45 applications of broad-spectrum insecticides were applied per season. Worse, tomatoes were being rejected by importers because of high pesticide residues and farmers entered the pesticide treadmill of secondary pests, such as the leafminer fly, *Liriomyza sativae*, created from previously minor pest species when their predators were killed by the pesticides targeted against the original pest. The solution was an IPM program combining mating disruption of tomato pin worm using the female sex pheromone, (*E*)-4-tridecenyl acetate, with largely biological control of other moth pests. The reduction of damage in IPM plots (with mating disruption) was due to fewer moth eggs laid and a higher percentage of parasitism of moth eggs by parasitoid wasps. Lower costs and lower crop losses gave increased profits. In addition, pesticide resistance risks were reduced, and there would be health benefits for farm workers, their families, and the environment.

Insects are not the only pests that could be controlled with mating disruption or related techniques. Parasitic nematodes, major pests in agriculture and causes of human and animal disease, are among the many nematodes that use acarosides, a group of nematode-specific small molecule pheromones (Chapters 1 and 2) (Choe *et al.* 2012). Apart from acarosides, pairing behavior in the disease-causing nematode *Schistosoma mansoni* is mediated by pheromones, which include a small peptide (Ghaleb *et al.* 2006). The sex pheromone of the soybean cyst nematode, *Heterodera glycines*, vanillic acid, has been used in successful small-scale field trials of mating disruption (Meyer *et al.* 1997).

12.2.3 "Lure and kill" (attracticide) and mass trapping

The aim of "lure and kill" or mass trap pest control is to reduce the pest population by attracting pests with pheromones and then either trapping, sterilizing, or killing responding individuals (see also Section 12.2.7) (El-Sayed *et al.* 2006, 2009). With lure and kill (also called attracticide), pest animals are attracted to the pheromone source and pick up an effective dose of insecticides, sterilants, or insect pathogens. Mass trapping confines the animals in a trap. Pheromones may be combined with visual targets and other semiochemicals such as host plant odors. Much less synthetic pheromone is required than for mating disruption and this may sometimes be a crucial financial factor (El-Sayed *et al.* 2009).

Both lure and kill and mass trapping rely on the specificity of pheromones to attract only members of the target pest species (though with bark beetle pheromones, some highly specialised predators may be attracted; Chapter 11). This means that the whole crop or forest does not need to be sprayed, thus helping to save beneficial insects that would normally be killed with area-wide pesticide sprayings.

Effective attracticide or mass trapping depends on highly attractive synthetic pheromone lures, low initial population densities, ideally attraction of females, a slow rate of population increase (longer lived species), ability to attract a high proportion of the population, and limited dispersal (El-Sayed *et al.* 2009). Large curculionid weevils have all these characteristics. For example, the curculionid beetle pest of banana, *Cosmopolites sordidus*, can be trapped out using traps baited with the male-produced aggregation pheromone, which attracts both sexes (Alpizar *et al.* 2012). The treatment reduced banana corm damage from 20 to 30% under insecticide control to less than 10% with pheromone control. The pheromone treatments cost US$185 per hectare per year but increased yield by US$4,240 per year per hectare. The successful trapping out of a curculionid pest carrying a nematode disease of palm trees is shown in Figure 12.2. Ticks can also be controlled using lure and kill, either on host animals or on resting locations (Allan 2010).

Attracticide approaches are particularly effective for controlling invasive species such as tephritid fruit flies because lure and kill is increasingly effective as pest population densities drop, ultimately taking them

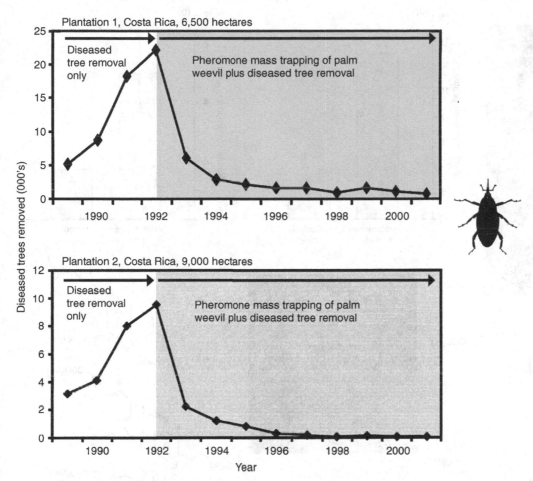

Figure 12.2 The nematode *Bursaphelenchus cocophilus*, which causes red ring disease of oil palms can be controlled by mass trapping its vector, the weevil *Rhynchophorus palmarum* (silhouette). In two commercial Costa Rican plantations, rising disease incidence (shown by the number of diseased palms that had to be removed) was halted in 1992 by the start of mass trapping of the weevils. The traps were baited with the male aggregation pheromone 6-methylhept-2-en-4-ol and sugar cane. Figure adapted from Oehlschlager *et al.* (2002). Weevil silhouette, adapted from CABI www.plantwise.org, creative commons.

below viable Allee densities (Chapter 4) (El-Sayed *et al.* 2009; Suckling *et al.* 2012a).

Mass trapping using pheromones could control invasive fish species. For example, the sea lamprey, *Petromyzon marinus*, a parasitic agnathan fish, which has devastated the Great Lakes fisheries in North America, is currently controlled with relatively non-specific toxic chemicals. Pheromones to lure ovulated females to traps or to unsuitable streams could be more effective and have much lower environmental impact

(Chung-Davidson *et al.* 2011). The lures, at concentrations down to 10^{-13} M, could be either synthetic male sex pheromone (Johnson *et al.* 2009,. 2012; Luehring *et al.* 2011) or synthetic larval (migratory) pheromone (Fine & Sorensen 2008; Meckley *et al.* 2012; Sorensen & Hoye 2007) (Figure 12.3). For species for which we do not have the synthetic pheromone, we may be able to use the animal itself to release the pheromone in the trap, such as common carp implanted with prostaglandin $F_{2\alpha}$ (Lim & Sorensen 2012).

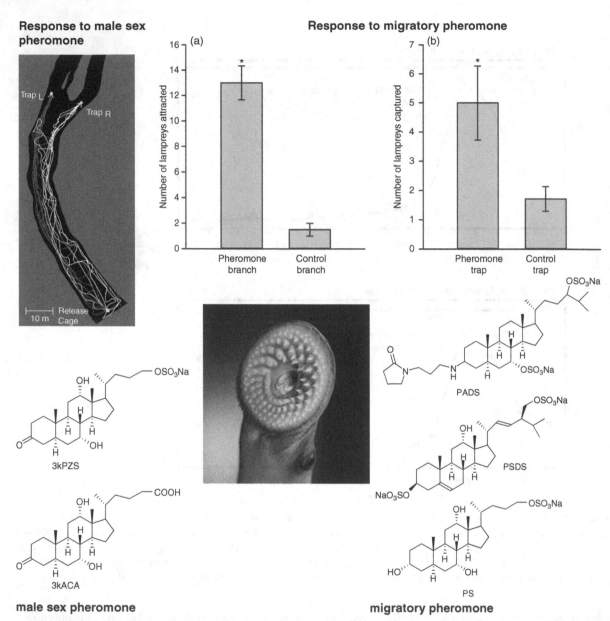

Response to male sex pheromone

Response to migratory pheromone

male sex pheromone

migratory pheromone

Figure 12.3 Pheromones offer the possibility of controlling sea lampreys, *Petromyzon marinus*, which have had a catastrophic effect on Great Lakes fisheries in North America. (Left) Traps baited with a component of the male sex pheromone, 3-keto-petromyzonal sulfate (3kPZS), capture all females when compared with unbaited traps (Johnson *et al.* 2009). Observed up-river movements of individual ovulated females trapped when 3kPZS was applied at 10^{-11}M, 10^{-12}M, or 10^{-13}M in a randomly selected trap and when control solvent was applied in the other (Trap L and Trap R). Gray lines illustrate ovulated females entering the left trap when the left trap was baited with 3kPZS. White lines illustrate ovulated females entering the right trap when it was baited with 3kPZS. 3kPZS is attractive but other molecules given off by males cause attracted females to stay and start nest preparation (Johnson *et al.* 2012; Luehring *et al.* 2011). Among these molecules could be another component of the male pheromone, 3-keto allocholic acid (3kACA), though its role in female behavior is not yet clear.

(Top right) Both male and female adult lampreys are attracted by low concentrations of the multicomponent lamprey larval migratory pheromone equivalent to 5×10^{-13} mol/l of petromyzonamine disulfate (PADS) (the other components are petromyzonol sulfate (PS) and petromyzosterol disulfate (PSDS)). (a) Mean ± 1 SEM of the number of adult sea lampreys attracted into the branch of

Tadpoles of the cane toad, *Rhinella marina* (formerly known as *Bufo marinus*), an invasive species spreading in tropical Australia, eliminate intra-specific competitors by locating and consuming newly laid eggs. The tadpoles find the eggs by searching for species-specific bufadienolide toxins (especially, bufogenins) and it is possible that these could be used to bait traps for lure and kill (though this would be more of a kairomone than pheromone) (Crossland *et al.* 2012). Another possibility is use of its tadpole alarm cue, which reduced toad tadpole survival rates by 50% and body mass at metamorphosis by 20% (Hagman & Shine 2009).

12.2.4 Using deterrent odors

Another approach to pest control uses the responses of animals to their own pheromones or to those of their predators to "persuade" animals to go somewhere else to lay eggs, nest, or feed.

12.2.4.1 Host-marking pheromones and insect herbivores

Synthetic host-marking pheromones (HMPs) (sometimes called oviposition deterring pheromones, ODPs) could be used to protect crops by persuading female insects that the plants or fruit are already occupied with competing eggs. Flies, beetles, and butterflies use HMPs (Chapter 4). Aluja *et al.* (2009b) reduced damage to tropical plums and mangos from the fruitfly *Anastrepha obliqua* by between 50 to 95% through spraying with a shared HMP in extracts of the feces of a related species, *A. ludens*, or simplified synthetic analogues of the *A. ludens* HMP (Figure 4.4) (Edmunds *et al.* 2010). As with other deterrent treatments it may be necessary to trap out females or offer untreated "sacrificial" trap rows, which are later destroyed, so that females can find

somewhere to lay their eggs (see Section 12.2.7) (Aluja *et al.* 2009a). Without this, despite the pheromone, they may eventually lay eggs on the HMP-protected crop.

12.2.4.2 Manipulating conspecific scent marking to prevent colonization

Scent marking forms a very important part of territorial behavior in mammals. However, in most species the scent marks do not act as "keep out" signals (Chapter 5). Nonetheless, for some species these may have some application as control agents. For example, beavers, *Castor canadensis*, which have recovered their numbers in the USA to reach pest status, may be deterred by beaver scent marks added by the experimenter so long as there are other sites available to colonize (Chapter 5) (Campbell-Palmer & Rosell 2010; Welsh & Müller-Schwarze 1989). A greater understanding of badger, *Meles meles*, scent-marking behavior and territories could help interpretation of the perturbation effects of culling badgers as part of experiments on transmission of bovine tuberculosis (McDonald *et al.* 2008; Roper 2010).

12.2.4.3 Predator odors to control vertebrate herbivores

Many vertebrate herbivores, such as deer and rodents, are repelled by the odors of predatory carnivores, in particular urine, gland secretions, and droppings (reviewed by Müller-Schwarze 2006, p. 397 ff.). Many of the active compounds are sulfur containing, perhaps metabolites of meat digestion, and from the carnivore anal glands. Snowshoe hares fed less on pine seedlings treated with a sulfur compound (3-propyl-1,2-dithiolane) from stoat anal glands (*Mustela erminea*) (Sullivan & Crump 1984).

Figure 12.3 (cont.)
the Trout River treated with pheromone (main effects, analysis of variance, $P < 0.0001$) and (b) captured by a baited trap in that branch of the river (vs. an adjacent, unbaited trap) (main effects, t test, $P < 0.05$) (Wagner *et al.* 2006). Later experiments suggest there may be further components to be identified (Meckley *et al.* 2012). Figures, left Johnson *et al.* (2009) and right Wagner *et al.* (2006). Lamprey image, Great Lakes Fishery Commission.

12.2.5 Alarm pheromones and cues

Synthetic alarm pheromones of aphids (Chapter 8) have the potential to increase the effectiveness of conventional pesticides or biological control agents such as the fungal pathogen *Verticillium lecanii* as the increased activity of the aphids in response to their alarm pheromone causes them to contact more insecticide or fungal spores (Dewhirst *et al.* 2010; Pickett *et al.* 1992) (see also Section 12.2.8). Similarly, desiccant formulations are more effective against the bed bug, *Cimex lectularius*, if their alarm pheromone is added, increasing movements of the insects (Benoit *et al.* 2009).

12.2.6 Trail pheromones

Invasive ant species are notoriously difficult to control. The specificity of trail pheromones and low toxicity of pheromones might allow control of the invasive ants but avoid impacts both on native ant species and other organisms. For example, the main component, (*Z*)-9-hexadecenal, of the trail pheromone of the Argentine ant, *Linepithema humile*, a highly invasive species, has been used experimentally to give up to 90% disruption of visible Argentine ant trails for two weeks (Suckling *et al.* 2010). A combined treatment of synthetic trail pheromone and insecticidal bait was effective at reducing populations of *L. humile* in a year-long trial (Sunamura *et al.* 2011). Preliminary experiments showed trail following by the invasive fire ant, *Solenopsis invicta*, could be disrupted by aerosol delivery of its trail pheromone, (*Z,E*)-α-farnesene (Suckling *et al.* 2012b).

12.2.7 Push–pull strategies

Push–pull strategies or stimulo-deterrent diversionary strategies (SDDS) seek to manipulate the behavior of pests and/or their natural enemies so that pests are repelled or deterred from a resource (such as a crop or domestic animal) (the push) and toward other areas such as traps or trap crops (Figure 12.4) (Cook *et al.* 2007). Among the cues that can be manipulated are semiochemicals including HMP (push) (Section 12.2.4.1) and

mosquito oviposition pheromones (pull) (Chapter 4) (Seenivasagan & Vijayaraghavan 2010). Many insect predators and parasitoids use herbivore-induced plant volatiles or the pheromones of their prey as cues to locate the prey (Chapter 11). The techniques can be used as part of IPM strategies for a wide range of targets, from forestry beetles, crop pests, and also for manipulating the behavior of insect vector species such as mosquitoes and tsetse flies (Khan *et al.* 2012; Pickett *et al.* 2010; Seenivasagan & Vijayaraghavan 2010).

12.2.8 Plant breeding and biotechnology for self-protecting plants

Theoretically, plants could be bred or genetically engineered to be self-protecting, able to produce their own mating disruption levels of moth sex pheromones or high concentrations of host-marking pheromones on their fruits. Inspired by wild potato species producing the aphid alarm pheromone (*E*)-β-farnesene (Chapter 11), Beale *et al.* (2006) developed transgenic *Arabidopsis thaliana* plants expressing an (*E*)-β-farnesene synthase gene cloned from *Mentha × piperita* to cause emission of pure (*E*)-β-farnesene. The plants' (*E*)-β-farnesene elicited potent alarm and repellent responses by the aphid, *Myzus persicae*, and an arrestant response by the aphid's parasitoid wasp, *Diaeretiella rapae*. Using the same transgenic plant system, de Vos *et al.* (2010) showed that the aphids became habituated to the plant-produced (*E*)-β-farnesene. However, while no longer repelled by the plant, the aphids also no longer responded to aphid-produced (*E*)-β-farnesene alarm pheromone, making them more vulnerable to ladybird predators.

12.2.9 Understanding barnacle settlement and how to stop it

Barnacles and other encrusting ("fouling") organisms cause costly drag and slow down ships (Aldred & Clare 2008). However, biocidal antifouling coatings cause significant environmental problems and many are now banned (Rittschof 2009). The search for non-toxic alternatives has renewed interest in the cues that barnacle

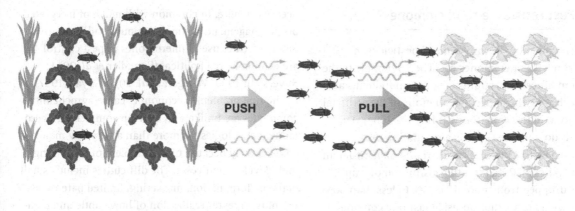

- **Visual distractions**
- **Non-host volatiles**
- **Anti-aggregation pheromones**
- **Alarm pheromones**
- **Oviposition deterrents**
- **Antifeedants**

- **Visual stimulants**
- **Host volatiles**
- **Aggregation pheromones**
- **Sex pheromones**
- **Oviposition stimulants**
- **Gustatory stimulants**

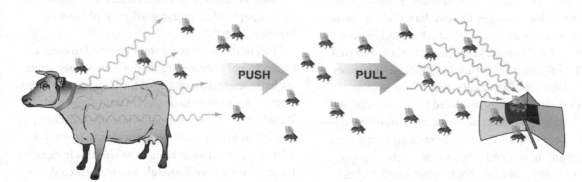

Figure 12.4 The push–pull strategy: diagrammatic representation of the components and generalized mode of action. This can be used with crop plants (above) or to protect livestock (below) or humans from blood sucking pests. Figure adapted from Cook *et al.* (2007).

larvae use when settling from the plankton into aggregations (Chapter 4) (Clare 2011). However, there are challenges in converting this knowledge into commercially applicable molecules to block settlement, as a broad spectrum and long life are needed (Aldred & Clare 2008; Rittschof 2009).

12.2.10 Primer effects of pheromones

We may in future be able to use the primer effects of pheromones (Chapter 6) to control important social insects that are pests, for example by suppressing

reproduction in fire ants, *Solenoposis invicta* (Le Conte & Hefetz 2008), or termites (Costa-Leonardo & Haifig 2010). The main component of the locust's adult cohesion pheromone, phenylacetonitrile, might offer a way of solitarizing hoppers that are starting to gregarize and the molecule may also lead to higher mortality (Chapter 4) (Bashir & Hassanali 2010).

The pheromone primer effects on reproduction in many rodents and fish (Chapter 9) also suggest that there should be rich possibilities for pest management, but I am not aware of commercially available treatments.

12.3 Pest resistance to pheromones?

Like pest control with conventional pesticides, mating disruption with pheromones could be vulnerable to the evolution of resistant strains of pests – individuals able to mate despite the high concentrations of the artificial pheromone in the crop (Cardé & Haynes 2004). The first well documented case is a population of the smaller tea tortrix moth, *Adoxophyes honmai*, in Japan (Mochizuki *et al.* 2002): mating disruption of trap catches dropped from more than 96% to less than 50% after 10 years of disruption using just one component of its pheromone. Mating disruption was restored when its full four-component blend was used. The mechanism of resistance seems to be both a change in the blend produced by females and a broadening of male responses to a wider range of blend ratios (Mochizuki *et al.* 2008). Theoretically, resistance could also come from changes in behaviors, such as the use of different mating sites, use of other mating cues, or selection for the dispersal of mated females (McNeil 1992). Resistance to mating disruption pheromones could also come from males responding to other chemical cues (already produced by the females and perceived by males but not currently part of the pheromone complex) in place of the original components. We already have good evidence that natural populations of insects have evolved regional differences in the pheromone blend used by a species (Chapters 3 and 11), which shows that change is possible. An illustration of the potential for signal change is the mutation in pheromone blend of the cabbage looper moth, *Trichoplusia ni*, observed by chance in a laboratory culture (Section 3.12.3.1) (Cardé & Haynes 2004). What we can predict is that insects are likely to surprise us.

12.4 Commercialization: problems and benefits of pheromones

Given the power of pheromones for insect pest control there still remains the question of why they have not been used more. In common with much of integrated pest management, the most serious challenges to getting greater use of pheromones in pest control are economic and political (Chandler *et al.* 2011; Silverstein 1990; Witzgall *et al.* 2010; Wyatt 1997).

The first problem is economic. Large agrochemical companies are unlikely to develop commercial pheromone technologies for more than a very few major crops, largely because of the problem of recovering high development costs. The difficulties include small markets, formulation, marketing, limited patentability, plus a greater realization of how subtle and complex pheromones can be, and competition with the companies' own conventional products (Arn 1990; Silverstein 1990). The specificity of pheromones, which is their strength for minimal impact, is a commercial disadvantage as each species has its own pheromone, and this situation is further complicated by geographical races using different pheromone blends (Chapter 3).

Most of the small companies involved in commercialization of pheromones do not have the resources for basic research, so much of the research and extension work will need to be government supported. However, it is ironic that politicians are legislating for reductions in conventional pesticide use (see below) while they continue to reduce the worldwide capacity for scientific research into alternative pest control methods at government and academic institutions (Witzgall *et al.* 2010).

Gaining acceptance for pheromone-based pest control, like the adoption of other new IPM methods, requires considerable effort. Adoption will be more likely if farmers find that there must be change, for example in response to resistance problems with conventional pesticides. Getting farmer acceptance of ultimately better but less certain pest control methods is hard. Pheromones need as much selling as pesticides, perhaps even more, if they are to compete successfully with heavily advertised pesticides.

Costs of active ingredients have been a limiting factor in some mating disruption schemes (Witzgall *et al.* 2010). Cheaper off-blends could be used where effective (Witzgall *et al.* 2010). The use of biotechnology and bacterial fermentation could bring down costs of pheromone synthesis. An alternative approach is to find higher plants that naturally synthesize compounds that can be used as precursors and so offer alternative, cheap, and renewable resources for production, for example, of pathogen vector pheromones in the countries where they are needed most (Pickett *et al.* 2010).

Second, political changes at the level of consumers and governments will affect the development of pheromones in pest control (Chandler *et al.* 2011). Changing consumer attitudes to pesticide use are already improving the climate for alternatives to conventional pest control and the use of pheromones is likely to benefit. Government influences through legislation include promising solutions such as the Swedish tax on pesticides, which provides money for researching pheromones and other alternatives to pesticides. Making it easier to register pheromone products would also make a big difference (Chandler *et al.* 2011). Legislation restricting pesticide use is increasingly important. For example, tougher legislation has dramatically changed the patterns of pesticide use in California (Trumble 1997). Government subsidy for pheromone treatments, such as those to wine-grape growers in Germany and Switzerland, can also help (Arn & Louis 1997).

Summary

Each insect pheromone symposium since the 1960s has emphasized the potential for pest control and that is now being realised, albeit slowly. However, we should remember that it is still just 50 years since the first chemical identification of a pheromone, for a moth, and we have made massive strides since then (Wyatt 2009). So far the greatest successes have been with moth and beetle pheromones. The central role of pheromones in the biology of most insects, crustaceans, fish, and mammals suggests there is still a vast potential for intervention with semiochemicals: we have only just begun.

Further reading

Witzgall *et al.* (2010) give a good review of applied pheromone use in insect pest management worldwide, from codling moths on apples in the USA to stem borers of brinjal in Bangladesh and India. Chapters in Allison and Cardé (2014) give further background on using moth pheromones. Pheromones are important in most insects, not just moths, and the chapters in Hardie and Minks (1999) cover these sometimes neglected groups.

Khan *et al.* (2012) discuss practical examples of push–pull cropping used by smallholder farmers in East Africa to tackle stemborers, striga weed, and poor soil fertility (see www.push-pull.net).

Chandler *et al*. (2011) discuss the national and international regulatory frameworks that can encourage development of pheromones and other alternatives to conventional pesticides.

For molecular structures see sites such as www.chemspider.com, which allows you to search by common name and shows synonyms as well as the systematic names. See also the Appendix for a short guide to the terminology (available for free download from the website associated with this book). Many pheromones are illustrated on Pherobase www.pherobase. com (El-Sayed 2013).

13 | On the scent of human attraction: human pheromones?

The sense of smell plays a key role in our lives, although we seem to value it the least of our senses, until it is lost.[1] Humans are just one of more than 300 living species of primate, and many primates, from lemurs to monkeys, use chemical communication extensively (Dixson 2012). Chemical information is still important to us, even if we and our nearest relatives, the great apes (the bonobos, chimpanzees, gorillas, and orang-utans), do not appear to use odor communication as much as other primates (Table 13.1) (Dixson 2012; Shepherd 2004; Stevenson 2010). Like other mammals and our fellow primates, humans and the great apes have many sebaceous and apocrine glands giving off a cloud of odor molecules that, with the help of bacterial breakdown, envelop the body with a complex, and individually distinctive, volatile label.

Pheromones might be hidden among the cloud of odor molecules. As we are mammals, it is likely that we do have pheromones (species-wide chemical signals), as most mammals studied so far appear to. However, there is little robust evidence for the existence of human pheromones (though there is early evidence that there might be a pheromone that babies respond to). The claims made on the Internet for irresistible "human pheromones" are baseless: no human pheromones of any kind have been chemically identified yet.

Until relatively recently, smell was thought to be unimportant in humans, largely because of the small relative size of the olfactory bulb in the human brain compared with other mammals. However, the size of olfactory bulbs and the number of neurons they contain may be less important than the extensive central olfactory brain regions in humans, which process the olfactory input (Shepherd 2004, 2012). These dedicated olfactory regions include the olfactory cortex, olfactory tubercle, entorhinal cortex, parts of the amygdala and the hypothalamus, and the medial and lateral orbitofrontal cortex. While relatively speaking we have smaller olfactory bulbs than mice, our olfactory bulb is larger than the entire brain of a mouse and the orbitofrontal region is huge by comparison (Figure 13.1) (Shepherd 2012, p. 112). Even though bulb size is not the critical factor, far from olfaction diminishing in human evolution, an increase in the size of the olfactory bulbs and supporting brain areas appears to have occurred as we diverged from common ancestors with the Neanderthals (Bastir et al. 2011).

Smell works by the interaction of odor molecules with olfactory receptors (ORs) in the nose. When an odor molecule binds to an OR in the membrane of an olfactory sensory neuron, the excited neuron sends a message to the brain (see Chapter 9 for a full explanation). The specificity of smell detection comes from the range of molecules an OR binds with. So one might expect that the more different ORs, the better the sense of smell. However, the human sense of smell is much better than we might imagine from a comparison of the number of functional (working) olfactory receptors (ORs) in mice (~1000) and humans (~370): despite smaller numbers of different ORs we have retained the ability to recognize a spectrum of odor molecules as broad as mice (Shepherd 2004, 2005; Zhang & Firestein 2002). Humans are just as good at distinguishing odors as squirrel monkeys (with ~1000 functional ORs) and pigtail macaques (with ~700), both species with relatively larger olfactory bulbs compared with humans (Laska et al. 2005).

---

[1] Oliver Sacks quotes a man who lost his sense of smell after a head injury: "when I lost [my sense of smell] it was like being struck blind. Life lost a good deal of its savor – one doesn't realize how much "savor" is smell. You *smell* people, you *smell* books, you *smell* the city, you *smell* the spring – maybe not consciously, but as a rich, unconscious background to everything else. My whole world was suddenly radically poorer." (Sacks 1987, p. 159)

Table 13.1 **Occurrence of specialized scent glands, scent-marking displays, and a functioning vomeronasal organ (VNO) in primates.**

Primate group	Number of genera studied	Functional VNO in adults	% Genera with specialized skin scent glands	Main positions of skin scent glands	% Genera that mark using glands	% Genera that mark using urine
Prosimians (e.g., lemurs)	17	Yes	94	Circumgenital/ ventral thorax/ shoulder/wrist	94	76
New World monkeys (e.g., marmosets)	16	Yes	94	Circumgenital/ ventral thorax/ throat	88	69
Old World monkeys (e.g., baboons, mangabeys)	14	No	14	Few have skin glands. Sternal (chest): (*Hylobates*, *Symphalangus*, *Mandrillus*)	14	0
Apes and man	5	No	100	Sternal: *Pongo*. Axillary (armpit): (*Pan, Gorilla, Homo*)	0	0

After Dixson (1998) Table 7.4, updated with information from Tables 7.4 and 7.5 in Dixson (2012). Vomeronasal information from Preti and Wysocki (1999), Dulac and Torello (2003), and Bhatnagar and Smith (2010).

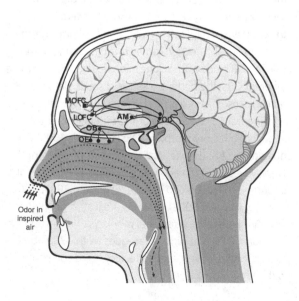

Odor in inspired air

Figure 13.1 Brain systems involved in human smell perception during orthonasal olfaction (sniffing in). Air flows indicated by dotted lines. As we breathe out, the retronasal olfactory flavor system integrates olfaction with many other inputs including taste, texture, and vision, leading to wide-ranging brain activity. The whole mouse brain is about the size of our olfactory bulb. AM, amygdala; LOFC, lateral orbitofrontal cortex; MOFC, medial orbitofrontal cortex; OB, olfactory bulb; OC, olfactory cortex; OE, olfactory epithelium. Caption and figure adapted from Shepherd (2006).

Box 13.1 **Cultural and social aspects of odors**

Smell is not just a biological and psychological phenomenon: smell is a cultural, hence a social and historical phenomenon (Classen *et al.* 1994, p. 3).

Different societies may have different universes of odor. These, in turn, give rise to different osmologies (words in different languages for smells that also reflect cultural significances) (see Classen *et al.* 1994 for a fascinating discussion).

Body odors are important to people in all societies. Added perfumes can mark us as members of a group. Seductive perfumes are used everywhere: in Melanesia, men wear a strong musky aromatic leaf during dances and, of course, perfume advertisements fill Western media (for more discussion of perfumes see Classen *et al.* 1994; Herz 2011; Schaal & Porter 1991; Schilling *et al.* 2010; Soo & Stevenson 2007; Stoddart 1990).

Over the last 300 years, cultural norms in the Western world have moved toward intense suppression and masking of natural body odors and control over overt appreciation (Classen *et al.* 1994; Corbin 1996; Soo & Stevenson 2007). However, these attitudes are not entirely recent: both Catullus and Ovid, Latin poets in ancient Rome, advised both men and women not to keep a smelly goat in their armpits when looking for a partner (Stoddart 1990). Smell can be political: those too poor to keep clean easily have been looked down on as the "great unwashed."

Still, natural body odors can hold a special place in sexual interaction. Some contact (personal) advertisements ask partners to remain unwashed. Napoleon is said to have written to Empress Josephine saying "Don't wash. I'm coming," two weeks before they would meet again (Ackerman 1990, p. 9).

In humans, as in other mammals, it seems that many odor preferences are learned and thus individual and cultural experience are important factors (see Stevenson 2010). While there is predisposition to respond to some classes of odor, much of the young baby's response to odors is learned (Schaal 2012). Soussignan *et al.* (1997) used the facial expressions of newly born babies (neonates) to judge their responses to odors (for example, smile or disgust). The responses of neonates to some odors are very different from those of adults and this continues into early childhood. For example, young children up to the age of four to five years tolerate odors such as butyric acid (rancid butter odor), which older children and adults find unpleasant or disgusting. Learned, socially appropriate, responses to odors (and thus foods) can mean that treasured national delicacies may be disgusting to visitors who have not learned to like them (see Deutsch & Murakhver 2012 for many examples such as Icelandic fermented shark or English Stilton blue cheese).

The learning of preferences has implications for attempts to study human pheromones. Like so many aspects of our behavior, our current condition is so strongly influenced by social evolution that it is hard to disentangle cultural influences from the biological. Even a modern cross-cultural approach may not yield the answers, as no present-day society shows the

Box 13.1 (cont.)

ancestral patterns of odor production and response. The challenge is to discover the biological odors and responses that are potentially common to all humans. The idea that odors might affect our emotions or subconscious, and not be entirely under our control, is slightly alarming to modern sensibilities (but paradoxically, it is also what perfume advertisements and commercial "pheromone" websites entice us with).

Our sophisticated enjoyment of food flavors is largely due to the subtlety and breadth of our sense of smell (olfaction) (Shepherd 2006, 2012). Many of our personal food likes and dislikes are due to our individual olfactory and gustatory (taste) worlds as well as experience: each of us has a unique combination of olfactory and gustatory (taste) receptor variants, so we sense the world differently (Box 9.1) (Section 13.4) (Newcomb *et al.* 2012).

It is not surprising, given the subtlety of odor discriminations we can make, that smell appears to be important in human culture (Box 13.1) as well as mate choice and courtship (in some languages, to kiss means to smell) (Penn & Potts 1998a). In this chapter, I start with recognition of family members by smell and the possible role of smell in mate choice for compatibility. I then look at our sense of smell and the odor molecules we give off, before exploring what we know about human pheromones. I end the chapter with what we will need to do to discover human pheromones.

13.1 Olfactory cues to recognition: signature mixtures

For a species once thought to have evolved a diminished sense of smell in favor of vision, we are surprisingly good at recognizing familiar humans by smell (Lenochova & Havlíček 2008; Schaal & Porter 1991). A clue to the individuality of our odors is given by the ease with which dogs are able to tell us apart by smell (unless we are identical twins on identical diets). The highly individual odors are our variable chemical profiles, not species-wide pheromones (Chapter 1) (Figure 1.1) (Wyatt 2010). We remember a subset of the whole smell (chemical) profile of another person as the *signature mixture* of that individual. A key difference between pheromones and signature mixtures is that in all animals so far investigated it seems that signature mixtures need to be learned, whereas, generally, pheromones do not. "Signature" is used as it implies individuality.

The typical experiment to investigate the ability of people to recognize, for example, the smell of family members, asks people to sniff a standard T-shirt or cotton pad that has been worn overnight by the "odor donor" (e.g., Roberts *et al.* 2005). The odor donors would have been given clean T-shirts/pads and instructions to minimize environmental odors such as scented soaps and strong smelling foods. Using T-shirts or odor extracts of various kinds makes it easier to standardize experiments and makes replicates possible. Odor donors do not have to be present during the testing. This removes the risk that responses will be influenced by the audio, visual, or behavioral characteristics of the odor donors.

Some of the most important work on human recognition by odor has been done on the interactions between mothers and newborn babies (neonates) (Mizuno 2011; Porter & Winberg 1999; Schaal *et al.* 2009). As well as mutual mother–baby recognition there may also be a mammary phero-mone common to all lactating mothers (see Section 13.5.1).

Newborn babies learn the individual odors of their mother in the first hours after birth, and by three days old will turn toward a pad carrying the odors of their mother's breast rather than odors of another lactating mother or a clean pad (Mizuno 2011; Porter & Winberg 1999). The baby also learns the individual odor of its mother's axillae (armpits), and will turn toward these odors in tests. Babies move their heads and arms less and suck more when exposed to their own mother's odors.

While the baby is breast feeding or sleeping, the mother is also learning its individual odors. Mothers can recognize their babies by smell alone within the first two to three days even if they have had only an hour or so of contact (Kaitz *et al.* 1987; Porter *et al.* 1983; Schaal *et al.* 1980). In the hours after giving birth many mothers rate the odors of their babies pleasing and find baby odors in general pleasing. In first-time mothers this, and success at recognizing the odors of their own baby, are positively correlated with their level of the adrenal hormone, cortisol (Fleming *et al.* 1997). Cortisol may simply be an easily measured marker for other, as yet unidentified, hormone changes underlying the changes in perception (Chapter 9). Olfactory recognition is not confined to the time after birth. Parents can recognize their older children by smell too (and the converse).

The comfort given by "security blankets" and other favorite "attachment objects" carried round by young children is largely due to their familiar smell (if washed, attachment objects may be rejected by younger children) (Ferdenzi *et al.* 2008). Young chil-dren with sleeping difficulties may be calmed by the axillary odors of their mother (Schaal 1988). Adults, too, get "olfactory comfort" from sniffing their part-ners' clothing during periods of separation (McBurney *et al.* 2006).

Other members of families can also recognize each other by smell (Porter *et al.* 1986). In a two-choice discrimination test, most fathers correctly recognized the T-shirt worn by their own infant, less than 72 hours after birth. Aunts and grandmothers also reliably identified their neonatal relatives by odor. Children aged between three to eight similarly were able to identify correctly T-shirts worn by their similar-aged siblings.

Early recognition of newborn infants by parents may be facilitated by the similarity of their odors to those of other family members to whom they are related (Schaal & Porter 1991). Many animals can use such "phenotype matching" (Chapter 1) to recognize genetically related but unfamiliar animals. This ability was tested in humans by asking strangers to match the T-shirt odors of children and parents, which they did successfully. It was assumed that if strangers could do this, family members would be able to. Cues such as diet in common between child and parent were ruled out as factors.

The memory for individual odors might be a long-lasting mechanism for social recognition (see also Box 13.2). Adult humans, for example, are able to identify their adult siblings correctly by odor even if they have not been in contact for up to 30 months (Schaal & Porter 1991). Olfactory memory of siblings may affect our choice of partner when adult (next section).

13.2 Choosing mates for genetic compatibility: avoiding kin and going for optimum difference?

Many kinds of animals avoid inbreeding, using learned olfactory cues to avoid mating with kin (Chapters 3 and 6). Olfactory cues described in the previous section might be part of the mechanism of the Westermarck ("kibbutz") effect in which we do not choose adult partners from among the people we have lived with when growing

Box 13.2 Odors and the "Proustian moment"

In our common experience, smells bring back powerful memories. Perhaps the most famous example in literature is Proust's account, in his novel *Swann's Way* (the first volume of *Remembrance of Things Past*, published in 1913), of the way the taste and smell of a biscuit dipped in tea triggered intense joy and long-forgotten memories of the narrator's childhood. Gilbert (2008, p. 189 ff.) notes that many nineteenth-century authors, including Wendell Holmes and Nathaniel Hawthorne in America and Louis-François Ramond de Carbonnières in France, made the association of smells and memory long before Proust.

However, memory of odors may be like memory of anything else – subject to fading, distortion, and misinterpretation (Gilbert 2008, p. 189 ff.). It is not clear that odor memory is a separate system from verbal and visual memories (Herz 2009b, 2012; White 2009; Wilson & Stevenson 2006). It may be simply that odor memories can surprise us because we don't recall consciously remembering them at the time (Gilbert 2008, p. 201).

Laboratory researchers struggle to create realistic long-term odor memory tasks. An opportunity for more natural "field tests" of memory associations was offered by synthetic odors used in the Jorvik museum in York, UK, to make an exhibit of tenth-century Viking life smellier (Aggleton & Waskett 1999). In the study, people out shopping in a city some 200 miles from York were asked if they had visited the museum and, if yes, were asked what they could recall. Up to six years after their visit to the museum, people were able to correctly answer more questions about the museum's contents if the synthetic odors used by the museum were presented during the interview. When originally visiting the museum, the people could have had no idea that years later they would be asked these questions.

My feeling is that we can still keep the idea of a Proustian moment but we should not give it a unique significance over memories prompted by our other senses.

up (see Rantala & Marcinkowska 2011). Again the odor cues would be signature mixtures, different in every person, not pheromones (Chapter 1) (Wyatt 2010).

Among the odors used by vertebrates to avoid kin are ones associated with the enormously variable major histocompatibility complex (MHC) of the immune system (see Chapter 3 for more detail). Mice can distinguish otherwise genetically identical mice differing at only one MHC locus. Humans can distinguish the odors of these same mice (Gilbert *et al.* 1986). Humans also have an MHC immune system and rats can learn to distinguish the urine odor of people who differ at the MHC

(Ferstl *et al.* 1992), which in humans is called human leukocyte antigen (HLA) because the molecules are on leukocytes, a type of white blood cell. As humans have both variability in MHC odors and the ability to detect them, do we use them in mate choice? We might expect so, based on studies of MHC-associated mate choice in five species of non-human primates, though it should be noted that not all species showed disassortative mating for the MHC (choosing mates dissimilar to themselves) (Setchell & Huchard 2010).

A first investigation of MHC and human mate choice examined the responses of Swiss female students to the odor of cotton T-shirts that had been worn by male students, using the kind of protocol for T-shirt smell testing described in the previous section (Wedekind *et al.* 1995). Wedekind and colleagues also tissue-typed the women and men to determine their MHC group and compared the preferences when the female had a similar or dissimilar MHC to the male who had worn the T-shirt. Women not taking the oral contraceptive pill (the "pill") preferred the odor of T-shirts previously worn by MHC-dissimilar men. In contrast, women who were taking the pill preferred T-shirts worn by MHC-similar men. This result was not due to perceptual differences of the intensity of odors. Preferred odors by women both taking and not taking the pill were rated just as intense as less preferred odors; this finding is important because strong odors often repel in this kind of experiment. When asked who the odor of the T-shirt reminded them of, women were more likely to say "of a current or previous mate" if the T-shirt had the odor of an MHC-dissimilar male.

A second experiment, designed slightly differently, but with the same protocol for T-shirts, tested the responses of both genders to male and female odors (Wedekind & Füri 1997). Women not on the pill and men preferred the odor of MHC-dissimilar individuals, and the preference was negatively correlated with the degree of MHC similarity between the smeller and the T-shirt wearer. The preference seemed to be for difference in MHC, not for any particular complementary MHC combination. Again, in contrast, women taking the pill preferred MHC-similar odors. Roberts *et al.* (2008) did not find a preference for MHC-dissimilar males by non-pill-using women but in a longitudinal study they did find a significant shift to a preference for MHC-similar men after starting pill use, which was not reflected among the control group of non-pill users.

Wedekind *et al.*'s (1995) experiments have been criticized (see Hedrick & Loeschcke 1996, and Wedekind's reply) but Wedekind was cautious in his interpretation of the experiments. He argued only that they show that females and males can discriminate MHC-related odors. The "reminding of past or present

partners" data suggests (but no more) that these preferences are acted on.

In a study of perfumes and MHC, Milinski and Wedekind (2001) found that people with a given MHC tended to choose the same perfume ingredients, and that these were different from those chosen by people with a different MHC. This was only true when the subjects were asked to choose a perfume for themselves, not when asked what perfume would they like a partner to wear (see Box 13.1 on perfumes and Box 13.3 for other ways we use smell).

As Wedekind and Füri (1997) conclude, "no one smells good to everybody: it depends on who is sniffing whom, and it is related to their respective MHCs."

13.2.1 Is this reflected in real-life choice of partners?

We would expect that if partners chose each other to be different at the MHC, there would be fewer same-MHC type matings than expected at random. However, to demonstrate this statistically can be difficult, not least because the high polymorphism in the MHC makes nearly all mating combinations rare (Hedrick 1999). Non-random mating for MHC has been found in some mouse experiments (Chapter 3), but the evidence in human populations is contradictory. Havlíček and Roberts (2009) reviewed ten studies of MHC similarity in established human couples and observed that most reported studies showed no significant disassortative mating choice for the MHC in natural human populations. The one study that did show disassortative choice was of 411 couples in a Hutterite population (a North American group isolated by religious beliefs), in which couples were less likely to share MHC haplotypes (genotype) than expected, even after statistically controlling for cultural incest taboos (Ober *et al.* 1997).

An analysis of the publicly available HapMap datasets of genetic markers was reported to show non-random, disassortative mate choice based on the MHC in some populations, but this is disputed by other researchers (for original references and the debate see Derti & Roth 2012; Laurent & Chaix 2012).

Box 13.3 Putting human odors to use: applications so far

Changing, masking, or getting rid of human odors is big business. In 2001, about US$6 billion was spent by US consumers on soaps and deodorants and US$4 billion was spent on putting smells back on (perfumes, see Box 13.1) (Gilbert & Firestein 2002). However, there are many other uses for human odors and our reactions to them, even if pheromones have not been identified yet.

Forensic uses

Our individually distinctive smells can be exploited for detective work. It has long been known that dogs can distinguish people by odor, only identical twins confuse them (humans have similar capabilities, Section 13.1) (Roberts *et al.* 2005). Dogs have remarkable abilities to track human scent trails (Chapter 10).

Matching trace odors at a crime scene or on a weapon, left by handprints for example, to a unique suspect is more open to challenge (Brisbin *et al.* 2000). The first problem is that the traces of odor are too small to be independently tested using current technologies such as gas chromatography. Another doubt is whether dogs can generalize from a handprint, for example, to the odors of other parts of the same person. Hamsters cannot generalize across odors from different parts of another hamster unless they are familiar with the animal and have learned to associate the various odors with the individual (Chapter 6) (Johnston & Jernigan 1994). Dogs may have similar difficulties. Brisbin and Austad (1991) found that dogs could not correctly distinguish their handler's elbow scent from a stranger's hand scent on objects. The training of dogs might inadvertently train them to respond to hand scents rather than to those of other parts of the body. Settle *et al.* (1994) found average success rates for a forensic task (comparing hand odor with an odor trace from another part of the same person's body) greater than 80% but this leaves a high error rate. Schoon (1997) developed new ways of presenting odor choices to the dogs that may improve the reliability of their decisions. Nonetheless, Brisbin *et al.* (2000) remind us to be cautious in accepting criminal evidence based on dog identification results alone.

Medical uses of smell

The growing understanding of olfaction in mother–neonate interactions has numerous potential clinical applications (Schaal *et al.* 2004). For example, amniotic or other odors could help pre-term babies thrive better in incubators. Olfactory cues are key to successful initiation of breast feeding and so washing of the mother's breast can be unhelpful; cleanliness is not necessarily a virtue. Similarly, close olfactory contact between mother and neonate in the hours after birth may be important for both to learn the odor signature(s) of the other, at a time when the mother and baby may be physiologically primed to learn each other's odors (Chapter 9) (Mizuno 2011). In the future, if a human mammary pheromone is identified, a synthetic version could perhaps

Box 13.3 **(cont.)**

help babies start breast feeding as failure can lead to dehydration and failure to thrive (Section 13.5.1) (Doucet *et al.* 2012).

Disease, either as a result of infection or metabolic disorder, may cause unusual odors. For example, cholera, *Vibrio cholera*, gives the patient's feces a sweetish smell, possibly from the volatiles dimethyl disulfide and *p*-menth-1-en-8-ol (Shirasu & Touhara 2011). Smelling the patient can help in the diagnosis of many diseases and has a long history, going back to Hippocrates around BCE 400 and later the Arabian physician Avicenna among others (Penn & Potts 1998a; Shirasu & Touhara 2011; Wilson & Baietto 2011). Today, a smart doctor can still make a preliminary diagnosis by smell of many rare metabolic diseases of newborn babies, such as phenylketonuria with its musty smell of phenylacetic acid, and start life-saving therapy while the diagnosis is being confirmed by slower laboratory tests (Doty 2001). Sadly, the use of smells by doctors for diagnosis is fast becoming a lost art.

Dogs can be trained to detect odors associated with particular diseases (Moser & McCulloch 2010). Electronic noses are likely to be more scalable for mass screening and non-invasive diagnosis of disease from breath and other odor samples (Shirasu & Touhara 2011; Wilson & Baietto 2011). However, though many volatile organic compounds have been claimed as potential biomarkers, only a few compounds have been consistently demonstrated and approved for clinical applications (Kwak & Preti 2011). Among the problems is the large individual variation in the concentrations of biomarkers in diseased and/or healthy subjects.

For some other diseases, early diagnosis can come from *loss* of the patient's sense of smell. Loss of this sense can be due to many causes, including simple congestion of the nose from a cold. However, patients with several neurodegenerative diseases, including early stages of Parkinson's disease and Alzheimer's disease, often show partial or complete loss of smell, distinguishing these from other neurological diseases with similar early symptoms (Doty 2009; Hawkes & Doty 2009). Olfactory dysfunction is present in approximately 90% of early-stage Parkinson's disease cases and can precede the onset of motor symptoms by years (Doty 2012a,b). The mechanisms responsible for olfactory dysfunction are currently unknown. The smell loss can be revealed by using standardized olfactory tests, including "scratch and sniff" strips with microencapsulated odors (Doty 2009; Hawkes & Doty 2009).

"Aromatherapy" and similar uses of odors

Many claims have been made for "aromatherapy" with plant and other odors, such as peppermint and lavender, but experimental results are contradictory and overall the evidence is weak (Herz 2009a; Howard & Hughes 2008). Responses may be largely due to expectation: for example, relaxation responses, including skin galvanic measures, to lavender oil depended on what the subjects were primed by the experimenter to expect from the odors (Howard & Hughes 2008).

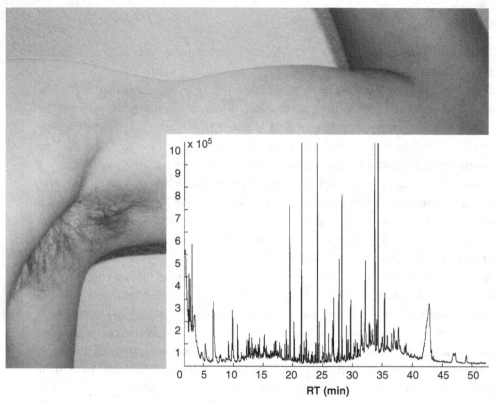

Figure 13.2 Gas chromatograph (GC) trace of a human axillary (armpit) sweat sample taken from a single person by stir-bar sorptive extraction (SBSE) (Chapter 2). Each of the more than 700 peaks in the trace is one or more molecules. Across similar samples from almost 200 people, the researchers described thousands of peaks: everyone's profile was different. Hidden among the peaks are the signature mixtures learned by other humans (and dogs) for individual recognition and, potentially, some molecules might be human pheromones. RT = retention time in minutes. Y-axis, amount. From Xu *et al.* (2007). Armpit photo by L. L. Freitas, Creative Commons licence.

One selective advantage for choosing a dissimilar partner is that there is a higher successful implantation rate if the fetus and mother differ in MHC (Chapter 3) (Beydoun & Saftlas 2005; Penn 2002). This was revealed for a human population by an analysis of pregnancy outcomes in the Hutterite population mentioned above (Ober 1999).

13.3 What molecules do humans give off?

The odor molecules we give off reflect our internal physiological and metabolic state in a complex interaction between our genetic make-up and the environment, including diet. The odors carry messages that give clues to our age, sex, reproductive maturity, and health status (Chapter 3). Our smell changes at puberty. Sebaceous and apocrine glands (see below) are secondary sexual characteristics that fully develop then, giving adult odors. The glands are androgen sensitive and are equivalent to the sexually selected apocrine and sebaceous glands in other mammals (Chapter 3).

We give off hundreds if not thousands of different volatile molecules (Figure 13.2). Many of these molecules are produced by bacteria and other micro-organisms on the skin, acting on largely odorless

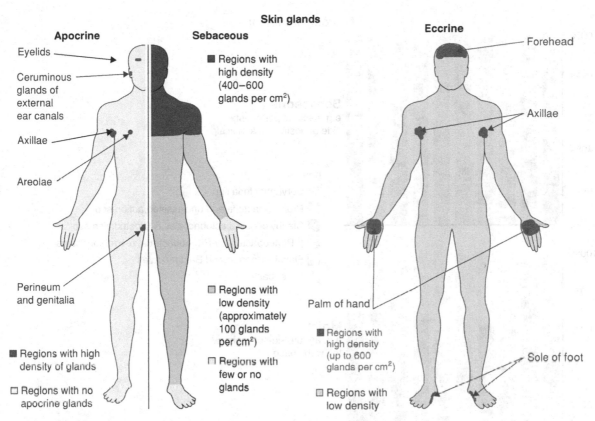

Figure 13.3 Distribution of apocrine, sebaceous, and eccrine glands on the human skin surface. The axillae stand out as having high densities of all three types of gland. Figure adapted from Wilson (2008).

secretions (Archie & Theis 2011; see reviews by Gautschi *et al.* 2007; Schilling *et al.* 2010; Smallegange *et al.* 2011). The characteristic odors of different parts of our body come from the interaction between secretions from different glands (Figure 13.3 and see below) and the characteristic microbiome, largely bacteria and yeasts, that lives on our skin in different areas (Figure 13.4) (Costello *et al.* 2009; Grice & Segre 2011; Human Microbiome Project Consortium 2012). The complexity, stability, and identity of the skin microbiome is highly dependent on the microenvironment of the sampled site and is influenced by levels of temperature, moisture, and oxygen. For example, hairy, moist armpits lie a short distance from smooth dry forearms, but these two niches are likely as ecologically dissimilar as rainforests are to deserts (Grice *et al.* 2009). As well as

the large variation of species and types of bacteria over the body, there are inter-person differences in microbiome that are consistent over time (Figure 13.4) (Grice *et al.* 2009; Human Microbiome Project Consortium 2012).

There is still much to learn about the molecules given off from the different parts of the human body, including the genitals. The type of systematic work to narrow down which areas of the body produce smells important for our behavior has not been carried out yet. Instead, olfactory behavioral research on humans has been largely focused on the axillae (armpits), in part because the axillary glands are unique to humans and the great apes (Table 13.1).

Male gorillas are reported to have a distinctive pungent odor different from females (Dixson 2012, p. 253).

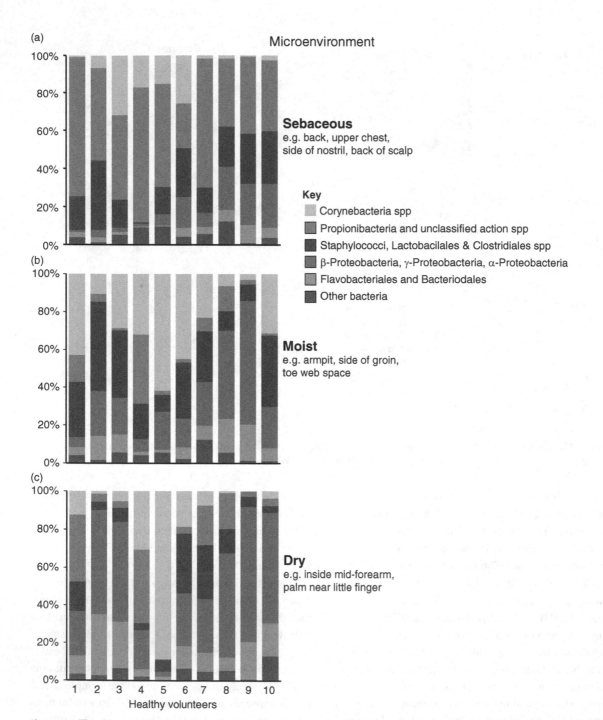

Figure 13.4 The skin microbiome, largely bacteria, varies between microenvironments on the body and also between individual humans. The microenvironment depends on glandular secretions and the local moisture and oxygen regime. The relative abundance of the most abundant bacterial groups associated with representative microenvironments is shown for each of ten

There are reports suggesting that the sexes in humans can be distinguished in the same way. However, experimental tests of people's ability to distinguish male and female humans by smell show that people often make errors, with stronger body smells frequently assumed to have come from males even whenthis was not the case (Doty 2010, p. 132 ff.; Doty *et al.* 1978). Women did not find men's armpit smells more attractive than women's and nor did men find women's armpit smells more attractive than men's. There are some differences in the mix of molecules between male and female armpits but there seem to be no unique and exclusive marker molecules to distinguish the sexes (Penn *et al.* 2007). The relative quantities of two odorless precursors have been reported to differ between men and women, potentially leading to women's sweat smelling more sulfurous onion-like and men's more cheesy rancid (see below) (Troccaz *et al.* 2009). Preliminary olfactory preference experiments suggest that sexual orientation might affect both odor production and preferences but there are likely to be many cultural factors so the results are hard to interpret (Martins *et al.* 2005).

13.3.1 Sebaceous and apocrine sweat glands are major sources of odor precursors

A number of gland systems secrete molecules to the skin surface: the sebaceous glands and two kinds of sweat glands, the apocrine and eccrine (to which a third has been added recently, apoeccrine, with characteristics of both; Wilke *et al.* 2007) (see also Labows & Preti 1992; Smallegange *et al.* 2011) (see Figure 13.5) (Table 13.2). Eccrine sweating occurs in response to thermal stress. In addition, the eccrine and apocrine sweat glands secrete in emotional sweating prompted by anxiety, fear, pain, or sexual arousal (Wilke *et al.* 2007). Emotional

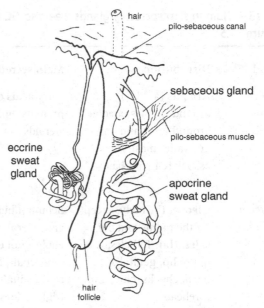

Figure 13.5 Diagram of a human hair follicle and associated glandular structures. Figure adapted from Stoddart (1990), redrawn from Montagna and Parakkal (1974).

sweating from apocrine glands in the axillary region does not occur before puberty.

The sebaceous glands are found over most of the body, especially the upper part (Figure 13.3). There may be up to 400 to 900 glands/cm^2 on the upper chest, back, scalp, face, and forehead. Their secretions are rich in lipids. The two main kinds of sweat glands produce quite different secretions. The apocrine glands secrete only small volumes of material (about 1 to 10 µl/day per gland), but are the most important for odor production. The apocrine glands are found primarily around the nipples, genitals, and in the armpit (axillary) region (Figure 13.3). They secrete into the hair follicles above the sebaceous gland, at times of emotion, stress, or sexual arousal.

Figure 13.4 (cont.)
healthy volunteers (who included both males and females, ages 20 to 41, with a variety of self-reported ethnicities). The sebaceous sites have many propionibacteria species, whereas in moist sites (which may also have sebaceous secretions, as in the armpit) the propionibacteria are joined by large numbers of staphylococci species. Dry sites may have many proteobacteria species, but as always there are variations in these proportions between people. (For more details see Grice & Segre 2011; Grice *et al.* 2009; Human Microbiome Project Consortium 2012). Data from Grice *et al.* (2009).

Table 13.2 Human skin secretory glands (see also GC trace in Figure 13.2; diagram of a human hair follicle in Figure 13.5).

Gland	Distribution	Main secretions	Control	Notes
Apocrine sweat gland	Armpits (axillae), around nipples (areolae), around anus and on genitalia, mostly associated with hair follicles	Fatty acids on carrier proteins, lipids, steroids	Continuous but especially if emotion/stress/ sexual arousal	Secretes at and after puberty, androgen dependent
Sebaceous gland	All over the body, except on the palms and the soles. Hair follicles, upper lip, genitalia, head, cheeks, chin, and forehead	Sebum (lipids such as cholesterol, cholesterol esters, long-chain fatty acids, squalene, triglycerides)	Continuous secretion	Secretes at and after puberty, androgen dependent. Lubricates and protects skin surface and hair.
Eccrine sweat gland	Whole body surface, most numerous in armpits, and on forehead, palms, and soles	Primarily water, with small traces of salts, proteins, amino acids, urea, ammonia, and lactic acid	Thermoregulation, emotion/stress	

Based on Stoddart (1990), Labows and Preti (1992), and Smallegange *et al.* (2011). The apoeccrine sweat gland, a third type with characteristics of both as its name suggests, has been identified recently (Wilke *et al.* 2007).

Eccrine sweat glands are distributed over the entire body (Figure 13.3). Each human adult has about three million eccrine sweat glands, capable of secreting up to 12 liters of very dilute aqueous fluid per day, for thermoregulation. Eccrine sweat does not contribute much to odor. In what we colloquially describe as "sweating" in hot weather or with hard physical exertion, almost all of the secretion comes from the eccrine sweat glands (hence the saying, "honest sweat doesn't stink"). However, the moisture does help to create the damp conditions needed for the growth of bacteria.

Most research on human odors so far has focused on apocrine secretions. Doty (2010, p. 126) reminds us that sebaceous secretions should not be forgotten as in most other mammals the sebaceous glands, rather than apocrine glands, are most important in chemical communication (and differ most between species) (Chapter 3).

13.3.2 Bacterial metabolism produces the odors

The apocrine and sebum secretions in the axillae are odorless when first secreted, but the characteristic odors emerge if incubated with skin bacteria, largely *Staphylococci* and *Corynebacteria* (Gautschi *et al.* 2007; James *et al.* 2004; Schilling *et al.* 2010). The springy hair of the armpit provides a vast surface area for the bacteria to live on and an evaporating surface for odors when the axillae are exposed by lifting the arms – not unlike the hair pencils of male moths and tarsal-gland hair tufts of deer (Chapters 1 and 3).

Shaving the armpit clear cuts the habitat and greatly increases the length of time before odors build up (Kohoutová *et al.* 2012; Shelley *et al.* 1953). A simple sniff test of each man's armpits showed that the unshaved, hairy axilla recovered its typical odor after 6 hours, whereas the shaved armpit only regained a noticeable odor after 24 hours (Shelley *et al.* 1953).

Antiperspirants control odor by reducing the flow from eccrine glands so that there is less moisture and nutrients for bacterial growth. Further ways of controlling odor include deodorants with various modes of action. Some change the pH to make it less favorable for bacterial growth; others kill the bacteria, inhibit their enzymes, mask the odor or, in a new tactic, give them alternative precursor molecules that they will break down into molecules we find sweet smelling (Gautschi *et al.* 2007; Schilling *et al.* 2010).

Chemical analysis combined with organoleptic (human sniffing sensory-based) studies and work on the skin bacteria has identified the molecules that give the characteristic smell to the armpits of about 80% of the world's population (see below) (Figure 13.6)

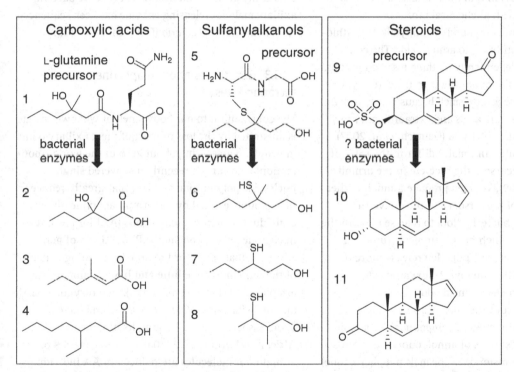

Figure 13.6 Three odorous chemical classes identified in human sweat: carboxylic acids, sulfanylalkanols, and steroids. They are secreted into the armpit as odorless precursors.

Carboxylic acids: the precursor *N*-α-3-hydroxy-3-methylhexanoyl-(L)-glutamine (1) is converted by bacterial axillary malodor releasing enzyme (AMRE) in bacteria such as *Corynebacterium striatum* into 3-hydroxy-3-methylhexanoic acid (2). Other carboxylic acids include 3-methyl-2-hexenoic acid (3), and 4-ethyloctanoic acid = "goat acid" (4). Sulfanylalkanols: the precursor (*R*) or (*S*)-[1-(2-hydroxyethyl)-1-methylbutyl]-(L)-cysteinylglycine (5) is converted in a number of bacterial enzyme steps into its respective enantiomer of 3-methyl-3-sulfanylhexan-1-ol (6) (see Appendix). Other sulfanylakanols include 3-sulfanyl-1-hexanol (7), and 2-methyl-3-sulfanyl-1-butanol (8).

Steroids: it is postulated that precursors such as dehydroepiandrosterone sulphate (9) are converted by bacterial enzymes to steroids such as 5α-androst-16-en-3-ol (10) (androstenol), 5α-androst-16-en-3-one (11) (androstenone). Caption and figures adapted from data in Gower *et al.* (1997), Schilling *et al.* (2010), and Troccaz *et al.* (2009).

(reviews: Gautschi *et al.* 2007; Schilling *et al.* 2010). These include various carboxylic acids, among them the "goat-acid," 4-ethyloctanoic acid (which gives what we perceive as the strong smell of male goats in the breeding season) (Zeng *et al.* 1991). (For some background on chemistry see the Appendix). Much of the smell comes from one of the most abundant molecules, 3-hydroxy-3-methyl-hexanoic acid, which also has a very low odor threshold of 0.0044 ng/l (Gautschi *et al.* 2007). There are also many other molecules, related to the hexenoic and hexanoic acids, which vary greatly between individuals in quantities and so might contribute to our individual chemical profiles.

A second class of compounds, sulfanylalkanols (thio-alcohols), also contributes to armpit odor. The compounds, found independently by three research groups, are present in smaller quantities than the carboxylic acids but add to the armpit smell because of their high volatility relative to the acids and extraordinarily low odor threshold values of 1 pg/l (Gautschi *et al.* 2007).

Caucasian females and males differ in the ratio of acid and thiol precursors they secrete in the armpit (Troccaz *et al.* 2009). Women produce a much higher proportion of thiol precursors and thus have the potential for their bacterial flora to release more of the sulfanylalkanols, which have a tropical fruit- and onion-like odor (Figure 13.6). Men release more of the precursor of (*E*)/(*Z*)-3-methyl-2-hexenoic acid, a molecule with a cheesy, rancid odor.

Only a small part of the odor of armpits appears to come from a third class of compounds, the steroids identified in early studies of armpit odors, 5α-androst-16-en-3-one ("androstenone," the male pig pheromone) and 5α-androst-16-en-3α-ol ("androstenol") (Gautschi *et al.* 2007) (see below for further discussion). These are secreted as odorless androgens: largely dehydro-epiandrosterone and its sulfate.

The odorless precursors of the smelly molecules described above are secreted in the apocrine secretions. Enzymatic activity by skin bacteria releases hexenoic acid from its carrier protein, apolipoprotein D, the major protein in axilla secretions (Preti & Leyden 2010). Other carboxylic acids, the sulfanylalkanols

and the steroids are similarly produced by bacteria from odorless precursors (Gautschi *et al.* 2007; Gower *et al.* 1997; Schilling *et al.* 2010). Bacterial breakdown of lipids from the sebaceous gland secretions also contribute short-chain (C_2 to C_5) volatile fatty acids to the odor (James *et al.* 2004).

The smell of human armpits is not a general bacterial decomposition. Instead, the unique odor is the result of the particular molecules we synthesize and actively secrete with their carrier molecules, transformed by particular enzymes of bacteria that characteristically grow in our armpits. A specific bacterial axillary malodor releasing enzyme has been sequenced from some *Corynebacteria* (Gautschi *et al.* 2007).

13.3.3 Genetic variation in apocrine secretion rates

The contribution to our body smell of what we secrete is illustrated by the loss of most of these axillary odors in people, representing about 20% of the world's population, who carry a recently discovered single-nucleotide polymorphism (SNP) that greatly reduces the amount secreted by the armpit apocrine glands. Individuals who are homozygous (AA) for the single-nucleotide polymorphism (SNP) 538G→A of gene *ABCC11*, that codes for the non-functioning version of the protein, secrete little and have a faint, acidic armpit odor (Martin *et al.* 2010a). Heterozygotes (GA) and the wild-type (GG) secrete more and have the odors associated with the armpits of Caucasians and Africans (Martin *et al.* 2010a). All three classes of armpit odor molecules are reduced in AA individuals. The reduction in smell is not due to differences in skin bacteria (Martin *et al.* 2010a).

The SNP was first noticed as the cause of white, dry earwax in AA individuals (see below). Almost 95% of East Asians, in China and Korea for example, are homozygous (AA) for the SNP 538G→A (Yoshiura *et al.* 2006). The SNP is rare (0 to 3%) in Caucasians and Africans. Intermediate frequencies (30 to 50%) are found in populations in Southeast Asia, the Pacific islands, and people of Native North

American ancestry. The distribution pattern in populations and the small amount of recombination around the gene suggests the SNP originated relatively recently in Northeast Asia and then spread rapidly, especially in the Asia-Pacific region. Its spread in North America suggests the SNP pre-dates the migration around 12,000 years ago from Northeast Asia to North America (Dillehay 2003).

Martin *et al.* (2010a) wonder if the rapid selection for a loss-of-function SNP to over 95% in large populations suggests a strong positive selection in mate choice for low-odorant partners with the SNP. People with AA have been known to comment on how strong GA and GG individuals smell. However, as Martin *et al.* (2010a) note, the gene is expressed in many tissues apart from the armpits and ears and it is not known if the selective advantage of the SNP is the dry, white earwax phenotype (the first discovered phenotype of the SNP; Yoshiura *et al.* 2006), loss of body odor, or an as yet undiscovered phenotype due to loss of *ABCC11* function in another tissue. It seems, in any event, that being less smelly is not a disadvantage in finding a partner. Perhaps the molecules we associate with Caucasian and African armpits are not important for sex.

13.4 Our sense of smell: perception of odors

About 1% of our genome is devoted to olfactory receptor genes, which form the largest gene superfamily in the human genome (see Chapter 9 for more detailed discussion of receptors and olfaction). Humans have about 855 olfactory receptor (OR) genes, of which ~370 are functional, about 60 have alleles coding for working and for disabled versions of an OR, and the rest are pseudogenes (reviews Hasin-Brumshtein *et al.* 2009; Zhang & Firestein 2009). Our brain combines the inputs from the functional ORs into a combinatorial code to recognize a potentially infinite number of odorants (Chapter 9). The human olfactory system has retained the ability to recognize a broad, if perhaps less discriminating, spectrum of chemicals while using about a third of the mouse's number of functioning ORs (Shepherd 2004, 2006; Zhang & Firestein 2002).

It has been suggested that much of the loss of functional ORs came after the divergence of Old World monkeys and hominoids from the New World monkeys, with a relaxation on selection on olfactory genes due to the acquisition of full trichromatic color vision (Gilad *et al.* 2004, 2007). However, with an analysis of the fuller primate sequences now available, Matsui *et al.* (2010) suggest the color vision hypothesis is not supported as there is no sudden loss of OR genes after the divergence and instead there was a similar gradual loss in both lineages (Figure 13.7).

Humans have a vestigial vomeronasal organ (VNO) but it is not a functional sense organ: there are no sensory cells, no functioning olfactory receptors, no connection to the brain, and no accessory olfactory bulb (to which a VNO would connect) (Chapter 9) (reviews Bhatnagar & Smith 2010; Meredith 2001; Wysocki & Preti 2009). Although vomeronasal nerves do appear in early embryos, and seem to have a role guiding gonadotropin-releasing hormone (GnRH) neurons to the basal forebrain, by 17 to 19 weeks the VNO could not be distinguished in the human fetus, leaving only a vestigial structure on the nasal septum (Kjaer & Hansen 1996). The molecular evidence suggests that more than 95% of human vomeronasal receptor 1 (V1R) genes, and all V2R genes, are non-functional pseudogenes (Chapter 9) (Mombaerts 2004). In addition, the crucial gene *TRPC2*, coding for the ion channel protein for transduction of V1R signals, is a non-functioning pseudogene in humans and Old World primates (Liman & Innan 2003). Intriguingly one human V1R is expressed in the main olfactory epithelium (Rodriguez *et al.* 2000) but how it would work without TRPC2 is not known.

A VNO is not a critical requirement for pheromone communication, though this was argued for some years (Baxi *et al.* 2006). While many sex pheromones in mice, other rodents, and elephants, for example, are mediated by the VNO, other pheromones in mice are sensed through the main olfactory system (MOE) (Chapter 9). In many other mammals, including sheep, sexual pheromones act via the MOE.

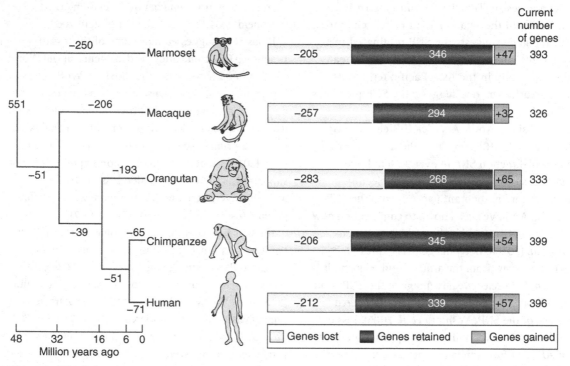

Figure 13.7 Olfactory receptor gene evolution in primates. Changes in the olfactory receptor gene repertoire in five primate species, based on cross-genomic comparisons (Matsui *et al.* 2010). The common ancestor of primates had at least 551 olfactory receptor genes (left). Many olfactory receptor genes have been lost along the five branches to modern primates (numbers of gene losses are shown on each branch). On the right, the number of functional olfactory receptor genes in modern primate species is shown. Humans have lost 212 of the original 551 olfactory receptor genes (white bars), retained 339 olfactory receptor genes (dark gray bars), and gained 57 olfactory receptor genes through gene duplication and divergence (light gray bars), giving a current total of 396 genes. The rate of change in human olfactory receptor genes is similar to the rate in other primate lineages. Original figure and data Matsui *et al.* (2010). This figure and caption adapted from Bendesky and Bargmann (2011).

13.4.1 Variation in human odor perception

Each one of us has an almost unique olfactory view of the world: each of us smells a different odor landscape as our unique combinations of olfactory receptors give each of us a different patchwork of small "odor blind-nesses" for some molecules and heightened odor sensitivities for others (Hasin-Brumshtein *et al.* 2009; Newcomb *et al.* 2012). Comparing 51 olfactory receptor (OR) genes, which had both functioning and inactive alleles, Menashe *et al.* (2003) found almost every one of 189 individuals tested had effectively a different olfactory genome. Olfaction, with potential differences in more than ~400 ORs, gives us hundreds of ways we each differ in what we can smell. This compares with the small number of varieties of color blindness from variation in our eye's four different opsin receptor proteins. However, the combinatorial nature of smell and broad tuning of most receptors means that the effects of most of the OR differences between us will be subtle.

The variations come first from SNPs, which can either turn the gene into a non-functioning pseudogene or change its sensitivity or interaction with odorants. Further variation comes from gene copy number variation (CNV) (duplication and deletions of OR genes). With a larger set of human genomes and looking at all OR genes, Olender *et al.* (2012) found almost 6,000 genetic variations, including SNPs, small insertion/deletions, and/or copy number variants, in 413 OR genes. Each of us has different combinations of these alleles (more than two-thirds of functional human OR loci can have either intact or inactivated alleles). A minority, 57, of human OR genes show evidence of selection against non-functional variants of those ORs (Olender *et al.* 2012).

Our responses to odor molecules can reveal some of the variation in our ORs. Observations from the nineteenth century onwards showed that some people, despite an otherwise good sense of smell, could not smell certain odor molecules or kinds of molecules at concentrations readily perceived by other individuals (Amoore 1977). Such individuals were said to have a specific anosmia or hyposmia for that molecule (or type of molecule). Specific anosmia concerns perception of a particular molecule, in contrast to anosmia, which is a complete lack of the sense of smell (Doty 2009). Complete anosmia is quite rare – about 2% in Western adults under 65 years old, commonly as a result of infection or head injury – and even rarer as a genetic disorder.

Though the emphasis in the literature has often been on specific anosmias, for most odorants they are just one end of a bell-shaped normal distribution for the lowest smell thresholds of individuals, covering several orders of magnitude of concentration (Figure 13.8) (Hasin-Brumshtein *et al.* 2009). For some odorants the distribution might be broader, bimodal, or even trimodal. Such distributions could result from multiple ORs with different affinities, but might also stem in part from environmental factors such as previous exposure to a molecule (Section 13.5.2.1) (Hasin-Brumshtein *et al.* 2009).

For many individuals, "specific anosmia" might be better described as simply hyposmia (Figure 13.8). At the other end of the distribution, hyperosmia, some individuals can detect the odor much better than the average. Specific anosmia probably implies a rare mutation of the highest affinity OR for a molecule. Specific hyperosmia is observed where the highest affinity OR is only rarely intact (so that people with the intact OR stand out from the rest of the population)

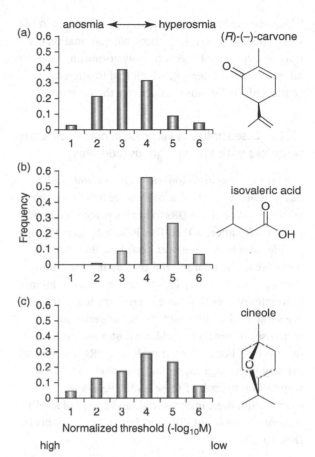

Figure 13.8 Testing the thresholds of the same individuals for a number of different molecules. The thresholds vary because the people have different combinations of olfactory receptor (OR) variants. Shown are threshold histograms for the odorants: (a) (*R*)-(–)-carvone, which has a spearmint smell; (b) isovaleric acid, which has a sweaty smell (Section 13.4.2); and (c) cineole, which has a eucalyptus smell (Menashe *et al.* 2007). The distribution of threshold concentrations in a population is typically a bell-shaped curve spanning several orders of magnitude of concentrations, with the extremes of this curve denoting anosmia and hyperosmia. For some odorants the distribution might be broader, bimodal, or even trimodal, as in the case of androstenone (Bremner *et al.* 2003). Such modified distributions might indicate the involvement of multiple ORs with different affinities, but can also stem in part from environmental factors (Hawkes & Doty 2009; Menashe *et al.* 2007).

Caption and figure adapted from Hasin-Brumshtein *et al.* (2009) with data from Menashe *et al.* (2007).

(Hasin-Brumshtein *et al.* 2009). The differences in the smells people perceive, in threshold and quality, can now be explained by genetic polymorphisms in an odorant receptor gene(s), which lead to altered perception of the ligands that activate the receptor.

13.4.2 Case studies of variation in human olfactory receptors with known ligands (odorants)

We know the ligands (odorants) for a number of human ORs, and for a few of these we have been able to relate variation in the ORs to human perception of odor (Olender & Lancet 2012). This is done by genotyping people after they have been tested for their smell response to the odorant and then by in vitro testing responses of the likely ORs to this and other odorants. Examples of ORs that have been identified in this way are some of the different ORs that underlie our responses to isovaleric acid, androstenone, and (Z)-3-hexen-1-ol. For each odor molecule, ORs with special affinity have been found but more than one OR is responding to each of these molecules (as we would expect from the combinatorial mechanism of smell). Androstenone is not different from other odorants in this regard.

13.4.2.1 Isovaleric acid

About 6% of people are very sensitive to isovaleric acid (3-methylbutanoic acid), characteristic of human foot odor (and cheese). The hypersensitivity is explained by variation in the OR gene *OR11H7P*: hypersensitive individuals usually had at least one copy of the intact allele of this OR (Menashe *et al.* 2007). When the intact allele of *OR11H7P* was expressed in vitro it responded to isovaleric acid but the pseudogenized allele (with a single nucleotide polymorphism SNP), which most people have, did not. Menashe *et al.* (2007) note that other ORs also respond in a lesser way to isovaleric acid (as would be expected) and that general sensitivity to smells is also under genetic control.

13.4.2.2 Androstenone

People differ widely in how they describe the smell of the steroid androstenone (5α-androst-16-en-3-one), ranging from offensive ("sweaty, urinous"), to pleasant ("sweet, floral"), to odorless (by those with "specific anosmia") (Wysocki & Beauchamp 1984). Keller *et al.* (2007) screened 335 human ORs in vitro and found that the olfactory receptor OR7D4 gave the strongest response to androstenone and the related odorous steroid androstadienone (androsta-4,16-dien-3-one) but did not respond to a panel of 64 other odors. A common variant of this receptor, WM, contains two linked SNPs, resulting in two amino acid substitutions (R88→W, T133→M, hence WM) that greatly reduced its response to androstenone and androstadienone in vitro. The RT version of the polymorphism has the sensitive response in vitro. Other rarer SNPs were also found, including P79→L, with a complete lack of in vitro response to androstenone and androstadienone, and S84→N, which was much more sensitive to both odors than the RT genotype.

Keller *et al.* (2007) then genotyped 391 volunteers for OR74D after psychophysical olfactory testing of their smell thresholds and how they verbally rated the smells. Individuals with RT/WM or WM/WM genotypes as a group were less sensitive to androstenone and androstadienone and found both odors less unpleasant than people with the RT/RT genotype. These results were the same for men and women. The RT/RT genotype would account for many individuals describing androstenone as offensive ("sweaty, urinous"), the RT/WM or WM/WM genotypes as pleasant ("sweet, floral"). Individuals with WM or other SNPs such as P79L would be, depending on steroid concentration, more likely to describe the molecule as odorless. The low frequencies of WM and P79→L match a more detailed examination of "specific anosmia" to androstenone, which found only ~2 to 6% of healthy young people were actually complete non-sensers of androstenone (rather than the ~30% often described as "specifically anosmic" to this molecule) (Bremner *et al.* 2003).

The olfactory receptor OR7D4 was the most sensitive OR tested but it was *not* the only receptor that responded to androstenone and androstadienone in vitro (Keller *et al.* 2007). The authors remind us that the response to these steroids, as for other odor molecules, is combinatorial across many receptors (Chapter 9) and this explains why the genotypic variation in OR7D4 only accounted for 19 to 39 % of the variation in valence (pleasantness or unpleasantness) and intensity perception of the steroids. The likely involvement of a number of receptors is supported by twin studies on olfactory perception of androstenone and other odorants (Knaapila *et al.* 2012).

13.4.2.3 (*Z*)-3-hexen-1-ol

Most people smell (*Z*)-3-hexen-1-ol, also known as *cis*-3-hexen-1-ol, as "green grassy" or the smell of "cut grass." There is a normally distributed variation in our ability to smell it, spanning two orders of magnitude (McRae *et al.* 2012). It is a key flavor in many fresh fruits such as raspberries as well as drinks such as white wines. McRae *et al.* (2012) identified three olfactory receptors that could respond to *cis*-3-hexen-1-ol. Single-nucleotide polymorphisms in one of these, OR2J3, could explain about 25% of the variation of human responses to (*Z*)-3-hexen-1-ol.

13.5 Human pheromones?

So far, no human pheromones have been chemically identified and shown to work in a rigorous bioassay. Sadly, there is no good evidence for a human pheromone to make the wearer irresistible to potential partners, despite many patented compounds and the advertising that comes up with any Internet search on "pheromones." However, there is one candidate phenomenon, which is potentially pheromone mediated, with a good bioassay: human babies' response to areolar mammary gland secretions. I deal with this first.

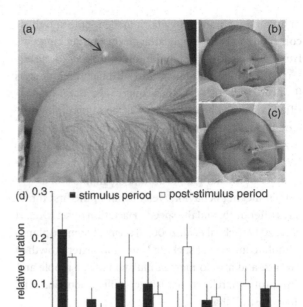

Figure 13.9 Human babies respond to secretions of the areolar glands from any lactating mother. This may be a candidate secretion for pheromonal activity. (a) Areola of a lactating mother (day 3 after giving birth) with the areolar glands giving off their secretion (arrow). The areolar glands combine sebaceous and milk producing glands. (b) and (c) Sleeping newborns' responses to the secretion of the areolar gland; (b) lip pursing; (c) putting out the tongue.
(d) Mean (± SE) relative durations of newborns' responses during and after presentation of various olfactory stimuli. During and after periods were each ten seconds long. (Abbreviations: f: familiar; nf: non-familiar; AG: secretions of areolar glands; S: sebum; HM: human milk; cow M: cow milk; FM: formula milk; van: vanillin; fM: habitual milk (mother's milk in breast-fed and formula milk in bottle-fed infants); n = 19). Caption and figure adapted from Doucet *et al.* (2009).

13.5.1 Mammary pheromone?

Like rabbit pups, human babies remember the individual signature mixture of their mother and can distinguish her from other lactating mothers (Section 13.1). In addition, like rabbit pups (Chapter 1), babies also respond to their mother's secretions (Figure 13.9) (Doucet *et al.* 2009). These secretions, from the skin glands in the areola around the nipple, may have a role in guiding the human newborn to the nipple and thus contribute to early nipple attachment and sucking. Human babies also respond in the same way to these secretions from unfamiliar, unrelated lactating mothers (Doucet *et al.* 2009). As these molecules, as yet unidentified, appear to be common to all lactating mothers, I would classify them as pheromones, though the researchers are still cautious at this stage (Chapter 1) (Wyatt 2010). There is a positive correlation between the number of areolar glands around the nipple and a baby's sucking behavior, weight gain in the first three days after birth, and the speed of lactation onset (Doucet *et al.* 2012; Schaal *et al.* 2006). It seems that more glands stimulate more avid sucking leading to faster growth of infants, and also to more stimulation of the nipple and thus prolactin release that triggers milk production after the colostral phase.

13.5.2 Human sex pheromones

If we were looking at humans as we would any other kind of animal, we would argue that the development of scent glands in particular places such as the groin and armpits at puberty would point to these glands being sexually selected (Chapter 3) (Section 13.3). In many mammals, females can use odor to choose dominant males, whose high androgen levels have produced the largest and most active odor glands (Chapter 3). Theoretically, human females might be attracted to dominant males with the smelliest armpits and other scent-producing areas (Section 13.3). However, recent work on the genetics of human variation in body odor suggests that, if anything, smelly armpits are not an essential part of reproductive success in humans, at least in the last 15,000 to 20,000 years (though it should be said that other molecules could be involved) (Section 13.3).

13.5.2.1 The trouble with steroidal "putative human pheromones": no evidence yet

Work on human sex pheromones is not in a healthy state: numerous studies are done on molecules with

no rigorous biological basis for choosing these molecules rather than others (Doty 2010; Wysocki & Preti 2009).

We do not have biological phenomena, potentially mediated by sex pheromones, which are sufficiently well defined to allow for an unambiguous bioassay, the first step in identifying biologically active molecules (Chapter 2). Without such bioassays, it is not surprising that we do not have any molecules with pheromonal activity in humans resulting from peer-reviewed "gold standard" bioassay-led work including molecular identification, synthesis, and confirmatory bioassay (Chapter 2).

Instead, there have been two waves of research looking at steroidal molecules as "putative human pheromones" without a solid experimental justification for choosing them: first, in the 1970s and 1980s, looking at two pig pheromone molecules androstenone and androstenol; then, from the late 1990s, two different steroids, another pig steroid Δ4,16-androstadien-3-one (androstadienone) and estra-1, 3, 5(10),16-tetraen-3-ol (estratetraenol), took their place (tabulated in Havlíček *et al.* 2010). In addition, in a separate strand of research, fatty acid "copulins" were proposed and later dropped in the 1970s (Section 13.5.4).

The first wave of pre-1990s studies of "human pheromones" looked at 5α-androst-16-en-3-one (androstenone) or 5α-androst-16-en-3α-ol (androstenol).[2] The molecules were proposed as "putative human pheromones" because of the coincidence of being a known pig sex pheromone and being found in human armpits (but see Sections 13.3 and 13.5.2.2). Perhaps as important was their commercial availability in aerosol cans as Boarmate™ (Chapter 12). The experiments included, for example, the rating of photographs while subjects were exposed to androstenol and choice of waiting room seats sprayed with androstenone. However, there were severe weaknesses in the experimental designs of these early

---

[2] In the literature, 5α-androst-16-en-3α-ol is abbreviated variously as 5α-androstenol or 3α-androstenol.

studies (see Doty 2010, pp. 141–53). A more natural experiment on the attractiveness of androstenol showed no effects (Black & Biron 1982).

The second wave came in the 1990s, after one set of researchers declared that some steroid molecules isolated from human skin were human pheromones (MontiBloch *et al.* 1998), based upon the false assumption that humans have a functional vomeronasal organ (Section 13.4) (Wysocki & Preti 2009). The molecules this time included androstadienone and estratetraenol, though the bioassays and procedures that led to these molecules being identified were never published in the scientific literature.

The switch to androstadienone and estratetraenol as stimuli seems to have started with, among other papers, Jacob and McClintock (2000), who looked at psychological and mood effects of the two molecules. The only reason for using these particular molecules was the claimed activity (above) on a supposed human VNO.

A large literature (reviewed by Doty 2010; Havlíček *et al.* 2010) reports experiments on largely these same compounds, androstadienone and/or estratetraenol, often at the concentrations used by Jacob and McClintock (2000), which have become almost traditional, and rarely at physiological concentrations (see below). The types of experiment vary and include psychological and physiological measures as well as high-tech fMRI and PET scans. Some of the better experiments are well carried out, by good scientists, using double-blind designs for example, and they show how humans respond to these molecules at these (non-physiological) concentrations. However, the whole literature is based on testing molecules that have no demonstrated biological validity as possible pheromones.

As there are no good bioassay grounds for researching androstenone, androstenol, androstadienone, or estratetraenol as pheromones, circumstantial evidence is used to argue that these molecules are special and thus worth testing. For example, it is argued variously that more is produced by one sex than the other and that perception of these molecules differs between the sexes and is (uniquely) different from other molecules. These claims are not well supported (Doty 2010; Wysocki & Preti 2009). I assess the claims below.

13.5.2.2 Sex differences in production?

In one of the most thorough investigations of individual and sex differences in axillary molecules, in almost 200 people over a ten-week period, none of the steroids featured or appeared as different between the genders (Penn *et al.* 2007). However, it should be noted that the extraction and analysis were not optimized for steroids (Penn, pers comm.).

In some studies, more androstenone was found in males' armpits than females' (e.g., 5 to 1019 pmol/24h in males and 1 to 17 pmol/24h in females, but the values overlapped and one normal female had 551 pmol/24h (Gower *et al.* 1985)). Androstadienone is not male specific and nor do males necessarily have more than females: in the one study that makes the comparison, the one female subject had more androstadienone than six of the seven males tested (Gower *et al.* 1994).

Wide variations between individual males for androstenone, androstenol, and androstadienone were found by Nixon *et al.* (1988): only 14 of 24 men had androstenone in axillary hair (quantities 0 to 6113 pmol/24h), and only 13 had androstadienone (quantities 0 to 433 pmol/24h). There was no relationship between the age of the donor and the presence of the steroids. Variation between males could be consistent with a signaling role but the overlap with female values is unexpected.

Estratetraenol is reportedly only known from the urine of pregnant women (Thysen *et al.* 1968) so it is strange that it has been proposed as a putative pheromone produced by women in general.

13.5.2.3 Perception

Much has been made of supposed differences between perception of "putative human pheromones" and other molecules. It has been suggested that "putative pheromones" were unlike other molecules because of their specific anosmias, induced increases in sensitivity, and differences in thresholds between the sexes. All these observations turn out to be common for all sorts of other molecules: none of them singles out "putative human pheromones." There is nothing special about these steroidal molecules.

Specific anosmias are not unique to steroids

While the first studies were on androstenone, "specific anosmias" are found for all sorts of molecules (Section 13.4). More are likely to be revealed as more human ORs are linked to their ligands.

Induced sensitivity is not unique to the steroids

An increase in sensitivity, after repeated brief exposures, in individuals initially unable to smell an odor was first shown with androstenone (perhaps in part because of the interest in this molecule) (Wysocki et al. 1989). However, it is now clear that the phenomenon occurs with other molecules such as benzaldehyde and citralva (a terpenoid: 3,7-dimethyl-2,6-octadienenitrile), quite unrelated to "putative human pheromones," and in individuals with average initial sensitivity (Dalton et al. 2002). Men and women started with the same sensitivities but only women of reproductive age showed the increased sensitivity, of up to 3 to 6 log units, to the test molecules with repeated exposure. There was no evidence of threshold fluctuations through women's monthly hormonal cycles. The mechanisms of induced sensitivity are yet to be resolved.

Differences in olfactory thresholds between the sexes do not support a pheromone role

Women, on average, are more sensitive to some odorants, although the sex differences are not large (Doty & Cameron 2009). While some studies suggest women are more sensitive than men to androstenone, many researchers have not shown such a difference between the sexes (see Bremner et al. 2003). Women have a lower threshold than men for a wide range of non-steroid molecules of all kinds including acetone, 1-butanol, 2-methyl,3-mercapto-butanol, citral, ethanol, and 1-hexanol (Doty & Cameron 2009), so if a greater female sensitivity for androstenone or other steroids was found it would not necessarily be biologically significant.

There is one intriguing exception to the pattern of greater female sensitivity: males are more sensitive than females to bourgeonal, a ligand for the human olfactory protein OR1D2 expressed on sperm as well as in the nose and important for sperm chemotaxis (Olsson & Laska 2010). How the greater male sensitivity has evolved is not known but it could be a pleiotropic effect because of upregulation of OR1D2 in sperm.

While there may be some variation in women's smell sensitivity over the menstrual cycle, it is also found in women on the pill (Doty & Cameron 2009). Sensitivities to sound and other stimuli also vary similarly. Doty and Cameron (2009) conclude that simple associations between circulating levels of gonadal hormones and measures of olfactory function are rarely present.

Non-physiological quantities of steroids are used in experiments

Pheromone concentration does matter. Olfactory receptor responses are usually concentration dependent and different receptors may be stimulated at different concentrations (Chapter 9). For example rabbit mammary pheromone elicits behavioral responses from pups only within a limited range of concentrations (from 2.5×10^{-9} to 2.5×10^{-5} g/ml) (Coureaud et al. 2004).

The quantities of androstadienone used in numerous studies far exceed levels found on humans (Table 3.2 in Havlíček et al. 2010). As Wysocki and Preti (2009) point out, the now traditional concentrations (250 μM) of androstadienone used by Jacob and McClintock (2000) are more than 1,000 times above the reported axillary concentrations (nanomolar, Gower *et al.* 1994). In fMRI and PET studies Savic and others (e.g., Savic *et al.* 2001) have used the headspace above 200 milligrams of solid crystalline androstadienone, up to a million times more than the nanogram amounts in human armpits. The results tell us little about the biological roles of these molecules.

13.5.3 Studies using armpit or other secretions

These are better as they use natural concentrations of the naturally occurring (but unknown) molecules (for review see Lundström & Olsson 2010). However, I think there are some unanswered questions about whether the sweat from people in different circumstances really differs in specific ways. For example, there are experiments looking at the effect of smelling sweat from people after exposure to sexually arousing films (Zhou & Chen 2008). Other experiments have used sweat from people exposed to scary films, parachute jumps, or exam stress (e.g., Prehn-Kristensen *et al.* 2009). I wonder if the sweat produced in the sexual and scary situations might actually be very similar – simply the result of emotional release from the apocrine glands (Wilke *et al.* 2007)? I am not aware of experiments that compare these two "kinds" of sweat directly.

Female tears are reported in one study to affect men's blood testosterone levels but the biological significance or relevance is not clear (Gelstein *et al.* 2011).

13.5.4 Female pheromone advertisement of ovulation?

In some mammals such as rats, dogs, and sheep, hormone cycles lead females to be sexually receptive around ovulation, during limited discrete times known as estrus, which is signaled by changes in behavior, smells, and visual signs. In some primate species, olfactory signals are important, in others there are visual signals such as brightly colored sexual skin produced during the most fertile period. However, for

Old World monkeys, apes, and humans, estrus is not an appropriate term as females are sexually receptive throughout the cycle (for a detailed discussion see Dixson 2009; 2012, p. 477). These species have been termed "concealed ovulators" (Sillén-Tullberg & Møller 1993). Dixson (2012, p. 477 ff.) argues many primate species described as "concealed ovulators," do have rhythmic changes in sexual activity through the ovarian cycle that are signaled by olfactory cues, subtle visual cues, or result from increases in female initiation of mating at those times. Thus to males of these species, for example vervets, marmosets, and tamarins, ovulation is not concealed (even if it is from the human observer).

Is ovulation in humans really concealed from the male (and the woman herself?). There are no obvious visual signs but work in the 1970s suggested that cyclic changes in vaginal odor might be a signal in rhesus monkeys and, since we are also primates, perhaps in humans too. In many monkey species, a male will sniff or lick the female's genitalia before mating (Lenochova & Havlíček 2008). In the early 1970s sex pheromones in rhesus monkeys were reported. The pheromones, named "copulins," were said to be estrogen-dependent short-chain aliphatic acids, found in the vagina of the female around ovulation, which attracted males and stimulated male sexual behavior (Michael & Keverne 1970). The same fatty acids are found in the vaginal secretions of some women. However, other researchers (Goldfoot 1981; Goldfoot *et al.* 1976) questioned these results: olfactory cues were not required for rhesus monkey male response, the highest concentrations of the aliphatic acids in rhesus monkey females occurred after the fertile period, and adding the natural or synthetic putative "copulin" mixtures to ovariectomized females did not increase mating by males (see reviews by Dixson 2012, p. 481 ff.; Doty 2010, pp. 57–9; Stoddart 1990). Olfactory cues are not ruled out but, on current evidence, a key releaser role for vaginal odors is not supported. The earliest experiments implicating fatty acids (by Michael and colleagues) used conditioning responses with a small number of rhesus monkey males, and may have been inadvertently testing other learned responses to olfactory cues (see Michael & Keverne 1970). The story is a complicated one – not surprising since the sexual behavior of primates is highly complex. Nonetheless, to date there is no evidence that "copulins" exist as human pheromones. The only double-blind placebo-controlled study with 62 human couples over three months showed no increase in sexual activity when the synthetic aliphatic acid mixture "copulins" rather than controls (water, alcohol, or a perfume) had been applied by the woman to her chest at bed-time (Morris & Udry 1978). Disturbingly, the early Michael papers on "copulins" are still sometimes quoted as evidence of human pheromones without mention of the associated debate.

Are there any odor cues to human ovulation? In a test of human responses to vaginal odors, Doty et al. (1975) found that although the mean results showed a pattern, there was so much individual variation between cycles that odors were unlikely to be a reliable way to identify the most fertile phase. However, as they admit, their "sniff" test of odors by a panel of volunteers in a laboratory setting might be rather out of context. More objective physiological responses of males might be a better way to investigate the question.

Odor changes in women's axillary secretions and mouth odors across the menstrual cycle have been reported (Preti & Wysocki 1999). Singh and Bronstad (2001) suggest that men may be able to distinguish odors from women in their ovulatory (late follicular) and non-ovulatory (luteal phases). Havlícek et al. (2006) found similar results.

13.5.5 Symmetry and odor

It has been proposed that animals might use body symmetry as an indicator of mate quality, but the idea remains controversial (see Chapter 3). In humans, it has been suggested that the T-shirt smell of more

symmetrical men is preferred by women (references in Gangestad & Thornhill 1998; Van Dongen 2011). However, a meta-analysis suggests substantial publication bias, leading Van Dongen (2011) to conclude that the current literature likely overstates the evolutionary importance of symmetry in determining human attractiveness.

13.5.6 Primer effects

Pheromone primer effects have long-term effects, for example on reproduction in mice and sheep (Chapter 9). One human example might be the report that male axillary secretions can alter the timing of the next luteinizing hormone peak of women (Preti et al. 2003). If odors from other humans did have effects on neuroendocrine processes, there would be the distant possibility of applications for control of fertility and contraception. However, evidence for pheromone primer effects in humans, including menstrual synchrony, remains controversial.

13.5.6.1 Menstrual synchrony?

A pioneering paper suggested that women who live together develop synchronized menstrual cycles (McClintock 1971). However, while some later studies have found synchrony, many others have failed to do so (for references see Doty 2010, p. 168 ff.; Yang & Schank 2006). Methodological critiques have argued that studies reporting synchrony have systematic errors of various kinds, including errors in establishing the dates of cycle starts (see Doty 2010, p. 168 ff.; Yang & Schank 2006). Differences in cycle length between women make the convergence of cycles needed for synchrony unlikely. Paradoxically, to a casual observer synchrony may appear to occur because cycle length differences between women lead to repeated cycle convergences and subsequent divergences by chance. Currently it seems that, on balance, menstrual synchrony in humans has not been established and there is similar doubt about its occurrence in other mammals (Chapter 9).

Given the doubts about the phenomenon of menstrual synchrony in humans, it may be too early to search for a pheromone. Stern and McClintock (1998) reported that armpit secretions produced by women at different stages in their menstrual cycle either speeded up or slowed down the cycles of other women. Schank (2006) reviews this and other studies of possible pheromone effects on menstrual cycles and concludes that none are demonstrated.

13.6 Where next with human pheromones?

Perhaps we need to start again. None of the "putative human pheromones" comes from bioassay-led work that follows the gold standards for identifying pheromones (Chapter 2). The much studied "putative human pheromones," androstadienone and estratetraenol, have little experimental validity because the basis for choosing these rather than some of the other 700 or more molecules in our armpits is founded on "VNO research" (but we do not have a VNO). In addition, the quantities/concentrations tested, at up to a million times the natural ones, tell us nothing about the possible normal functions of the molecules tested so far.

I think the best way forward will be to go back to first principles – looking for phenomena that appear to be affected by species-wide odors to establish the bioassays and then looking for the molecules involved, but from scratch. It is what we would do with any other animal.

We do not even know which parts of the body produce odors important for human behavior. While some of the odors produced by humans are known, in particular from the armpits, the exploration of the odors we give off is far from complete. The human odors we seem to notice are characteristic ones of armpits and feet. However, the smells we find most noticeable might not be pheromones or otherwise important in communication. The choice of armpits as a focus may be partly because they are less embarrassing to study than other parts of the body. If we really want to understand human olfaction, we must lose all taboos about investigating some "embarrassing" sources of odorants. For example, genital and perianal smells are produced by parts of our bodies with many secretory glands and the hair to support bacteria (Figure 13.3). These odor sources might be important at short range but these are rarely described and discussed.

The isolation of real human pheromones will need to be guided by bioassays (Chapter 2). The challenge will be formidable, as it is for studies of other mammal pheromones, because out of the hundreds of potential molecules (Figure 13.2), it could be some of the smaller peaks that are the signal, perhaps active only in combination (Chapter 1). Some of the androgens could be involved despite their low concentrations, but there is no reason to suppose the pheromones will necessarily include androgens rather than other molecules.

Choosing which peaks to investigate will be a problem. I wonder if pheromone research could be helped by the development of human metabolomics for medical diagnosis (Chapter 2) (for an introduction, see Nicholson & Lindon 2008). Metabolomics characterizes and quantifies all the molecules produced by an organism, and for medical applications it compares healthy and ill subjects to find "biomarkers" of disease. I suggest candidate pheromone components might be pinpointed by looking for molecules that appear with the hormonally driven changes at puberty, when the apocrine glands become active and the sebaceous glands secrete more. Do some of the molecules start to differ between the sexes at puberty?

Another challenge for human pheromone research is that of designing rigorous experiments that eliminate other cues and variables. As well as the complexity of mammal odors that being a mammal brings, like other mammals humans learn and associate. This makes us doubly difficult as experimental subjects. The greatest need is for well controlled experiments with physiological concentrations of molecules, run double blind so that neither subject nor experimenter knows which treatment is being tested (Chapter 2). Among the other challenges is cultural conditioning of responses to odors (Box 13.1). We know from other mammals, such as the hamster, that responses depend on a complex interaction of previous experience and the context of the message (Chapter 9); we should expect human responses to be just as complex.

Progress will only come from good bioassays and we seem as far from this as ever. Perhaps most promising in the shorter term is the possible human mammary pheromone, with a candidate glandular secretion and a biologically relevant bioassay of baby response (Section 13.5.1) (Doucet *et al.* 2009).

Olfaction was the last of the senses to have its mechanisms properly understood, celebrated by the Nobel Prize awarded to Richard Axel and Linda Buck in 2004. There is increasing recognition that olfaction is medically important, at the beginning of life and, as the key component of flavor, important for food intake leading to obesity and, toward the end of life, to loss of appetite as olfactory abilities fade (Shepherd 2012). With the new tools from genomics, molecular biology, and chemistry, and developments in behavioral techniques, it is surely time to greatly increase funding for olfactory research, including serious new investigations of possible human pheromones.

Summary: To smell is human

We each produce a heady mixture of odors. There is a paradox in the enormous interest in human odors compared to the relatively limited work on them; if only we knew as much about humans as we do about moths and mice! Reliably demonstrated effects include parent–offspring recognition and, possibly, choice of genetically compatible partners. No human pheromones have been identified to date. We do not have a functioning VNO but having one is not a pre-requisite for us to have pheromones.

Most of our secretions, such as those from the apocrine glands, are odorless when produced and only gain their smells when metabolized by particular kinds of skin bacteria. Underarm and other hair gives them a habitat.

We each have a unique sense of smell as our individual combination of functional olfactory receptor genes gives us a unique patchwork of small "odor blindnesses" for some molecules and heightened odor sensitivities for others.

Even though we might try hard to forget it, we are mammals first and foremost and, as mammals, odors are likely to be important means of communication. There is surely an interesting world of olfactory communication to investigate. No doubt there will be some surprises as we learn more.

Further reading

The way our brain creates flavor from smell and taste is explored in a popular but scholarly book by Shepherd (2012). Hasin-Brumshtein *et al.* (2009) review olfactory receptor variation in humans. A brief review of human variation in smell and taste is given by Newcomb *et al.* (2012). Clinical aspects of smell including diagnosis of neurodegenerative diseases are covered by Hawkes and Doty (2009).

In his two books, Dixson (2009, 2012) describes primate sexuality, including that of humans, from an evolutionary point of view. I find the skeptical analysis by Doty (2010) of work on human "pheromones" is helpful and illuminating, though I disagree with his ultimate conclusion that pheromones in mammals are a myth.

Three popular accounts of our senses are by Ackerman (1990), who discusses all of the senses; by Gilbert (2008), who entertainingly describes our olfactory world, including a skeptical view of autobiographical olfactory memory and an account of cinema's flirtation with Smell-O-Vision and AromaRama; and by Herz (2007), with a more psychological view. Classen *et al.* (1994) discuss the sociology and anthropology of smell and are strong on history and social context.

Patrick Süskind's imaginative novel *Perfume* (1986) is chilling but compelling. It has a sinister main character, Grenouille, who has an extraordinarily sensitive sense of smell and memory for odors, but is odorless himself. *Perfume* has been translated into many languages and made into a film (2006, director Tom Tykwer).

For links to websites offering entry points to good sources of information about the chemical ecology of humans see www.zoo.ox.ac.uk/group/pheromones.

For molecular structures see sites such as www.chemspider.com, which allows you to search by common name and shows synonyms as well as the systematic names. See also the Appendix for a short guide to the terminology (available for free download from the website associated with this book). Many pheromones are illustrated on Pherobase www.pherobase.com (El-Sayed 2013).

An introduction to chemical terms for non-chemists

The chemical names for semiochemicals can seem complicated to a biologist. One of my favorites is the termite sex and trail pheromone (1*E*,5*E*,9*E*,12*R*)-1,5,9-trimethyl-12-(1-methylethenyl)-1,5,9-cyclotetradecatriene (not surprisingly better known by its common name, neocembrene) (Chapter 7) (Bordereau & Pasteels 2011).

Neocembrene

However, getting the molecule right in every detail can be essential for getting a response and the systematic naming protocols developed by chemists allow us to describe a particular molecule unambiguously. The aim of this short appendix is to explain the basics of chemical names. Some of the ideas and examples for this appendix come from chapters by Stevens in Howse *et al.* (1998).

We need precise specification of molecules because chemosensory receptor proteins are activated by the particular characteristics of ligand molecules. These characteristics include a ligand's functional groups, charge, and size as well as its shape in three dimensions, hydrophobicity, and flexibility (Chapter 9) (Kato & Touhara 2009; Leal 2013; Reisert & Restrepo 2009). This potential range of molecular characteristics, in biosynthesis and reception, has been acted on by natural selection so these variations are vitally important for responses to semiochemicals. Each of the possible variations between molecules described below can be important biologically, as in each case the molecule is different and it may stimulate a different range of olfactory receptors (Chapter 9). The same

considerations of molecular characteristics apply to the selectivity of enzymes catalyzing the synthesis of biological signal molecules.

Organic molecules are based on a chain of carbon and attached hydrogen atoms. The carbon backbone forms a zigzag because of the tetrahedral arrangement of the four carbon bonds (these angles are important for other characteristics of the molecular shape, see Section A.2). The hydrogen atoms attached to the carbon backbone lie in two planes, above and below the paper (represented here by bonds as solid wedges (the plane above) and as dashed wedges (the plane below)):

The structure is often simplified to show just the carbon backbone:

When other atoms such as oxygen or nitrogen, or other functional groups, are added to the chain or substituted for hydrogens or carbons, the chemical nature of the molecule changes (see Table A.1 for common functional groups).

The naming of compounds tells the reader the length of the carbon chain and what and where the important functional groups occur. The name also indicates the number and position of carbon double bonds (C=C). For example, the formal name for one component of the

Table A.1 Prefixes and suffixes for common functional groups. Adapted from Howse *et al.* (1998).

Functional group	Formula	Prefix	Suffix
Alcohol	-OH	Hydroxy-	-ol
Aldehyde	-CH=O	Formyl-	-al
Amine	-NH$_2$	Amino-	-amine
Carboxylic acid	-COOH	Carboxy-	-oic acid
Ester	-COOR	*R*-oxycarbonyl-	-*R*-oate
Ketone	>C=O	Oxo-	-one

pheromone of the pink bollworm, *Pectinophora gossypiella*, is (*Z,E*)-7,11-hexadecandienyl acetate (the other component is the *Z,Z* isomer; see Section A.2.2.1). The "dien" tells us that there are two double bonds, the "7,11" that these occur at carbons 7 and 11:

(*Z,E*)-7,11-hexadecandienyl acetate

Isomers

Having the formula for a molecule, e.g., C$_2$H$_6$O, is just the beginning. This molecular formula could describe many molecules. Isomers are groups of compounds that have the same atoms, in the same molecular formula, but have different structures and shapes. They may have different physical and chemical properties too. The word "isomer" comes from the Greek (*isos* = "equal" (same), méros = "part").

There are two main different kinds of isomers: **structural** (constitutional) isomers have the same atoms connected in different ways, whereas **stereoisomers** have atoms connected in the same way but they differ in the arrangement of atoms in space. Isomers exclude any temporary arrangements simply

due to the molecule rotating as a whole or around particular bonds.

A.1 Structural (constitutional) isomers

Structural (constitutional) isomers have the atoms connected in different ways, creating either different functional groups and/or different shaped molecules.

A.1.1 Chain isomers

These arise because carbon chains can branch. For example there are two isomers of butane, C$_4$H$_{10}$. One has the carbons in a "straight chain," the other is branched.

A.1.2 Functional group isomers

Atoms with the same molecular formula can be connected in ways that produce different functional groups, giving quite different chemical properties to the molecule. For example, a molecular formula C$_2$H$_6$O could be ethanol (an alcohol) or methoxymethane (an ether):

Ethanol   Methoxymethane ("dimethyl ether")

A.1.3 Positional isomers

Positional isomers share the same molecular formula but differ, for example, in the position of a functional group. Heptan-2-ol and heptan-3-ol, with the –OH group on the second or third carbon respectively:

Heptan-2-ol

Heptan-3-ol

or the position of a double bond, for example (Z)-9-tetradecen-1-ol acetate and (Z)-11-tetradecen-1-ol acetate, the pheromone components of the summer fruit tortrix moth, *Adoxophyes orana*:

The highly variable cuticular hydrocarbons of ants, used in colony odors for nestmate recognition, include many different isomers using various positions of double bonds and/or the position and number of methyl ($-CH_3$) groups (Chapters 1 and 6) (Martin *et al.* 2008b).

A.2 Stereoisomers (spatial isomers)

Stereoisomers are compounds with the same atoms (the same molecular formula) and the same order of connecting the atoms together (connectivity), but with different spatial orientations of the atoms (configurations), changing the shape of the molecule. There are several types of stereoisomer. These are described in the following sections and the nomenclature used is shown in Table A.2. Stereoisomers are usually treated by receptors as different molecules (so proper chemical identification of semiochemical molecules must include stereochemistry, Mori 2007).

If the stereoisomers are mirror images of each other, they are called enantiomers. If they are not mirror images of each other then they are called diastereoisomers. Geometric isomers are a subclass of diastereoisomers.

A.2.1 Enantiomers (mirror image molecules)

Chiral molecules are mirror images of each other and, just like your hands, cannot be superimposed on

Table A.2 **Naming of isomers: the meanings of the letters and symbols. The development of chemical nomenclature to precisely describe molecules has left us with a number of different naming schemes, including some that are synonymous and others that appear similar but are based on different principles.**

Symbols	Synonym	Type of isomer
E, Z		Geometrical spatial arrangement around double bond
l, d	–, +	Optical isomers (*laevo, dextro* rotation of polarized light)
L, D		Configurations of a sugar or amino acid, based on the absolute configuration of glyceraldehyde (Fischer nomenclature)
S, R		Different absolute configurations of stereocenters (asymmetric centers)

each other (chiral comes from the Greek word *cheir*, meaning "hand"). The two mirror images are called **enantiomers**. The ability to distinguish between a pair of enantiomers requires a chiral agent; a glove is a chiral agent that distinguishes right and left hands. Shoes do the same for feet. Enzymes and receptors are just such systems in biology. Nature is inherently chiral (Mori 2007).

The chirality is a consequence of the tetrahedral geometry of carbon atoms: if different groups are attached to each of the four bonds of one of the carbon atoms, the molecule can be made in two different ways (or as two enantiomers, see diagram below). A central carbon atom with four *different* groups attached is called a stereogenic or asymmetric center (sometimes called a chiral center). An asymmetric center can be in one of the two forms, *R* or *S* (see below) (Table A.2). It is called asymmetric because there is no way to split the molecule in half to give two equal halves. For example, butan-2-ol has four different groups attached to the central carbon:

$$CH_3 - \overset{\overset{\displaystyle H}{|}}{\underset{\underset{\displaystyle OH}{|}}{C^*}} - CH_2 - CH_3$$

The asymmetric carbon (chiral center) is marked with a star.

And represented spatially:

In this and other diagrams, solid wedges ▶ represent bonds coming out of the page toward you and dashed wedges [ıııı·] (or dotted lines) represent bonds going away from you behind the plane of the paper. Lines – are bonds in the plane of the paper.

In the upper diagram on the next page, a symmetrical molecule A with two different groups attached to the carbon is contrasted with an asymmetric molecule C with four different groups.

One way to see this in a molecule is to imagine the page of this book as a plane of symmetry and look along it (lower diagram, next page).

A special property of chiral compounds, which led to the discovery of the phenomenon, is optical activity. Solutions of pure enantiomers rotate the plane of polarization of plane-polarized light passing through them: opposite enantiomers rotate it in opposite directions. The plane of polarization is rotated to the left by l-molecules, and to the right by d-molecules. The abbreviations come from Latin: l (*laevo*, for "left") and d (*dextro*, "right"). The letters l and d have been replaced by (−) and (+) respectively (Table A.2). A different naming system uses D and L.[1]

A **racemic mixture** or **racemate** is an equal mix of the two enantiomers. As they cancel each other out optically, such solutions do not rotate the plane of polarization of plane-polarized light, so they are optically inactive. Most chemically synthesized enantiomeric compounds, unless special steps are used in synthesis or purification, are racemic mixtures. Molecules produced by enzymes tend to be one enantiomer or the other.

The +/− naming system is based on the observation of the direction in which polarized light is shifted but

---

[1] An example is L–kynurenine, an amino acid used by masu salmon, *Oncorhynchus masou*, as a female sex pheromone. D and L are used to define the absolute configuration of a sugar or of an amino acid, based on the absolute configuration of glyceraldehyde (Fischer nomenclature). D, L (absolute configuration) should not be confused with d, l (rotation of the plane of polarization of plane-polarized light). The rules for naming are not important here, but biologically the differences between the different naming systems can be important (see Howse *et al.* 1998, p. 152).

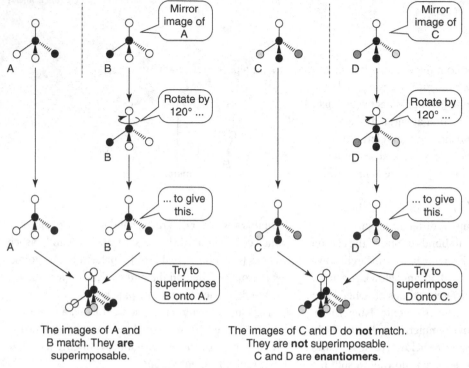

The images of A and B match. They **are** superimposable.

The images of C and D do **not** match. They are **not** superimposable. C and D are **enantiomers**.

Molecule A and its mirror image (B) can be superimposed on each other. Molecule C, with four different groups around the central carbon, cannot be superimposed on its mirror image D. Molecules C and D are enantiomers. Figure from Crowe and Bradshaw (2010).

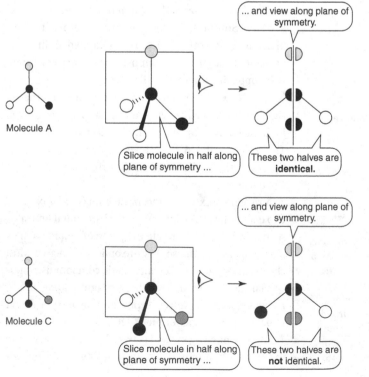

Molecule A is symmetrical: it can be divided in half along the plane of the page to generate two half molecules, which are identical. However, the enantiomer, Molecule C, is asymmetric. It cannot be divided in half along the plane of the page, or along any other plane to generate two identical half molecules. Caption and figure from Crowe and Bradshaw (2010).

does not tell us the actual position of all atoms in space around a molecule. This is done by the **absolute configuration**, which uses a rather complicated set of rules that allow the chemist to describe the absolute configuration at each of the asymmetric centers as either R (from *rectus* = "right") and S (from *sinister* = "left"). As the R and S are defined according to a set of nomenclatural rules, these do not necessarily predict which way the whole molecule will shift polarized light (+/−). The full names of the carvone enantiomers are (R)-(−)-carvone and its mirror image, (S)-(+)-carvone.

An example of stereoselectivity in a mammalian pheromone is dehydro-*exo*-brevicomin, with two chiral centers, which comes in R, R and S, S forms; only the R, R form is active biologically and only this form is produced by male mice (Novotny *et al.* 1995):

(*R*,*R*)-3,4-dehydro-
exo-brevicomin

(*S*,*S*)-3,4-dehydro-
exo-brevicomin

We can smell a difference between the enantiomers of many but not all molecules that have them (Laska 2004). An example of our sensitivity to enantiomers is the way we can distinguish (−)-carvone, which gives us the smell of spearmint, from its mirror image, (+)-carvone, perceived by us as the spicy smell of caraway seed:

(−)-carvone
smell: spearmint

(+)-carvone
caraway

Another pair of enantiomers that we can distinguish are molecules produced by bacteria from odorless precursors secreted in the armpits of 80% of the world's population: one enantiomer, (S)-3-methyl-3-sulfanylhexan-1-ol, is responsible for the oniony armpit smell of sweat:

(*S*)-3SH

(*R*)-3SH

By contrast, the other enantiomer (R)-3-methyl-3-sulfanylhexan-1-ol, smells fruity, of grapefruit (Chapter 13) (Troccaz *et al.* 2009).

We and other organisms can distinguish different enantiomers of molecules because of the chiral characteristics of chemosensory receptors. In those cases where we can distinguish different enantiomers, they are stimulating different combinations of receptors, giving the signal to the brain that they are different smells (Chapter 9).

A.2.2 Diastereoisomers

When a molecule has more than one asymmetric center, a second form of stereoisomerism is possible, diastereoisomerism (stereoisomers that are not mirror images, so are not enantiomers). Each of the asymmetric centers can again be in one of the two forms, R and S. If there are two asymmetric centers there will be $2^2 = 4$ stereoisomers. Some may be mirror image molecules, others will not be (diastereoisomers).

For example, one component of the male sex pheromone of the cerambycid grape borer beetle, *Xylotrechus pyrrhoderus*, 2,3-octandiol occurs in four forms: (2*R*,3*R*), (2*R*,3*S*), (2*S*,3*S*), and (2*S*,3*R*):

Some of these molecules represent mirror images of each other, for example, (2*S*,3*S*) and (2*R*,3*R*) and are enantiomers. Other pairs of isomers, such as (2*R*,3*R*) and (2*S*,3*R*) are not mirror images and these are called diastereoisomers, that is, they are stereoisomers that are not enantiomers. These are common

The four optical isomers of 2, 3-octandiol

The four isomers of 2,3-octandiol. Only one, (2S,3S), is active as a pheromone component for the grape borer beetle, *Xylotrechus pyrrhoderus* (Sakai *et al.* 1984). Insects that use chiral pheromones typically produce and respond to either a single stereoisomer or to a species-specific blend of only some of the possible stereoisomers (Mori 2007).

in branched compounds. If there are three asymmetric centers in the molecule there will be eight isomers (2^3), as the number of optical isomers is 2^N where N is the number of asymmetric centers. This complicates the work of both the synthetic chemist and biologist as only some isomers (or combinations of them) may be active as pheromones.

A.2.2.1 Geometrical isomers

Geometrical isomers, also known as *cis–trans* or *E–Z* isomers, are a particular form of diastereoisomer. During biosynthesis of many moth pheromones, precursors have particular hydrogens removed, catalyzed by specific desaturase enzymes, to leave double bonds (see e.g., Figure 3.15). The double bond makes the carbon chain rigid at that point. Because of the angles of the chain, there are two versions of the molecule depending on the way molecules are arranged around the double bond, either on the same side (*Z*) or opposite sides (*E*) (from the German, *Zusammen* = "together" and *Entgegen* = "opposite"). Like other isomers, these two forms are different chemical compounds, often with slightly different chemical and physical properties. In simple molecules, the older terms *cis* and *trans* are roughly equivalent to *Z* and *E* respectively (though not strictly so by modern, more systematic naming rules better suited to more complicated structures, see www.chemguide.co.uk/basicorg/isomerism/ez.html).

This pheromone, one component of the pink bollworm moth, *Pectinophora gossypiella*, female pheromone has two double bonds, one in each orientation:

(*Z,E*)-7,11-hexadecandienyl acetate.

Further reading

There is more about naming conventions for organic molecules at www.chemguide.co.uk/basicorg/convmenu.html and isomers at www.chemguide.co.uk/basicorg/isomermenu.html. These are websites written and maintained by Clark (2009).

Crowe and Bradshaw (2010) introduce chemistry for the biosciences.

The Leffingwell website www.leffingwell.com/chirality/chirality2.htm, has molecular structures of odor molecules and additional information on olfactory thresholds for these.

Mori (2007) describes the importance of chirality for pheromones, giving many examples of the ways that chirality determines specificity of pheromones in animals as varied as elephants and beetles.

For the structures of many pheromones see Pherobase www.pherobase.com (El-Sayed 2013).

For the structure and naming of organic molecules try www.chemspider.com of the Royal Society of Chemistry. It allows you to search by common name and shows synonyms as well as the systematic names.

This Appendix can be downloaded from www.cambridge.org/pheromones.

REFERENCES

Abbott, D H, Saltzman, W, Schultz-Darken, N J & Tannenbaum, P L (1998) Adaptations to subordinate status in female marmoset monkeys. *Comp Biochem Physiol C Pharmacol Toxicol Endocrinol* **119**: 261–74.

Abel, R L, Maclaine, J S, Cotton, R *et al.* (2010) Functional morphology of the nasal region of a hammerhead shark. *Comp Biochem Physiol A Mol Integr Physiol* **155**: 464–75.

Ackerman, D (1990) *A Natural History of the Senses*. New York: Random House (Phoenix Pbk).

Adams, E S & Traniello, J F A (1981) Chemical interference competition by *Monomorium minimum* (Hymenoptera, Formicidae). *Oecologia* **51**: 265–70.

Aeschlimann, P B, Haberli, M A, Reusch, T B H, Boehm, T & Milinski, M (2003) Female sticklebacks *Gasterosteus aculeatus* use self-reference to optimize MHC allele number during mate selection. *Behav Ecol Sociobiol* **54**: 119–26.

Aggio, J & Derby, C D (2011) Chemical communication in lobsters. In Breithaupt, T & Thiel, M (eds.) *Chemical Communication in Crustaceans*. pp. 239–56. New York: Springer.

Aggleton, J P & Waskett, L (1999) The ability of odours to serve as state-dependent cues for real-world memories: can Viking smells aid the recall of Viking experiences? *Br J Psychol* **90**: 1–7.

Akino, T (2008) Chemical strategies to deal with ants: a review of mimicry, camouflage, propaganda, and phytomimesis by ants (Hymenoptera: Formicidae) and other arthropods. *Myrmecol News* **11**: 173–81.

Akino, T, Yamamura, K, Wakamura, S & Yamaoka, R (2004) Direct behavioral evidence for hydrocarbons as nestmate recognition cues in *Formica japonica* (Hymenoptera: Formicidae). *Appl Entomol Zool* **39**: 381–7.

Al Abassi, S, Birkett, M A, Pettersson, J, Pickett, J A & Woodcock, C M (1998) Ladybird beetle odour identified and found to be responsible for attraction between adults. *Cell Mol Life Sci* **54**: 876–9.

Alaux, C & Robinson, G (2007) Alarm pheromone induces immediate–early gene expression and slow behavioral response in honey bees. *J Chem Ecol* **33**: 1346–50.

Alaux, C, Le Conte, Y, Adams, H A *et al.* (2009a) Regulation of brain gene expression in honey bees by brood pheromone. *Genes Brain Behav* **8**: 309–19.

Alaux, C, Sinha, S, Hasadsri, L *et al.* (2009b) Honey bee aggression supports a link between gene regulation and behavioral evolution. *Proc Natl Acad Sci USA* **106**: 15400–5.

Alaux, C, Maisonnasse, A & Le Conte, Y (2010) Pheromones in a superorganism: from gene to social regulation. In Gerald, L (ed.) *Pheromones*. pp. 401–23. London: Academic Press.

Alberts, A C (1990) Chemical-properties of femoral gland secretions in the desert iguana, *Dipsosaurus dorsalis*. *J Chem Ecol* **16**: 13–25.

Alberts, A C (1992) Constraints on the design of chemical communication-systems in terrestrial vertebrates. *Am Nat* **139**: 62–89.

Albone, E S (1984) *Mammalian Semiochemistry: the Investigation of Chemical Signals between Mammals*. Chichester: John Wiley.

Alcock, J (1982) Natural selection and communication in bark beetles. *Fla Entomol* **65**: 17–32.

Alcock, J (2009) *Animal Behaviour. An Evolutionary Approach*, 9th edn. Sunderland, MA: Sinuaer.

Aldred, N & Clare, A S (2008) The adhesive strategies of cyprids and development of barnacle-resistant marine coatings. *Biofouling* **24**: 351–63.

Aldrich, J R (1995) Chemical communication in the true bugs and parasitoid exploitation. In Cardé, R T & Bell, W J (eds.) *Chemical Ecology of Insects 2*. pp. 318–63. London: Chapman and Hall.

Aldrich, J R (1999) Predators. In Hardie, J & Minks, A K (eds.) *Pheromones of Non-lepidopteran Insects associated with Agricultural Plants.* pp. 357–81. Wallingford, Oxon: CAB International.

Aldrich, J R, Kochansky, J R & Abrams, C B (1984) Attractant for a beneficial insect and its parasitoids: pheromone of the predatory spined soldier bug, *Podisus maculiventris* (Hemiptera: Pentatomidae). *Environ Entomol* **13**: 1031–6.

Aldrich, J R, Neal, J W, Oliver, J E & Lusby, W R (1991) Chemistry vis-a-vis maternalism in lace bugs (Heteroptera, Tingidae) – alarm pheromones and exudate defense in *Corythucha* and *Gargaphia* species. *J Chem Ecol* **17**: 2307–22.

Allan, R A, Elgar, M A & Capon, R J (1996) Exploitation of an ant chemical alarm signal by the zodariid spider *Habronestes bradleyi* Walckenaer. *Proc R Soc B* **263**: 69–73.

Allan, S A (2010) Chemical ecology of tick–host interactions. In Takken, W & Knols, B G J (eds.) *Olfaction in Vector–Host Interactions.* pp. 327. Wageningen: Wageningen Academic Publishers.

Allee, W C (1931) *Animal Aggregations: a Study in General Sociology.* Chicago: Chicago University Press.

Allen, C E, Zwaan, B J & Brakefield, P M (2011) Evolution of sexual dimorphism in the Lepidoptera. *Annu Rev Entomol* **56**: 445–64.

Allison, J D & Cardé, R T (eds.) (2014) *Pheromone Communication in Moths: Evolution, Behavior and Application.* Berkeley, CA: University of California Press.

Alpizar, D, Fallas, M, Oehlschlager, A & Gonzalez, L (2012) Management of *Cosmopolites sordidus* and *Metamasius hemipterus* in banana by pheromone-based mass trapping. *J Chem Ecol* **38**: 245–52.

Aluja, M, Leskey, T C & Vincent, C (eds.) (2009a) *Biorational Tree Fruit Pest Management.* Wallingford: CABI Publishing.

Aluja, M, Diaz-Fleischer, F, Boller, E F *et al.* (2009b) Application of feces extracts and synthetic analogues of the host marking pheromone of *Anastrepha ludens* significantly reduces fruit infestation by *A. obliqua* in tropical plum and mango backyard orchards. *J Econ Entomol* **102**: 2268–78.

Alves, H, Rouault, JD, Kondoh, Y *et al.* (2010) Evolution of cuticular hydrocarbons of Hawaiian Drosophilidae. *Behav Genet* **40**: 694–705.

Amoore, J E (1977) Specific anosmia and the concept of primary odors. *Chem Senses* **2**: 267–81.

Andersson, J, Borg-Karlson, A K & Wiklund, C (2003) Antiaphrodisiacs in pierid butterflies: a theme with variation! *J Chem Ecol* **29**: 1489–99.

Andersson, J, Borg-Karlson, A K & Wiklund, C (2004) Sexual conflict and anti-aphrodisiac titre in a poly-androus butterfly: male ejaculate tailoring and absence of female control. *Proc R Soc B* **271**: 1765–70.

Andersson, M (1994) *Sexual Selection.* Princeton: Princeton University Press.

Andersson, M & Iwasa, Y (1996) Sexual selection. *Trends Ecol Evol* **11**: 53–8.

Andersson, M & Simmons, L W (2006) Sexual selection and mate choice. *Trends Ecol Evol* **21**: 296–302.

Andrade, M C B & Roitberg, B D (1995) Rapid response to intraclonal selection in the pea aphid (*Acyrthosiphon pisum*). *Evol Ecol* **9**: 397–410.

Anon (2008) General guidelines for authors for submission of manuscripts that contain identifications and syntheses of compounds. *J Chem Ecol* **34**: 984–6.

Anon (2009) New SPME guidelines. *J Chem Ecol* **35**: 1383.

Anon (2010) General guidelines for authors for submission of manuscripts that contain molecular biological content. *J Chem Ecol* **36**: 1288–92.

Anton, S, Dufour, M C & Gadenne, C (2007) Plasticity of olfactory-guided behaviour and its neurobiological basis: lessons from moths and locusts. *Entomol Exp Appl* **123**: 1–11.

Anton, S, Evengaard, K, Barrozo, R B, Anderson, P & Skals, N (2011) Brief predator sound exposure elicits behavioral and neuronal long-term sensitization in the olfactory system of an insect. *Proc Natl Acad Sci USA* **108**: 3401–5.

Aoki, S (1977) *Colophina clematis* (Homoptera, Pemphigidae), an aphid species with soldiers. *Kontyu* **45**: 276–82.

Apanius, V, Penn, D, Slev, P R, Ruff, L R & Potts, W K (1997) The nature of selection on the major histocompatibility complex. *Crit Rev Immunol* **17**: 179–224.

Appelt, C W & Sorensen, P W (2007) Female goldfish signal spawning readiness by altering when and where they release a urinary pheromone. *Anim Behav* **74**: 1329–38.

Arak, A & Enquist, M (1993) Hidden preferences and the evolution of signals. *Phil Trans R Soc B* **340**: 207–13.

Arakaki, N (1989) Alarm pheromone eliciting attack and escape responses in the sugar-cane woolly aphid, *Ceratovacuna lanigera* (Homoptera, Pemphigidae). *J Ethol* **7**: 83–90.

Arakaki, N (1990) Colony defense by first instar nymphs and dual function of alarm pheromone in the sugar cane woolly aphid, *Ceratovacuna lanigera*. In Veeresh, G K, Mallik, B & Viraktamath, C A (eds.) *Social Insects and the Environment*. pp. 299–300. Bombay: Oxford University Press.

Arakaki, N, Wakamura, S & Yasuda, T (1996) Phoretic egg parasitoid, *Telenomus euproctidis* (Hymenoptera, Scelionidae), uses sex-pheromone of tussock moth *Euproctis taiwana* (Lepidoptera, Lymantriidae) as a kairomone. *J Chem Ecol* **22**: 1079–85.

Arakaki, N, Wakamura, S, Yasuda, T & Yamagishi, K (1997) Two regional strains of a phoretic egg parasitoid, *Telenomus euproctidis* (Hymenoptera: Scelionidae), that use different sex pheromones of two allopatric tussock moth species as kairomones. *J Chem Ecol* **23**: 153–61.

Arakaki, N, Yamazawa, H & Wakamura, S (2011) The egg parasitoid *Telenomus euproctidis* (Hymenoptera: Scelionidae) uses sex pheromone released by immobile female tussock moth *Orgyia postica* (Lepidoptera: Lymantriidae) as kairomone. *Appl Entomol Zool* **46**: 195–200.

Arakawa, H, Cruz, S & Deak, T (2011) From models to mechanisms: odorant communication as a key determinant of social behavior in rodents during illness-associated states. *Neurosci Biobehav Rev* **35**: 1916–28.

Arathi, H S, Shakarad, M & Gadagkar, R (1997) Factors affecting the acceptance of alien conspecifics on nests of the primitively eusocial wasp, *Ropalidia marginata* (Hymenoptera: Vespidae). *J Insect Behav* **10**: 343–53.

Arcese, P (1999) Effect of auxiliary males on territory ownership in the oribi and the attributes of multi-male groups. *Anim Behav* **57**: 61–71.

Archie, E A & Theis, K R (2011) Animal behaviour meets microbial ecology. *Anim Behav* **82**: 425–36.

Arn, H (1990) Pheromones: prophesies, economics, and the ground swell. In Ridgeway, R L, Silverstein, R M & Inscoe, M N (eds.) *Behavior-modifying Chemicals for Insect Management*. pp. 717–22. New York: Marcel Dekker.

Arn, H & Louis, F (1997) Mating disruption in European vineyards. In Cardé, R T & Minks, A K (eds.) *Pheromone Research: New Directions*. pp. 377–82. New York: Chapman and Hall.

Arnold, A P (2009) The organizational–activational hypothesis as the foundation for a unified theory of sexual differentiation of all mammalian tissues. *Horm Behav* **55**: 570–8.

Arnqvist, G (2006) Sensory exploitation and sexual conflict. *Phil Trans R Soc B* **361**: 375–86.

Arnqvist, G & Rowe, L (2005) *Sexual Conflict*. Princeton: Princeton University Press.

Arvedlund, M, McCormick, M I, Fautin, D G & Bildsøe, M (1999) Host recognition and possible imprinting in the anemonefish *Amphiprion melanopus* (Pisces: Pomacentridae). *Mar Ecol Prog Ser* **188**: 207–18.

Atema, J (1986) Review of sexual selection and chemical communication in the lobster, *Homarus americanus*. *Can J Fish Aquat Sci* **43**: 2283–90.

Atema, J (1995) Chemical signals in the marine environment: dispersal, detection, and temporal signal analysis. In Eisner, T & Meinwald, J (eds.) *Chemical Ecology: the Chemistry of Biotic Interaction*. pp. 147–59. Washington, DC: National Academy of Sciences.

Atema, J (2012) Aquatic odour dispersal fields: opportunities and limits of detection, communication, and navigation. In Brönmark, C & Hansson, L-A (eds.) *Chemical Ecology in Aquatic Systems*. pp. 1–18. Oxford: Oxford University Press.

Atema, J & Steinbach, M A (2007) Chemical communication and social behavior of the lobster *Homarus americanus* and other decapod Crustacea. In Duffy, J E & Thiel, M (eds.) *Evolutionary Ecology of Social and Sexual Systems: Crustaceans as Model Organisms*. pp. 115–44. Oxford & New York: Oxford University Press.

Avila, F W, Sirot, L K, Laflamme, B A, Rubinstein, C D & Wolfner, M F (2011) Insect seminal fluid proteins: identification and function. *Annu Rev Entomol* **56**: 21–40.

Axel, R (2005) Scents and sensibility: a molecular logic of olfactory perception (Nobel Lecture). *Angew Chem Int Ed* **44**: 6111–27.

Bagley, K R, Goodwin, T E, Rasmussen, L E L & Schulte, B A (2006) Male African elephants, *Loxodonta africana*, can distinguish oestrous status via urinary signals. *Anim Behav* 71: 1439–45.

Bagnères, A-G & Lorenzi, M C (2010) Chemical deception/mimicry using cuticular hydrocarbons. In Blomquist, G J & Bagnères, A-G (eds.) *Insect Hydrocarbons: Biology, Biochemistry, and Chemical Ecology.* pp. 282–324. Cambridge: Cambridge University Press.

Bagøien, E & Kiørboe, T (2005) Blind dating – mate finding in planktonic copepods. I. Tracking the pheromone trail of *Centropages typicus. Mar Ecol Prog Ser* 300: 105–15.

Baker, C F, Montgomery, J C & Dennis, T E (2002) The sensory basis of olfactory search behavior in banded kokopu (*Galaxias fasciatus*). *J Comp Physiol A* 188: 553–60.

Baker, T C (2002) Mechanism for saltational shifts in pheromone communication systems. *Proc Natl Acad Sci USA* 99: 13368–70.

Baker, T C (2008) Balanced olfactory antagonism as a concept for understanding evolutionary shifts in moth sex pheromone blends. *J Chem Ecol* 34: 971–81.

Baker, T C (2011) Insect pheromones: useful lessons for crustacean pheromone programs? In Breithaupt, T & Thiel, M (eds.) *Chemical Communication in Crustaceans.* pp. 531–50. New York: Springer.

Baker, T C, Cossé, A A & Todd, J L (1998) Behavioral antagonism in the moth *Helicoverpa zea* in response to pheromone blends of three sympatric heliothine moth species is explained by one type of antennal neuron. *Ann N Y Acad Sci* 855: 511–13.

Bakker, J, Pierman, S & Gonzalez-Martinez, D (2010) Effects of aromatase mutation (ArKO) on the sexual differentiation of kisspeptin neuronal numbers and their activation by same versus opposite sex urinary pheromones. *Horm Behav* 57: 390–5.

Barata, E N, Hubbard, P C, Almeida, O G, Miranda, A & Canário, A V M (2007) Male urine signals social rank in the *Mozambique tilapia* (Oreochromis mossambicus). *BMC Biol* 5: 54.

Barata, E N, Serrano, R M, Miranda, A, *et al.* (2008a) Putative pheromones from the anal glands of male blennies attract females and enhance male reproductive success. *Anim Behav* 75: 379–89.

Barata, E N, Fine, J M, Hubbard, P C *et al.* (2008b) A sterol-like odorant in the urine of Mozambique tilapia males likely signals social dominance to females. *J Chem Ecol* 34: 438–49.

Barbero, F, Bonelli, S, Thomas, J A, Balletto, E & Schonrogge, K (2009) Acoustical mimicry in a predatory social parasite of ants. *J Exp Biol* 212: 4084–90.

Bargmann, C I (2006a) Comparative chemosensation from receptors to ecology. *Nature* 444: 295–301.

Bargmann, C I (2006b) Chemosensation in *C. elegans*. In The C. elegans Research Community (ed.) *WormBook: The Online Review of C. elegans Biology [Internet].* doi/10.1895/wormbook.1891.1123.1891. Pasadena, CA: WormBook.

Barki, A, Jones, C & Karplus, I (2011) Chemical communication and aquaculture of decapod crustaceans: needs, problems, and possible solutions. In Breithaupt, T & Thiel, M (eds.) *Chemical Communication in Crustaceans.* pp. 485–506. New York: Springer.

Barrozo, R B, Jarriault, D, Simeone, X *et al.* (2010) Mating-induced transient inhibition of responses to sex pheromone in a male moth is not mediated by octopamine or serotonin. *J Exp Biol* 213: 1100–6.

Barth, F G, Hrncir, M & Jarau, S (2008) Signals and cues in the recruitment behavior of stingless bees (Meliponini). *J Comp Physiol A* 194: 313–27.

Bartlet, R J (2010) Volatile hydrocarbon pheromones from beetles. In Blomquist, G J & Bagnères, A-G (eds.) *Insect Hydrocarbons: Biology, Biochemistry, and Chemical Ecology.* pp. 448–76. Cambridge: Cambridge University Press.

Bashir, M & Hassanali, A (2010) Novel cross-stage solitarising effect of gregarious-phase adult desert locust (*Schistocerca gregaria* (Forskål)) pheromone on hoppers. *J Insect Physiol* 56: 640–5.

Basil, J A, Hanlon, R T, Sheikh, S I & Atema, J (2000) Three-dimensional odor tracking by *Nautilus pompilius. J Exp Biol* 203: 1409–14.

Bastir, M, Rosas, A, Gunz, P *et al.* (2011) Evolution of the base of the brain in highly encephalized human species. *Nat Commun* 2: 588.

Bateman, A & Logan, D W (2010) Time to underpin Wikipedia wisdom. *Nature* 468: 765.

Bateson, P & Mameli, M (2007) The innate and the acquired: useful clusters or a residual distinction from folk biology? *Dev Psychobiol* 49: 818–31.

Bathellier, B, Gschwend, O & Carleton, A (2010) Temporal coding in olfaction. In Menini, A (ed.) *The Neurobiology of Olfaction.* Boca Raton, FL.: CRC Press. Available online at www.ncbi.nlm.nih.gov/books/NBK55968/.

Bauer, R T (2011) Chemical communication in decapod shrimps: the influence of mating and social systems on the relative importance of olfactory and contact pheromones. In Breithaupt, T & Thiel, M (eds.) *Chemical Communication in Crustaceans.* pp. 277–96. New York: Springer.

Baum, M J (2012) Contribution of pheromones processed by the main olfactory system to mate recognition in female mammals. *Front Neuroanat* 6: 20.

Baum, M J & Kelliher, K R (2009) Complementary roles of the main and accessory olfactory systems in mammalian mate recognition. *Annu Rev Physiol* 71: 141–60.

Baxi, K N, Dorries, K M & Eisthen, H L (2006) Is the vomeronasal system really specialized for detecting pheromones? *Trends Neurosci* 29: 1–7.

Beale, M H, Birkett, M A, Bruce, T J A *et al.* (2006) Aphid alarm pheromone produced by transgenic plants affects aphid and parasitoid behavior. *Proc Natl Acad Sci USA* 103: 10509–13.

Beauchamp, G K, Doty, R L, Moulton, D G & Mugford, R A (1976) The pheromone concept in mammalian chemical communication: a critique. In Doty, R L (ed.) *Mammalian Olfaction, Reproductive Processes, and Behavior.* pp. 143–60. New York: Academic Press.

Becker, S D & Hurst, J L (2009) Female behaviour plays a critical role in controlling murine pregnancy block. *Proc R Soc B* 276: 1723–9.

Beckers, R, Deneubourg, J L & Goss, S (1993) Modulation of trail laying in the ant *Lasius niger* (Hymenoptera, Formicidae) and its role in the collective selection of a food source. *J Insect Behav* 6: 751–9.

Bedell, V M, Westcot, S E & Ekker, S C (2011) Lessons from morpholino-based screening in zebrafish. *Brief Funct Genomics* 10: 181–8.

Beggs, K T & Mercer, A R (2009) Dopamine receptor activation by honey bee queen pheromone. *Curr Biol* 19: 1206–9.

Belanger, R M & Moore, P A (2006) The use of the major chelae by reproductive male crayfish (*Orconectes rusticus*) for discrimination of female odours. *Behaviour* 143: 713–31.

Bell, W J (1991) *Searching Behaviour. The Behavioural Ecology of Finding Resources.* London: Chapman and Hall.

Bell, W J, Roth, L M & Nalepa, C A (2007) *Cockroaches: Ecology, Behavior, and Natural History.* Baltimore: Johns Hopkins University Press.

Bellés, X (2010) Beyond *Drosophila*: RNAi in vivo and functional genomics in insects. *Annu Rev Entomol* 55: 111–28.

Ben-Shahar, Y, Dudek, N L & Robinson, G E (2004) Phenotypic deconstruction reveals involvement of manganese transporter malvolio in honey bee division of labor. *J Exp Biol* 207: 3281–8.

Ben-Shaul, Y, Katz, L C, Mooney, R & Dulac, C (2010) In vivo vomeronasal stimulation reveals sensory encoding of conspecific and allospecific cues by the mouse accessory olfactory bulb. *Proc Natl Acad Sci USA* 107: 5172–7.

Bendesky, A & Bargmann, C I (2011) Genetic contributions to behavioural diversity at the gene–environment interface. *Nat Rev Genet* 12: 809–20.

Bengtsson, B O & Löfstedt, C (2007) Direct indirect selection in moth pheromone evolution: population genetical simulations of asymmetric sexual interactions. *Biol J Linn Soc* 90: 117–23.

Benoit, J B, Phillips, S A, Croxall, T J *et al.* (2009) Addition of alarm pheromone components improves the effectiveness of desiccant dusts against *Cimex lectularius.* *J Med Entomol* 46: 572–9.

Benton, R (2011) Decision making: singin' in the brain. *Neuron* 69: 399–401.

Benton, R, Vannice, K S, Gomez-Diaz, C & Vosshall, L B (2009) Variant ionotropic glutamate receptors as chemosensory receptors in *Drosophila.* *Cell* 136: 149–62.

Bentz, B J, Régnière, J, Fettig, C J *et al.* (2010) Climate change and bark beetles of the western United States and Canada: direct and indirect effects. *Bioscience* 60: 602–13.

Berec, L, Angulo, E & Courchamp, F (2007) Multiple Allee effects and population management. *Trends Ecol Evol* 22: 185–91.

Berg, H C (2004) *E. coli in Motion.* New York: Springer.

Bergstrom, C T & Lachmann, M (1998) Signaling among relatives. III. Talk is cheap. *Proc Natl Acad Sci USA* 95: 5100.

Bergstrom, C T, Számadó, S & Lachmann, M (2002) Separating equilibria in continuous signalling games. *Phil Trans R Soc B* 357: 1595–606.

Bertness, M D, Leonard, G H, Levine, J M & Bruno, J F (1999) Climate-driven interactions among rocky intertidal organisms caught between a rock and a hot place. *Oecologia* 120: 446–50.

Beydoun, H & Saftlas, A (2005) Association of human leucocyte antigen sharing with recurrent spontaneous abortions. *Tissue Antigens* 65: 123–35.

Bhatnagar, K & Smith, T (2010) The human vomeronasal organ. Part VI: A nonchemosensory vestige in the context of major variations of the mammalian vomeronasal organ. *Curr Neurobiol* 1: 1–9.

Billen, J (2006) Signal variety and communication in social insects. *Proc Neth Entomol Soc Meet* 17: 9–25 available at http://tinyurl.com/billen2006.

Billen, J & Morgan, E D (1998) Pheromone communication in social insects: sources and secretions. In Vander Meer, R K, Breed, M D, Espelie, K E & Winston, M L (eds.) *Pheromone Communication in Social Insects: Ants, Wasps, Bees, and Termites.* pp. 3–33. Boulder, CO: Westview Press.

Billeter, J C, Rideout, E J, Dornan, A J & Goodwin, S F (2006) Control of male sexual behavior in *Drosophila* by the sex determination pathway. *Curr Biol* 16: R766–R776.

Billeter, J C, Atallah, J, Krupp, J J, Millar, J G & Levine, J D (2009) Specialized cells tag sexual and species identity in *Drosophila melanogaster. Nature* 461: 987–91.

Birch, M C, Poppy, G M & Baker, T C (1990) Scents and eversible scent structures of male moths. *Annu Rev Entomol* 35: 25–58.

Bisulco, S & Slotnick, B (2003) Olfactory discrimination of short chain fatty acids in rats with large bilateral lesions of the olfactory bulbs. *Chem Senses* 28: 361–70.

Black, S & Biron, C (1982) Androstenol as a human pheromone: no effect on perceived physical attractiveness. *Behav Neural Biol* 34: 326–30.

Blomquist, G J (2010) Structure and analysis of insect hydrocarbons. In Blomquist, G J & Bagnères, A-G (eds.) *Insect Hydrocarbons: Biology, Biochemistry, and Chemical Ecology.* pp. 19–34. Cambridge: Cambridge University Press.

Blomquist, G J & Bagnéres, A G (eds.) (2010) *Insect Hydrocarbons: Biology, Biochemistry, and Chemical Ecology.* Cambridge: Cambridge University Press.

Blomquist, G J, Teran, R F, Aw, M *et al.* (2010) Pheromone production in bark beetles. *Insect Biochem Mol Biol* 40: 699–712.

Blum, M S (1974) Pheromonal bases of social manifestations in insects. In Birch, M C (ed.) *Pheromones.* pp. 190–99. Amsterdam: North-Holland.

Blum, M S (1982) Pheromonal bases of insect sociality: communications, conundrums and caveats. *Colloques de l'INRA* 7: 149–62.

Blum, M S (1985) Alarm pheromones. In Kerkut, GA & Gilbert, LI (eds.) *Comprehensive Insect Physiology, Biochemistry and Pharmacology.* pp. 193–224. Oxford: Pergamon Press.

Blum, M S (1996) Semiochemical parsimony in the arthropoda. *Annu Rev Entomol* 41: 353–74.

Boake, C R B (1991) Coevolution of senders and receivers of sexual signals: genetic coupling and genetic correlations. *Trends Ecol Evol* 6: 225–7.

Boehm, U, Zou, Z H & Buck, L B (2005) Feedback loops link odor and pheromone signaling with reproduction. *Cell* 123: 683–95.

Bolnick, D I & Fitzpatrick, B M (2007) Sympatric speciation: models and empirical evidence. *Annu Rev Ecol Evol Syst* 38: 459–87.

Bonabeau, E, Theraulaz, G, Deneubourg, J L, Aron, S & Camazine, S (1997) Self-organization in social insects. *Trends Ecol Evol* 12: 188–93.

Bonabeau, E, Theraulaz, G, Deneubourg, JL *et al.* (1998) A model for the emergence of pillars, walls and royal chambers in termite nests. *Phil Trans R Soc B* 353: 1561–76.

Bonadonna, F & Sanz-Aguilar, A (2012) Kin recognition and inbreeding avoidance in wild birds: the first evidence for individual kin-related odour recognition. *Anim Behav* 84: 509–13.

Bond, A L (2011) Why ornithologists should embrace and contribute to Wikipedia. *Ibis* 153: 640–1.

Bonthuis, P J, Cox, K H, Searcy, B T *et al.* (2010) Of mice and rats: key species variations in the sexual differentiation of brain and behavior. *Front Neuroendocrinol* 31: 341–58.

Boone, C K, Six, D L & Raffa, K F (2008) The enemy of my enemy is still my enemy: competitors add to predator load of a tree-killing bark beetle. *Agric For Entomol* **10**: 411–21.

Booth, D W & Signoret, J P (1992) Olfaction and reproduction in ungulates. In Milligan, S R (ed.) *Oxford Reviews of Reproduction*. pp. 263–301. Oxford: Oxford University Press.

Boppré, M (1990) Lepidoptera and pyrrolizidine alkaloids – exemplification of complexity in chemical ecology. *J Chem Ecol* **16**: 165–85.

Bordereau, C & Pasteels, J M (2011) Pheromones and chemical ecology of dispersal and foraging in termites. In Bignell, D E, Roisin, Y & Lo, N (eds.) *Biology of Termites: a Modern Synthesis*, 2nd edn. pp. 279–320. Dordrecht: Springer.

Bos, D H, Williams, R N, Gopurenko, D, Bulut, Z & Dewoody, J A (2009) Condition-dependent mate choice and a reproductive disadvantage for MHC-divergent male tiger salamanders. *Mol Ecol* **18**: 3307–15.

Bos, N & d'Ettorre, P (2012) Recognition of social identity in ants. *Front Psychol* **3**: 83.

Bossert, W H & Wilson, E O (1963) The analysis of olfactory communication among animals. *J Theor Biol* **5**: 443–69.

Bourke, A F G (2011a) The validity and value of inclusive fitness theory. *Proc R Soc B* **278**: 3313–20.

Bourke, A F G (2011b) *Principles of Social Evolution*. Oxford: Oxford University Press.

Bousquet, F, Nojima, T, Houot, B *et al.* (2012) Expression of a desaturase gene, desat1, in neural and nonneural tissues separately affects perception and emission of sex pheromones in *Drosophila*. *Proc Natl Acad Sci USA* **109**: 249–54.

Boydston, E E, Morelli, T L & Holekamp, K E (2001) Sex differences in territorial behavior exhibited by the spotted hyena (Hyaenidae, *Crocuta crocuta*). *Ethology* **107**: 369–85.

Bradbury, J W & Vehrencamp, S L (2011) *Principles of Animal Communication*, 2nd edn. Sunderland, MA: Sinauer.

Bradshaw, J W S, Baker, R & Howse, P E (1975) Multicomponent alarm pheromones of the weaver ant. *Nature* **258**: 230–1.

Bradshaw, J W S, Baker, R & Howse, P E (1979) Multicomponent alarm pheromones in the mandibular glands of the African weaver ant, *Oecophylla longinoda*. *Physiol Entomol* **4**: 15–25.

Brandstaetter, A S & Kleineidam, C J (2011) Distributed representation of social odors indicates parallel processing in the antennal lobe of ants. *J Neurophysiol* **106**: 2437–49.

Brandstaetter, A S, Endler, A & Kleineidam, C J (2008) Nestmate recognition in ants is possible without tactile interaction. *Naturwissenschaften* **95**: 601–8.

Brandstaetter, A S, Rossler, W & Kleineidam, C J (2011) Friends and foes from an ant brain's point of view – neuronal correlates of colony odors in a social insect. *PLoS ONE* **6**: e21383.

Brashares, J S & Arcese, P (1999a) Scent marking in a territorial African antelope: I. The maintenance of borders between male oribi. *Anim Behav* **57**: 1–10.

Brashares, J S & Arcese, P (1999b) Scent marking in a territorial African antelope: II. The economics of marking with faeces. *Anim Behav* **57**: 11–17.

Brechbühl, J, Klaey, M & Broillet, M-C (2008) Grueneberg ganglion cells mediate alarm pheromone detection in mice. *Science* **321**: 1092–5.

Brechbühl, J, Luyet, G, Moine, F, Rodriguez, I & Broillet, M-C (2011) Imaging pheromone sensing in a mouse vomeronasal acute tissue slice preparation. *J Vis Exp* e3311.

Breed, M D (1998a) Chemical cues in kin recognition: criteria for identification, experimental approaches, and the honey bee as an example. In Vander Meer, R K, Breed, M D, Espelie, K E & Winston, M L (eds.) *Pheromone Communication in Social Insects: Ants, Wasps, Bees, and Termites*. pp. 57–78. Boulder, CO: Westview Press.

Breed, M D (1998b) Recognition pheromones of the honey bee. *Bioscience* **48**: 463–70.

Breed, M D & Buchwald, R (2009) Cue diversity and social recognition. In Gadau, J & Fewell, J H (eds.) *Organization of Insect Societies: From Genome to Sociocomplexity*. pp. 173–94. Cambridge, MA: Harvard University Press.

Breed, M D, Stiller, T M & Moor, M J (1988) The ontogeny of kin discrimination cues in the honey bee, *Apis mellifera*. *Behav Genet* **18**: 439–48.

Breed, M D, Garry, M F, Pearce, A N *et al.* (1995) The role of wax comb in honey-bee nestmate recognition. *Anim Behav* **50**: 489–96.

Breed, M D, Perry, S & Bjostad, L B (2004a) Testing the blank slate hypothesis: why honey bee colonies accept young bees. *Insectes Soc* 51: 12–16.

Breed, M D, Guzmán-Novoa, E & Hunt, G J (2004b) Defensive behavior of honey bees: organization, genetics, and comparisons with other bees. *Annu Rev Entomol* 49: 271–98.

Breed, M D, Cook, C & Krasnec, M O (2012) Cleptobiosis in social insects. *Psyche* 2012 doi:10.1155/2012/484765.

Breithaupt, T & Thiel, M (eds.) (2011) *Chemical Communication in Crustaceans*. New York: Springer.

Breithaupt, T & Hardege, J (2012) Pheromones mediating sex and dominance in aquatic animals. In Brönmark, C & Hansson, L-A (eds.) *Chemical Ecology in Aquatic Systems*. pp. 39–56. Oxford: Oxford University Press.

Bremner, E A, Mainland, J D, Khan, R M & Sobel, N (2003) The prevalence of androstenone anosmia. *Chem Senses* 28: 423–32.

Brennan, P A (2009) Outstanding issues surrounding vomeronasal mechanisms of pregnancy block and individual recognition in mice. *Behav Brain Res* 200: 287–94.

Brennan, P A (2010) Pheromones and mammalian behavior. In Menini, A (ed.) *The Neurobiology of Olfaction*. pp. 157. Boca Raton, FL: CRC Press. Available online at www.ncbi.nlm.nih.gov/books/NBK55973/.

Brennan, P A & Kendrick, K M (2006) Mammalian social odours: attraction and individual recognition. *Phil Trans R Soc B* 361: 2061–78.

Brennan, P A & Zufall, F (2006) Pheromonal communication in vertebrates. *Nature* 444: 308–15.

Brenner, S (2002) Life sentences: Detective Rummage investigates. *Genome Biol* 3: 1–1013.

Bretman, A, Westmancoat, J D, Gage, M J G & Chapman, T (2011) Males use multiple, redundant cues to detect mating rivals. *Curr Biol* 21: 617–22.

Brisbin, I L & Austad, S N (1991) Testing the individual odour theory of canine olfaction. *Anim Behav* 42: 63–9.

Brisbin, I L, Austad, S N & Jacobson, S K (2000) Canine detectives: the nose knows – or does it? *Science* 290: 1093.

Brodmann, J, Twele, R, Francke, W *et al.* (2009) Orchid mimics honey bee alarm pheromone in order to attract hornets for pollination. *Curr Biol* 19: 1368–72.

Brönmark, C & Hansson, L-A (eds.) (2012) *Chemical Ecology in Aquatic Systems*. Oxford: Oxford University Press.

Brown, G E & Smith, R J F (1998) Acquired predator recognition in juvenile rainbow trout (*Oncorhynchus mykiss*): conditioning hatchery-reared fish to recognize chemical cues of a predator. *Can J Fish Aquat Sci* 55: 611–17.

Brown, G E, Chivers, D P & Smith, R J F (1997) Differential learning rates of chemical versus visual cues of a northern pike by fathead minnows in a natural habitat. *Environ Biol Fishes* 49: 89–96.

Brown, R E, Roser, B & Singh, P B (1989) Class I and class II regions of the major histocompatibility complex both contribute to individual odors in congenic inbred strains of rats. *Behav Genet* 19: 659–74.

Bruschini, C, Cervo, R & Turillazzi, S (2010) Pheromones in social wasps. In Gerald, L (ed.) *Pheromones*. pp. 447–92. London: Academic Press.

Buck, L B (2005) Unraveling the sense of smell (Nobel Lecture). *Angew Chem Int Ed* 44: 6128–40.

Buck, L B & Axel, R (1991) A novel multigene family may encode odorant receptors – a molecular-basis for odor recognition. *Cell* 65: 175–87.

Buckley, S H, Tregenza, T & Butlin, R K (2003) Transitions in cuticular composition across a hybrid zone: historical accident or environmental adaptation? *Biol J Linn Soc* 78: 193–201.

Būda, V, Mozūraitis, R, Kutra, J & Borg-Karlson, A-K (2012) *p*-Cresol: a sex pheromone component identified from the estrous urine of mares. *J Chem Ecol* 38: 811–13.

Buesching, C D, Stopka, P & Macdonald, D W (2003) The social function of allo-marking in the European badger (*Meles meles*). *Behaviour* 140: 965–80.

Burgener, N, Dehnhard, M, Hofer, H & East, M L (2009) Does anal gland scent signal identity in the spotted hyaena? *Anim Behav* 77: 707–15.

Burghardt, G M, Bartmess-LeVasseur, J N, Browning, S A *et al.* (2012) Perspectives – Minimizing observer bias in behavioral studies: a review and recommendations. *Ethology* 118: 511–17.

Burke, R D (1984) Pheromonal control of metamorphosis in the Pacific sand dollar, *Dendraster excentricus. Science* 225: 442–3.

Burkholder, W E (1982) Reproductive biology and communication among grain storage and warehouse beetles. *J Ga Entomol Soc* 17 (II. suppl.): 1–10.

Burton, J L & Franks, N R (1985) The foraging ecology of the army ant *Eciton rapax* – an ergonomic enigma. *Ecol Entomol* 10: 131–41.

Buschinger, A (2009) Social parasitism among ants: a review (Hymenoptera: Formicidae). *Myrmecol News* 12: 219–35.

Butenandt, A, Beckmann, R, Stamm, D & Hecker, E (1959) Uber den sexual-lockstoff des seidenspinners *Bombyx mori* – reindarstellung und konstitution. *Zeitschrift Fur Naturforschung Part B-Chemie Biochemie Biophysik Biologie Und Verwandten Gebiete* 14: 283–4.

Butler, C (1623) *The Feminine Monarchie, Or the Historie of Bees*, 2nd edn. John Haviland. Available Google Books http://tinyurl.com/butler-1623-feminine.

Butlin, R K & Ritchie, M G (1989) Genetic coupling in mate recognition systems – what is the evidence. *Biol J Linn Soc* 37: 237–46.

Butlin, R K & Trickett, A J (1997) Can population genetic simulations help to interpret pheromone evolution? In Cardé, R T & Minks, A K (eds.) *Pheromone Research: New Directions* pp. 548–62. New York: Chapman and Hall.

Byers, J A (1992) Optimal fractionation and bioassay plans for isolation of synergistic chemicals: the subtractive-combination method. *J Chem Ecol* 18: 1603–21.

Byers, J A (2005) A cost of alarm pheromone production in cotton aphids, *Aphis gossypii. Naturwissenschaften* 92: 69–72.

Byers, J A & Zhang, Q (2011) Chemical ecology of bark beetles in regard to search and selection of host trees. In Liu, T & Kang, L (eds.) *Recent Advances in Entomological Research: from Molecular Biology to Pest Management.* pp. 150–90. Beijing and Berlin: Higher Education Press and Springer.

Calenbuhr, V & Deneubourg, J L (1992) A model for osmotropotactic orientation. 1. *J Theor Biol* 158: 359–93.

Calenbuhr, V, Chretien, L, Deneubourg, J L & Detrain, C (1992) A model for osmotropotactic orientation. 2. *J Theor Biol* 158: 395–407.

Camazine, S, Deneubourg, J-L, Franks, N R *et al.* (2001) *Self-organization in Biological Systems*. Princeton: Princeton University Press.

Cammaerts, M C & Cammaerts, R (1980) Food recruitment strategies of the ant *Myrmica sabuleti* and *Myrmica ruginodis. Behav Processes* 5: 251–70.

Campagna, S, Mardon, J, Celerier, A & Bonadonna, F (2012) Potential semiochemical molecules from birds: a practical and comprehensive compilation of the last 20 years studies. *Chem Senses* 37: 3–25.

Campbell-Palmer, R & Rosell, F (2010) Conservation of the Eurasian beaver *Castor fiber*: an olfactory perspective. *Mammal Rev* 40: 293–312.

Cardé, R T & Baker, T C (1984) Sexual communication with pheromones. In Bell, W J & Cardé, R T (eds.) *Chemical Ecology of Insects*. pp. 355–86. London: Chapman & Hall.

Cardé, R T & Haynes, K F (2004) Structure of the pheromone communication channel in moths. In: Cardé, R & Millar, J G (eds.) *Advances in Insect Chemical Ecology*. pp. 283–332. Cambridge: Cambridge University Press.

Cardé, R T & Mafra-Neto, A (1997) Mechanisms of flight of male moths to pheromone. In Cardé, R T & Minks, A K (eds.) *Pheromone Research: New Directions*. pp. 275–90. New York: Chapman and Hall.

Cardé, R T & Minks, A K (1995) Control of moth pests by mating disruption – successes and constraints. *Annu Rev Entomol* 40: 559–85.

Cardé, R T & Willis, M A (2008) Navigational strategies used by insects to find distant, wind-borne sources of odor. *J Chem Ecol* 34: 854–66.

Cardé, R T, Cardé, A M & Girling, R D (2012) Observations on the flight paths of the day flying moth *Virbia lamae* during periods of mate location: do males have a strategy for contacting the pheromone plume? *J Anim Ecol* 81: 268–76.

Cardwell, J R, Stacey, N E, Tan, E S P, McAdam, D S O & Lang, S L C (1995) Androgen increases olfactory receptor response to a vertebrate sex pheromone. *J Comp Physiol A* 176: 55–61.

Carey, A F & Carlson, J R (2011) Insect olfaction from model systems to disease control. *Proc Natl Acad Sci USA* 108: 12987–95.

Carlin, N F & Hölldobler, B (1987) The kin recognition system of carpenter ants (*Camponotus* spp). 2. Larger colonies. *Behav Ecol Sociobiol* 20: 209–17.

Caro, S P & Balthazart, J (2010) Pheromones in birds: myth or reality? *J Comp Physiol A* 196: 751–66.

Carroll, J F, Mills, G D & Schmidtmann, E T (1996) Field and laboratory responses of adult *Ixodes scapularis* (Acari: Ixodidae) to kairomones produced by white tailed deer. *J Med Entomol* 33: 640–4.

Carson, C, Birkett, M A, Logan, J G *et al.* (2010) Novel use of stir bar sorptive extraction (SBSE) as a tool for isolation of oviposition site attractants for gravid *Culex quinquefasciatus*. *Bull Entomol Res* 100: 1–7.

Carter, C S & Getz, L L (1993) Monogamy and the prairie vole. *Sci Am* 268: 100–6.

Carter, C S & Roberts, R L (1997) The psychobiological basis of cooperative breeding in rodents. In Solomon, N G (ed.) *Cooperative Breeding in Mammals*. pp. 231–66. Cambridge: Cambridge University Press.

Caspers, B A, Schroeder, F C, Franke, S, Streich, W J & Voigt, C C (2009) Odour-based species recognition in two sympatric species of sac-winged bats (*Saccopteryx bilineata*, *S. leptura*): combining chemical analyses, behavioural observations and odour preference tests. *Behav Ecol Sociobiol* 63: 741–9.

Chamero, P, Marton, T F, Logan, D W *et al.* (2007) Identification of protein pheromones that promote aggressive behaviour. *Nature* 450: 899–902.

Chamero, P, Leinders-Zufall, T & Zufall, F (2012) From genes to social communication: molecular sensing by the vomeronasal organ. *Trends Neurosci* 35: 597–606.

Chandler, D, Bailey, A S, Tatchell, G M *et al.* (2011) The development, regulation and use of biopesticides for integrated pest management. *Phil Trans R Soc B* 366: 1987–98.

Chapman, R F (1998) *The Insects. Structure and Function*, 4th edn. Cambridge: Cambridge University Press.

Chapman, T (2008) The soup in my fly: evolution, form and function of seminal fluid proteins. *PLoS Biol* 6: e179.

Charlesworth, D & Willis, J H (2009) The genetics of inbreeding depression. *Nat Rev Genet* 10: 783–96.

Charra, R, Datiche, F, Casthano, A *et al.* (2012) Brain processing of the mammary pheromone in newborn rabbits. *Behav Brain Res* 226: 179–88.

Chemineau, P (2011) A foresight reflection on sustainable methods for controlling mammalian farm animal reproduction. *Trop Subtrop Agroecosyst* [Online], 15. Available: www.veterinaria.uady.mx/ojs/index.php/TSA/article/view/1344/639 [Accessed 6 June 2013].

Chen, Z-F, Matsumura, K, Wang, H *et al.* (2011) Toward an understanding of the molecular mechanisms of barnacle larval settlement: a comparative transcriptomic approach. *PLoS ONE* 6: e22913.

Chivers, D P, Brown, G E & Ferrari, M C O (2012) The evolution of alarm substances and disturbance cues in aquatic animals. In Brönmark, C & Hansson, L-A (eds) *Chemical Ecology in Aquatic Systems*. pp. 127–39. Oxford: Oxford University Press.

Choe, A, von Reuss, SH, Kogan, D *et al.* (2012) Ascaroside signaling is widely conserved among nematodes. *Curr Biol* 22: 772–80.

Choleris, E, Clipperton-Allen, A E, Phan, A & Kavaliers, M (2009) Neuroendocrinology of social information processing in rats and mice. *Front Neuroendocrinol* 30: 442–59.

Choleris, E, Pfaff, D W & Kavaliers, M (eds.) (2013) *Oxytocin, Vasopressin and Related Peptides in the Regulation of Behavior*. Cambridge: Cambridge University Press.

Christensen, T A (2005) Making scents out of spatial and temporal codes in specialist and generalist olfactory networks. *Chem Senses* 30: i283–4.

Christensen, T A & Hildebrand, J G (2002) Pheromonal and host-odor processing in the insect antennal lobe: how different? *Curr Opin Neurobiol* 12: 393–9.

Christy, J H (2011) Timing of hatching and release of larvae by brachyuran crabs: patterns, adaptive significance and control. *Integr Comp Biol* 51: 62–72.

Chung-Davidson, Y-W, Huertas, M & Li, W (2011) A review of research in fish pheromones. In Breithaupt, T & Thiel, M (eds.) *Chemical Communication in Crustaceans*. pp. 467–82. New York: Springer.

Clare, A S (2011) Toward a characterization of the chemical cue to barnacle gregariousness. In Breithaupt, T & Thiel, M (eds.) *Chemical Communication in Crustaceans*. pp. 431–50. New York: Springer.

Clark, C J (2012) The role of power versus energy in courtship: what is the 'energetic cost' of a courtship display? *Anim Behav* **84**: 269–77.

Clark, J & Edexcel (2009) *Edexcel International GCSE Chemistry*. Harlow: Pearson Education. Supported by the website www.chemguide.co.uk.

Clarke, I J (2011) Control of GnRH secretion: one step back. *Front Neuroendocrinol* **32**: 367–75.

Classen, C, Honer, D & Synott, A (1994) *Aroma. The Cultural History of Smell*. London: Routledge.

Clavijo McCormick, A, Unsicker, S B & Gershenzon, J (2012) The specificity of herbivore-induced plant volatiles in attracting herbivore enemies. *Trends Plant Sci* **17**: 303–10.

Clegg, J M & Barlow, C A (1982) Escape behavior of the pea aphid *Acyrthosiphon pisum* (Harris) in response to alarm pheromone and vibration. *Can J Zool* **60**: 2245–52.

Clément, J L & Bagnères, A G (1998) Nestmate recognition in termites. In Vander Meer, R K, Breed, M D, Espelie, K E & Winston, M L (eds.) *Pheromone Communication in Social Insects*, pp. 126–55. Boulder, CD: Westview Press.

Cleveland, A, Verde, E A & Lee, R W (2011) Nutritional exchange in a tropical tripartite symbiosis: direct evidence for the transfer of nutrients from anemonefish to host anemone and zooxanthellae. *Mar Biol* **158**: 589–602.

Clifford, K T, Gross, L, Johnson, K *et al.* (2003) Slime-trail tracking in the predatory snail, *Euglandina rosea*. *Behav Neurosci* **117**: 1086.

Clutton-Brock, T (2007) Sexual selection in males and females. *Science* **318**: 1882–5.

Clutton-Brock, T (2009) Structure and function in mammalian societies. *Phil Trans R Soc B* **364**: 3229–42.

Clutton-Brock, T & McAuliffe, K (2009) Female mate choice in mammals. *Q Rev Biol* **84**: 3–27.

Clyne, J D & Miesenböck, G (2009) Postcoital finesse. *Neuron* **61**: 491–3.

Clyne, P J, Warr, C G, Freeman, M R *et al.* (1999) A novel family of divergent seven-transmembrane proteins: candidate odorant receptors in *Drosophila. Neuron* **22**: 327–38.

Cohen, L, Rothschild, G & Mizrahi, A (2011) Multisensory integration of natural odors and sounds in the auditory cortex. *Neuron* **72**: 357–69.

Colazza, S, Peri, E, Salerno, G & Conti, E (2010) Host searching by egg parasitoids: exploitation of host chemical cues. In Consoli, F L, Parra, J R P & Zucchi, R A (eds.) *Egg Parasitoids in Agroecosystems with Emphasis on Trichogramma*. pp. 97–147. Dordrecht: Springer.

Colledge, W H (2009) Kisspeptins and GnRH neuronal signalling. *Trends Endocrinol Metab* **20**: 115–21.

Conner, W E (ed.) (2009) *Tiger Moths and Woolly Bears: Behavior, Ecology, and Evolution of the Arctiidae*. Oxford: Oxford University Press.

Conner, W E & Best, B A (1988) Biomechanics of the release of sex pheromone in moths: effects of body posture on local airflow. *Physiol Entomol* **13**: 15–20.

Conner, W E & Weller, S J (2004) A quest for alkaloids: the curious relationship between tiger moths and plants containing pyrrolizidine alkaloids. In Cardé, R & Millar, J G (eds.) *Advances in Insect Chemical Ecology*. pp. 248–82. Cambridge: Cambridge University Press.

Conner, W E, Eisner, T, Vander Meer, R K, Guerrero, A & Meinwald, J (1980) Sex attractant of an arctiid moth (*Utetheisa ornatrix*): a pulsed chemical signal. *Behav Ecol Sociobiol* **7**: 55–63.

Conrad, T, Paxton, R J, Barth, F G, Francke, W & Ayasse, M (2010) Female choice in the red mason bee, *Osmia rufa* (L.) (Megachilidae). *J Exp Biol* **213**: 4065–73.

Consuegra, S & Garcia de Leaniz, C (2008) MHC-mediated mate choice increases parasite resistance in salmon. *Proc R Soc B* **275**: 1397–403.

Cook, S M, Khan, Z R & Pickett, J A (2007) The use of push–pull strategies in integrated pest management. *Annu Rev Entomol* **52**: 375–400.

Corbin, A (1996) *The Foul and the Fragrant: Odour and the Social Imagination*. London: Papermac.

Cornwallis, C K & Uller, T (2010) Towards an evolutionary ecology of sexual traits. *Trends Ecol Evol* **25**: 145–52.

Costa-Leonardo, A M & Haifig, I (2010) Pheromones and exocrine glands in Isoptera. In Gerald, L (ed.) *Pheromones*. pp. 521–49. London: Academic Press.

Costello, E K, Lauber, C L, Hamady, M *et al.* (2009) Bacterial community variation in human body habitats across space and time. *Science* 326: 1694–17.

Courchamp, F, Berec, L & Gascoigne, J (2008) *Allee Effects in Ecology and Conservation*. Oxford: Oxford University Press.

Coureaud, G, Langlois, D, Sicard, G & Schaal, B (2004) Newborn rabbit responsiveness to the mammary pheromone is concentration-dependent. *Chem Senses* 29: 341–50.

Coureaud, G, Charra, R, Datiche, F *et al.* (2010) A pheromone to behave, a pheromone to learn: the rabbit mammary pheromone. *J Comp Physiol A* 196: 779–90.

Couvillon, M J, Caple, J P, Endsor, S L *et al.* (2007) Nestmate recognition template of guard honeybees (*Apis mellifera*) is modified by wax comb transfer. *Biol Lett* 3: 228–30.

Cox, J P L (2008) Hydrodynamic aspects of fish olfaction. *J R Soc Interface* 5: 575.

Coyne, J A & Elwyn, S (2006) Does the *desaturase-2* locus in *Drosophila melanogaster* cause adaptation and sexual isolation? *Evolution* 60: 279–91.

Coyne, J A & Orr, H A (2004) *Speciation*. Sunderland, MA: Sinauer.

Craven, B A, Paterson, E G & Settles, G S (2010) The fluid dynamics of canine olfaction: unique nasal airflow patterns as an explanation of macrosmia. *J R Soc Interface* 7: 933–43.

Cremer, S, d'Ettorre, P, Drijfhout, F P *et al.* (2008) Imperfect chemical female mimicry in males of the ant *Cardiocondyla obscurior*. *Naturwissenschaften* 95: 1101–5.

Crespi, B J (2005) Social sophistry: logos and mythos in the forms of cooperation. *Ann Zool Fenn* 42: 569–71.

Crimaldi, J P (2012) The role of structured stirring and mixing on gamete dispersal and aggregation in broadcast spawning. *J Exp Biol* 215: 1031–9.

Crossland, M R, Haramura, T, Salim, A A, Capon, R J & Shine, R (2012) Exploiting intraspecific competitive mechanisms to control invasive cane toads (*Rhinella marina*). *Proc R Soc B* 279: 3436–42.

Crowe, J & Bradshaw, T (2010) *Chemistry for the Biosciences: the Essential Concepts*, 2nd edn. Oxford: Oxford University Press.

Crozier, R H (1986) Genetic clonal recognition abilities in marine invertebrates must be maintained by selection for something else. *Evolution* 40: 1100–1.

Crozier, R H (1987) Genetic aspects of kin recognition: concepts, models, and synthesis. In Fletcher, D J C & Michener, C D (eds.) *Kin Recognition in Animals*. pp. 55–73. New York: Wiley.

Crozier, R H, Newey, P S, Schluns, E A & Robson, S K A (2010) A masterpiece of evolution – *Oecophylla* weaver ants (Hymenoptera: Formicidae). *Myrmecol News* 13: 57–71.

Cummins, S F & Bowie, J H (2012) Pheromones, attractants and other chemical cues of aquatic organisms and amphibians. *Nat Prod Rep* 29: 642–58.

Cummins, S F & Degnan, B M (2010) Sensory sea slugs: towards decoding the molecular toolkit required for a mollusc to smell. *Commun Integr Biol* 3: 423–6.

Cummins, S F, Xie, F, de Vries, M R *et al.* (2007) *Aplysia* temptin – the 'glue' in the water-borne attractin pheromone complex. *FEBS J* 274: 5425–37.

Cummins, S F, Boal, J G, Buresch, K C *et al.* (2011) Extreme aggression in male squid induced by a beta-MSP-like pheromone. *Curr Biol* 21: 322–7.

Czaczkes, T J & Ratnieks, F L W (2012) Pheromone trails in the Brazilian ant *Pheidole oxyops*: extreme properties and dual recruitment action. *Behav Ecol Sociobiol*: 1149–56.

Czaczkes, T J, Grüter, C, Jones, S M & Ratnieks, F L W (2012) Uncovering the complexity of ant foraging trails. *Commun Integr Biol* 5: 78–80.

d'Ettorre, P & Heinze, J (2005) Individual recognition in ant queens. *Curr Biol* 15: 2170–4.

d'Ettorre, P & Lenoir, A (2010) Nestmate recognition in ants. In Lach, L, Parr, C L & Abbott, K L (eds.) *Ant Ecology*. pp. 194–209. Oxford: Oxford University Press.

d'Ettorre, P & Moore, A J (2008) Chemical communication and the coordination of social interactions in insects. In d'Ettorre, P & Hughes, D P (eds.) *Sociobiology of Communication: an Interdisciplinary Perspective*. pp. 81–96. Oxford: Oxford University Press.

Dahanukar, A & Ray, A (2011) Courtship, aggression and avoidance: pheromones, receptors and neurons for social behaviors in *Drosophila*. *Fly* 5: 58–63.

Dalton, P, Doolittle, N & Breslin, P A S (2002) Gender-specific induction of enhanced sensitivity to odors. *Nat Neurosci* 5: 199–200.

Dani, F R (2006) Cuticular lipids as semiochemicals in paper wasps and other social insects. *Ann Zool Fenn* 43: 500–14.

Dani, F R, Jones, G R, Corsi, S *et al.* (2005) Nestmate recognition cues in the honey bee: differential importance of cuticular alkanes and alkenes. *Chem Senses* 30: 477–89.

Dani, F R, Michelucci, E, Francese, S *et al.* (2011) Odorant-binding proteins and chemosensory proteins in pheromone detection and release in the silkmoth *Bombyx mori. Chem Senses* 36: 335–44.

Darwin, C (1871) *The Descent of Man and Selection in Relation to Sex.* London: John Murray.

David, C T, Kennedy, J S & Ludlow, A R (1983) Finding of a sex-pheromone source by gypsy moths released in the field. *Nature* 303: 804–6.

Davies, N B, Krebs, J R & West, S A (2012) *An Introduction to Behavioural Ecology*, 4th edn. Chicester: Wiley-Blackwell

Davis, E C (2007) Investigation in the laboratory of mucous trail detection in the terrestrial pulmonate snail *Mesodon thyroidus* (Say, 1817) (Mollusca: Gastropoda: Polygyridae). *Am Malacol Bull* 22: 157–64.

Dawkins, M S (2007) *Observing Animal Behaviour: Design and Analysis of Quantitative Data.* Oxford: Oxford University Press.

Dawkins, R (1976) *The Selfish Gene.* Oxford: Oxford University Press.

Dawkins, R (1982) *The Extended Phenotype.* San Francisco, CA: WH Freeman.

de Brito-Sanchez, M G, Deisig, N, Sandoz, J C & Giurfa, M (2008) Neurobiology of olfactory communication in the honeybee. In d'Ettorre, P & Hughes, D P (eds.) *Sociobiology of Communication: an Interdisciplinary Perspective.* pp. 119–38. Oxford: Oxford University Press.

de Bruyne, M & Baker, T C (2008) Odor detection in insects: volatile codes. *J Chem Ecol* 34: 882–97.

de Vos, M, Cheng, W Y, Summers, H E, Raguso, R A & Jander, G (2010) Alarm pheromone habituation in *Myzus persicae* has fitness consequences and causes extensive gene expression changes. *Proc Natl Acad Sci USA* 107: 14673–8.

DeBose, J & Nevitt, G (2008) The use of odors at different spatial scales: comparing birds with fish. *J Chem Ecol* 34: 867–81.

Dehnhard, M (2011) Mammal semiochemicals: understanding pheromones and signature mixtures for better zoo animal husbandry and conservation. *Int Zoo Yearb* 45: 1–25.

delBarco-Trillo, J & Ferkin, M H (2004) Male mammals respond to a risk of sperm competition conveyed by odours of conspecific males. *Nature* 431: 446–9.

delBarco-Trillo, J, Burkert, B A, Goodwin, T E & Drea, C M (2011) Night and day: the comparative study of strepsirrhine primates reveals socioecological and phylogenetic patterns in olfactory signals. *J Evol Biol* 24: 82–98.

delBarco-Trillo, J, Sacha, C R, Dubay, G R & Drea, C M (2012) *Eulemur*, me lemur: the evolution of scent-signal complexity in a primate clade. *Phil Trans R Soc B* 367: 1909–22.

delBarco-Trillo, J, Harelimana, I H, Goodwin, T E & Drea, C M (2013) Chemical differences between voided and bladder urine in the aye-aye (*Daubentonia madagascariensis*): implications for olfactory communication studies. *Am J Primatol* 75: 695–702.

Delgadillo, J A, Gelez, H, Ungerfeld, R, Hawken, P A R & Martin, G B (2009) The 'male effect' in sheep and goats – revisiting the dogmas: pheromonal communication in higher vertebrates and its implication for reproductive function. *Behav Brain Res* 200: 304–14.

Deneubourg, J L, Goss, S, Franks, N & Pasteels, J M (1989) The blind leading the blind – modeling chemically mediated army ant raid patterns. *J Insect Behav* 2: 719–25.

Denny, M W (1993) *Air and Water.* Princeton: Princeton University Press.

Derby, C D & Sorensen, P W (2008) Neural processing, perception, and behavioral responses to natural chemical stimuli by fish and crustaceans. *J Chem Ecol* 34: 898–914.

Derti, A & Roth, F P (2012) Response to "MHC-dependent mate choice in humans: Why genomic patterns from the HapMap European American data set support the hypothesis" (DOI: 10.1002/bies.201100150). *Bioessays* 34: 576–7.

Dethier, V G (1987) Sniff, flick, and pulse: an appreciation of intermittency. *Proc Am Philos Soc* 131: 159–76.

Detrain, C & Deneubourg, J-L (1997) Scavenging by *Pheidole pallidula*: a key for understanding decision-making systems in ants. *Anim Behav* 53: 537–47.

Detrain, C & Deneubourg, J-L (2006) Self-organized structures in a superorganism: do ants "behave" like molecules? *Phys Life Rev* 3: 162–87.

Deutsch, J & Murakhver, N (eds.) (2012) *They Eat That? A Cultural Encyclopedia of Weird and Exotic Food from Around the World*. Santa Barbara, CA: ABC-CLIO.

Deutsch, J C & Nefdt, R J C (1992) Olfactory cues influence female choice in two lek-breeding antelopes. *Nature* 356: 596–8.

Dewhirst, S Y, Pickett, J A & Hardie, J (2010) Aphid pheromones. In Gerald, L (ed.) *Pheromones*. pp. 551–74. London: Academic Press.

Dhawale, A K, Hagiwara, A, Bhalla, U S, Murthy, V N & Albeanu, D F (2010) Non-redundant odor coding by sister mitral cells revealed by light addressable glomeruli in the mouse. *Nat Neurosci* 13: 1404–12.

Dicke, M & Baldwin, I T (2010) The evolutionary context for herbivore-induced plant volatiles: beyond the 'cry for help'. *Trends Plant Sci* 15: 167–75.

Dicke, M & Sabelis, M W (1988) Infochemical terminology: based on cost-benefit analysis rather than origin of compounds? *Funct Ecol* 2: 131–9.

Dickson, B J (2008) Wired for sex: the neurobiology of *Drosophila* mating decisions. *Science* 322: 904–9.

Dietemann, V, Liebig, J, Hölldobler, B & Peeters, C (2005) Changes in the cuticular hydrocarbons of incipient reproductives correlate with triggering of worker policing in the bulldog ant *Myrmecia gulosa*. *Behav Ecol Sociobiol* 58: 486–96.

Diggle, S P, Gardner, A, West, S A & Griffin, A S (2007) Evolutionary theory of bacterial quorum sensing: when is a signal not a signal? *Phil Trans R Soc B* 362: 1241–9.

Dill, L M, Fraser, A H G & Roitberg, B D (1990) The economics of escape behavior in the pea aphid, *Acyrthosiphon pisum*. *Oecologia* 83: 473–8.

Dillehay, T D (2003) Tracking the first Americans. *Nature* 425: 23–4.

Dixson, A F (1998) *Primate Sexuality. Comparative Studies of the Prosimians, Monkeys, Apes, and Human Beings*. Oxford: Oxford University Press.

Dixson, A F (2009) *Sexual Selection and the Origins of Human Mating Systems*. Oxford: Oxford University Press.

Dixson, A F (2012) *Primate Sexuality: Comparative Studies of the Prosimians, Monkeys, Apes, and Human Beings*, 2nd edn. Oxford: Oxford University Press.

Dixson, D L, Munday, P L & Jones, G P (2010) Ocean acidification disrupts the innate ability of fish to detect predator olfactory cues. *Ecol Lett* 13: 68–75.

do Nascimento, R R, Morgan, E D, Billen, J et al. (1993) Variation with caste of the mandibular gland secretion in the leaf-cutting ant – *Atta sexdens rubropilosa*. *J Chem Ecol* 19: 907–18.

Doall, M H, Colin, S P, Strickler, J R & Yen, J (1998) Locating a mate in 3D: the case of *Temora longicornis*. *Phil Trans R Soc B* 353: 681–9.

Domingue, M J, Haynes, K F, Todd, J L & Baker, T C (2009) Altered olfactory receptor neuron responsiveness is correlated with a shift in behavioral response in an evolved colony of the cabbage looper moth, *Trichoplusia ni*. *J Chem Ecol* 35: 405–15.

Doney, S C, Fabry, V J, Feely, R A & Kleypas, J A (2009) Ocean acidification: the other CO_2 problem. *Ann Rev Mar Sci* 1: 169–92.

Doney, S C, Ruckelshaus, M, Emmett Duffy, J et al. (2012) Climate change impacts on marine ecosystems. *Ann Rev Mar Sci* 4: 11–37.

Dopman, E B, Robbins, P S & Seaman, A (2010) Components of reproductive isolation between north American pheromone strains of the European corn borer. *Evolution* 64: 881–902.

Dorries, K M, Adkins-Regan, E & Halpern, B P (1997) Sensitivity and behavioral responses to the pheromone androstenone are not mediated by the vomeronasal organ in domestic pigs. *Brain Behav Evol* 49: 53–62.

Doty, R L (2001) Olfaction. *Annu Rev Psychol* 52: 423–52.

Doty, R L (2009) The olfactory system and its disorders. *Semin Neurol* 29: 74–81.

Doty, R L (2010) *The Great Pheromone Myth*. Baltimore, MD: Johns Hopkins University Press.

Doty, R L (2012a) Olfaction in Parkinson's disease and related disorders. *Neurobiol Dis* 46: 527–52.

Doty, R L (2012b) Olfactory dysfunction in Parkinson disease. *Nat Rev Neurol* 8: 329–39.

Doty, R L & Cameron, E L (2009) Sex differences and reproductive hormone influences on human odor perception. *Physiol Behav* 97: 213–28.

Doty, R L, Ford, M, Preti, G & Huggins, G R (1975) Changes in the intensity and pleasantness of human vaginal odors during the menstrual cycle. *Science* 190: 1316–18.

Doty, R L, Orndorff, M M, Leyden, J & Kligman, A (1978) Communication of gender from human axillary odors: relationship to perceived intensity and hedonicity. *Behav Biol* 23: 373–80.

Doucet, S, Soussignan, R, Sagot, P & Schaal, B (2009) The secretion of areolar (Montgomery's) glands from lactating women elicits selective, unconditional responses in neonates. *PLoS ONE* 4: e7579.

Doucet, S, Soussignan, R, Sagot, P & Schaal, B (2012) An overlooked aspect of the human breast: areolar glands in relation with breastfeeding pattern, neonatal weight gain, and the dynamics of lactation. *Early Hum Dev* 88: 119–28.

Døving, K B & Lastein, S (2009) The alarm reaction in fishes – odorants, modulations of responses, neural pathways. *Ann N Y Acad Sci* 1170: 413–23.

Drea, C M, Boulet, M, delBarco-Trillo, J *et al.* (2013) The "secret" in secretions: methodological considerations in deciphering primate olfactory communication. *Am J Primatol* 75: 621–42.

Dreanno, C, Kirby, R R & Clare, A S (2006a) Smelly feet are not always a bad thing: the relationship between cyprid footprint protein and the barnacle settlement pheromone. *Biol Lett* 2: 423–5.

Dreanno, C, Matsumura, K, Dohmae, N *et al.* (2006b) An α2-macroglobulin-like protein is the cue to gregarious settlement of the barnacle *Balanus amphitrite*. *Proc Natl Acad Sci USA* 103: 14396–401.

Drickamer, L C (1992) Estrous female house mice discriminate dominant from subordinate males and sons of dominant from sons of subordinate males by odor cues. *Anim Behav* 43: 868–70.

Drickamer, L C (1995) Rates of urine excretion by house mouse (*Mus domesticus*) – differences by age, sex, social-status, and reproductive condition. *J Chem Ecol* 21: 1481–93.

Droney, D C (2003) Females lay fewer eggs for males with greater courtship success in a lekking *Drosophila*. *Anim Behav* 65: 371–8.

Dronnet, S, Lohou, C, Christides, J P & Bagnères, A G (2006) Cuticular hydrocarbon composition reflects genetic relationship among colonies of the introduced termite *Reticulitermes santonensis* Feytaud. *J Chem Ecol* 32: 1027–42.

Drury, J P (2010) Immunity and mate choice: a new outlook. *Anim Behav* 79: 539–45.

Duarte, A, Weissing, F J, Pen, I & Keller, L (2011) An evolutionary perspective on self-organized division of labor in social insects. *Annu Rev Ecol Evol Syst* 42: 91–110.

Duffy, J E & Macdonald, K S (2010) Kin structure, ecology and the evolution of social organization in shrimp: a comparative analysis. *Proc R Soc B* 277: 575–84.

Duistermars, B J, Chow, D M & Frye, M A (2009) Flies require bilateral sensory input to track odor gradients in flight. *Curr Biol* 19: 1301–7.

Dukas, R (2008) Evolutionary biology of insect learning. *Annu Rev Entomol* 53: 145–60.

Dulac, C & Torello, A T (2003) Molecular detection of pheromone signals in mammals: from genes to behaviour. *Nat Rev Neurosci* 4: 551–62.

Durand, N, Carot-Sans, G, Chertemps, T *et al.* (2010) A diversity of putative carboxylesterases are expressed in the antennae of the noctuid moth *Spodoptera littoralis*. *Insect Mol Biol* 19: 87–97.

Dusenbery, D B (1989) Calculated effect of pulsed pheromone release on range of attraction. *J Chem Ecol* 15: 971–8.

Dusenbery, D B (1992) *Sensory Ecology. How Organisms Acquire and Respond to Information*. New York: WH Freeman.

Dusenbery, D B (2009) *Living at Micro Scale: The Unexpected Physics of Being Small*. Cambridge, MA: Harvard University Press.

Dusenbery, D B & Snell, T W (1995) A critical body size for use of pheromones in mate location. *J Chem Ecol* 21: 427–38.

Dussourd, D E, Harvis, C A, Meinwald, J & Eisner, T (1991) Pheromonal advertisement of a nuptial gift by a male moth (*Utetheisa ornatrix*). *Proc Natl Acad Sci USA* 88: 9224–7.

Dussutour, A & Beekman, M (2009) How to tell your mates. In Jarau, S & Hrncir, M (eds.) *Food Exploitation by Social Insects: Ecological, Behavioral and Theoretical Approaches*. pp. 115–34. Boca Raton, FL: CRC Press.

East, M L & Dehnhard, M (eds.) (2013) *Chemical Signals in Vertebrates 12*. New York: Springer.

Eberhard, W G (2009) Postcopulatory sexual selection: Darwin's omission and its consequences. *Proc Natl Acad Sci USA* 106: 10025–32.

Edison, A S (2009) *Caenorhabditis elegans* pheromones regulate multiple complex behaviors. *Curr Opin Neurobiol* 19: 378–88.

Edmunds, A J F, Aluja, M, Diaz-Fleischer, F, Patrian, B & Hagmann, L (2010) Host marking pheromone (HMP) in the Mexican fruit fly *Anastrepha ludens*. *Chimia* 64: 37–42.

Edward, D A & Chapman, T (2011) The evolution and significance of male mate choice. *Trends Ecol Evol* 26: 647–54.

Eggert, A K & Müller, J K (1997) Biparental care and social evolution in burying beetles: lessons from the larder. In Choe, J C & Crespi, B J (eds.) *The Evolution of Social Behavior in Insects and Arachnids*. pp. 216–36. Cambridge: Cambridge University Press.

Eggert, A K & Sakaluk, S K (1995) Female-coerced monogamy in burying beetles. *Behav Ecol Sociobiol* 37: 147–53.

Eisner, T (2003) *For Love of Insects*. Cambridge, MA: Belknap Press of Harvard University Press.

Eisner, T & Meinwald, J (1995) Defense-mechanisms of arthropods. 129. The chemistry of sexual selection. *Proc Natl Acad Sci USA* 92: 50–5.

Eisner, T & Meinwald, J (2003) Alkaloid-derived pheromones and sexual selection in Lepidoptera. In Blomquist, G J & Vogt, R G (eds.) *Insect Pheromone Biochemistry and Molecular Biology: the Biosynthesis and Detection of Insect Pheromones and Plant Volatiles*. pp. 341–68. New York: Academic Press.

Eisthen, H L (2002) Why are olfactory systems of different animals so similar? *Brain Behav Evol* 59: 273–93.

Eizaguirre, C & Lenz, T L (2010) Major histocompatability complex polymorphism: dynamics and consequences of parasite-mediated local adaptation in fishes. *J Fish Biol* 77: 2023–47.

Eizaguirre, C, Lenz, T L, Sommerfeld, R D *et al.* (2011) Parasite diversity, patterns of MHC II variation and olfactory based mate choice in diverging three-spined stickleback ecotypes. *Evol Ecol* 25: 605–22.

Ekerholm, M & Hallberg, E (2005) Primer and short-range releaser pheromone properties of premolt female urine from the shore crab *Carcinus maenas*. *J Chem Ecol* 31: 1845–64.

El-Sayed, A M (2013) The Pherobase: database of pheromones and semiochemicals. [Online]. Available: www.pherobase.com [Accessed 6 June 2013].

El-Sayed, A M, Suckling, D M, Wearing, C H & Byers, J A (2006) Potential of mass trapping for long-term pest management and eradication of invasive species. *J Econ Entomol* 99: 1550–64.

El-Sayed, A M, Suckling, D M, Byers, J A, Jang, E B & Wearing, C H (2009) Potential of "lure and kill" in long-term pest management and eradication of invasive species. *J Econ Entomol* 102: 815–35.

Elgar, M A & Allan, R A (2004) Predatory spider mimics acquire colony-specific cuticular hydrocarbons from their ant model prey. *Naturwissenschaften* 91: 143–7.

Eliyahu, D, Ross, K, Haight, K, Keller, L & Liebig, J (2011) Venom alkaloid and cuticular hydrocarbon profiles are associated with social organization, queen fertility status, and queen genotype in the fire ant *Solenopsis invicta*. *J Chem Ecol* 37: 1242–54.

Eliyahu, D, Nojima, S, Santangelo, R G *et al.* (2012) Unusual macrocyclic lactone sex pheromone of *Parcoblatta lata*, a primary food source of the endangered red-cockaded woodpecker. *Proc Natl Acad Sci USA* 109: E490–E496.

Elliott, J K, Mariscal, R N & Roux, K H (1994) Do anemonefishes use molecular mimicry to avoid being stung by host anemones. *J Exp Mar Biol Ecol* 179: 99–113.

Endler, A, Liebig, J, Schmitt, T *et al.* (2004) Surface hydrocarbons of queen eggs regulate worker reproduction in a social insect. *Proc Natl Acad Sci USA* 101: 2945–50.

Endler, J A & Basolo, A L (1998) Sensory ecology, receiver biases and sexual selection. *Trends Ecol Evol* 13: 415–20.

Epple, G, Belcher, A M, Kuderling, I *et al.* (1993) Making sense out of scents – species-differences in scent glands, scent marking behavior and scent mark composition in the Callitrichidae. In Rylands, A B (ed.) *Marmosets and Tamarins: Systematics, Behaviour and Ecology*. pp. 123–51. Oxford: Oxford University Press.

Espelie, K E, Gamboa, G J, Grudzien, T A & Bura, E A (1994) Cuticular hydrocarbons of the paper wasp, *Polistes fuscatus* – a search for recognition pheromones. *J Chem Ecol* 20: 1677–87.

Estrada, C, Yildizhan, S, Schulz, S & Gilbert, L E (2010) Sex-specific chemical cues from immatures facilitate the evolution of mate guarding in *Heliconius* butterflies. *Proc R Soc B* 277: 407–13.

Estrada, C, Schulz, S, Yildizhan, S & Gilbert, L E (2011) Sexual selection drives the evolution of antiaphrodisiac pheromones in butterflies. *Evolution* 65: 2843–54.

Evans, C S & Goy, R W (1968) Social behaviour and reproductive cycles in captive ring-tailed lemurs (*Lemur catta*). *J Zool* 156: 181–97.

Evans, I, Thornton, H, Chalmers, I & Glasziou, P (2011) *Testing Treatments: Better Research for Better Healthcare*, 2nd edn. London: Pinter and Martin.

Everaerts, C, Farine, J P, Cobb, M & Ferveur, J F (2010) *Drosophila* cuticular hydrocarbons revisited: mating status alters cuticular profiles. *PLoS ONE* 5: e9607.

Fabre, J H (1911) *Social Life in the Insect World*. Translated by B Miall. London: Fisher Unwin.

Fang, S, Ting, C T, Lee, C R *et al.* (2009) Molecular evolution and functional diversification of fatty acid desaturases after recurrent gene duplication in *Drosophila*. *Mol Biol Evol* 26: 1447–56.

Farbman, A I (1992) *Cell Biology of Olfaction*. Cambridge: Cambridge University Press.

Fatouros, N E, Dicke, M, Mumm, R, Meiners, T & Hilker, M (2008) Foraging behavior of egg parasitoids exploiting chemical information. *Behav Ecol* 19: 677–89.

Faulkes, C G & Abbott, D H (1993) Evidence that primer pheromones do not cause social suppression of reproduction in male and female naked mole-rats (*Heterocephalus glaber*). *J Reprod Fertil* 99: 225–30.

Faulkes, C G & Bennett, N C (2009) Reproductive skew in African mole rats: behavioural and physiological mechanisms to maintain high skew. In Hager, R & Jones, C B (eds.) *Reproductive Skew: Proximate and Ultimate Causes*. pp. 369–96. Cambridge: Cambridge University Press.

Fedina, T Y & Lewis, S M (2008) An integrative view of sexual selection in *Tribolium* flour beetles. *Biol Rev* 83: 151–71.

Feener, D H, Jacobs, L F & Schmidt, J O (1996) Specialized parasitoid attracted to a pheromone of ants. *Anim Behav* 51: 61–6.

Felix, M-A & Duveau, F (2012) Population dynamics and habitat sharing of natural populations of *Caenorhabditis elegans* and *C. briggsae*. *BMC Biol* 10: 59.

Ferdenzi, C, Coureaud, G, Camos, V & Schaal, B (2008) Human awareness and uses of odor cues in everyday life: results from a questionnaire study in children. *Int J Behav Dev* 32: 422–6.

Ferkin, M H (2011) Odor-related behavior and cognition in meadow voles, *Microtus pennsylvanicus* (Arvicolidae, Rodentia). *Folia Zool* 60: 262–76.

Ferkin, M H & Pierce, A A (2007) Perspectives on overmarking: is it good to be on top? *J Ethol* 25: 107–16.

Ferkin, M H, Sorokin, E S, Renfroe, M W & Johnston, R E (1994) Attractiveness of male odors to females varies directly with plasma testosterone concentration in meadow voles. *Physiol Behav* 55: 347–53.

Ferkin, M H, Sorokin, E S, Johnston, R E & Lee, C J (1997) Attractiveness of scents varies with protein content of the diet in meadow voles. *Anim Behav* 53: 133–41.

Ferner, M C & Weissburg, M J (2005) Slow-moving predatory gastropods track prey odors in fast and turbulent flow. *J Exp Biol* 208: 809–19.

Ferrari, M C O, Wisenden, B D & Chivers, D P (2010) Chemical ecology of predator–prey interactions in aquatic ecosystems: a review and prospectus. *Can J Zool* 88: 698–724.

Ferrero, D M, Lemon, J K, Fluegge, D *et al.* (2011) Detection and avoidance of a carnivore odor by prey. *Proc Natl Acad Sci USA* 108: 11235–40.

Ferstl, R, Eggert, F, Westphal, E, Zavazava, N & MullerRuchholtz, W (1992) MHC-related odors in humans. In Doty, R L & Müller-Schwarze, D (eds.) *Chemical Signals in Vertebrates VI*. pp. 206–11. New York: Plenum Press.

Ferveur, J-F (2005) Cuticular hydrocarbons: their evolution and roles in *Drosophila* pheromonal communication. *Behav Genet* 35: 279–95.

Ferveur, J-F (2007) Elements of courtship behavior in *Drosophila*. In North, G & Greenspan, R J (eds.) *Invertebrate Neurobiology*. pp. 405–36. Cold Spring Harbor, NY Cold Spring Harbor Laboratory Press.

Ferveur, J-F (2010) *Drosophila* female courtship and mating behaviors: sensory signals, genes, neural structures and evolution. *Curr Opin Neurobiol* 20: 764–9.

Ferveur, J-F & Cobb, M (2010) Behavioral and evolutionary roles of cuticular hydrocarbons in *Diptera*. In Blomquist, G J & Bagnères, A-G (eds.) *Insect Hydrocarbons: Biology, Biochemistry, and Chemical Ecology*. pp. 325–43. Cambridge: Cambridge University Press.

Ferveur, J-F, Cobb, M, Boukella, H & Jallon, J M (1996) World-wide variation in *Drosophila melanogaster* sex-pheromone behavioral effects, genetic bases and potential evolutionary consequences. *Genetica* 97: 73–80.

Fiedler, K (2012) The host genera of ant-parasitic Lycaenidae butterflies: a review. *Psyche* 2012: doi:10.1155/2012/153975.

Fine, J M & Sorensen, P W (2008) Isolation and biological activity of the multi-component sea lamprey migratory pheromone. *J Chem Ecol* 34: 1259–67.

Fine, J M, Vrieze, L A & Sorensen, P W (2004) Evidence that petromyzontid lampreys employ a common migratory pheromone that is partially comprised of bile acids. *J Chem Ecol* 30: 2091–110.

Fischer-Tenhagen, C, Wetterholm, L, Tenhagen, B A & Heuwieser, W (2011) Training dogs on a scent platform for oestrus detection in cows. *Appl Anim Behav Sci* 131: 63–70.

Fischman, B J, Woodard, S H & Robinson, G E (2011) Molecular evolutionary analyses of insect societies. *Proc Natl Acad Sci USA* 108: 10847–54.

Fisher, H S, Swaisgood, R & Fitch-Snyder, H (2003) Countermarking by male pygmy lorises (*Nycticebus pygmaeus*): do females use odor cues to select mates with high competitive ability? *Behav Ecol Sociobiol* 53: 123–30.

Fitzgerald, T D (1995) *The Tent Caterpillars*. Ithaca, NY: Cornell University Press.

Fitzgerald, T D & Gallagher, E M (1976) A chemical trail factor from the silk of the eastern tent caterpillar *Malacosoma americanum* (Lepidoptera: Lasiocamidae). *J Chem Ecol* 2: 564–74.

Flanagan, K A, Webb, W & Stowers, L (2011) Analysis of male pheromones that accelerate female reproductive organ development. *PLoS ONE* 6: e16660.

Flecke, C & Stengl, M (2009) Octopamine and tyramine modulate pheromone-sensitive olfactory sensilla of the hawkmoth *Manduca sexta* in a time-dependent manner. *J Comp Physiol A* 195: 529–45.

Fleischer, J & Breer, H (2010) The Grueneberg ganglion: a novel sensory system in the nose. *Histol Histopathol* 25: 909–15.

Fleischer, J, Breer, H & Strotmann, J (2009) Mammalian olfactory receptors. *Front Cell Neurosci* 3: 9.

Fleming, A S, Steiner, M & Corter, C (1997) Cortisol, hedonics, and maternal responsiveness in human mothers. *Horm Behav* 32: 85–98.

Folstad, I & Karter, A J (1992) Parasites, bright males, and the immunocompetence handicap. *Am Nat* 139: 602–22.

Font, E, Barbosa, D, Sampedro, C & Carazo, P (2012) Social behavior, chemical communication, and adult neurogenesis: studies of scent mark function in *Podarcis* wall lizards. *Gen Comp Endocrinol* 177: 9–17.

Ford, N B (1986) The role of pheromone trails in the sociobiology of snakes. In Duvall, D (ed.) *Chemical Signals in Vertebrates 4*. pp. 261–78. New York: Plenum Publishing.

Ford, N B & Low, J R (1984) Sex pheromone source location by garter snakes: a mechanism for detection of direction in non-volatile trails. *J Chem Ecol* 10: 1193–9.

Forseth, R R & Schroeder, F C (2011) NMR-spectroscopic analysis of mixtures: from structure to function. *Curr Opin Chem Biol* 15: 38–47.

Forward, R B (2009) Larval biology of the crab *Rhithropanopeus harrisii* (Gould): a synthesis. *Biol Bull* 216: 243.

Foster, K R (2010) Social behavior in microorganisms. In Székely, T, Moore, A J & Komdeur, J (eds.) *Social Behaviour: Genes, Ecology and Evolution*. pp. 331–56. Cambridge: Cambridge University Press.

Foster, S P & Johnson, C P (2011) Signal honesty through differential quantity in the female-produced sex pheromone of the moth *Heliothis virescens*. *J Chem Ecol* 37: 717–23.

Fourcassié, V, Dussutour, A & Deneubourg, J-L (2010) Ant traffic rules. *J Exp Biol* 213: 2357–63.

Fraenkel, G S & Gunn, D L (1940) *The Orientation of Animals. Kineses, Taxes and Compass Reactions*. Oxford: Clarendon Press.

Francke, D L, Harmon, J P, Harvey, C T & Ives, A R (2008) Pea aphid dropping behavior diminishes foraging efficiency of a predatory ladybeetle. *Entomol Exp Appl* **127**: 118–24.

Francke, W & Schulz, S (2010) Pheromones in terrestrial invertebrates. In Mander, L & Lui, H-W (eds.) *Comprehensive Natural Products II Chemistry and Biology. Vol 4.* pp. 153–223. Oxford: Elsevier.

Frank, D, Beauchamp, G & Palestrini, C (2010) Systematic review of the use of pheromones for treatment of undesirable behavior in cats and dogs. *J Am Vet Med Assoc* **236**: 1308–16.

Franks, N R, Gomez, N, Goss, S & Deneubourg, J L (1991) The blind leading the blind in army ant raid patterns – testing a model of self-organization (Hymenoptera, Formicidae). *J Insect Behav* **4**: 583–607.

Free, J B (1987) *Pheromones of Social Bees.* London: Chapman and Hall.

French, J A (1997) Proximate regulation of singular breeding in callitrichid primates. In Solomon, N G & French, J A (eds.) *Cooperative Breeding in Mammals.* pp. 34–75. Cambridge: Cambridge University Press.

Friedrich, R W (2011) Olfactory neuroscience: beyond the bulb. *Curr Biol* **21**: R438 – R440.

Frostig, R D (ed.) (2009) *In Vivo Optical Imaging of Brain Function*, 2nd edn. Boca Raton, FL: CRC Press.

Frye, M A (2010) Multisensory systems integration for high-performance motor control in flies. *Curr Opin Neurobiol* **20**: 347–52.

Fujii, T, Ito, K, Tatematsu, M *et al.* (2011) Sex pheromone desaturase functioning in a primitive *Ostrinia* moth is cryptically conserved in congeners' genomes. *Proc Natl Acad Sci USA* **108**: 7102–6.

Funasaka, N, Yoshioka, M & Fujise, Y (2010) Features of the ocular Harderian gland in three balaenopterid species based on anatomical, histological and histochemical observations. *Mammal Study* **35**: 9–15.

Futuyama, D J (2009) *Evolution*, 2nd edn. Sunderland, MA: Sinauer.

Gabirot, M, López, P & Martín, J (2011) Interpopulational variation in chemosensory responses to selected steroids from femoral secretions of male lizards, *Podarcis hispanica*, mirrors population differences in chemical signals. *Chemoecology* **22**: 65–73.

Gadagkar, R (2009) Interrogating an insect society. *Proc Natl Acad Sci USA* **106**: 10407–14.

Gadau, J & Fewell, J H (eds.) (2009) *Organization of Insect Societies: from Genome to Sociocomplexity.* Cambridge, MA: Harvard University Press.

Gadau, J, Helmkampf, M, Nygaard, S *et al.* (2012) The genomic impact of 100 million years of social evolution in seven ant species. *Trends Genet* **28**: 14–21.

Gaillard, I, Rouquier, S, Pin, J P *et al.* (2002) A single olfactory receptor specifically binds a set of odorant molecules. *Eur J Neurosci* **15**: 409–18.

Galef, B G & Buckley, L L (1996) Use of foraging trails by Norway rats. *Anim Behav* **51**: 765–71.

Galizia, C G & Rössler, W (2010) Parallel olfactory systems in insects: anatomy and function. *Annu Rev Entomol* **55**: 399–420.

Galizia, C G & Vetter, R S (2004) Optical methods for analyzing odor-evoked activity in the insect brain. In Christensen, T A (ed.) *Methods in Insect Sensory Neuroscience.* pp. 345–88. Boca Raton, FL: CRC Press.

Gamboa, G J (2004) Kin recognition in eusocial wasps. *Ann Zool Fenn* **41**: 789–808.

Gamboa, G J, Grudzien, T A, Espelie, K E & Bura, E A (1996) Kin recognition pheromones in social wasps: combining chemical and behavioural evidence. *Anim Behav* **51**: 625–9.

Gangestad, S W & Thornhill, R (1998) Menstrual cycle variation in women's preferences for the scent of symmetrical men. *Proc R Soc B* **265**: 927–33.

Gao, Q & Chess, A (1999) Identification of candidate *Drosophila* olfactory receptors from genomic DNA sequence. *Genomics* **60**: 31–9.

Gardiner, J M & Atema, J (2007) Sharks need the lateral line to locate odor sources: rheotaxis and eddy chemotaxis. *J Exp Biol* **210**: 1925–34.

Gardiner, J M & Atema, J (2010) The function of bilateral odor arrival time differences in olfactory orientation of sharks. *Curr Biol* **20**: 1187–91.

Gardner, A & West, S A (2007) Social evolution: the decline and fall of genetic kin recognition. *Curr Biol* **17**: R810–R812.

Gardner, A & West, S A (2010) Greenbeards. *Evolution* **64**: 25–38.

Garner, S R, Bortoluzzi, R N, Heath, D D & Neff, B D (2010) Sexual conflict inhibits female mate choice for major histocompatibility complex dissimilarity in Chinook salmon. *Proc R Soc B* **277**: 885–94.

Gascoigne, J, Berec, L, Gregory, S & Courchamp, F (2009) Dangerously few liaisons: a review of mate-finding Allee effects. *Popul Ecol* 51: 355–72.

Gaskett, A C (2007) Spider sex pheromones: emission, reception, structures, and functions. *Biol Rev* 82: 26–48.

Gaskett, A C (2011) Orchid pollination by sexual deception: pollinator perspectives. *Biol Rev* 86: 33–75.

Gaskett, A C, Winnick, C G & Herberstein, M E (2008) Orchid sexual deceit provokes ejaculation. *Am Nat* 171: E206–E212.

Gasparini, C & Pilastro, A (2011) Cryptic female preference for genetically unrelated males is mediated by ovarian fluid in the guppy. *Proc R Soc B* 278: 2495–501.

Gaudry, Q, Nagel, K I & Wilson, R I (2012) Smelling on the fly: sensory cues and strategies for olfactory navigation in *Drosophila*. *Curr Opin Neurobiol* 22: 216–22.

Gautschi, M, Natsch, A & Schröder, F (2007) Biochemistry of human axilla malodor and chemistry of deodorant ingredients. *CHIMIA* 61: 27–32.

Gelstein, S, Yeshurun, Y, Rozenkrantz, L *et al.* (2011) Human tears contain a chemosignal. *Science* 331: 226–30.

Gemeno, C & Schal, C (2004) Sex pheromones of cockroaches. In Cardé, R & Millar, J G (eds.) *Advances in Insect Chemical Ecology.* pp. 179–247. Cambridge: Cambridge University Press.

Gemeno, C, Yeargan, K V & Haynes, K F (2000) Aggressive chemical mimicry by the bolas spider *Mastophora hutchinsoni*: identification and quantification of a major prey's sex pheromone components in the spider's volatile emissions. *J Chem Ecol* 26: 1235–43.

Gemeno, C, Snook, K, Benda, N & Schal, C (2003) Behavioral and electrophysiological evidence for volatile sex pheromones in *Parcoblatta* wood cockroaches. *J Chem Ecol* 29: 37–54.

Getty, T (2006) Sexually selected signals are not similar to sports handicaps. *Trends Ecol Evol* 21: 83–8.

Getz, W (1991) The honey bee as a model kin recognition system. In Hepper, P G, (ed.) *Kin Recognition.* pp. 358–412. Cambridge: Cambridge University Press.

Ghaleb, A, Atwood III, J, Morales-Montor, J & Damian, R (2006) A 3 kDa peptide is involved in the chemoattraction in vitro of the male *Schistosoma mansoni* to the female. *Microbes Infect* 8: 2367–75.

Gibson, R W & Pickett, J A (1983) Wild potato repels aphids by release of aphid alarm pheromone. *Nature* 302: 608–9.

Gilad, Y, Wiebel, V, Przeworski, M, Lancet, D & Paabo, S (2004) Loss of olfactory receptor genes coincides with the acquisition of full trichromatic vision in primates. *PLoS Biol* 2: 120–5.

Gilad, Y, Wiebe, V, Przeworski, M, Lancet, D & Paabo, S (2007) Correction: Loss of olfactory receptor genes coincides with the acquisition of full trichromatic vision in primates (Vol 2, pg 120, 2004). *PLoS Biol* 5: 1383.

Gilbert, A N (2008) *What the Nose Knows: the Science of Scent in Everyday Life.* New York: Crown.

Gilbert, A N & Firestein, S (2002) Dollars and scents: commercial opportunities in olfaction and taste. *Nat Neurosci* 5: 1043–5.

Gilbert, A N, Yamazaki, K, Beauchamp, GK & Thomas, L (1986) Olfactory discrimination of mouse strains (*Mus musculus*) and major histocompatibility types by humans (*Homo sapiens*). *J Comp Psychol* 100: 262–5.

Gilley, D C, Kuzora, J M & Thom, C (2012) Hydrocarbons emitted by waggle-dancing honey bees stimulate colony foraging activity by causing experienced foragers to exploit known food sources. *Apidologie* 43: 85–94.

Gillingham, M A F, Richardson, D S, Lovlie, H *et al.* (2009) Cryptic preference for MHC-dissimilar females in male red junglefowl, *Gallus gallus*. *Proc R Soc B* 276: 1083–92.

Gillott, C (2003) Male accessory gland secretions: modulators of female reproductive physiology and behavior. *Annu Rev Entomol* 48: 163–84.

Glastad, K M, Hunt, B G & Goodisman, M A D (2013) Evidence of a conserved functional role for DNA methylation in termites. *Insect Mol Biol* 22: 143–54.

Gleason, J M, James, R A, Wicker-Thomas, C & Ritchie, M G (2009) Identification of quantitative trait loci function through analysis of multiple cuticular hydrocarbons differing between *Drosophila simulans* and *Drosophila sechellia* females. *Heredity* 103: 416–24.

Godfray, H C J (1994) *Parasitoids: Behavioral and Evolutionary Ecology.* Princeton: Princeton University Press.

Goldfoot, D A (1981) Olfaction, sexual-behavior, and the pheromone hypothesis in rhesus monkeys – a critique. *Am Zool* 21: 153–64.

Goldfoot, D A, Kravetz, M A, Goy, R W & Freeman, S K (1976) Lack of effect of vaginal lavages and aliphatic acids on ejaculatary responses in rhesus monkeys. *Horm Behav* 7: 1–27.

Gomez-Diaz, C, Reina, J H, Cambillau, C & Benton, R (2013) Ligands for pheromone-sensing neurons are not conformationally activated odorant binding proteins. *PLoS Biol* 11: e1001546.

Gomez-Marin, A & Louis, M (2012) Active sensation during orientation behavior in the *Drosophila* larva: more sense than luck. *Curr Opin Neurobiol* 22: 208–15.

Gomez-Marin, A, Stephens, G J & Louis, M (2011) Active sampling and decision making in *Drosophila* chemotaxis. *Nat Commun* 2: 441.

Gorman, M L (1976) A mechanism for individual recognition by odour in *Herpestes auropunctatus*. *Anim Behav* 24: 141–6.

Gorman, M L & Mills, M G L (1984) Scent marking strategies in hyaenas (Mammalia). *J Zool* 202: 535–47.

Gorman, M L & Stone, R D (1990) Mutual avoidance by European moles *Talpa europaea*. In Macdonald, D W, Müller-Schwarze, D & Natynczuk, S E (eds.) *Chemical Signals in Vertebrates 5.* pp. 367–77. Oxford: Oxford Science Publications.

Gosling, L M (1982) A reassessment of the function of scent marking in territories. *Z Tierpsychol* 60: 89–118.

Gosling, L M (1990) Scent marking by resource holders: alternative mechanisms for advertising the cost of competition. In Macdonald, D W, Müller-Schwarze, D & Natynczuk, S E (eds.) *Chemical Signals in Vertebrates 5.* pp. 315–28. Oxford: Oxford Science Publications.

Gosling, L M & McKay, H V (1990) Competitor assessment by scent matching – an experimental test. *Behav Ecol Sociobiol* 26: 415–20.

Gosling, L M & Roberts, S C (2001) Scent-marking by male mammals: cheat-proof signals to competitors and mates. *Adv Study Behav* 30: 169–217.

Gosling, L M, Atkinson, N W, Dunn, S & Collins, S A (1996) The response of subordinate male mice to scent marks varies in relation to their own competitive ability. *Anim Behav* 52: 1185–91.

Gosling, L M, Roberts, S C, Thornton, E A & Andrew, M J (2000) Life history costs of olfactory status signalling in mice. *Behav Ecol Sociobiol* 48: 328–32.

Gotzek, D & Ross, K G (2007) Genetic regulation of colony social organization in fire ants: an integrative overview. *Q Rev Biol* 82: 201–26.

Gotzek, D & Ross, K G (2009) Current status of a model system: the gene *Gp-9* and its association with social organization in fire ants. *PLoS ONE* 4: e7713.

Gould, F, Estock, M, Hillier, N K *et al.* (2010) Sexual isolation of male moths explained by a single pheromone response QTL containing four receptor genes. *Proc Natl Acad Sci USA* 107: 8660–5.

Goulson, D (2009) The use of scent marks by foraging bumble bees. In Jarau, S & Hrncir, M (eds.) *Food Exploitation by Social Insects: Ecological, Behavioral and Theoretical Approaches.* Boca Raton, FL: CRC Press.

Gowaty, P A, Drickamer, L C & Schmid-Holmes, S (2003) Male house mice produce fewer offspring with lower viability and poorer performance when mated with females they do not prefer. *Anim Behav* 65: 95–103.

Gower, D B, Bird, S, Sharma, P & House, F R (1985) Axillary 5-alpha-androst-16-en-3-one in men and women – relationships with olfactory acuity to odorous 16-androstenes. *Experientia* 41: 1134–6.

Gower, D B, Holland, K T, Mallet, A I, Rennie, P J & Watkins, W J (1994) Comparison of 16-androstene steroid concentrations in sterile apocrine sweat and axillary secretions – interconversions of 16-androstenes by the axillary microflora – a mechanism for axillary odor production in man. *J Steroid Biochem Mol Biol* 48: 409–18.

Gower, D B, Mallet, A I, Watkins, W J, Wallace, L M & Calame, J P (1997) Capillary gas chromatography with chemical ionization negative ion mass spectrometry in the identification of odorous steroids formed in metabolic studies of the sulphates of androsterone, DHA and 5 alpha-androst-16-en-3 beta-ol with human axillary bacterial isolates. *J Steroid Biochem Mol Biol* 63: 81–9.

Grafen, A (1990a) Biological signals as handicaps. *J Theor Biol* 144: 517–46.

Grafen, A (1990b) Do animals really recognize kin? *Anim Behav* 39: 42–54.

Granero, A M, Sanz, J M G, Gonzalez, F J E *et al.* (2005) Chemical compounds of the foraging recruitment pheromone in bumblebees. *Naturwissenschaften* 92: 371–4.

Grassé, P-P (1959) La reconstruction du nid et les coordinations inter-individuelles chez *Bellicoitermes natalenis* et *Cubitermes* sp. La théorie de la stigmergie: essai d'interprétation des termites constructeurs. *Insectes Soc* 6: 41–83.

Grassé, P-P (1984) *Termitologia, fondation des sociétés. Construction. Tome II.* Paris: Masson.

Grasso, D A, Sledge, M F, Le Moli, F, Mori, A & Turillazzi, S (2005) Nest-area marking with faeces: a chemical signature that allows colony-level recognition in seed harvesting ants (Hymenoptera, Formicidae). *Insectes Soc* 52: 36–44.

Gray, S & Hurst, J L (1995) The effects of cage cleaning on aggression within groups of male laboratory mice. *Anim Behav* 49: 821–6.

Greenberg, L (1979) Genetic component of bee odor in kin recognition. *Science* 206: 1095–7.

Greene, M J (2010) Cuticular hydrocarbon cues in the formation and maintenance of insect social groups. In Blomquist, G J & Bagnères, A-G (eds.) *Insect Hydrocarbons: Biology, Biochemistry, and Chemical Ecology.* pp. 244–53. Cambridge: Cambridge University Press.

Greene, M J & Gordon, D M (2003) Cuticular hydrocarbons inform task decisions. *Nature* 423: 32.

Greenfield, M D (2002) *Signalers and Receivers: Mechanisms and Evolution of Arthropod Communication.* Oxford: Oxford University Press.

Greenfield, M D (2006) Honesty and deception in animal signals. In Lucas, J R & Simmons, L W (eds.) *Essays in Animal Behaviour: Celebrating 50 Years of Animal Behaviour.* pp. 281–98. Burlington, MA: Academic Press.

Grether, G F (2010) The evolution of mate preferences, sensory biases, and indicator traits. *Adv Study Behav* 41: 35–76.

Grice, E A & Segre, J A (2011) The skin microbiome. *Nat Rev Microbiol* 9: 244–53.

Grice, E A, Kong, H H, Conlan, S *et al.* (2009) Topographical and temporal diversity of the human skin microbiome. *Science* 324: 1190–2.

Gries, R, Khaskin, G, Gries, G *et al.* (2002) (*Z, Z*)-4,7-Tridecadien-(S)-2-yl acetate: sex pheromone of douglas-fir cone gall midge, *Contarinia oregonensis*. *J Chem Ecol* 28: 2283–97.

Griffith, L C & Ejima, A (2009) Courtship learning in *Drosophila melanogaster*: diverse plasticity of a reproductive behavior. *Learn Mem* 16: 743–50.

Griggio, M, Biard, C, Penn, D & Hoi, H (2011) Female house sparrows "count on" male genes: experimental evidence for MHC-dependent mate preference in birds. *BMC Evol Biol* 11: 44.

Grillet, M, Everaerts, C, Houot, B *et al.* (2012) Incipient speciation in *Drosophila melanogaster* involves chemical signals. *Scientific Reports* 2: 224.

Gronenberg, W & Riveros, A J (2009) Social brains and behavior, past and present. In Gadau, J & Fewell, J H (eds.) *Organization of Insect Societies: from Genome to Sociocomplexity.* pp. 377–401. Cambridge, MA: Harvard University Press.

Groot, A T, Horovitz, J L, Hamilton, J *et al.* (2006) Experimental evidence for interspecific directional selection on moth pheromone communication. *Proc Natl Acad Sci USA* 103: 5858–63.

Groot, A T, Estock, M L, Horovitz, J L *et al.* (2009) QTL analysis of sex pheromone blend differences between two closely related moths: insights into divergence in biosynthetic pathways. *Insect Biochem Mol Biol* 39: 568–77.

Grosjean, Y, Rytz, R, Farine, J-P *et al.* (2011) An olfactory receptor for food-derived odours promotes male courtship in *Drosophila*. *Nature* 478: 236–40.

Grosmaitre, X, Santarelli, L C, Tan, J, Luo, M & Ma, M (2007) Dual functions of mammalian olfactory sensory neurons as odor detectors and mechanical sensors. *Nat Neurosci* 10: 348–54.

Grozinger, C M (2013) Honey bee pheromones In Graham, J (ed.) *The Hive and the Honey Bee.* Hamilton, IL: Dadant & Sons Inc.

Grozinger, C M & Robinson, G E (2007) Pheromone-mediated gene expression in the honey bee brain. *J Comp Physiol A* 193: 461–70.

Grozinger, C M, Sharabash, N M, Whitfield, C W & Robinson, G E (2003) Pheromone-mediated gene expression in the honey bee brain. *Proc Natl Acad Sci USA* 100: 14519–25.

Grozinger, C M, Fischer, P & Hampton, J E (2007a) Uncoupling primer and releaser responses to pheromone in honey bees. *Naturwissenschaften* 94: 375–9.

Grozinger, C M, Fan, Y, Hoover, S E R & Winston, M L (2007b) Genome wide analysis reveals differences in brain gene expression patterns associated with caste and reproductive status in honey bees (*Apis mellifera*). *Mol Ecol* 16: 4837–48.

Grüter, C, Menezes, C, Imperatriz-Fonseca, V L & Ratnieks, F L W (2012) A morphologically specialized soldier caste improves colony defense in a neotropical eusocial bee. *Proc Natl Acad Sci USA* 109: 1182–6.

Guerra Sanz, J M & Roldán Serrano, A (2008) Influence of honey bees brood pheromone on the production of triploid watermelon. In Pitrat, M, (ed.) *Cucurbitaceae 2008, Proc IXth EUCARPIA genetics & breeding Cucurbitaceae.* pp. 385–9. Avignon (France): INRA.

Guerrieri, F J & d'Ettorre, P (2008) The mandible opening response: quantifying aggression elicited by chemical cues in ants. *J Exp Biol* 211: 1109–13.

Guilford, T & Dawkins, M S (1991) Receiver psychology and the evolution of animal signals. *Anim Behav* 42: 1–14.

Guilford, T & Dawkins, M S (1993) Receiver psychology and the design of animal signals. *Trends Neurosci* 16: 430–6.

Guilford, T, Nicol, C, Rothschild, M & Moore, B P (1987) The biological roles of pyrazines – evidence for a warning odor function. *Biol J Linn Soc* 31: 113–28.

Guo, C C, Hwang, J S & Fautin, D G (1996) Host selection by shrimps symbiotic with sea anemones: a field survey and experimental laboratory analysis. *J Exp Mar Biol Ecol* 202: 165–76.

Gutiérrez-Castellanos, N, Martínez-Marcos, A, Martínez-García, F & Lanuza, E (2010) Chemosensory function of the amygdala. In Gerald, L (ed.) *Pheromones.* pp. 165–96. London: Academic Press.

Haberer, W, Schmitt, T, Peschke, K, Schreier, P & Müller, J (2008) Ethyl 4-methyl heptanoate: a male-produced pheromone of *Nicrophorus vespilloides. J Chem Ecol* 34: 94–8.

Hafernik, J & Saul-Gershenz, L (2000) Beetle larvae cooperate to mimic bees. *Nature* 405: 35–6.

Haga, S, Hattori, T, Sato, T *et al.* (2010) The male mouse pheromone ESP1 enhances female sexual receptive behaviour through a specific vomeronasal receptor. *Nature* 466: 118–22.

Hagelin, J C (2007) The citrus-like scent of crested auklets: reviewing the evidence for an avian olfactory ornament. *J Ornithol* 148: S195–S201.

Hagelin, J C & Jones, I L (2007) Bird odors and other chemical substances: a defense mechanism or overlooked mode of intraspecific communication? *Auk* 124: 741–61.

Hager, R & Jones, C B (eds.) (2009) *Reproductive Skew in Vertebrates: Proximate and Ultimate Causes.* Cambridge: Cambridge University Press.

Hagman, M & Shine, R (2009) Larval alarm pheromones as a potential control for invasive cane toads (*Bufo marinus*) in tropical Australia. *Chemoecology* 19: 211–17.

Hallem, E A & Carlson, J R (2006) Coding of odors by a receptor repertoire. *Cell* 125: 143–60.

Halpern, M & Martinez-Marcos, A (2003) Structure and function of the vomeronasal system: an update. *Prog Neurobiol* 70: 245–318.

Halpin, Z T (1986) Individual odors among mammals – origins and functions. *Adv Study Behav* 16: 39–70.

Hamdani, E H & Døving, K B (2007) The functional organization of the fish olfactory system. *Prog Neurobiol* 82: 80–6.

Hamilton, W D (1964) The genetical evolution of social behaviour. I and II. *J Theor Biol* 7: 1–32.

Hamilton, W D (1971) Geometry for the selfish herd. *J Theor Biol* 31: 295–311.

Hamilton, W D (1987) Kinship, recognition, disease, and intelligence: constraints of social evolution. In Itô, Y (ed.) *Animal Societies: Theories and Facts.* pp. 88–102. Tokyo: Japan Science Society Press.

Hangartner, W (1967) Spezifität und inaktivierung des spurpheromons von *Lasius fuliginosus* Latr. und orientierung der arbeiterinnen in duftfeld. *ZeitverglPhysiol* 57: 103–36.

Hanin, O, Azrielli, A, Applebaum, S W & Rafaeli, A (2012) Functional impact of silencing the *Helicoverpa armigera* sex-peptide receptor on female reproductive behaviour. *Insect Mol Biol* 21: 161–7.

Hansson, B S & Stensmyr, M C (2011) Evolution of insect olfaction. *Neuron* 72: 698–711.

Hanus, R, Vrkoslav, V, Hrdý, I, Cvacka, J & Sobotník, J (2010) Beyond cuticular hydrocarbons: evidence of proteinaceous secretion specific to termite kings and queens. *Proc R Soc B* 277: 995–1002.

Hardege, J D (1999) Nereidid polychaetes as model organisms for marine chemical ecology. *Hydrobiologia* 402: 145–61.

Hardege, J D & Terschak, J A (2011) Identification of crustacean sex pheromones. In Breithaupt, T & Thiel, M (eds.) *Chemical Communication in Crustaceans.* pp. 373–92. New York: Springer.

Hardege, J D, Bartels-Hardege, H, Muller, C T & Beckmann, M (2004) Peptide pheromones in female *Nereis succinea. Peptides* 25: 1517–22.

Hardege, J D, Rotchell, J M, Terschak, J & Greenway, G M (2011a) Analytical challenges and the development of biomarkers to measure and to monitor the effects of ocean acidification. *Trends Analyt Chem* 30: 1320–6.

Hardege, J D, Bartels-Hardege, H, Fletcher, N *et al.* (2011b) Identification of a female sex pheromone in *Carcinus maenas*. *Mar Ecol Prog Ser* 436: 177–89.

Hardie, J & Minks, A K (eds.) (1999) *Pheromones of Non-lepidopteran Insects associated with Agricultural Plants*. Wallingford, Oxon: CAB International.

Hardy, S, Legagneux, V, Audic, Y & Paillard, L (2010) Reverse genetics in eukaryotes. *Biol Cell* 102: 561–80.

Harris, M O & Foster, S P (1995) Behavior and integration. In Cardé, R T & Bell, W J (eds.) *Chemical Ecology of Insects 2*. pp. 3–46. London: Chapman and Hall.

Hart, A C & Chao, M Y (2010) From odors to behaviors in Caenorhabditis elegans. In Menini, A (ed.) *The Neurobiology of Olfaction*. Boca Raton, FL: CRC Press. Available online at www.ncbi.nlm.nih.gov/books/NBK55983.

Harter, J (1979) *Animals. 1419 Copyright-free Illustrations of Mammals etc.* New York: Dover.

Häsemeyer, M, Yapici, N, Heberlein, U & Dickson, B J (2009) Sensory neurons in the *Drosophila* genital tract regulate female reproductive behavior. *Neuron* 61: 511–18.

Hasin-Brumshtein, Y, Lancet, D & Olender, T (2009) Human olfaction: from genomic variation to phenotypic diversity. *Trends Genet* 25: 178–84.

Hassanali, A, Njagi, P G N & Bashir, M O (2005) Chemical ecology of locusts and related acridids. *Annu Rev Entomol* 50: 223–45.

Hassanali, A, Nyandat, E, Obenchain, F A, Otieno, D A & Galun, R (1989) Humidity effects on response of *Argas persicus* (Oken) to guanine, an assembly pheromone of ticks. *J Chem Ecol* 15: 791–3.

Haupt, S S, Sakurai, T, Namiki, S, Kazawa, T & Kanzaki, R (2010) Olfactory information processing in moths. In Menini, A (ed.) *The Neurobiology of Olfaction*. Boca Raton, FL: CRC Press. Available online at www.ncbi.nlm.nih.gov/books/NBK55976/.

Havlíček, J & Roberts, S C (2009) MHC-correlated mate choice in humans: a review. *Psychoneuroendocrinology* 34: 497–512.

Havlíček, J, Dvoráková, R, Bartoš, L & Flegr, J (2006) Non advertized does not mean concealed: body odour changes across the human menstrual cycle. *Ethology* 112: 81–90.

Havlíček, J, Murray, A K, Saxton, T K & Roberts, S C (2010) Current issues in the study of androstenes in human chemosignaling. In Gerald, L (ed.) *Pheromones*. pp. 47–81. London: Academic Press.

Hawken, P & Martin, G (2012) Sociosexual stimuli and gonadotropin-releasing hormone/luteinizing hormone secretion in sheep and goats. *Domest Anim Endocrinol* 43: 85–94.

Hawkes, C H & Doty, R L (2009) *The Neurology of Olfaction*. Cambridge: Cambridge University Press.

Hawkins, R D, Hon, G C & Ren, B (2010) Next-generation genomics: an integrative approach. *Nat Rev Genet* 11: 476–86.

Hay, M E (2009) Marine chemical ecology: chemical signals and cues structure marine populations, communities, and ecosystems. *Ann Rev Mar Sci* 1: 193–212.

Hayden, S, Bekaert, M, Crider, T A *et al.* (2010) Ecological adaptation determines functional mammalian olfactory subgenomes. *Genome Res* 20: 1–9.

Hayes, R A, Richardson, B J & Wyllie, S G (2003) To fix or not to fix: the role of 2-phenoxyethanol in rabbit, *Oryctolagus cuniculus*, chin gland secretion. *J Chem Ecol* 29: 1051–64.

Haynes, K F & Millar, J G (eds.) (1998) *Methods in Chemical Ecology. Volume 2. Bioassay Methods*. London: Chapman & Hall.

Haynes, K F, Gemeno, C, Yeargan, K V, Millar, J G & Johnson, K M (2002) Aggressive mimicry of moth pheromones by a bolas spider: how does this specialist predator attract more than one species of prey? *Chemoecology* 12: 99–105.

He, J, Ma, L M, Kim, S, Nakai, J & Yu, C R (2008) Encoding gender and individual information in the mouse vomeronasal organ. *Science* 320: 535–8.

Hebets, E A & Papaj, D R (2005) Complex signal function: developing a framework of testable hypotheses. *Behav Ecol Sociobiol* 57: 197–214.

Hedin, P A, Hardee, D D, Thompson, A C & Gueldner, R C (1974) An assessment of the lifetime biosynthesis potential of the male boll weevil. *J Insect Physiol* 20: 1707–12.

Hedrick, P W (1999) Balancing selection and MHC. *Genetica* 104: 207–14.

Hedrick, P W & Loeschcke, V (1996) MHC and mate selection in humans? *Trends Ecol Evol* 11: 24.

Hefetz, A (2007) The evolution of hydrocarbon pheromone parsimony in ants (Hymenoptera: Formicidae) – interplay of colony odor uniformity and odor idiosyncrasy. *Myrmecol News* 10: 59–68.

Hefetz, A, Bergström, G & Tengo, J (1986) Species, individual and kin specific blends in Dufour's gland secretions of halictine bees – chemical evidence. *J Chem Ecol* 12: 197–208.

Heinze, J & d'Ettorre, P (2009) Honest and dishonest communication in social Hymenoptera. *J Exp Biol* 212: 1775–9.

Helantera, H, Lee, Y R, Drijfhout, F P & Martin, S J (2011) Genetic diversity, colony chemical phenotype, and nest mate recognition in the ant *Formica fusca*. *Behav Ecol* 22: 710–16.

Helgason, A, Palsson, S & Guthbjartsson, D F (2008) An association between the kinship and fertility of human couples. *Science* 319: 813.

Hendrichs, J, Katsoyannos, B I, Wornoayporn, V & Hendrichs, M A (1994) Odor-mediated foraging by yellowjacket wasps (Hymenoptera, Vespidae) – predation on leks of pheromone-calling Mediterranean fruit-fly males (Diptera, Tephritidae). *Oecologia* 99: 88–94.

Hendrichs, M A & Hendrichs, J (1998) Perfumed to be killed: interception of Mediterranean fruit fly (Diptera: Tephritidae) sexual signaling by predatory foraging wasps (Hymenoptera: Vespidae). *Ann Entomol Soc Am* 91: 228–34.

Hensch, T K (2004) Critical period regulation. *Annu Rev Neurosci* 27: 549–79.

Hepper, P G & Wells, D L (2005) How many footsteps do dogs need to determine the direction of an odour trail? *Chem Senses* 30: 291–8.

Herb, B R, Wolschin, F, Hansen, K D *et al.* (2012) Reversible switching between epigenetic states in honeybee behavioral subcastes. *Nat Neurosci.* 15: 1371–3.

Herz, R S (2007) *The Scent of Desire: Discovering our Enigmatic Sense of Smell.* New York: William Morrow/HarperCollins.

Herz, R S (2009a) Aromatherapy facts and fictions: a scientific analysis of olfactory effects on mood, physiology and behavior. *Int J Neurosci* 119: 263–90.

Herz, R S (2009b) Symposium overview. Basic processes in human olfactory cognition: current findings and future directions. *Ann N Y Acad Sci* 1170: 313–17.

Herz, R S (2011) Perfume. In Gottfried, J A (ed.) *Neurobiology of Sensation and Reward.* Boca Raton, FL: CRC Press. Available from: www.ncbi.nlm.nih.gov/books/NBK92802.

Herz, R S (2012) Odor memory and the special role of associative learning. In Zucco, G M, Schaal, B & Herz, R S (eds.) *Olfactory Cognition: from Perception and Memory to Environmental Odours and Neuroscience.* pp. 95–114. Amsterdam: John Benjamins.

Hesselschwerdt, J, Tscharner, S, Necker, J & Wantzen, K (2009) A local gammarid uses kairomones to avoid predation by the invasive crustaceans *Dikerogammarus villosus* and *Orconectes limosus*. *Biol Invasions* 11: 2133–40.

Hettyey, A, Hegyi, G, Puurtinen, M *et al.* (2010) Mate choice for genetic benefits: time to put the pieces together. *Ethology* 116: 1–9.

Heuskin, S, Verheggen, F, Haubruge, E, Wathelet, J P & Lognay, G (2011) The use of semiochemical slow-release devices in integrated pest management strategies. *Biotechnol Agron Soc Environ* 15: 459–70.

Hildebrand, J G & Shepherd, G M (1997) Mechanisms of olfactory discrimination: converging evidence for common principles across phyla. *Annu Rev Neurosci* 20: 595–631.

Hill, G E (2011) Condition-dependent traits as signals of the functionality of vital cellular processes. *Ecol Lett* 14: 625–34.

Hillier, N K & Vickers, N J (2011) Hairpencil volatiles influence interspecific courtship and mating between two related moth species. *J Chem Ecol* 37: 1127–36.

Himuro, C, Yokoi, T & Matsuura, K (2011) Queen-specific volatile in a higher termite *Nasutitermes takasagoensis* (Isoptera: Termitidae). *J Insect Physiol* 57: 962–5.

Hine, E, McGuigan, K & Blows, M W (2011) Natural selection stops the evolution of male attractiveness. *Proc Natl Acad Sci USA* **108**: 3659–64.

Hobbs, N J & Ferkin, M H (2011) Dietary protein content affects the response of meadow voles, *Microtus pennsylvanicus*, to over-marks. *Acta Ethol* **14**: 57–64.

Hoffmeister, T S & Roitberg, B D (2002) Evolutionary ecology of oviposition marking pheromones. In Hilker, M & Meiners, T (eds.) *Chemoecology of Insect Eggs and Egg Deposition*. pp. 319–47. Berlin: Blackwell.

Hofstetter, R W, Gaylord, M L, Martinson, S & Wagner, M R (2012) Attraction to monoterpenes and beetle-produced compounds by syntopic *Ips* and *Dendroctonus* bark beetles and their predators. *Agric For Entomol* **14**: 207–15.

Högland, J & Alatalo, R V (1995) *Leks*. Princeton: Princeton University Press.

Holekamp, K E & Dloniak, S M (2010) Intraspecific variation in the behavioral ecology of a tropical carnivore, the spotted hyena. *Adv Study Behav* **42**: 189–229.

Holland, D & Rice, W R (1998) Chase-away sexual selection: antagonistic seduction versus resistance. *Evolution* **52**: 1–7.

Hölldobler, B & Carlin, N F (1987) Anonymity and specificity in the chemical communication signals of social insects. *J Comp Physiol A* **161**: 567–81.

Hölldobler, B & Wilson, E O (1977) Weaver ants. *Sci Am* **237**: 146–54.

Hölldobler, B & Wilson, E O (1978) The multiple recruitment systems of the African weaver ant *Oecophylla longinoda* (Latreille) (Hymenoptera: Formicidae). *Behav Ecol Sociobiol* **3**: 19–60.

Hölldobler, B & Wilson, E O (1990) *The Ants*. Berlin: Springer.

Hölldobler, B & Wilson, E O (1994) *Journey to the Ants. A Story of Scientific Exploration*. Cambridge, MA: Harvard University Press.

Hölldobler, B & Wilson, E O (2009) *The Superorganism: the Beauty, Elegance, and Strangeness of Insect Societies*. New York: W.W. Norton.

Hölldobler, B, Stanton, R C & Markl, H (1978) Recruitment and food-retrieving behavior in *Novomessor* (Formicidae, Hymenoptera). I. Chemical signals. *Behav Ecol Sociobiol* **4**: 163–81.

Holman, L (2010) Queen pheromones: the chemical crown governing insect social life. *Commun Integr Biol* **3**: 558–60.

Holman, L, Dreier, S & d'Ettorre, P (2010a) Selfish strategies and honest signalling: reproductive conflicts in ant queen associations. *Proc R Soc B* **277**: 2007–15.

Holman, L, Jørgensen, C, Nielsen, J & d'Ettorre, P (2010b) Identification of an ant queen pheromone regulating worker sterility. *Proc R Soc B* **277**: 3793–800.

Holmes, W G (1986) Identification of paternal half-siblings by captive Belding ground-squirrels. *Anim Behav* **34**: 321–7.

Holmes, W G (2004) The early history of Hamiltonian-based research on kin recognition. *Ann Zool Fenn* **41**: 691–711.

Holmes, W G & Sherman, P W (1982) The ontogeny of kin recognition in 2 species of ground-squirrels. *Am Zool* **22**: 491–517.

Hong, W S, Chen, S X, Zhang, Q Y & Zheng, W Y (2006) Sex organ extracts and artificial hormonal compounds as sex pheromones to attract broodfish and to induce spawning of Chinese black sleeper (*Bostrichthys sinensis* Lacépède). *Aquac Res* **37**: 529–34.

Hoover, S E R, Keeling, C I, Winston, M L & Slessor, K N (2003) The effect of queen pheromones on worker honey bee ovary development. *Naturwissenschaften* **90**: 477–80.

Hoover, S E R, Oldroyd, B P, Wossler, T C & Winston, M L (2005) Anarchistic queen honey bees have normal queen mandibular pheromones. *Insectes Soc* **52**: 6–10.

Horne, T J & Ylönen, H (1998) Heritabilities of dominance-related traits in male bank voles (*Clethrionomys glareolus*). *Evolution* **52**: 894–9.

Horner, A J, Nickles, S P, Weissburg, M J & Derby, C D (2006) Source and specificity of chemical cues mediating shelter preference of Caribbean spiny lobsters (*Panulirus argus*). *Biol Bull* **211**: 128–39.

Horner, A J, Weissburg, M J & Derby, C D (2008) The olfactory pathway mediates sheltering behavior of Caribbean spiny lobsters, *Panulirus argus*, to conspecific urine signals. *J Comp Physiol A* **194**: 243–53.

Hosken, D J, Stockley, P, Tregenza, T & Wedell, N (2009) Monogamy and the battle of the sexes. *Annu Rev Entomol* **54**: 361–78.

Houck, L D (2009) Pheromone communication in amphibians and reptiles. *Annu Rev Physiol* **71**: 161–76.

House, P K, Vyas, A & Sapolsky, R (2011) Predator cat odors activate sexual arousal pathways in brains of *Toxoplasma gondii* infected rats. *PLoS ONE* **6**: e23277.

Hovestadt, T, Thomas, J A, Mitesser, O, Elmes, G W & Schönrogge, K (2012) Unexpected benefit of a social parasite for a key fitness component of its ant host. *Am Nat* **179**: 110–23.

Howard, L O & Fiske, W F (1911) *The Importation into the United States of the Parasites of the Gipsy Moth and the Brown-tail Moth. Bulletin 91.* Washington, DC: US Department of Agriculture, Bureau of Entomology.

Howard, R W & Akre, R D (1995) Propaganda, crypsis, and slave-making. In Cardé, R T & Bell, W J (eds.) *Chemical Ecology of Insects 2.* pp. 364–424. London: Chapman and Hall.

Howard, R W & Blomquist, G J (2005) Ecological, behavioral, and biochemical aspects of insect hydrocarbons. *Annu Rev Entomol* **50**: 371–93.

Howard, R W, McDaniel, C A & Blomquist, G J (1980) Chemical mimicry as an integrating mechanism: cuticular hydrocarbons of a termitophile and its host. *Science* **210**: 431–3.

Howard, S & Hughes, B M (2008) Expectancies, not aroma, explain impact of lavender aromatherapy on psychophysiological indices of relaxation in young healthy women. *Br J Health Psychol* **13**: 603–17.

Howe, N R & Harris, L G (1978) Transfer of the sea anemone pheromone, anthopleurine, by the nudibranch *Aedidia papillosa. J Chem Ecol* **4**: 551–61.

Howe, N R & Sheik, Y M (1975) Anthopleurine: a sea anemone alarm pheromone. *Science* **189**: 386–8.

Howse, P E, Stevens, I D R & Jones, O T (1998) *Insect Pheromones and their Use in Pest Management.* London: Chapman & Hall.

Hu, P J (2007) Dauer. In The C. elegans Research Community (ed.) *WormBook: The Online Review of C. elegans Biology [Internet].* doi/10.1895/wormbook.1891.1144.1891. Pasadena, CA: WormBook.

Hudson, R (1993) Olfactory imprinting. *Curr Opin Neurobiol* **3**: 548–52.

Hughes, M (1996) The function of concurrent signals: visual and chemical communication in snapping shrimp. *Anim Behav* **52**: 247–57.

Huigens, M E, Pashalidou, F G, Qian, M H *et al.* (2009) Hitch-hiking parasitic wasp learns to exploit butterfly antiaphrodisiac. *Proc Natl Acad Sci USA* **106**: 820.

Huigens, M E, de Swart, E & Mumm, R (2011) Risk of egg parasitoid attraction depends on anti-aphrodisiac titre in the large cabbage white butterfly *Pieris brassicae. J Chem Ecol* **37**: 364–7.

Human Microbiome Project Consortium (2012) Structure, function and diversity of the healthy human microbiome. *Nature* **486**: 207–14.

Hunter, J R & Hasler, A D (1965) Spawning association of the redfin shiner *Notropis umbratilis* and the green sunfish *Lepomis cyanellus. Copeia* **1965**: 265–81.

Hurd, P L (1995) Communication in discrete action-response games. *J Theor Biol* **174**: 217–22.

Hurd, P L (1997) Is signalling of fighting ability costlier for weaker individuals? *J Theor Biol* **184**: 83–8.

Hurst, J L (1993) The priming effects of urine substrate marks on interactions between male house mice, *Mus musculus domesticus* Schwarz and Schwarz. *Anim Behav* **45**: 55–81.

Hurst, J L (2009) Female recognition and assessment of males through scent. *Behav Brain Res* **200**: 295–303.

Hurst, J L & Beynon, R J (2004) Scent wars: the chemobiology of competitive signalling in mice. *Bioessays* **26**: 1288–98.

Hurst, J L & Beynon, R J (2008) Chemical communication in societies of rodents. In d'Ettorre, P & Hughes, DP (eds.) *Sociobiology of Communication: an Interdisciplinary Perspective.* pp. 97–117. Oxford: Oxford University Press.

Hurst, J L & Beynon, R J (2013) Rodent urinary proteins used in scent communication. In East, M L & Dehnhard, M (eds.) *Chemical Signals in Vertebrates 12.* pp. 117–33. New York: Springer.

Hurst, J L, Beynon, R J, Roberts, S C & Wyatt, T D (eds.) (2008) *Chemical Signals in Vertebrates 11.* New York: Springer.

Hutchison, L V & Wenzel, B M (1980) Olfactory guidance in procellariiforms. *Condor* **82**: 314–19.

Iino, Y & Yoshida, K (2009) Parallel use of two behavioral mechanisms for chemotaxis in *Caenorhabditis elegans. J Neurosci* **29**: 5370–80.

Imai, T, Sakano, H & Vosshall, L B (2010) Topographic mapping – the olfactory system. *Cold Spring Harb Perspect Biol* 2: a001776.

Inoshita, T, Martin, J R, Marion-Poll, F & Ferveur, J F (2011) Peripheral, central and behavioral responses to the cuticular pheromone bouquet in *Drosophila melanogaster* males. *PLoS ONE* 6: e19770.

Ishida, Y & Leal, W S (2005) Rapid inactivation of a moth pheromone. *Proc Natl Acad Sci USA* 102: 14075–9.

Isogai, Y, Si, S, Pont-Lezica, L *et al.* (2011) Molecular organization of vomeronasal chemoreception. *Nature* 478: 241–5.

Ivy, T M, Weddle, C B & Sakaluk, S K (2005) Females use self-referent cues to avoid mating with previous mates. *Proc R Soc B* 272: 2475–8.

Iyengar, V K & Eisner, T (1999a) Heritability of body mass, a sexually selected trait, in an arctiid moth (*Utetheisa ornatrix*). *Proc Natl Acad Sci USA* 96: 9169–71.

Iyengar, V K & Eisner, T (1999b) Female choice increases offspring fitness in an arctiid moth (*Utetheisa ornatrix*). *Proc Natl Acad Sci USA* 96: 15013–16.

Iyengar, V K, Reeve, H K & Eisner, T (2002) Paternal inheritance of a female moth's mating preference. *Nature* 419: 830–2.

Izard, M K (1983) Pheromones and reproduction in domestic animals. In Vandenbergh, J G (ed.) *Pheromones and Reproduction in Mammals.* pp. 253–85. New York: Academic Press.

Jackson, D E & Ratnieks, F L W (2006) Communication in ants. *Curr Biol* 16: R570–R574.

Jackson, D E, Holcombe, M & Ratnieks, F L W (2004) Trail geometry gives polarity to ant foraging networks. *Nature* 432: 907–9.

Jackson, D E, Martin, S J, Holcombe, M & Ratnieks, F L W (2006) Longevity and detection of persistent foraging trails in Pharaoh's ants, *Monomorium pharaonis* (L.). *Anim Behav* 71: 351–9.

Jackson, D E, Martin, S J, Ratnieks, F L W & Holcombe, M (2007) Spatial and temporal variation in pheromone composition of ant foraging trails. *Behav Ecol* 18: 444–50.

Jacob, S & McClintock, M K (2000) Psychological state and mood effects of steroidal chemosignals in women and men. *Horm Behav* 37: 57–78.

Jallon, J M & David, J R (1987) Variations in cuticular hydrocarbons among the 8 species of the *Drosophila melanogaster* subgroup. *Evolution* 41: 294–302.

James, A, Plank, M J & Edwards, A M (2011) Assessing Lévy walks as models of animal foraging. *J R Soc Interface* 8: 1233–47.

James, A, Casey, J, Hylands, D & Mycock, G (2004) Fatty acid metabolism by cutaneous bacteria and its role in axillary malodour. *World J Microbiol Biotechnol* 20: 787–93.

Jarau, S (2009) Chemical communication during food exploitation in stingless bees. In Jarau, S & Hrncir, M (eds.) *Food Exploitation by Social Insects: Ecological, Behavioral and Theoretical Approaches.* pp. 223–49. Boca Raton, FL: CRC Press.

Jarau, S, Schulz, C M, Hrncir, M *et al.* (2006) Hexyl decanoate, the first trail pheromone compound identified in a stingless bee, *Trigona recursa. J Chem Ecol* 32: 1555–64.

Jarriault, D, Barrozo, R B, Pinto, C J D *et al.* (2009) Age-dependent plasticity of sex pheromone response in the moth, *Agrotis ipsilon*: combined effects of octopamine and juvenile hormone. *Horm Behav* 56: 185–91.

Jeanson, R, Dussutour, A & Fourcassié, V (2012) Key factors for the emergence of collective decision in invertebrates. *Front Neurosci* 6: 121.

Jefferis, G S X E & Livet, J (2012) Sparse and combinatorial neuron labelling. *Curr Opin Neurobiol* 22: 101–10.

Jefferis, G S X E, Potter, C J, Chan, A I *et al.* (2007) Comprehensive maps of *Drosophila* higher olfactory centers: spatially segregated fruit and pheromone representation. *Cell* 128: 1187–203.

Jenkins, S R, Marshall, D & Fraschetti, S (2009) Settlement and recruitment. In Whal, M (ed.) *Marine Hard Bottom Communities: Patterns, Dynamics, Diversity, and Change.* pp. 177–90. Dordrecht: Springer.

Jeong, S, Rokas, A & Carroll, S B (2006) Regulation of body pigmentation by the Abdominal-B Hox protein and its gain and loss in *Drosophila* evolution. *Cell* 125: 1387–99.

Johansson, B G & Jones, T M (2007) The role of chemical communication in mate choice. *Biol Rev* 82: 265–89.

John, L, Aguilar, I, Ayasse, M & Jarau, S (2012) Nest-specific composition of the trail pheromone of the stingless bee *Trigona corvina* within populations. *Insectes Soc* **59**: 527–32.

Johnson, M E & Atema, J (2005) The olfactory pathway for individual recognition in the American lobster *Homarus americanus*. *J Exp Biol* **208**: 2865–72.

Johnson, N S & Li, W M (2010) Understanding behavioral responses of fish to pheromones in natural freshwater environments. *J Comp Physiol A* **196**: 701–11.

Johnson, N S, Yun, S S, Thompson, H T, Brant, C O & Li, W (2009) A synthesized pheromone induces upstream movement in female sea lamprey and summons them into traps. *Proc Natl Acad Sci USA* **106**: 1021–6.

Johnson, N S, Yun, S S, Buchinger, T J & Li, W (2012) Multiple functions of a multi-component mating pheromone in sea lamprey *Petromyzon marinus*. *J Fish Biol* **80**: 538–54.

Johnston, C E (1994) Nest association in fishes – evidence for mutualism. *Behav Ecol Sociobiol* **35**: 379–83.

Johnston, R E (1998) Pheromones, the vomeronasal system, and communication – from hormonal responses to individual recognition. *Ann N Y Acad Sci* **855**: 333–48.

Johnston, R E (2003) Chemical communication in rodents: from pheromones to individual recognition. *J Mammal* **84**: 1141–62.

Johnston, R E (2005) Communication by mosaic signals: individual recognition and underlying neural mechanisms. In Mason, R T, LeMaster, M P & Müller-Schwarze, D (eds.) *Chemical Signals in Vertebrates 10*. pp. 269–82. New York, NY: Springer.

Johnston, R E (2008) Individual odors and social communication: individual recognition, kin recognition, and scent over-marking. *Adv Study Behav* **38**: 439–505.

Johnston, R E & Jernigan, P (1994) Golden hamsters recognize individuals, not just individual scents. *Anim Behav* **48**: 129–36.

Johnston, R E & Rasmussen, K (1984) Individual recognition of female hamsters by males: role of chemical cues and of the olfactory and vomeronasal systems. *Physiol Behav* **33**: 95–104.

Johnston, R E & Robinson, T A (1993) Cross-species discrimination of individual odors by hamsters (Muridae: *Mesocricetus auratus*, *Phodopus campbelli*). *Ethology* **94**: 317–25.

Johnston, R E, Derzie, A, Chiang, G, Jernigan, P & Lee, H C (1993) Individual scent signatures in golden hamsters: evidence for specialization of function. *Anim Behav* **45**: 1061–70.

Jones, A G & Ratterman, N L (2009) Mate choice and sexual selection: what have we learned since Darwin? *Proc Natl Acad Sci USA* **106**: 10001–8.

Jones, T M, Quinnell, R J & Balmford, A (1998) Fisherian flies: benefits of female choice in a lekking sandfly. *Proc R Soc B* **265**: 1651–7.

Jordan, N R, Mwanguhya, F, Furrer, R D *et al.* (2011) Scent marking in wild banded mongooses: 2. Intrasexual overmarking and competition between males. *Anim Behav* **81**: 43–50.

Judd, T M & Sherman, P W (1996) Naked mole-rats recruit colony mates to food sources. *Anim Behav* **52**: 957–69.

Jumper, G Y & Baird, R C (1991) Location by olfaction – a model and application to the mating problem in the deep-sea hatchetfish *Argyropelecus hemigymnus*. *Am Nat* **138**: 1431–58.

Jurenka, R A, Haynes, K F, Adlof, R O, Bengtsson, M & Roelofs, W L (1994) Sex-pheromone component ratio in the cabbage-looper moth altered by a mutation affecting the fatty-acid chain-shortening reactions in the pheromone biosynthetic-pathway. *Insect Biochem Mol Biol* **24**: 373–81.

Kaib, M (1999) Termites. In Hardie, J & Minks, A K (eds.) *Pheromones of Non-lepidopteran Insects Associated with Agricultural Plants*. pp. 329–53. Wallingford: CAB International.

Kaib, M, Husseneder, C, Epplen, C, Epplen, J T & Brandl, R (1996) Kin-biased foraging in a termite. *Proc R Soc B* **263**: 1527–32.

Kaib, M, Jmhasly, P, Wilfert, L *et al.* (2004) Cuticular hydrocarbons and aggression in the termite *Macrotermes subhyalinus*. *J Chem Ecol* **30**: 365–85.

Kaissling, K-E (1987) *R. H. Wright Lectures on Insect Olfaction* Burnaby, BC, Canada: Simon Fraser University.

Kaissling, K-E (1998) Olfactory transduction in moths: I. Generation of receptor potentials and nerve impulses. In Taddei-Ferretti, C & Musio, C (eds.) *From Structure to Information in Sensory Systems*. pp. 93–112. Singapore: World Scientific.

Kaissling, K-E (2009a) Olfactory perireceptor and receptor events in moths: a kinetic model revised. *J Comp Physiol A* 195: 895–922.

Kaissling, K-E (2009b) The sensitivity of the insect nose: the example of *Bombyx mori*. In Gutiérrez, A & Marco, S (eds.) *Biologically Inspired Signal Processing for Chemical Sensing.* pp. 45–52. Berlin: Springer.

Kaitz, M, Good, A, Rokem, A M & Eidelman, A I (1987) Mothers' recognition of their newborns by olfactory cues. *Dev Psychobiol* 20: 587–91.

Kalbe, M, Eizaguirre, C, Dankert, I *et al.* (2009) Lifetime reproductive success is maximized with optimal major histocompatibility complex diversity. *Proc R Soc B* 276: 925–34.

Kalberer, N M, Reisenman, C E & Hildebrand, J G (2010) Male moths bearing transplanted female antennae express characteristically female behaviour and central neural activity. *J Exp Biol* 213: 1272–80.

Kamakura, M (2011) Royalactin induces queen differentiation in honeybees. *Nature* 473: 478–83.

Kamio, M & Derby, C D (2011) Approaches to a molecular identification of sex pheromones in blue crabs. In Breithaupt, T & Thiel, M (eds.) *Chemical Communication in Crustaceans.* pp. 393–412. New York: Springer.

Kamio, M, Reidenbach, M A & Derby, C D (2008) To paddle or not: context dependent courtship display by male blue crabs, *Callinectes sapidus. J Exp Biol* 211: 1243.

Kanaujia, S & Kaissling, K E (1985) Interactions of pheromone with moth antennae – adsorption, desorption and transport. *J Insect Physiol* 31: 71–81.

Kaneshiro, K Y (1989) The dynamics of sexual selection and founder effects in species formation. In Giddings, L V, Kaneshiro, K Y & Anderson, W W (eds.) *Genetics, Speciation, and the Founder Principle.* pp. 279–96. Oxford: Oxford University Press.

Kaneshiro, K Y (2006) Dynamics of sexual selection in the Hawaiian Drosophilidae: a paradigm for evolutionary change. *Proc Hawaiian Entomol Soc* 38: 1–19.

Kaneshiro, K Y & Boake, C R B (1987) Sexual selection and speciation – issues raised by Hawaiian *Drosophila. Trends Ecol Evol* 2: 207–12.

Karlson, P & Lüscher, M (1959) 'Pheromones': a new term for a class of biologically active substances. *Nature* 183: 55–6.

Karpati, Z, Dekker, T & Hansson, B S (2008) Reversed functional topology in the antennal lobe of the male European corn borer. *J Exp Biol* 211: 2841–8.

Karpati, Z, Olsson, S, Hansson, B S & Dekker, T (2010) Inheritance of central neuroanatomy and physiology related to pheromone preference in the male European corn borer. *BMC Evol Biol* 10: 286.

Kasparov, S (2011) The many facets of optogenetics. *Exp Physiol* 96: 1–3.

Kathrin, S (2012) Just follow your nose: homing by olfactory cues in ants. *Curr Opin Neurobiol* 22: 231–5.

Kato, A & Touhara, K (2009) Mammalian olfactory receptors: pharmacology, G protein coupling and desensitization. *Cell Mol Life Sci* 66: 3743–53.

Kaupp, U B (2010) Olfactory signalling in vertebrates and insects: differences and commonalities. *Nat Rev Neurosci* 11: 188–200.

Kausrud, K L, Gregoire, J C, Skarpaas, O *et al.* (2011) Trees wanted – dead or alive! Host selection and population dynamics in tree-killing bark beetles. *PLoS ONE* 6: e18274.

Kavaliers, M, Choleris, E & Pfaff, D W (2005) Genes, odours and the recognition of parasitized individuals by rodents. *Trends Parasitol* 21: 423–9.

Kay, L M, Beshel, J, Brea, J *et al.* (2009) Olfactory oscillations: the what, how and what for. *Trends Neurosci* 32: 207–14.

Keeling, C I, Plettner, E & Slessor, K N (2004) Hymenopteran semiochemicals. *Top Curr Chem* 239: 133–77.

Keil, T A (1992) Fine structure of a developing insect olfactory organ: morphogenesis of the silkmoth antenna. *Microsc Res Tech* 22: 351–71.

Keller, A & Vosshall, L B (2007) Influence of odorant receptor repertoire on odor perception in humans and fruit flies. *Proc Natl Acad Sci USA* 104: 5614–19.

Keller, A, Zhuang, H Y, Chi, Q Y, Vosshall, L B & Matsunami, H (2007) Genetic variation in a human odorant receptor alters odour perception. *Nature* 449: 468–73.

Keller, L & Nonacs, P (1993) The role of queen pheromones in social insects – queen control or queen signal. *Anim Behav* 45: 787–94.

Keller, L & Reeve, H K (1994) Partitioning of reproduction in animal societies. *Trends Ecol Evol* 9: 98–102.

Keller, L & Surette, M G (2006) Communication in bacteria: an ecological and evolutionary perspective. *Nat Rev Microbiol* 4: 249–58.

Keller, M, Baum, M J, Brock, O, Brennan, P A & Bakker, J (2009) The main and the accessory olfactory systems interact in the control of mate recognition and sexual behavior. *Behav Brain Res* 200: 268–76.

Kelliher, K R, Spehr, M, Li, X H, Zufall, F & Leinders-Zufall, T (2006) Pheromonal recognition memory induced by TRPC2-independent vomeronasal sensing. *Eur J Neurosci* 23: 3385–90.

Kelly, C A, Norbutus, A J, Lagalante, A F & Iyengar, V K (2012) Male courtship pheromones as indicators of genetic quality in an arctiid moth (*Utetheisa ornatrix*). *Behav Ecol* 23: 1009–14.

Kelly, C D & Jennions, M D (2011) Sexual selection and sperm quantity: meta-analyses of strategic ejaculation. *Biol Rev* 86: 863–84.

Kelly, D R (1996) When is a butterfly like an elephant? *Chem Biol* 3: 595–602.

Kempenaers, B (2007) Mate choice and genetic quality: a review of the heterozygosity theory. *Adv Study Behav* 37: 189–278.

Kendrick, K M, DaCosta, A P C, Broad, K D *et al.* (1997) Neural control of maternal behaviour and olfactory recognition of offspring. *Brain Res Bull* 44: 383–95.

Kennedy, J S (1986) Some current issues in orientation to odour sources. In Payne, T L, Birch, M C & Kennedy, C E J (eds.) *Mechanisms in Insect Olfaction*. pp. 1–25. New York: Oxford University Press.

Kennedy, J S (1992) *The New Anthropomorphism*. Cambridge: Cambridge University Press.

Kepecs, A, Uchida, N & Mainen, Z F (2006) The sniff as a unit of olfactory processing. *Chem Senses* 31: 167–79.

Kerr, J N D & Nimmerjahn, A (2012) Functional imaging in freely moving animals. *Curr Opin Neurobiol* 22: 45–53.

Khan, Z R, Midega, C A O, Pittchar, J, Bruce, T J A & Pickett, J A (2012) 'Push–pull' revisited: the process of successful deployment of a chemical ecology based pest management tool. In Gurr, G M, Wratten, S D, Snyder, W E & Read, D M Y (eds.) *Biodiversity and Insect Pests*. pp. 259–75. Chichester: John Wiley.

Kiemnec-Tyburczy, K M, Woodley, S K, Feldhoff, P W, Feldhoff, R C & Houck, L D (2011) Dermal application of courtship pheromones does not influence receptivity in female red-legged salamanders (*Plethodon shermani*). *J Herpetol* 45: 169–73.

Kikuta, S, Sato, K, Kashiwadani, H *et al.* (2010) Neurons in the anterior olfactory nucleus pars externa detect right or left localization of odor sources. *Proc Natl Acad Sci USA* 107: 12363–8.

Kilner, R M & Langmore, N E (2011) Cuckoos versus hosts in insects and birds: adaptations, counter-adaptations and outcomes. *Biol Rev* 86: 836–52.

Kiørboe, T (2011) What makes pelagic copepods so successful? *J Plankton Res* 33: 677.

Kirkendall, L R, Kent, D S & Raffa, K A (1997) Interactions among males, females and offspring in bark and ambrosia beetles: the significance of living in tunnels for the evolution of social behavior. In Choe, J C & Crespi, B J (eds.) *The Evolution of Social Behavior in Insects and Arachnids*. pp. 181–215. Cambridge: Cambridge University Press.

Kjaer, I & Hansen, B F (1996) The human vomeronasal organ: prenatal developmental stages and distribution of luteinizing hormone-releasing hormone. *Eur J Oral Sci* 104: 34–40.

Klarica, J, Bittner, L, Pallua, J *et al.* (2011) Near-infrared imaging spectroscopy as a tool to discriminate two cryptic *Tetramorium* ant species. *J Chem Ecol* 37: 549–52.

Kleineidam, C J & Rossler, W (2009) Adaptations in the olfactory system of social Hymenoptera. In Gadau, J & Fewell, J H (eds.) *Organization of Insect Societies: from Genome to Sociocomplexity*. pp. 195–219. Cambridge, MA: Harvard University Press.

Kleineidam, C J, Rössler, W, Hölldobler, B & Roces, F (2007) Perceptual differences in trail-following leaf-cutting ants relate to body size. *J Insect Physiol* 53: 1233–41.

Kloppenburg, P & Mercer, A (2008) Serotonin modulation of moth central olfactory neurons. *Annu Rev Entomol* 53: 179–90.

Knaapila, A, Zhu, G, Medland, S E *et al.* (2012) A genome-wide study on the perception of the odorants androstenone and galaxolide. *Chem Senses* 37: 541–52.

Knöpfel, T & Boyden, E S (eds.) (2012) *Optogenetics: Tools for Controlling and Monitoring Neuronal Activity*. Amsterdam: Elsevier.

Kobayakawa, K, Kobayakawa, R, Matsumoto, H *et al.* (2007) Innate versus learned odour processing in the mouse olfactory bulb. *Nature* 450: 503-8.

Kocher, S & Grozinger, C (2011) Cooperation, conflict, and the evolution of queen pheromones. *J Chem Ecol* 37: 1263-75.

Kock, D, Ruther, J & Sauer, K P (2007) A male sex pheromone in a scorpionfly. *J Chem Ecol* 33: 1249-56.

Koehl, M A R (2006) The fluid mechanics of arthropod sniffing in turbulent odor plumes. *Chem Senses* 31: 93-105.

Koehl, M A R (2011) Hydrodynamics of sniffing by crustaceans. In Breithaupt, T & Thiel, M (eds.) *Chemical Communication in Crustaceans.* pp. 85-102. New York: Springer.

Koene, J M & ter Maat, A (2001) "Allohormones": a class of bioactive substances favoured by sexual selection. *J Comp Physiol A* 187: 323-6.

Koene, J M & ter Maat, A (2002) The distinction between pheromones and allohormones – reply. *J Comp Physiol A* 188: 163-4.

Kohatsu, S, Koganezawa, M & Yamamoto, D (2011) Female contact activates male-specific interneurons that trigger stereotypic courtship behavior in *Drosophila. Neuron* 69: 498-508.

Kohoutová, D, Rubešová, A & Havlíček, J (2012) Shaving of axillary hair has only a transient effect on perceived body odor pleasantness. *Behav Ecol Sociobiol* 66: 569-81.

Koivula, M & Viitala, J (1999) Rough-legged buzzards use vole scent marks to assess hunting areas. *J Avian Biol* 30: 329-32.

Kokko, H (2005) Treat 'em mean, keep 'em (sometimes) keen: evolution of female preferences for dominant and coercive males. *Evol Ecol* 19: 123-35.

Kokko, H & Rankin, D J (2006) Lonely hearts or sex in the city? Density-dependent effects in mating systems. *Phil Trans R Soc B* 361: 319-34.

Kokko, H & Wong, B B M (2007) What determines sex roles in mate searching? *Evolution* 61: 1162-75.

Kokko, H, Jennions, M D & Brooks, R (2006) Unifying and testing models of sexual selection. *Annu Rev Ecol Evol Syst* 37: 43-66.

Kolliker, M, Chuckalovcak, J P, Haynes, K F & Brodie, E D (2006) Maternal food provisioning in relation to condition-dependent offspring odours in burrower bugs (*Sehirus cinctus*). *Proc R Soc B* 273: 1523-8.

Korb, J & Hartfelder, K (2008) Life history and development a framework for understanding developmental plasticity in lower termites. *Biol Rev* 83: 295-313.

Korb, J, Weil, T, Hoffmann, K, Foster, K R & Rehli, M (2009) A gene necessary for reproductive suppression in termites. *Science* 324: 758.

Kotiaho, J S (2001) Costs of sexual traits: a mismatch between theoretical considerations and empirical evidence. *Biol Rev* 76: 365-76.

Kou, R, Chang, H W, Chen, S C & Ho, H Y (2009) Suppression pheromone and cockroach rank formation. *Naturwissenschaften* 96: 691-701.

Kozak, G M, Head, M L & Boughman, J W (2011) Sexual imprinting on ecologically divergent traits leads to sexual isolation in sticklebacks. *Proc R Soc B* 278: 2604-10.

Krause, E T, Krüger, O, Kohlmeier, P & Caspers, B A (2012) Olfactory kin recognition in a songbird. *Biol Lett* 8: 327-9.

Krautwurst, D (2008) Human olfactory receptor families and their odorants. *Chemistry & Biodiversity* 5: 842-52.

Kreher, S A, Mathew, D, Kim, J & Carlson, J R (2008) Translation of sensory input into behavioral output via an olfactory system. *Neuron* 59: 110-24.

Kristoffersen, L, Hansson, B S, Anderbrant, O & Larsson, M C (2008) Aglomerular hemipteran antennal lobes – basic neuroanatomy of a small nose. *Chem Senses* 33: 771-8.

Kroiss, J, Lechner, K & Strohm, E (2010) Male territoriality and mating system in the European beewolf *Philanthus triangulum* F. (Hymenoptera: Crabronidae): evidence for a "hotspot" lek polygyny. *J Ethol* 28: 295-304.

Kronauer, D J C & Pierce, N E (2011) Myrmecophiles. *Curr Biol* 21: R208-R209.

Kronforst, M R, Young, L G, Kapan, D D *et al.* (2006) Linkage of butterfly mate preference and wing color preference cue at the genomic location of wingless. *Proc Natl Acad Sci USA* 103: 6575-80.

Kruuk, H (1972) *The Spotted Hyena. A Study of Predation and Social Behavior.* Chicago: Chicago University Press.

Kruuk, H (1989) *The Social Badger. Ecology and Behaviour of a Group-living Carnivore (Meles meles).* Oxford: Oxford University Press.

Kruuk, H, Gorman, M & Leitch, A (1984) Scent-marking with the subcaudal gland by the European badger, *Meles meles* L. *Anim Behav* 32: 899–907.

Kubli, E & Bopp, D (2012) Sexual behavior: how sex peptide flips the postmating switch of female flies. *Curr Biol* 22: R520–R522.

Kuebler, L S, Kelber, C & Kleineidam, C J (2010) Distinct antennal lobe phenotypes in the leaf-cutting ant (*Atta vollenweideri*). *J Comp Neurol* 518: 352–65.

Kunert, G, Otto, S, Röse, U, Gershenzon, J & Weisser, W (2005) Alarm pheromone mediates production of winged dispersal morphs in aphids. *Ecol Lett* 8: 596–603.

Kwak, J & Preti, G (2011) Volatile disease biomarkers in breath: a critique. *Curr Pharm Biotechnol* 12: 1067–74.

Kwak, J, Opiekun, M C, Matsumura, K *et al.* (2009) Major histocompatibility complex-regulated odortypes: peptide-free urinary volatile signals. *Physiol Behav* 96: 184–8.

Kwak, J, Willse, A, Preti, G, Yamazaki, K & Beauchamp, GK (2010) In search of the chemical basis for MHC odourtypes. *Proc R Soc B* 277: 2417–25.

Labows, J N & Preti, G (1992) Human semiochemicals. In Van Toller, S & Dodd, G H (eds.) *Fragrance: the Psychology and Biology of Perfume.* pp. 69–90. London: Elsevier Science.

Lacey, E A & Sherman, P W (2005) Redefining eusociality: concepts, goals and levels of analysis. *Ann Zool Fenn* 42: 573–77.

Lachmann, M, Szamado, S & Bergstrom, C T (2001) Cost and conflict in animal signals and human language. *Proc Natl Acad Sci USA* 98: 13189.

Lamunyon, C W & Eisner, T (1993) Postcopulatory sexual selection in an arctiid moth (*Utetheisa ornatrix*). *Proc Natl Acad Sci USA* 90: 4689–92.

Lamunyon, C W & Eisner, T (1994) Spermatophore size as determinant of paternity in an arctiid moth (*Utetheisa ornatrix*). *Proc Natl Acad Sci USA* 91: 7081–4.

Landolt, P J (1997) Sex attractant and aggregation pheromones of male phytophagous insects. *Am Entomol* 43: 12–22.

Landolt, P J, Reed, H C & Heath, R R (1992) Attraction of female papaya fruit-fly (Diptera, Tephritidae) to male pheromone and host fruit. *Environ Entomol* 21: 1154–9.

Landolt, P J, Molina, O H, Heath, R R *et al.* (1996) Starvation of cabbage looper moths (Lepidoptera: Noctuidae) increases attraction to male pheromone. *Ann Entomol Soc Am* 89: 459–65.

LaPorte, J (2002) Must signals handicap? *Monist* 85: 86–104.

Larsson, M C & Svensson, G P (2009) Pheromone monitoring of rare and threatened insects: exploiting a pheromone–kairomone system to estimate prey and predator abundance. *Conserv Biol* 23: 1516–25.

Laska, M (2004) Olfactory discrimination ability of human subjects for enantiomers with an isopropenyl group at the chiral center. *Chem Senses* 29: 143–52.

Laska, M, Genzel, D & Wieser, A (2005) The number of functional olfactory receptor genes and the relative size of olfactory brain structures are poor predictors of olfactory discrimination performance with enantiomers. *Chem Senses* 30: 171–5.

Lassance, J-M (2010) Journey in the *Ostrinia* world: from pest to model in chemical ecology. *J Chem Ecol* 36: 1155–69.

Lassance, J-M & Löfstedt, C (2009) Concerted evolution of male and female display traits in the European corn borer, *Ostrinia nubilalis*. *BMC Biol* 7: 10.

Lassance, J-M, Groot, A T, Liénard, M A *et al.* (2010) Allelic variation in a fatty-acyl reductase gene causes divergence in moth sex pheromones. *Nature* 466: 486–9.

Lassance, J-M, Bogdanowicz, S M, Wanner, K W, Löfstedt, C & Harrison, R G (2011) Gene genealogies reveal differentiation at sex pheromone olfactory receptor loci in pheromone strains of the European corn borer, *Ostrinia nubilalis*. *Evolution* 65: 1583–93.

Lassance, J-M, Liénard, M A, Antony, B, *et al.* (2013) Functional consequences of sequence variation in the pheromone biosynthetic gene *pgFAR* for *Ostrinia* moths. *Proc Natl Acad Sci USA* 110: 3967–72.

Lastein, S, Hamdani, E H & Døving, K B (2014) Olfactory discrimination of pheromones. In Sorensen, P W & Wisenden, B D (eds.) *Fish Pheromones and Related Conspecific Chemical Cues.* Chichester: Wiley-Blackwell.

Laughlin, J D, Ha, T S, Jones, D N M & Smith, D P (2008) Activation of pheromone-sensitive neurons is mediated by conformational activation of pheromone-binding protein. *Cell* 133: 1255–65.

Laurence, B R & Pickett, J A (1985) An oviposition attractant pheromone in *Culex quinquefasciatus* Say (Diptera: Culicidae). *Bull Entomol Res* 75: 283–90.

Laurent, R & Chaix, R (2012) HapMap European American genotypes are compatible with the hypothesis of MHC-dependent mate choice (response to DOI 10.1002/bies.201200023, Derti and Roth). *Bioessays* 34: 871–2.

Lawniczak, M K N, Barnes, A I, Linklater, J R *et al.* (2007) Mating and immunity in invertebrates. *Trends Ecol Evol* 22: 48–55.

Lawson, L P, Vander Meer, R K & Shoemaker, D (2012a) Male reproductive fitness and queen polyandry are linked to variation in the supergene Gp-9 in the fire ant *Solenopsis invicta*. *Proc R Soc B* 279: 3217–22.

Lawson, M J, Craven, B A, Paterson, E G & Settles, G S (2012b) A computational study of odorant transport and deposition in the canine nasal cavity: implications for olfaction. *Chem Senses* 37: 553–66.

Le Conte, Y & Hefetz, A (2008) Primer pheromones in social Hymenoptera. *Annu Rev Entomol* 53: 523–42.

Le Roux, A, Cherry, M I & Manser, M B (2008) The effects of population density and sociality on scent marking in the yellow mongoose. *J Zool* 275: 33–40.

Leal, W S (1999) Scarab beetles. In Hardie, J & Minks, A K (eds.) *Pheromones of Non-lepidopteran Insects Associated with Agricultural Plants*. pp. 51–68. Wallingford, Oxon: CAB International.

Leal, W S (2005) Pheromone reception. In Schulz, S (ed.) *Chemistry of Pheromones and Other Semiochemicals II*. pp. 1–36. Berlin: Springer.

Leal, W S (2013) Odorant reception in insects: roles of receptors, binding proteins, and degrading enzymes. *Annu Rev Entomol* 58: 373–91.

Leal, W S & Ishida, Y (2008) GP-9s are ubiquitous proteins unlikely involved in olfactory mediation of social organization in the red imported fire ant, *Solenopsis invicta*. *PLoS ONE* 3: e3762.

Leal, W S, Chen, A M, Ishida, Y *et al.* (2005) Kinetics and molecular properties of pheromone binding and release. *Proc Natl Acad Sci USA* 102: 5386–91.

Leary, G P, Allen, J E, Bunger, P L *et al.* (2012) Single mutation to a sex pheromone receptor provides adaptive specificity between closely related moth species. *Proc Natl Acad Sci USA* 109: 14081–6.

Lehner, P N (1996) *Handbook of Ethological Methods*, 2nd edn. Cambridge: Cambridge University Press.

Leinders-Zufall, T, Lane, A P, Puche, A C *et al.* (2000) Ultrasensitive pheromone detection by mammalian vomeronasal neurons. *Nature* 405: 792–6.

Leinders-Zufall, T, Brennan, P, Widmayer, P *et al.* (2004) MHC Class I peptides as chemosensory signals in the vomeronasal organ. *Science* 306: 1033–7.

Leinders-Zufall, T, Ishii, T, Mombaerts, P, Zufall, F & Boehm, T (2009) Structural requirements for the activation of vomeronasal sensory neurons by MHC peptides. *Nat Neurosci* 12: 1551–8.

Lelito, J, Myrick, A & Baker, T (2008) Interspecific pheromone plume interference among sympatric heliothine moths: a wind tunnel test using live, calling females. *J Chem Ecol* 34: 725–33.

Lenochova, P & Havlíček, J (2008) Human body odour individuality. In Hurst, J L, Beynon, R J, Roberts, S C & Wyatt, T D (eds.) *Chemical Signals in Vertebrates 11*. pp. 189–98. New York: Springer.

Lenoir, A, d'Ettorre, P, Errard, C & Hefetz, A (2001) Chemical ecology and social parasitism in ants. *Annu Rev Entomol* 46: 573–99.

Leonhardt, S D, Brandstaetter, A S & Kleineidam, C J (2007) Reformation process of the neuronal template for nestmate-recognition cues in the carpenter ant *Camponotus floridanus*. *J Comp Physiol A* 193: 993–1000.

Levesque, H M, Scaffidi, D, Polkinghorne, C N & Sorensen, P W (2011) A multi-component species identifying pheromone in the goldfish. *J Chem Ecol* 37: 219–27.

Levinson, A & Levinson, H (1995) Reflections on structure and function of pheromone glands in storage insect species. *Anz Schädlingsk, Pflanzen, Umweltschutz* 68: 99–118.

Levitan, D R & McGovern, T M (2005) The Allee effect in the sea. In Norse, E A & Crowder, L B (eds.) *Marine Conservation Biology: the Science of Maintaining the Sea's Biodiversity*. pp. 47–57. Washington DC: Island Press.

Lévy, F & Keller, M (2008) Neurobiology of maternal behavior in sheep. *Adv Study Behav* 38: 399–437.

Lévy, F & Keller, M (2009) Olfactory mediation of maternal behavior in selected mammalian species. *Behav Brain Res* 200: 336–45.

Lévy, F, Porter, R H, Kendrick, K M, Keverne, E B & Romeyer, A (1996) Physiological, sensory, and experiential factors of parental care in sheep. *Adv Study Behav* 25: 385–422.

Lewis, S M & Austad, S N (1994) Sexual selection in flour beetles – the relationship between sperm precedence and male olfactory attractiveness. *Behav Ecol* 5: 219–24.

Leypold, B G, Yu, C R, Leinders-Zufall, T et al. (2002) Altered sexual and social behaviors in TRP2 mutant mice. *Proc Natl Acad Sci USA* **99**: 6376–81.

Li, Q, Korzan, W J, Ferrero, D M et al. (2013) Synchronous evolution of an odor biosynthesis pathway and behavioral response. *Curr Biol* **23**: 11–20.

Liberles, S D & Buck, L B (2006) A second class of chemosensory receptors in the olfactory epithelium. *Nature* **442**: 645–50.

Liberles, S D, Horowitz, L F, Kuang, D H et al. (2009) Formyl peptide receptors are candidate chemosensory receptors in the vomeronasal organ. *Proc Natl Acad Sci USA* **106**: 9842–7.

Lichtman, J W, Livet, J & Sanes, J R (2008) A technicolour approach to the connectome. *Nat Rev Neurosci* **9**: 417–22.

Liebig, J, Eliyahu, D & Brent, C S (2009) Cuticular hydrocarbon profiles indicate reproductive status in the termite *Zootermopsis nevadensis*. *Behav Ecol Sociobiol* **63**: 1799–807.

Liebig, J (2010) Hydrocarbon profiles indicate fertility and dominance status in ant, bee, and wasp colonies. In Blomquist, G J & Bagnères, A-G (eds.) *Insect Hydrocarbons: Biology, Biochemistry, and Chemical Ecology*. pp. 254–81. Cambridge: Cambridge University Press.

Liénard, M A, Strandh, M, Hedenström, E, Johansson, T & Löfstedt, C (2008) Key biosynthetic gene subfamily recruited for pheromone production prior to the extensive radiation of Lepidoptera. *BMC Evol Biol* **8**: 270.

Liénard, M A, Hagstrom, A K, Lassance, J-M & Löfstedt, C (2010) Evolution of multicomponent pheromone signals in small ermine moths involves a single fatty-acyl reductase gene. *Proc Natl Acad Sci USA* **107**: 10955–60.

Lihoreau, M & Rivault, C (2009) Kin recognition via cuticular hydrocarbons shapes cockroach social life. *Behav Ecol* **20**: 46–53.

Lihoreau, M, Zimmer, C & Rivault, C (2008) Mutual mate choice: when it pays both sexes to avoid inbreeding. *PLoS ONE* **3**: e3365.

Lihoreau, M, Costa, J & Rivault, C (2012) The social biology of domiciliary cockroaches: colony structure, kin recognition and collective decisions. *Insectes Soc* **59**: 445–52.

Lim, H & Sorensen, P W (2012) Common carp implanted-with prostaglandin F2α release a sex pheromone complex that attracts conspecific males in both the laboratory and field. *J Chem Ecol* **38**: 127–34.

Liman, E R & Innan, H (2003) Relaxed selective pressure on an essential component of pheromone transduction in primate evolution. *Proc Natl Acad Sci USA* **100**: 3328–32.

Lin, C P (2006) Social behaviour and life history of membracine treehoppers. *J Nat Hist* **40**: 1887–907.

Lin, D Y, Zhang, S Z, Block, E & Katz, L C (2005) Encoding social signals in the mouse main olfactory bulb. *Nature* **434**: 470–7.

Lin, D Y, Shea, S D & Katz, L C (2006) Representation of natural stimuli in the rodent main olfactory bulb. *Neuron* **50**: 937–49.

Lindauer, M & Kerr, W E (1958) Die gegenseitige Verständigung bei den stachellosen Bienen. *J Comp Physiol A* **41**: 405–34.

Lindsay, S M (2009) Ecology and biology of chemoreception in polychaetes. *Zoosymposia* **2**: 339–67.

Linn, C E & Roelofs, W L (1989) Response specificity of male moths to multicomponent pheromones. *Chem Senses* **14**: 421–37.

Linn, C E, Campbell, M G & Roelofs, W L (1987) Pheromone components and active spaces: what do moths smell and where do they smell it? *Science* **237**: 650–2.

Linn, C E, O'Connor, M & Roelofs, W (2003) Silent genes and rare males: a fresh look at pheromone blend response specificity in the European corn borer moth, *Ostrinia nubilalis*. *J Insect Sci* **3**: 15.

Liu, S, Zhao, B & Bonjour, E (2011) Host marking and host discrimination in phytophagous insects. In Liu, T & Kang, L (eds.) *Recent Advances in Entomological Research: from Molecular Biology to Pest Management*. pp. 73–85. Beijing and Berlin: Higher Education Press and Springer.

Liu, Y B & Haynes, K F (1992) Filamentous nature of pheromone plumes protects integrity of signal from background chemical noise in cabbage-looper moth, *Trichoplusia ni*. *J Chem Ecol* **18**: 299–307.

Lledo, P M, Alonso, M & Grubb, M S (2006) Adult neurogenesis and functional plasticity in neuronal circuits. *Nat Rev Neurosci* **7**: 179–93.

Locatello, L, Mazzoldi, C & Rasotto, M (2002) Ejaculate of sneaker males is pheromonally inconspicuous in the black goby, *Gobius niger* (Teleostei, Gobiidae). *J Exp Zool* **293**: 601–5.

Lockery, S R (2011) The computational worm: spatial orientation and its neuronal basis in *C. elegans*. *Curr Opin Neurobiol* **21**: 782–90.

Löfstedt, C (1990) Population variation and genetic control of pheromone communication systems in moths. *Entomol Exp Appl* **54**: 199–218.

Löfstedt, C (1993) Moth pheromone genetics and evolution. *Phil Trans R Soc B* **340**: 167–77.

Löfstedt, C, Vickers, N J, Roelofs, W L & Baker, T C (1989) Diet related courtship success in the oriental fruit moth, *Grapholita molesta* (Tortricidae). *Oikos* **55**: 402–8.

Löfstedt, C, Herrebout, W M & Menken, S B J (1991) Sex pheromones and their potential role in the evolution of reproductive isolation in small ermine moths (Yponomeutidae). *Chemoecology* **2**: 20–8.

Logan, D W, Marton, T F & Stowers, L (2008) Species specificity in major urinary proteins by parallel evolution. *PLoS ONE* **3**: e3280.

Logan, D W, Sandal, M, Gardner, P P, Manske, M & Bateman, A (2010) Ten simple rules for editing Wikipedia. *PLoS Comput Biol* **6**: e1000941.

Logan, D W, Brunet, J L, Webb *et al.* (2012) Learned recognition of maternal signature odors mediates the first suckling episode in mice. *Curr Biol* **22**: 1998–2007.

Lois, C & Groves, J O (2012) Genetics in non-genetic model systems. *Curr Opin Neurobiol* **22**: 79–85.

Lopez, F, Acosta, F J & Serrano, J M (1994) Guerrilla vs phalanx strategies of resource capture – growth and structural plasticity in the trunk trail system of the harvester ant *Messor barbarus*. *J Anim Ecol* **63**: 127–38.

Lorenzi, M C (2006) The result of an arms race: the chemical strategies of *Polistes* social parasites. *Ann Zool Fenn* **43**: 550–63.

Louis, M, Huber, T, Benton, R, Sakmar, T P & Vosshall, L B (2008) Bilateral olfactory sensory input enhances chemotaxis behavior. *Nat Neurosci* **11**: 187–99.

Lu, B, LaMora, A, Sun, Y, Welsh, M J & Ben-Shahar, Y (2012) ppk23-Dependent chemosensory functions contribute to courtship behavior in *Drosophila melanogaster*. *PLoS Genet* **8**: e1002587.

Luehring, M A, Wagner, C M & Li, W M (2011) The efficacy of two synthesized sea lamprey sex pheromone components as a trap lure when placed in direct competition with natural male odors. *Biol Invasions* **13**: 1589–97.

Lundström, J N & Olsson, M J (2010) Functional neuronal processing of human body odors. In Gerald, L (ed.) *Pheromones*. pp. 1–23. London: Academic Press.

Lundström, J N, Gordon, A R, Alden, E C, Boesveldt, S & Albrecht, J (2010) Methods for building an inexpensive computer-controlled olfactometer for temporally-precise experiments. *Int J Psychophysiol* **78**: 179–89.

Lürling, M (2012) Infodisruption: pollutants interfering with the natural chemical information conveyance in aquatic systems. In Brönmark, C & Hansson, L-A (eds.) *Chemical Ecology in Aquatic Systems*. pp. 250–71. Oxford: Oxford University Press.

Lyko, F, Foret, S, Kucharski, R *et al.* (2010) The honeybee epigenomes: differential methylation of brain DNA in queens and workers. *PLoS Biol* **8**: e1000506.

Ma, M (2010) Multiple olfactory subsystems convey various sensory signals. In Menini, A (ed.) *The Neurobiology of Olfaction*. Boca Raton, FL: CRC Press. Available online at www.ncbi.nlm.nih.gov/books/NBK55971.

Ma, W D, Miao, Z S & Novotny, M V (1999) Induction of estrus in grouped female mice (*Mus domesticus*) by synthetic analogues of preputial gland constituents. *Chem Senses* **24**: 289–93.

Macbeth, A H, Edds, J S & Young, W S (2009) Housing conditions and stimulus females: a robust social discrimination task for studying male rodent social recognition. *Nat Protoc* **4**: 1574–81.

Macdonald, D W (1985a) The rodents IV: suborder Hystricomorpha. In Brown, R E & Macdonald, D W (eds.) *Social Odours in Mammals*. pp. 480–506. Oxford: Oxford University Press.

Macdonald, D W (1985b) The carnivores: order Carnivora. In Brown, R E & Macdonald, D W (eds.) *Social Odours in Mammals*. pp. 619–722. Oxford: Oxford University Press.

Madsen, T, Shine, R, Loman, J & Håkansson, T (1992) Why do female adders copulate so frequently? *Nature* **355**: 440–1.

Maeno, K & Tanaka, S (2012) Adult female desert locusts require contact chemicals and light for progeny gregarization. *Physiol Entomol* **37**: 109–18.

Magro, A, Ducamp, C, Ramon-Portugal, F *et al.* (2010) Oviposition deterring infochemicals in ladybirds: the role of phylogeny. *Evol Ecol* 24: 251–71.

Mainland, J & Sobel, N (2006) The sniff is part of the olfactory percept. *Chem Senses* 31: 181–96.

Maisonnasse, A, Lenoir, J C, Beslay, D, Crauser, D & Le Conte, Y (2010) E-β-ocimene, a volatile brood pheromone involved in social regulation in the honey bee colony (*Apis mellifera*). *PLoS ONE* 5: e13531.

Malka, O, Karunker, I, Yeheskel, A, Morin, S & Hefetz, A (2009) The gene road to royalty – differential expression of hydroxylating genes in the mandibular glands of the honeybee. *FEBS J* 276: 5481–90.

Mallet, J (2008) Hybridization, ecological races and the nature of species: empirical evidence for the ease of speciation. *Phil Trans R Soc B* 363: 2971–86.

Malnic, B, Hirono, J, Sato, T & Buck, L B (1999) Combinatorial receptor codes for odors. *Cell* 96: 713–23.

Malnic, B, Gonzalez-Kristeller, D C & Gutiyama, L M (2010) Odorant receptors. In Menini, A (ed.) *The Neurobiology of Olfaction*. Boca Raton, FL: CRC Press. Available online at www.ncbi.nlm.nih.gov/books/NBK55985.

Mameli, M & Bateson, P (2011) An evaluation of the concept of innateness. *Phil Trans R Soc B* 366: 436–43.

Manning, A & Dawkins, M S (1998) *An Introduction to Animal Behaviour*, 5th edn. Cambridge: Cambridge University Press.

Manoli, D S, Meissner, G W & Baker, B S (2006) Blueprints for behavior: genetic specification of neural circuitry for innate behaviors. *Trends Neurosci* 29: 444–51.

Mardon, J, Saunders, S M, Anderson, M J, Couchoux, C & Bonadonna, F (2010) Species, gender, and identity: cracking petrels' sociochemical code. *Chem Senses* 35: 309–21.

Maresh, A, Gil, D R, Whitman, M C & Greer, C A (2008) Principles of glomerular organization in the human olfactory bulb – implications for odor processing. *PLoS ONE* 3: e2640.

Martin, A, Saathoff, M, Kuhn, F *et al.* (2010a) A functional ABCC11 allele is essential in the biochemical formation of human axillary odor. *J Invest Dermatol* 130: 529–40.

Martin, G B, Milton, J T B, Davidson, R H *et al.* (2004) Natural methods for increasing reproductive efficiency in small ruminants. *Anim Reprod Sci* 82: 231–45.

Martin, H (1965) Osmotropotaxis in the honey-bee. *Nature* 208: 59–63.

Martín, J & López, P (2008) Female sensory bias may allow honest chemical signaling by male Iberian rock lizards. *Behav Ecol Sociobiol* 62: 1927–34.

Martín, J & López, P (2010a) Condition-dependent pheromone signaling by male rock lizards: more oily scents are more attractive. *Chem Senses* 35: 253–62.

Martín, J & López, P (2010b) Pheromones and reproduction in reptiles. In Norris, D O & Lopez, K H (eds.) *Hormones and Reproduction of Vertebrates*. pp. 141–67. San Diego, CA: Academic Press.

Martin, J A & Wang, Z (2011) Next-generation transcriptome assembly. *Nat Rev Genet* 12: 671–82.

Martin, J P, Beyerlein, A, Dacks, AM *et al.* (2011a) The neurobiology of insect olfaction: Sensory processing in a comparative context. *Prog Neurobiol* 95: 427–47.

Martin, P & Bateson, P (2007) *Measuring Behaviour. An Introductory Guide*, 3rd edn. Cambridge: Cambridge University Press.

Martin, S J & Drijfhout, F P (2009a) Nestmate and task cues are influenced and encoded differently within ant cuticular hydrocarbon profiles. *J Chem Ecol* 35: 368–74.

Martin, S J & Drijfhout, F P (2009b) A review of ant cuticular hydrocarbons. *J Chem Ecol* 35: 1151–61.

Martin, S J, Châline, N G, Ratnieks, F L W & Jones, G R (2005) Searching for the egg-marking signal in honeybees. *J Negat Results* 2: 1–9.

Martin, S J, Helanterä, H & Drijfhout, F P (2008a) Evolution of species specific cuticular hydrocarbon patterns in *Formica* ants. *Biol J Linn Soc* 95: 131–40.

Martin, S J, Helanterä, H & Drijfhout, F P (2008b) Colony-specific hydrocarbons identify nest mates in two species of *Formica* ant. *J Chem Ecol* 34: 1072–80.

Martin, S J, Vitikainen, E, Helanterä, H & Drijfhout, F P (2008c) Chemical basis of nest-mate discrimination in the ant *Formica exsecta*. *Proc R Soc B* 275: 1271–78.

Martin, S J, Carruthers, J M, Williams, P H & Drijfhout, F P (2010b) Host specific social parasites (*Psithyrus*) indicate chemical recognition system in bumblebees. *J Chem Ecol* 36: 855–63.

Martin, S J, Helanterä, H & Drijfhout, F P (2011b) Is parasite pressure a driver of chemical cue diversity in ants? *Proc R Soc B* **278**: 496–503.

Martins, Y, Preti, G, Crabtree, C R *et al.* (2005) Preference for human body odors is influenced by gender and sexual orientation. *Psychol Sci* **16**: 694–701.

Mas, F & Kölliker, M (2008) Maternal care and offspring begging in social insects: chemical signalling, hormonal regulation and evolution. *Anim Behav* **76**: 1121–31.

Mason, R T (1993) Chemical ecology of the red-sided garter snake, *Thamnophis sirtalis parietalis. Brain Behav Evol* **41**: 261–8.

Mason, R T & Parker, M R (2010) Social behavior and pheromonal communication in reptiles. *J Comp Physiol A* **196**: 729–49.

Mateo, J M (2004) Recognition systems and biological organization: the perception component of social recognition. *Ann Zool Fenn* **41**: 729–45.

Mateo, J M (2009) The causal role of odours in the development of recognition templates and social preferences. *Anim Behav* **77**: 115–21.

Mateo, J M (2010) Self-referent phenotype matching and long-term maintenance of kin recognition. *Anim Behav* **80**: 929–35.

Mateo, J M & Johnston, R E (2000) Kin recognition and the 'armpit effect': evidence of self-referent phenotype matching. *Proc R Soc B* **267**: 695–700.

Mateo, J M & Johnston, R E (2003) Kin recognition by self-referent phenotype matching: weighing the evidence. *Anim Cogn* **6**: 73–6.

Mathis, K A & Philpott, S M (2012) Current understanding and future prospects of host selection, acceptance, discrimination, and regulation of phorid fly parasitoids that attack ants. *Psyche* 2012: doi:10.1155/2012/895424.

Mathuru, A S, Kibat, C, Cheong, W F *et al.* (2012) Chondroitin fragments are odorants that trigger fear behavior in fish. *Curr Biol* **22**: 538–44.

Matsui, A, Go, Y & Niimura, Y (2010) Degeneration of olfactory receptor gene repertories in primates: no direct link to full trichromatic vision. *Mol Biol Evol* **27**: 1192–200.

Matsumura, K, Nagano, M & Fusetani, N (1998) Purification of a larval settlement-inducing protein complex (SIPC) of the barnacle, *Balanus amphitrite. J Exp Zool* **281**: 12–20.

Matsuo, T, Sugaya, S, Yasukawa, J, Aigaki, T & Fuyama, Y (2007) Odorant-binding proteins OBP57d and OBP57e affect taste perception and host-plant preference in *Drosophila sechellia. PLoS Biol* **5**: e118.

Matsuura, K (2012) Multifunctional queen pheromone and maintenance of reproductive harmony in termite colonies. *J Chem Ecol* **38**: 746–54.

Matsuura, K, Himuro, C, Yokoi, T *et al.* (2010) Identification of a pheromone regulating caste differentiation in termites. *Proc Natl Acad Sci USA* **107**: 12963–8.

Maynard Smith, J (1991) Honest signalling: the Philip Sidney game. *Anim Behav* **42**: 1034–5.

Maynard Smith, J & Harper, D (1995) Animal signals: models and terminology. *J Theor Biol* **177**: 305–11.

Maynard Smith, J & Harper, D (2003) *Animal Signals.* Oxford: Oxford University Press.

Mays, H L & Hill, G E (2004) Choosing mates: good genes versus genes that are a good fit. *Trends Ecol Evol* **19**: 554–9.

McAllister, M K & Roitberg, B D (1987) Adaptive suicidal behaviour in pea aphids. *Nature* **328**: 797–9.

McBurney, D H, Shoup, M L & Streeter, S A (2006) Olfactory comfort: smelling a partner's clothing during periods of separation. *J Appl Soc Psychol* **36**: 2325–35.

McClintock, M K (1971) Menstrual synchrony and suppression. *Nature* **229**: 244–5.

McDonald, R A, Delahay, R J, Carter, S P, Smith, G C & Cheeseman, C L (2008) Perturbing implications of wildlife ecology for disease control. *Trends Ecol Evol* **23**: 53–6.

McGlone, J J & Morrow, J L (1988) Reduction of pig agonistic behavior by androstenone. *J Anim Sci* **66**: 880–4.

McGrath, P T, Xu, Y F, Ailion, M *et al.* (2011) Parallel evolution of domesticated *Caenorhabditis* species targets pheromone receptor genes. *Nature* **477**: 321–5.

McGraw, L A & Young, L J (2010) The prairie vole: an emerging model organism for understanding the social brain. *Trends Neurosci* **33**: 103–9.

McNeil, J N (1992) Evolutionary perspectives and insect pest control: an attractive blend for the deployment of semiochemicals in management systems. In Roitberg, B D & Isman, M B (eds.) *Insect Chemical Ecology: an Evolutionary Approach.* pp. 334–52. New York: Chapman and Hall.

McRae, J F, Mainland, J D, Jaeger, S R *et al.* (2012) Genetic variation in the odorant receptor OR2J3 is associated with the ability to detect the "grassy" smelling odor, cis-3-hexen-1-ol. *Chem Senses* **37**: 585–93.

Mead, K S & Caldwell, R (2011) Mantis shrimp: olfactory apparatus and chemosensory behavior. In Breithaupt, T & Thiel, M (eds.) *Chemical Communication in Crustaceans.* pp. 219–38. New York: Springer.

Mead, K S, Koehl, M A R & O'Donnell, M J (1999) Stomatopod sniffing: the scaling of chemosensory sensillae and flicking behavior with body size. *J Exp Mar Biol Ecol* **241**: 235–61.

Meaney, M J (2001) Nature, nurture, and the disunity of knowledge. *Ann N Y Acad Sci* **935**: 50–61.

Mebs, D (2009) Chemical biology of the mutualistic relationships of sea anemones with fish and crustaceans. *Toxicon* **54**: 1071–4.

Meckley, T D, Wagner, C M & Luehring, M A (2012) Field evaluation of larval odor and mixtures of synthetic pheromone components for attracting migrating sea lampreys in rivers. *J Chem Ecol* **38**: 1062–9.

Meinwald, J (2003) Understanding the chemistry of chemical communication: are we there yet? *Proc Natl Acad Sci USA* **100**: 14514–16.

Meinwald, J (2009) The chemistry of biotic interactions in perspective: small molecules take center stage. *J Org Chem* **74**: 1813–25.

Mellon, D (2012) Smelling, feeling, tasting and touching: behavioral and neural integration of antennular chemosensory and mechanosensory inputs in the crayfish. *J Exp Biol* **215**: 2163–72.

Melo, A & González-Mariscal, G (2010) Communication by olfactory signals in rabbits: its role in reproduction. In Gerald, L (ed.) *Pheromones.* pp. 351–71. London: Academic Press.

Menashe, I, Man, O, Lancet, D & Gilad, Y (2003) Different noses for different people. *Nat Genet* **34**: 143–4.

Menashe, I, Abaffy, T, Hasin, Y *et al.* (2007) Genetic elucidation of human hyperosmia to isovaleric acid. *PLoS Biol* **5**: e284.

Menini, A (ed.) (2010) *The Neurobiology of Olfaction.* Boca Raton, FL: CRC Press. Available online at www.ncbi.nlm.nih.gov/books/NBK55980.

Mercier, A & Hamel, J F (2010) Synchronized breeding events in sympatric marine invertebrates: role of behavior and fine temporal windows in maintaining reproductive isolation. *Behav Ecol Sociobiol* **64**: 1749–65.

Meredith, M (1998) Vomeronasal, olfactory, hormonal convergence in the brain – cooperation or coincidence? *Ann N Y Acad Sci* **855**: 349–61.

Meredith, M (2001) Human vomeronasal organ function: a critical review of best and worst cases. *Chem Senses* **26**: 433–45.

Meyer, S L F, Johnson, G, Dimock, M, Fahey, J W & Huettel, R N (1997) Field efficacy of *Verticillium lecanii,* sex pheromone, and pheromone analogs as potential management agents for soybean cyst nematode. *J Nematol* **29**: 282–8.

Michael, R P & Keverne, E B (1970) Primate sex pheromones of vaginal origin. *Nature* **225**: 84–5.

Miesenböck, G (2009) The optogenetic catechism. *Science* **326**: 395–9.

Miklos, G L G & Maleszka, R (2011) Epigenomic communication systems in humans and honey bees: from molecules to behavior. *Horm Behav* **59**: 399–406.

Milinski, M (2006) The major histocompatibility complex, sexual selection, and mate choice. *Annu Rev Ecol Evol Syst* **37**: 159–86.

Milinski, M & Wedekind, C (2001) Evidence for MHC-correlated perfume preferences in humans. *Behav Ecol* **12**: 140–9.

Milinski, M, Griffiths, S, Wegner, K M *et al.* (2005) Mate choice decisions of stickleback females predictably modified by MHC peptide ligands. *Proc Natl Acad Sci USA* **102**: 4414–18.

Millar, J G & Haynes, K F (eds.) (1998) *Methods in Chemical Ecology. Volume 1. Chemical Methods.* London: Chapman & Hall.

Miller, E J, Eldridge, M D B & Herbert, C A (2010a) Dominance and paternity in the tammar wallaby. In Coulson, G M & Eldridge, M D B (eds.) *Macropods: Biology of Kangaroos, Wallabies and Rat-kangaroos.* pp. 77–86. Collingwood, VIC: CSIRO Publishing.

Miller, J R, McGhee, P S, Siegert, P Y *et al.* (2010b) General principles of attraction and competitive attraction as revealed by large-cage studies of moths responding to sex pheromone. *Proc Natl Acad Sci USA* **107**: 22–7.

Mills, M G L, Gorman, M L & Mills, M E J (1980) The scent marking behaviour of the brown hyaena, *Hyaena brunea. S Afr J Zool* **15**: 240–8.

Minks, A K & Cardé, R T (1988) Disruption of pheromone communication in moths – is the natural blend really most efficacious. *Entomol Exp Appl* **49**: 25–36.

Mitchell, M D, McCormick, M I, Ferrari, M C O & Chivers, D P (2011) Coral reef fish rapidly learn to identify multiple unknown predators upon recruitment to the reef. *PLoS ONE* **6**: e15764.

Miura, T & Scharf, M (2011) Molecular basis underlying caste differentiation in termites. In Bignell, D, Roisin, Y & Lo, N (eds.) *Biology of Termites: a Modern Synthesis*, 2nd edn., pp. 211–53. Dordrecht: Springer.

Mizuno, K (2011) Infantile olfactory learning. In Preedy, V R, Watson, R R & Martin, C R (eds.) *Handbook of Behavior, Food and Nutrition*. pp. 119–32. New York: Springer.

Mochizuki, F, Fukumoto, T, Noguchi, H *et al.* (2002) Resistance to a mating disruptant composed of (Z)-11-tetradecenyl acetate in the smaller tea tortrix, *Adoxophyes honmai* (Yasuda)(Lepidoptera: Tortricidae). *Appl Entomol Zool* **37**: 299–304.

Mochizuki, F, Noguchi, H, Sugie, H, Tabata, J & Kainoh, Y (2008) Sex pheromone communication from a population resistant to mating disruptant of the smaller tea tortrix, *Adoxophyes honmai* Yasuda (Lepidoptera: Tortricidae). *Appl Entomol Zool* **43**: 293–8.

Molet, M, Chittka, L & Raine, N E (2009) Potential application of the bumblebee foraging recruitment pheromone for commercial greenhouse pollination. *Apidologie* **40**: 608–16.

Møller, A P & Thornhill, R (1998) Bilateral symmetry and sexual selection: a meta-analysis. *Am Nat* **151**: 174–92.

Mombaerts, P (2004) Genes and ligands for odorant, vomeronasal and taste receptors. *Nat Rev Neurosci* **5**: 263–78.

Mombaerts, P (2006) Axonal wiring in the mouse olfactory system. *Annu Rev Cell Dev Biol* **22**: 713–37.

Monaco, E L, Tallamy, D W & Johnson, R K (1998) Chemical mediation of egg dumping in the lace bug *Gargaphia solani* Heidemann (Heteroptera: Tingidae). *Anim Behav* **56**: 1491–5.

Mondor, E B & Roitberg, B D (2004) Inclusive fitness benefits of scent-marking predators. *Proc R Soc B* **271**: S341.

Monnin, T & Peeters, C (1999) Dominance hierarchy and reproductive conflicts among subordinates in a monogynous queenless ant. *Behav Ecol* **10**: 323–32.

Monnin, T, Malosse, C & Peeters, C (1998) Solid-phase microextraction and cuticular hydrocarbon differences related to reproductive activity in queenless ant *Dinoponera quadriceps*. *J Chem Ecol* **24**: 473–90.

Montagna, W & Parakkal, P F (1974) *The Structure and Function of Skin*. New York: Academic Press.

Montell, C (2009) A taste of the *Drosophila* gustatory receptors. *Curr Opin Neurobiol* **19**: 345–53.

Montgomery, J C, Diebel, C, Halstead, M B D & Downer, J (1999) Olfactory search tracks in the Antarctic fish *Trematomus bernacchii*. *Polar Biol* **21**: 151–4.

Montgomery, J C, Carton, G, Voigt, R, Baker, C & Diebel, C (2000) Sensory processing of water currents by fishes. *Phil Trans R Soc B* **355**: 1325.

MontiBloch, L, JenningsWhite, C & Berliner, D L (1998) The human vomeronasal system – a review. *Ann N Y Acad Sci* **855**: 373–89.

Moore, A J & Moore, P J (1999) Balancing sexual selection through opposing mate choice and male competition. *Proc R Soc B* **266**: 711–16.

Moore, A J, Reagan, N L & Haynes, K F (1995) Conditional signaling strategies – effects of ontogeny, social experience and social-status on the pheromonal signal of male cockroaches. *Anim Behav* **50**: 191–202.

Moore, B P, Brown, W V & Rothschild, M (1990) Methylalkylpyrazines in aposematic insects, their host plants and mimics. *Chemoecology* **1**: 43–51.

Moore, P J, Reagan-Wallin, N L, Haynes, K F & Moore, A J (1997) Odour conveys status on cockroaches. *Nature* **389**: 25.

Moreno-Rueda, G (2007) Is there empirical evidence for the cost of begging? *J Ethol* **25**: 215–22.

Morgan, E D (2008) Chemical sorcery for sociality: exocrine secretions of ants (Hymenoptera: Formicidae). *Myrmecol News* **11**: 79–90.

Morgan, E D (2009) Trail pheromones of ants. *Physiol Entomol* **34**: 1–17.

Mori, K (2007) Significance of chirality in pheromone science. *Biorg Med Chem* **15**: 7505–23.

Mori, K & Sakano, H (2011) How is the olfactory map formed and interpreted in the mammalian brain? *Annu Rev Neurosci* **34**: 467–99.

Morris, N M & Udry, R J (1978) Pheromonal influences on human sexual behavior: an experimental search. *J Biosoc Sci* 10: 147–57.

Moser, E & McCulloch, M (2010) Canine scent detection of human cancers: a review of methods and accuracy. *J Vet Behav* 5: 145–52.

Mucignat-Caretta, C, Caretta, A & Cavaggioni, A (1995) Acceleration of puberty onset in female mice by male urinary proteins. *J Physiol* 486: 517–22.

Mucignat-Caretta, C, Redaelli, M & Caretta, A (2012) One nose, one brain: contribution of the main and accessory olfactory system to chemosensation. *Front Neuroanat* 6: 46.

Muenz, T, Maisonnasse, A, Plettner, E, Le Conte, Y & Rössler, W (2012) Sensory reception of the primer pheromone ethyl oleate. *Naturwissenschaften* 99: 421–5.

Müller, C A & Manser, M B (2007) 'Nasty neighbours' rather than 'dear enemies' in a social carnivore. *Proc R Soc B* 274: 959.

Müller-Schwarze, D (2006) *Chemical Ecology of Vertebrates*. Cambridge: Cambridge University Press.

Müller-Schwarze, D, Altieri, R & Porter, N (1984) Alert odor from skin gland in deer. *J Chem Ecol* 10: 1707–29.

Munger, S D, Leinders-Zufall, T & Zufall, F (2009) Subsystem organization of the mammalian sense of smell. *Annu Rev Physiol* 71: 115–40.

Murata, K, Wakabayashi, Y, Sakamoto, K *et al.* (2011) Effects of brief exposure of male pheromone on multiple-unit activity at close proximity to kisspeptin neurons in the goat arcuate nucleus. *J Reprod Dev* 57: 197–202.

Murlis, J, Elkinton, JS & Cardé, R T (1992) Odor plumes and how insects use them. *Annu Rev Entomol* 37: 505–32.

Mylonas, C C, Fostier, A & Zanuy, S (2010) Broodstock management and hormonal manipulations of fish reproduction. *Gen Comp Endocrinol* 165: 516–34.

Nakada, T, Toyoda, F, Iwata, T *et al.* (2007) Isolation, characterization and bioactivity of a region-specific pheromone, [Val8]sodefrin from the newt Cynops pyrrhogaster. *Peptides* 28: 774–80.

Nakagawa, T & Vosshall, L B (2009) Controversy and consensus: noncanonical signaling mechanisms in the insect olfactory system. *Curr Opin Neurobiol* 19: 284–92.

Nakagawa, T, Pellegrino, M, Sato, K, Vosshall, L B & Touhara, K (2012) Amino acid residues contributing to function of the heteromeric insect olfactory receptor complex. *PLoS ONE* 7: e32372.

Nash, D R & Boomsma, J J (2008) Communication between hosts and social parasites. In d'Ettorre, P & Hughes, D P (eds.) *Sociobiology of Communication: An Interdisciplinary Perspective*. pp. 55–79. Oxford: Oxford University Press.

Nash, D R, Als, T D, Maile, R, Jones, G R & Boomsma, J J (2008) A mosaic of chemical coevolution in a large blue butterfly. *Science* 319: 88–90.

Natsch, A, Kuhn, F & Tiercy, JM (2010) Lack of evidence for HLA-linked patterns of odorous carboxylic acids released from glutamine conjugates secreted in the human axilla. *J Chem Ecol* 36: 837–46.

Nault, L R (1973) Alarm pheromones help aphids escape predators. *Ohio Report* 58: 16–17.

Nault, L R, Montgomery, M E & Bowers, W S (1976) Ant-aphid association: role of aphid alarm pheromone. *Science* 192: 1349–51.

Naumann, K, Winston, M L, Slessor, K N, Prestwich, G D & Latli, B (1992) Intra-nest transmission of aromatic honey-bee queen mandibular gland pheromone components – movement as a unit. *Can Entomol* 124: 917–34.

Naumann, K, Winston, M L & Slessor, K N (1993) Movement of honey bee (*Apis mellifera* L.) queen mandibular gland pheromone in populous and unpopulous colonies. *J Insect Behav* 6: 211–23.

Nehring, V, Evison, S E F, Santorelli, L A, d'Ettorre, P & Hughes, W O H (2011) Kin-informative recognition cues in ants. *Proc R Soc B* 278: 1942–8.

Nehring, V, Wyatt, T D & d'Ettorre, P (2014) Noise in chemical communication. In H Brumm (ed.) *Animal Communication and Noise. Animal Signals and Communication*, Vol. 2. New York: Springer.

Nei, M, Niimura, Y & Nozawa, M (2008) The evolution of animal chemosensory receptor gene repertoires: roles of chance and necessity. *Nat Rev Genet* 9: 951–63.

Nevitt, G A (2008) Sensory ecology on the high seas: the odor world of the procellariiform seabirds. *J Exp Biol* 211: 1706–13.

Nevitt, G A, Losekoot, M & Weimerskirch, H (2008) Evidence for olfactory search in wandering albatross, *Diomedea exulans. Proc Natl Acad Sci USA* 105: 4576–81.

Newcomb, R D, Xia, M B & Reed, D R (2012) Heritable differences in chemosensory ability among humans. *Flavour* 1: 9.

Newey, P S, Robson, S K A & Crozier, R H (2008) Near-infrared spectroscopy as a tool in behavioural ecology: a case study of the weaver ant, *Oecophylla smaragdina*. *Anim Behav* 76: 1727–33.

Newey, P S, Robson, S K A & Crozier, R H (2010) Weaver ants *Oecophylla smaragdina* encounter nasty neighbors rather than dear enemies. *Ecology* 91: 2366–72.

Nicholson, J K & Lindon, J C (2008) Systems biology: metabonomics. *Nature* 455: 1054–6.

Nie, Y, Swaisgood, R R, Zhang, Z *et al.* (2012) Giant panda scent-marking strategies in the wild: role of season, sex and marking surface. *Anim Behav* 84: 39–44.

Nieberding, C M, de Vos, H, Schneider, M V *et al.* (2008) The male sex pheromone of the butterfly *Bicyclus anynana*: towards an evolutionary analysis. *PLoS ONE* 3: e2751.

Nieh, J C, Contrera, F A L & Nogueira-Neto, P (2003) Pulsed mass recruitment by a stingless bee, *Trigona hyalinata*. *Proc R Soc B* 270: 2191.

Nieh, J C, Contrera, F A L, Yoon, R R, Barreto, L S & Imperatriz-Fonseca, V L (2004) Polarized short odor trail recruitment communication by a stingless bee, *Trigona spinipes*. *Behav Ecol Sociobiol* 56: 435–48.

Nixon, A, Mallet, A I & Gower, D B (1988) Simultaneous quantification of 5 odorous steroids (16-androstenes) in the axillary hair of men. *J Steroid Biochem Mol Biol* 29: 505–10.

Nodari, F, Hsu, F F, Fu, X Y *et al.* (2008) Sulfated steroids as natural ligands of mouse pheromone-sensing neurons. *J Neurosci* 28: 6407–18.

Noldus, L P J J, Potting, R P J & Barendregt, H E (1991) Moth sex-pheromone adsorption to leaf surface – bridge in time for chemical spies. *Physiol Entomol* 16: 329–44.

Nonacs, P & Hager, R (2011) The past, present and future of reproductive skew theory and experiments. *Biol Rev* 86: 271–98.

Nordlund, D A & Lewis, W J (1976) Terminology of chemical releasing stimuli in intraspecific and interspecific interactions. *J Chem Ecol* 2: 211–20.

Nosil, P & Schluter, D (2011) The genes underlying the process of speciation. *Trends Ecol Evol* 26: 160–7.

Novotny, M V (2003) Pheromones, binding proteins and receptor responses in rodents. *Biochem Soc Trans* 31: 117–22.

Novotny, M V, Harvey, S, Jemiolo, B & Alberts, J (1985) Synthetic pheromones that promote inter-male aggression in mice. *Proc Natl Acad Sci USA* 82: 2059–61.

Novotny, M V, Xie, T M, Harvey, S *et al.* (1995) Stereoselectivity in mammalian chemical communication – male-mouse pheromones. *Experientia* 51: 738–43.

Novotny, M V, Ma, W, Zidek, L & Daev, E (1999a) Recent biochemical insights into puberty acceleration, estrus induction and puberty delay in the house mouse. In Johnston, R E, Müller-Schwarze, D & Sorensen, P W (eds.) *Advances in Chemical Signals in Vertebrates.* pp. 99–116. New York: Kluwer Academic/Plenum Press.

Novotny, M V, Ma, W D, Wiesler, D & Zidek, L (1999b) Positive identification of the puberty-accelerating pheromone of the house mouse: the volatile ligands associating with the major urinary protein. *Proc R Soc B* 266: 2017–22.

Nufio, C R & Papaj, D R (2001) Host marking behavior in phytophagous insects and parasitoids. *Entomol Exp Appl* 99: 273–93.

Nyby, J G (2009) Adult house mouse (*Mus musculus*) ultrasonic calls: hormonal and pheromonal regulation. In Brudzynski, S M (ed.) *Handbook of Mammalian Vocalization.* pp. 303–10. Oxford: Academic Press.

O'Riain, M J & Jarvis, J U M (1997) Colony member recognition and xenophobia in the naked mole-rat. *Anim Behav* 53: 487–98.

Ober, C (1999) Studies of HLA, fertility and mate choice in a human isolate. *Hum Reprod Update* 5: 103–7.

Ober, C, Weitkamp, L R, Cox, N *et al.* (1997) HLA and mate choice in humans. *Am J Hum Genet* 61: 497–504.

Obin, M S & Vander Meer, R K (1989) Nestmate recognition in fire ants (*Solenopsis invicta* Buren) – do queens label workers? *Ethology* 80: 255–64.

Oboti, L, Schellino, R, Giachino, C *et al.* (2011) Newborn interneurons in the accessory olfactory bulb promote mate recognition in female mice. *Front Neurosci* 5: 10.3389/fnins.2011.00113.

Oehlschlager, A C, Chinchilla, C, Castillo, G & Gonzalez, L (2002) Control of red ring disease by mass trapping of *Rhynchophorus palmarum* (Coleoptera: Curculionidae). *Fla Entomol* 85: 507–13.

Olender, T & Lancet, D (2012) Evolutionary grass roots for odor recognition. *Chem Senses* 37: 581–4.

Olender, T, Waszak, S, Viavant, M *et al.* (2012) Personal receptor repertoires: olfaction as a model. *BMC Genomics* 13: 414.

Olsén, K H (2011) Effects of pollutants on olfactory mediated behaviors in fish and crustaceans. In Breithaupt, T & Thiel, M (eds.) *Chemical Communication in Crustaceans.* pp. 507–29. New York: Springer.

Olsson, M & Shine, R (1998) Chemosensory mate recognition may facilitate prolonged mate guarding by male snow skinks, *Niveoscincus microlepidotus. Behav Ecol Sociobiol* 43: 359–63.

Olsson, P & Laska, M (2010) Human male superiority in olfactory sensitivity to the sperm attractant odorant bourgeonal. *Chem Senses* 35: 427–32.

Olsson, S B, Kesevan, S, Groot, A T *et al.* (2010) *Ostrinia* revisited: evidence for sex linkage in European corn borer *Ostrinia nubilalis* (Hubner) pheromone reception. *BMC Evol Biol* 10: 285.

Ono, M, Igarashi, T, Ohno, E & Sasaki, M (1995) Unusual thermal defence by a honeybee against mass attack by hornets. *Nature* 377: 334–6.

Ophir, A G, Schrader, S B & Gillooly, J F (2010) Energetic cost of calling: general constraints and species-specific differences. *J Evol Biol* 23: 1564–9.

Oppelt, A, Spitzenpfeil, N, Kroiss, J & Heinze, J (2008) The significance of intercolonial variation of cuticular hydrocarbons for inbreeding avoidance in ant sexuals. *Anim Behav* 76: 1029–34.

Ortiz, C O, Etchberger, J F, Posy, S L *et al.* (2006) Searching for neuronal left/right asymmetry: genome wide analysis of nematode receptor-type guanylyl cyclases. *Genetics* 173: 131–49.

Ouyang, G, Vuckovic, D & Pawliszyn, J (2011) Nondestructive sampling of living systems using in vivo solid-phase microextraction. *Chem Rev* 111: 2784–814.

Owens, I P F, Rowe, C & Thomas, A L R (1999) Sexual selection, speciation and imprinting: separating the sheep from the goats. *Trends Ecol Evol* 14: 131–2.

Oyarzun, F X & Strathmann, R R (2011) Plasticity of hatching and the duration of planktonic development in marine invertebrates. *Integr Comp Biol* 51: 81–90.

Ozaki, M, Wada-Katsumata, A, Fujikawa, K *et al.* (2005) Ant nestmate and non-nestmate discrimination by a chemosensory sensillum. *Science* 309: 311–14.

Ozsolak, F & Milos, P M (2010) RNA sequencing: advances, challenges and opportunities. *Nat Rev Genet* 12: 87–98.

Page, J L, Dickman, B D, Webster, D R & Weissburg, M J (2011) Staying the course: chemical signal spatial properties and concentration mediate cross-stream motion in turbulent plumes. *J Exp Biol* 214: 1513–22.

Page, R E, Rueppell, O & Amdam, G V (2012) Genetics of reproduction and regulation of honeybee (*Apis mellifera* L.) social behavior. *Annu Rev Genet* 46: 97–119.

Pain, F, L'Heureux, B & Gurden, H (2011) Visualizing odor representation in the brain: a review of imaging techniques for the mapping of sensory activity in the olfactory glomeruli. *Cell Mol Life Sci* 68: 2689–709.

Palmer, A R (2000) Quasireplication and the contract of error: lessons from sex ratios, heritabilities and fluctuating asymmetry. *Annu Rev Ecol Syst* 31: 441–80.

Palmer, C A, Watts, R A, Gregg, R G *et al.* (2005) Lineage-specific differences in evolutionary mode in a salamander courtship pheromone. *Mol Biol Evol* 22: 2243–56.

Palmer, C A, Watts, R A, Houck, L D, Picard, A L & Arnold, S J (2007a) Evolutionary replacement of components in a salamander pheromone signaling complex: more evidence for phenotypic-molecular decoupling. *Evolution* 61: 202–15.

Palmer, C A, Hollis, D M, Watts, R A *et al.* (2007b) Plethodontid modulating factor, a hypervariable salamander courtship pheromone in the three-finger protein superfamily. *FEBS J* 274: 2300–10.

Palmer, C A, Watts, R A, Hastings, A P, Houck, L D & Arnold, S J (2010) Rapid evolution of plethodontid modulating factor, a hypervariable salamander courtship pheromone, is driven by positive selection. *J Mol Evol* 70: 427–40.

Pankiw, T (2004) Brood pheromone regulates foraging activity of honey bees (Hymenoptera: Apidae). *J Econ Entomol* 97: 748–51.

Parker, G A & Pizzari, T (2010) Sperm competition and ejaculate economics. *Biol Rev* 85: 897–934.

Parker, M R & Mason, R T (2012) How to make a sexy snake: estrogen activation of female sex pheromone in male red-sided garter snakes. *J Exp Biol* 215: 723–30.

Parker, M R, Young, B A & Kardong, K V (2008) The forked tongue and edge detection in snakes *Crotalus oreganus*: an experimental test. *J Comp Psychol* 122: 35–40.

Partan, S R & Marler, P (2005) Issues in the classification of multimodal communication signals. *Am Nat* 166: 231–45.

Pasteels, J M (2007) Chemical defence, offence and alliance in ants–aphids–ladybirds relationships. *Popul Ecol* 49: 5–14.

Pasteels, J M, Deneubourg, J L & Goss, S (1987) Self-organization mechanisms in ant societies (I): trail recruitment to newly discovered food sources. *Experimentia Supplementum* 54: 155–75.

Paterson, S & Pemberton, J M (1997) No evidence for major histocompatibility complex-dependent mating patterns in a free-living ruminant population. *Proc R Soc B* 264: 1813–19.

Peakall, R (1990) Responses of male *Zaspilothynnus trilobatus* Turner wasps to females and the sexually deceptive orchid it pollinates. *Funct Ecol* 4: 159–68.

Peakall, R, Ebert, D, Poldy, J *et al.* (2010) Pollinator specificity, floral odour chemistry and the phylogeny of Australian sexually deceptive *Chiloglottis* orchids: implications for pollinator-driven speciation. *New Phytol* 188: 437–50.

Peele, P, Salazar, I, Mimmack, M, Keverne, E B & Brennan, P A (2003) Low molecular weight constituents of male mouse urine mediate the pregnancy block effect and convey information about the identity of the mating male. *Eur J Neurosci* 18: 622–8.

Peeters, C & Liebig, J (2009) Fertility signaling as a general mechanism of regulating reproductive division of labor in ants. In Gadau, J & Fewell, J H (eds.) *Organization of Insect Societies: From Genome to Sociocomplexity.* pp. 220–42. Cambridge, MA: Harvard University Press.

Peeters, C, Monnin, T & Malosse, C (1999) Cuticular hydrocarbons correlated with reproductive status in a queenless ant. *Proc R Soc B* 266: 1323–7.

Pelosi, P (1996) Perireceptor events in olfaction. *J Neurobiol* 30: 3–19.

Pelosi, P, Zhou, J, Ban, L & Calvello, M (2006) Soluble proteins in insect chemical communication. *Cell Mol Life Sci* 63: 1658–76.

Pener, M P & Simpson, S J (2009) Locust phase polyphenism: an update. *Adv Insect Physiol* 36: 1–272.

Penn, D J (2002) The scent of genetic compatibility: sexual selection and the major histocompatibility complex. *Ethology* 108: 1–21.

Penn, D J & Frommen, J G (2010) Kin recognition: an overview of conceptual issues, mechanisms and evolutionary theory. In Kappeler, P M (ed.) *Animal Behaviour: Evolution and Mechanisms.* pp. 55–85. Heidelberg: Springer.

Penn, D J & Ilmonen, P (2005) Major histocompatibility complex (MHC). *eLS.* Chichester: John Wiley DOI: 10.1038/npg.els.0003986.

Penn, D J & Potts, W K (1998a) How do major histocompatibility complex genes influence odor and mating preferences? *Adv Immunol* 69: 411–36.

Penn, D J & Potts, W K (1998b) MHC-disassortative mating preferences reversed by cross-fostering. *Proc R Soc B* 265: 1299–306.

Penn, D J & Potts, W K (1998c) Chemical signals and parasite-mediated sexual selection. *Trends Ecol Evol* 13: 391–6.

Penn, D J, Oberzaucher, E, Grammer, K *et al.* (2007) Individual and gender fingerprints in human body odour. *J R Soc Interface* 4: 331–40.

Pereira, R, Sivinski, J & Teal, P E A (2010a) Influence of a juvenile hormone analog and dietary protein on male *Anastrepha suspensa* (Diptera: Tephritidae) sexual success. *J Econ Entomol* 103: 40–6.

Pereira, R, Sivinski, J, Teal, P & Brockmann, J (2010b) Enhancing male sexual success in a lekking fly (*Anastrepha suspensa* Diptera: Tephritidae) through a juvenile hormone analog has no effect on adult mortality. *J Insect Physiol* 56: 1552–7.

Pernal, S F, Baird, D S, Birmingham, A L *et al.* (2005) Semiochemicals influencing the host-finding behaviour of *Varroa destructor*. *Exp Appl Acarol* 37: 1–26.

Petrulis, A (2009) Neural mechanisms of individual and sexual recognition in Syrian hamsters (*Mesocricetus auratus*): pheromonal communication in higher vertebrates and its implication for reproductive function. *Behav Brain Res* 200: 260–7.

Pettis, J, Pankiw, T & Plettner, E (1999) Bees. In Hardie, J & Minks, A K (eds.) *Pheromones of Non-lepidopteran Insects Associated with Agricultural Plants*. pp. 429–50. Wallingford, Oxon: CAB International.

Pettis, R J, Erickson, B W, Forward, R B & Rittschof, D (1993) Superpotent synthetic tripeptide mimics of the mud-crab pumping pheromone. *Int J Pept Protein Res* 42: 312–19.

Pfennig, D W & Sherman, P W (1995) Kin recognition. *Sci Am* 272: 98–103.

Pham-Delègue, M, Trouiller, J, Caillaud, C, Roger, B & Masson, C (1993) Effect of queen pheromone on worker bees of different ages: behavioural and electrophysiological responses. *Apidologie* 24: 267–81.

Phelan, P L (1992) Evolution of sex pheromones and the role of asymmetric tracking. In Roitberg, B D & Isman, M B (eds.) *Insect Chemical Ecology: an Evolutionary Approach*. pp. 245–64. New York: Chapman and Hall.

Phelan, P L (1997) Evolution of mate-signalling in moths: phylogenetic considerations and predictions from the asymmetric tracking hypothesis. In Choe, J C & Crespi, B J (eds.) *The Evolution of Mating Systems in Insects and Arachnids*. pp. 240–56. Cambridge: Cambridge University Press.

Phelan, P L & Baker, T C (1986) Male-size-related courtship success and intersexual selection in the tobacco moth, *Ephestia elutella*. *Experientia* 42: 1291–3.

Phelan, P L & Baker, T C (1987) Evolution of male pheromones in moths – reproductive isolation through sexual selection. *Science* 235: 205–7.

Phillips, T W & Throne, J E (2010) Biorational approaches to managing stored-product insects. *Annu Rev Entomol* 55: 375–97.

Pickett, J A, Wadhams, L J, Woodcock, C M & Hardie, J (1992) The chemical ecology of aphids. *Annu Rev Entomol* 37: 67–90.

Pickett, J A, Birkett, M A, Dewhirst, S Y *et al.* (2010) Chemical ecology of animal and human pathogen vectors in a changing global climate. *J Chem Ecol* 36: 113–21.

Pierce, N E, Braby, M F, Heath, A *et al.* (2002) The ecology and evolution of ant association in the Lycaenidae (Lepidoptera). *Annu Rev Entomol* 47: 733–71.

Piertney, S B & Oliver, M K (2006) The evolutionary ecology of the major histocompatibility complex. *Heredity* 96: 7–21.

Pike, N & Foster, W A (2008) The ecology of altruism in a clonal insect. In Korb, J & Heinze, J (eds.) *Ecology of Social Evolution*. pp. 37–56. Berlin: Springer.

Pilkington, L J, Messelink, G, van Lenteren, J C & Le Mottee, K (2010) "Protected biological control" – biological pest management in the greenhouse industry. *Biol Control* 52: 216–20.

Pizzari, T & Bonduriansky, R (2010) Sexual behaviour: conflict, cooperation and co-evolution. In Székely, T, Moore, A & Komdeur, J (eds.) *Social Behaviour: Genes, Ecology and Evolution*. pp. 230–66. Cambridge: Cambridge University Press.

Pizzari, T & Snook, R R (2004) Sexual conflict and sexual selection: measuring antagonistic coevolution. *Evolution* 58: 1389–93.

Pizzolon, M, Giacomello, E, Marri, L *et al.* (2010) When fathers make the difference: efficacy of male sexually selected antimicrobial glands in enhancing fish hatching success. *Funct Ecol* 24: 141–8.

Plenderleith, M, Oosterhout, C, Robinson, R L & Turner, G F (2005) Female preference for conspecific males based on olfactory cues in a Lake Malawi cichlid fish. *Biol Lett* 1: 411–14.

Plettner, E, Slessor, K N, Winston, M L & Oliver, J E (1996) Caste-selective pheromone biosynthesis in honey-bees. *Science* 271: 1851–3.

Podjasek, J O, Bosnjak, L M, Brooker, D J & Mondor, E B (2005) Alarm pheromone induces a transgenerational wing polyphenism in the pea aphid, *Acyrthosiphon pisum*. *Can J Zool* 83: 1138–41.

Poiani, A (2006) Complexity of seminal fluid: a review. *Behav Ecol Sociobiol* 60: 289–310.

Polak, M (2008) The developmental instability-sexual selection hypothesis: a general evaluation and case study. *Evol Biol* 35: 208–30.

Porat, D & Chadwick-Furman, N (2004) Effects of anemonefish on giant sea anemones: expansion behavior, growth, and survival. *Hydrobiologia* 530: 513–20.

Porter, J, Craven, B, Khan, R M *et al.* (2007) Mechanisms of scent-tracking in humans. *Nat Neurosci* 10: 27–9.

Porter, M L, Blasic, J R, Bok, M J *et al.* (2012) Shedding new light on opsin evolution. *Proc R Soc B* 279: 3–14.

Porter, R H & Blaustein, A R (1989) Mechanisms and ecological correlates of kin recognition. *Sci Prog* **73**: 53–66.

Porter, R H & Winberg, J (1999) Unique salience of maternal breast odors for newborn infants. *Neurosci Biobehav Rev* **23**: 439–49.

Porter, R H, Tepper, V J & White, D M (1981) Experiential influences on the development of huddling preferences and sibling recognition in spiny mice. *Dev Psychobiol* **14**: 375–82.

Porter, R H, Cernoch, J M & McLaughlin, F J (1983) Maternal recognition of neonates through olfactory cues. *Physiol Behav* **30**: 151–4.

Porter, R H, Balogh, R D, Cernoch, J M & Franchi, C (1986) Recognition of kin through characteristic body odors. *Chem Senses* **11**: 389–95.

Porter, R H, McFadyen-Ketchum, S A & King, G A (1989) Underlying bases of recognition signatures in spiny mice, *Acomys cahirinus*. *Anim Behav* **37**: 638–44.

Poth, D, Wollenberg, K C, Vences, M & Schulz, S (2012) Volatile amphibian pheromones: macrolides from mantellid frogs from Madagascar. *Angew Chem Int Ed* **51**: 2187–90.

Poulin, G B (2011) A guide to using RNAi and other nucleotide-based technologies. *Brief Funct Genomics* **10**: 173–4.

Prehn-Kristensen, A, Wiesner, C, Bergmann, T O *et al.* (2009) Induction of empathy by the smell of anxiety. *PLoS ONE* **4**: e5987.

Prestwich, K N & Walter, T J (1981) Energetics of singing in crickets: effect of temperature in three trilling species (Orthoptera: Gryllidae). *J Comp Physiol B* **143**: 199–212.

Preti, G & Leyden, J J (2010) Genetic influences on human body odor: from genes to the axillae. *J Invest Dermatol* **130**: 344–6.

Preti, G & Wysocki, C J (1999) Human pheromones: releasers or primers, fact or myth. In Johnston, R E, Müller-Schwarze, D & Sorensen, P W (eds.) *Advances in Chemical Signals in Vertebrates*. pp. 315–32. New York: Kluwer Academic/Plenum Press.

Preti, G, Wysocki, C J, Barnhart, K T, Sondheimer, S J & Leyden, JJ (2003) Male axillary extracts contain pheromones that affect pulsatile secretion of luteinizing hormone and mood in women recipients. *Biol Reprod* **68**: 2107–13.

Pungaliya, C, Srinivasan, J, Fox, B *et al.* (2009) A shortcut to identifying small molecule signals that regulate behavior and development in *Caenorhabditis elegans*. *Proc Natl Acad Sci USA* **106**: 7708–13.

Puurtinen, M, Ketola, T & Kotiaho, J S (2009) The good-genes and compatible-genes benefits of mate choice. *Am Nat* **174**: 741–52.

Queller, D C & Strassmann, J E (2010) Evolution of complex societies. In Westneat, D F & Fox, C W (eds.) *Evolutionary Behavioral Ecology*. pp. 327–40. New York: Oxford University Press.

Quental, T B, Patten, M M & Pierce, N E (2007) Host plant specialization driven by sexual selection. *Am Nat* **169**: 830–6.

Quinet, Y & Pasteels, J M (1996) Spatial specialization of the foragers and foraging strategy in *Lasius fuliginosus* (Latreille) (Hymenoptera, Formicidae). *Insectes Soc* **43**: 333–46.

Qvarnström, A & Forsgren, E (1998) Should females prefer dominant males? *Trends Ecol Evol* **13**: 498–501.

Raffa, K F (2001) Mixed messages across multiple trophic levels: the ecology of bark beetle chemical communication systems. *Chemoecology* **11**: 49–65.

Raffa, K F & Dahlsten, D L (1995) Differential responses among natural enemies and prey to bark beetle pheromones. *Oecologia* **102**: 17–23.

Raffa, K F, Phillips, T W & Salom, S M (1993) Strategies and mechanisms of host colonization by bark beetles. In Schowalter, T D (ed.) *Beetle–Pathogens Interactions in Conifer Forests*. pp. 103–28. London: Academic Press.

Raffa, K F, Hobson, K R, LaFontaine, S & Aukema, B H (2007) Can chemical communication be cryptic? Adaptations by herbivores to natural enemies exploiting prey semiochemistry. *Oecologia* **153**: 1009–19.

Rajan, R, Clement, J P & Bhalla, U S (2006) Rats smell in stereo. *Science* **311**: 666–70.

Ramdya, P & Benton, R (2010) Evolving olfactory systems on the fly. *Trends Genet* **26**: 307–16.

Ramírez, S R, Eltz, T, Fujiwara, M K *et al.* (2011) Asynchronous diversification in a specialized plant-pollinator mutualism. *Science* **333**: 1742–6.

Ramm, S A, McDonald, L, Hurst, J L, Beynon, R J &
Stockley, P (2009) Comparative proteomics reveals
evidence for evolutionary diversification of rodent
seminal fluid and its functional significance in
sperm competition. *Mol Biol Evol* 26: 189–98.

Rantala, M J & Marcinkowska, U M (2011) The role of
sexual imprinting and the Westermarck effect in
mate choice in humans. *Behav Ecol Sociobiol* 65:
859–73.

Rasa, O A E (1973) Marking behaviour and its social significance in the African dwarf mongoose, *Helogale
undulata rufula. Z Tierpsychol* 32: 293–318.

Rasmussen, H B, Ganswindt, A, Douglas-Hamilton, I &
Vollrath, F (2008) Endocrine and behavioral changes
in male African elephants: linking hormone changes
to sexual state and reproductive tactics. *Horm Behav*
54: 539–48.

Rasmussen, L E L, Lee, T D, Zhang, A J,
Roelofs, W L & Daves, G D (1997) Purification,
identification, concentration and bioactivity of
(*Z*)-7-dodecen-1-yl acetate: sex pheromone of the
female Asian elephant, *Elephas maximus. Chem
Senses* 22: 417–37.

Rasmussen, L E L, Lazar, J & Greenwood, DR (2003)
Olfactory adventures of elephantine pheromones.
Biochem Soc Trans 31: 137–41.

Rasmussen, L E L, Krishnamurthy, V & Sukumar, R (2005)
Behavioural and chemical confirmation of the preo-
vulatory pheromone, (*Z*)-7-dodecenyl acetate, in wild
Asian elephants: its relationship to musth. *Behaviour*
142: 351–96.

Ratnieks, F L W & Wenseleers, T (2005) Policing insect
societies. *Science* 307: 54–6.

Ratnieks, F L W & Wenseleers, T (2008) Altruism in insect
societies and beyond: voluntary or enforced? *Trends
Ecol Evol* 23: 45–52.

Ratnieks, F L W, Foster, K R & Wenseleers, T (2006)
Conflict resolution in insect societies. *Annu Rev
Entomol* 51: 581–608.

Rechav, Y, Norval, R A I, Tannock, J & Colborne, J (1978)
Attraction of the tick *Ixodes neitzi* to twigs
marked by the klipspringer antelope. *Nature*
275: 310–11.

Reddy, G V P & Guerrero, A (2004) Interactions of insect
pheromones and plant semiochemicals. *Trends Plant
Sci* 9: 253–61.

Reddy, G V P & Guerrero, A (2010) New pheromones and
insect control strategies. In Gerald, L, (ed.)
Pheromones. pp. 493–519. London: Academic Press.

Reeve, H K & Sherman, P W (1993) Adaptation and the
goals of evolutionary research. *Q Rev Biol* 68: 1–32.

Regnier, F E & Wilson, E O (1969) The alarm-defense
system of the ant Lasius alienus. *J Insect Physiol* 15:
893–8.

Regnier, F E & Wilson, E O (1971) Chemical communication and 'propaganda' in slave-maker ants. *Science*
172: 267–9.

Reichle, C, Aguilar, I, Ayasse, M *et al.* (2013) Learnt
information in species-specific 'trail pheromone'
communication in stingless bees. *Anim Behav*
85: 225–32.

Reidenbach, M A & Koehl, M A R (2011) The spatial and
temporal patterns of odors sampled by lobsters and
crabs in a turbulent plume. *J Exp Biol* 214: 3138–53.

Reinhard, J & Kaib, M (1995) Interaction of pheromones
during food exploitation by the termite
Schedorhinotermes lamanianus. Physiol Entomol
20: 266–72.

Reinhard, J, Lacey, M J, Ibarra, F *et al.* (2002)
Hydroquinone: a general phagostimulating pheromone in termites. *J Chem Ecol* 28: 1–14.

Reisert, J & Restrepo, D (2009) Molecular tuning of odorant receptors and its implication for odor signal
processing. *Chem Senses* 34: 535–45.

Rekwot, P I, Ogwu, D, Oyedipe, E O & Sekoni, V O (2001)
The role of pheromones and biostimulation in animal reproduction. *Anim Reprod Sci* 65: 157–70.

Restrepo, D, Lin, W H, Salcedo, E, Yarnazaki, K &
Beauchamp, G (2006) Odortypes and MHC peptides:
complementary chemosignals of MHC haplotype?
Trends Neurosci 29: 604–9.

Restrepo, D, Doucette, W, Whitesell, J D, McTavish, T S &
Salcedo, E (2009) From the top down: flexible reading of a fragmented odor map. *Trends Neurosci* 32:
525–31.

Rettenmeyer, C W (1963) Behavioral studies of army ants.
Univ Kans Sci Bull 44: 281–465.

Rettenmeyer, C W, Rettenmeyer, M, Joseph, J &
Berghoff, S (2011) The largest animal association
centered on one species: the army ant *Eciton burch-
ellii* and its more than 300 associates. *Insectes Soc*
58: 281–92.

Reusch, T B H, Haberli, M A, Aeschlimann, P B & Milinski, M (2001) Female sticklebacks count alleles in a strategy of sexual selection explaining MHC polymorphism. *Nature* 414: 300–2.

Reynolds, A M (2010) Bridging the gulf between correlated random walks and Lévy walks: autocorrelation as a source of Lévy walk movement patterns. *J R Soc Interface* 7: 1753–8.

Reynolds, A M, Reynolds, D R, Smith, A D, Svensson, G P & Löfstedt, C (2007) Appetitive flight patterns of male *Agrotis segetum* moths over landscape scales. *J Theor Biol* 245: 141–9.

Rezával, C, Pavlou, H J, Dornan, A J *et al.* (2012) Neural circuitry underlying *Drosophila* female postmating behavioral responses. *Curr Biol* 22: 1155–65.

Richgels, P K & Rollmann, S M (2012) Genetic variation in odorant receptors contributes to variation in olfactory behavior in a natural population of *Drosophila melanogaster. Chem Senses* 37: 229–40.

Richter, S H, Garner, J P & Wurbel, H (2009) Environmental standardization: cure or cause of poor reproducibility in animal experiments? *Nat Meth* 6: 257–61.

Richter, S H, Garner, J P, Zipser, B *et al.* (2011) Effect of population heterogenization on the reproducibility of mouse behavior: a multi-laboratory study. *PLoS ONE* 6: e16461.

Ritchie, M G (2007) Sexual selection and speciation. *Annu Rev Ecol Evol Syst* 38: 79–102.

Rittschof, D (2009) Future trends in antifouling research. In Hellio, C & Yebra, D (eds.) *Advances in Marine Antifouling Coatings and Technologies.* pp. 725–48. Cambridge/Boca Raton, FL: Woodhead/CRC Press.

Rittschof, D & Cohen, J H (2004) Crustacean peptide and peptide-like pheromones and kairomones. *Peptides* 25: 1503–16.

Rivière, S, Challet, L, Fluegge, D, Spehr, M & Rodriguez, I (2009) Formyl peptide receptor-like proteins are a novel family of vomeronasal chemosensors. *Nature* 459: 574–7.

Roberts, M L, Buchanan, K L & Evans, M R (2004) Testing the immunocompetence handicap hypothesis: a review of the evidence. *Anim Behav* 68: 227–39.

Roberts, R L, Zullo, A, Gustafson, E A & Carter, C S (1996) Perinatal steroid treatments alter alloparental and affiliative behavior in prairie voles. *Horm Behav* 30: 576–82.

Roberts, R L, Williams, J R, Wang, A K & Carter, C S (1998) Cooperative breeding and monogamy in prairie voles: influence of the sire and geographical variation. *Anim Behav* 55: 1131–40.

Roberts, S A, Davidson, A J, McLean, L, Beynon, R J & Hurst, J L (2012) Pheromonal induction of spatial learning in mice. *Science* 338: 1462–5.

Roberts, S A, Simpson, D M, Armstrong, S D *et al.* (2010) Darcin: a male pheromone that stimulates female memory and sexual attraction to an individual male's odour. *BMC Biol* 8: 75.

Roberts, S C (2007) Scent marking. In Wolff, J O & Sherman, P W (eds.) *Rodent Societies: an Ecological and Evolutionary Perspective.* pp. 255–67. Chicago: Chicago University Press.

Roberts, S C (2012) On the relationship between scent-marking and territoriality in callitrichid primates. *Int J Primatol* 33: 749–61.

Roberts, S C & Gosling, L M (2003) Genetic similarity and quality interact in mate choice decisions by female mice. *Nat Genet* 35: 103–6.

Roberts, S C & Gosling, L M (2004) Manipulation of olfactory signaling and mate choice for conservation breeding: a case study of harvest mice. *Conserv Biol* 18: 548–56.

Roberts, S C & Lowen, C (1997) Optimal patterns of scent marks in klipspringer (*Oreotragus oreotragus*) territories. *J Zool* 243: 565–78.

Roberts, S C, Gosling, L M, Spector, T D *et al.* (2005) Body odor similarity in noncohabiting twins. *Chem Senses* 30: 651–6.

Roberts, S C, Gosling, L M, Carter, V & Petrie, M (2008) MHC-correlated odour preferences in humans and the use of oral contraceptives. *Proc R Soc B* 275: 2715–22.

Robertson, H M & Wanner, K W (2006) The chemoreceptor superfamily in the honey bee, *Apis mellifera*: expansion of the odorant, but not gustatory, receptor family. *Genome Res* 16: 1395.

Robinette, S L, Brutschweiler, R, Schroeder, F C & Edison, A S (2011) NMR in metabolomics and natural products research: two sides of the same coin. *Acc Chem Res* 45: 288–97.

Robinson, E J H (2009) Physiology as a caste-defining feature. *Insectes Soc* 56: 1–6.

Robinson, E J H, Jackson, D E, Holcombe, M &
Ratnieks, FLW (2005a) 'No entry' signal in ant foraging. *Nature* **438**: 442.

Robinson, G E, Grozinger, C M & Whitfield, C W (2005b)
Sociogenomics: social life in molecular terms. *Nat
Rev Genet* **6**: 257–70.

Rodriguez, I, Greer, C A, Mok, M Y & Mombaerts, P
(2000) A putative pheromone receptor gene
expressed in human olfactory mucosa. *Nat Genet*
26: 18–19.

Rodriguez-Saona, C R & Stelinski, L L (2009) Behavior-
modifying strategies in IPM: theory and practice. In
Peshwin, R & Dhawan, AK (eds.) *Integrated Pest
Management: Innovation-Development Process*. pp.
263–315. Dordrecht: Springer.

Roelofs, W L & Rooney, A P (2003) Molecular genetics and
evolution of pheromone biosynthesis in Lepidoptera.
Proc Natl Acad Sci USA **100**: 9179–84.

Roelofs, W L, Liu, W T, Hao, G X *et al.* (2002) Evolution of
moth sex pheromones via ancestral genes. *Proc Natl
Acad Sci USA* **99**: 13621–6.

Roessingh, P, Peterson, S C & Fitzgerald, T D (1988) The
sensory basis of trail following in some lepidopter-
ous larvae – contact chemoreception. *Physiol
Entomol* **13**: 219–24.

Roitberg, B D & Prokopy, R J (1981) Experience required
for pheromone recognition by the apple maggot fly.
Nature **292**: 540–1.

Roitberg, B D, Lauzon, C R, Opp, S B & Papaj, D (2009)
Functional and behavioural ecology of tree-fruit
pests: the four Fs of fruit flies (Diptera: Tephritidae).
In Aluja, M, Leskey, T C & Vincent, C (eds.)
Biorational Tree Fruit Pest Management. pp. 56–84.
Wallingford: CABI Publishing.

Rollmann, S M, Wang, P, Date, P *et al.* (2010) Odorant
receptor polymorphisms and natural variation in
olfactory behavior in *Drosophila melanogaster*.
Genetics **186**: 687–97.

Romantshik, O, Porter, R, Tillmann, V & Varendi, H (2007)
Preliminary evidence of a sensitive period for olfactory learning by human newborns. *Acta Paediatr* **96**:
372–6.

Romeyer, A, Porter, R H, Poindron, P *et al.* (1993)
Recognition of dizygotic and monozygotic twin
lambs by ewes. *Behaviour* **127**: 119–39.

Roper, T J (2010) *Badger*. London: Collins.

Roper, T J, Conradt, L, Butler, J *et al.* (1993) Territorial
marking with faeces in badgers (*Meles meles*) – a
comparison of boundary and hinterland latrine use.
Behaviour **127**: 289–307.

Roulston, T H, Buczkowski, G & Silverman, J (2003)
Nestmate discrimination in ants: effect of bioassay
on aggressive behavior. *Insectes Soc* **50**: 151–9.

Roux, O, Martin, J M, Ghomsi, N T & Dejean, A (2009) A
non-lethal water-based removal-reapplication
technique for behavioral analysis of cuticular compounds of ants. *J Chem Ecol* **35**: 904–12.

Rowley, A F, Vogan, C L, Taylor, G W & Clare, A S (2005)
Prostaglandins in non-insectan invertebrates: recent
insights and unsolved problems. *J Exp Biol* **208**:
3–14.

Roy, S, Macleod, I & Moore, N (2006) The use of scent
glands to improve the efficiency of mink (*Mustela
vison*) captures in the Outer Hebrides. *N Z J Zool* **33**:
267–71.

Russell, E M (1985) The metatherians: order Marsupialia.
In Brown, R E & Macdonald, D W (eds.) *Social Odours
in Mammals*. pp. 45–104. Oxford: Oxford University
Press.

Russo, C A, Takezaki, N & Nei, M (1995) Molecular phylogeny and divergence times of drosophilid species.
Mol Biol Evol **12**: 391–404.

Ruta, V, Datta, S R, Vasconcelos, M L *et al.* (2010) A
dimorphic pheromone circuit in *Drosophila* from
sensory input to descending output. *Nature* **468**:
686–90.

Ruther, J & Steidle, J L M (2002) "Allohormones": a new
class of bioactive substances or old wine in new
skins? *J Comp Physiol A* **188**: 161–2.

Ruxton, G D & Schaefer, H M (2011) Resolving current
disagreements and ambiguities in the terminology of
animal communication. *J Evol Biol* **24**: 2574–85.

Ruxton, G D & Sherratt, T N (2006) Aggregation, defence
and warning signals: the evolutionary relationship.
Proc R Soc B **273**: 2417.

Ryan, M J (1998) Sexual selection, receiver biases, and the
evolution of sex differences. *Science* **281**: 1999–
2003.

Ryan, M J, Bernal, X E & Rand, A S (2010) Female mate
choice and the potential for ornament evolution in
túngara frogs *Physalaemus pustulosus*. *Curr Zool*
56: 343–57.

Sachs, B D (1999) Airborne aphrodisiac odor from estrous rats: implication for pheromonal classification. In Johnston, R E, Müller-Schwarze, D & Sorensen, P W (eds.) *Advances in Chemical Signals in Vertebrates*. pp. 333–42. New York: Kluwer Academic/Plenum Press.

Sacks, O (1987) *The Man who Mistook his Wife for a Hat*. London: Duckworth.

Sakai, R, Fukuzawa, M, Nakano, R, Tatsuki, S & Ishikawa, Y (2009) Alternative suppression of transcription from two desaturase genes is the key for species-specific sex pheromone biosynthesis in two *Ostrinia* moths. *Insect Biochem Mol Biol* 39: 62–7.

Sakai, T, Nakagawa, Y, Takahashi, J, Iwabuchi, K & Ishii, K (1984) Isolation and identification of the male sex pheromone of the grape borer *Xylotrechus pyrrhoderus* Bates (Coleoptera: Cerambycidae). *Chem Lett* 1984: 263–4.

Saleh, N, Scott, A, Bryning, G & Chittka, L (2007) Distinguishing signals and cues: bumblebees use general footprints to generate adaptive behaviour at flowers and nest. *Arthropod Plant Interact* 1: 119–27.

Saltzman, W, Digby, L J & Abbott, D H (2009) Reproductive skew in female common marmosets: what can proximate mechanisms tell us about ultimate causes? *Proc R Soc B* 276: 389–99.

Sanchez-Andrade, G & Kendrick, K M (2009) The main olfactory system and social learning in mammals: pheromonal communication in higher vertebrates and its implication for reproductive function. *Behav Brain Res* 200: 323–35.

Sanchez-Gracia, A, Vieira, F G & Rozas, J (2009) Molecular evolution of the major chemosensory gene families in insects. *Heredity* 103: 208–16.

Sandoz, J-C (2011) Behavioural and neurophysiological study of olfactory perception and learning in honeybees. *Front Syst Neurosci* 5: 98.

Sandoz, J-C (2012) Olfaction in honey bees: from molecules to behavior honeybee neurobiology and behavior. In Galizia, C G, Eisenhardt, D & Giurfa, M (eds.) *Honeybee Neurobiology and Behavior*. pp. 235–52. Dordrecht: Springer.

Sandoz, J-C, Deisig, N, de Brito Sanchez, M G & Giurfa, M (2007) Understanding the logics of pheromone processing in the honeybee brain: from labeled-lines to across-fiber patterns. *Front Behav Neurosci* 1: 5.

Saraiva, J L, Gonçalves, D M & Oliveira, R F (2010) Environmental modulation of androgen levels and secondary sex characters in two populations of the peacock blenny *Salaria pavo*. *Horm Behav* 57: 192–7.

Saul-Gershenz, L & Millar, J (2006) Phoretic nest parasites use sexual deception to obtain transport to their host's nest. *Proc Natl Acad Sci USA* 103: 14039.

Savarit, F, Sureau, G, Cobb, M & Ferveur, J F (1999) Genetic elimination of known pheromones reveals the fundamental chemical bases of mating and isolation in *Drosophila*. *Proc Natl Acad Sci USA* 96: 9015–20.

Savic, I, Berglund, H, Gulyas, B & Roland, P (2001) Smelling of odorous sex hormone-like compounds causes sex-differentiated hypothalamic activations in humans. *Neuron* 31: 661–8.

Schaal, B (1988) Olfaction in infants and children – developmental and functional perspectives. *Chem Senses* 13: 145–90.

Schaal, B (2009) Social chemosignal. In Binder, M D, Hirokawa, N & Windhorst, U (eds.) *Encyclopedia of Neuroscience*. pp. 3756–9. Berlin: Springer.

Schaal, B (2012) Emerging chemosensory preferences. Another playground for the innate-acquired dichotomy in human cognition. In Zucco, G M, Schaal, B & Herz, R S (eds.) *Olfactory Cognition. From Perception and Memory to Environmental Odours and Neuroscience*. pp. 237–68. Amsterdam: John Benjamins.

Schaal, B & Porter, R H (1991) Microsmatic humans revisited – the generation and perception of chemical signals. *Adv Study Behav* 20: 135–99.

Schaal, B, Montagner, H, Hertling, E *et al.* (1980) Les stimulations olfactives dans les relations entre l'enfant et la mère. *Reprod Nutr Dev* 20: 843–58.

Schaal, B, Coureaud, G, Langlois, D *et al.* (2003) Chemical and behavioural characterization of the rabbit mammary pheromone. *Nature* 424: 68–72.

Schaal, B, Hummel, T & Soussignan, R (2004) Olfaction in the fetal and premature infant: functional status and clinical implications. *Clin Perinatol* 31: 261–85.

Schaal, B, Doucet, S, Sagot, P, Hertling, E & Soussignan, R (2006) Human breast areolae as scent organs: morphological data and possible involvement in maternal–neonatal coadaptation. *Dev Psychobiol* 48: 100–10.

Schaal, B, Coureaud, G, Doucet, S *et al.* (2009) Mammary olfactory signalisation in females and odor processing in neonates: ways evolved by rabbits and humans. *Behav Brain Res* 200: 346–58.

Schaefer, A T & Claridge-Chang, A (2012) The surveillance state of behavioral automation. *Curr Opin Neurobiol* 22: 170–6.

Schaefer, M L, Yamazaki, K, Osada, K, Restrepo, D & Beauchamp, G K (2002) Olfactory fingerprints for major histocompatibility complex-determined body odors II: relationship among odor maps, genetics, odor composition, and behavior. *J Neurosci* 22: 9513–21.

Schank, J (2006) Do human menstrual-cycle pheromones exist? *Hum Nat* 17: 448–70.

Schellinck, H M, Rooney, E & Brown, R E (1995) Odors of individuality of germ-free mice are not discriminated by rats in a habituation–dishabituation procedure. *Physiol Behav* 57: 1005–8.

Schellinck, H M, Cyr, D P & Brown, R E (2010) How many ways can mouse behavioral experiments go wrong? Confounding variables in mouse models of neurodegenerative diseases and how to control them. *Adv Study Behav* 41: 255–366.

Schiestl, F P, Ayasse, M, Paulus, H F *et al.* (1999) Orchid pollination by sexual swindle. *Nature* 399: 421–2.

Schiestl, F P, Peakall, R, Mant, J G *et al.* (2003) The chemistry of sexual deception in an orchid–wasp pollination system. *Science* 302: 437–8.

Schilling, B, Kaiser, R, Natsch, A & Gautschi, M (2010) Investigation of odors in the fragrance industry. *Chemoecology* 20: 135–47.

Schlechter-Helas, J, Schmitt, T & Peschke, K (2011) A contact anti-aphrodisiac pheromone supplied by the spermatophore in the rove beetle *Aleochara curtula*: mode of transfer and evolutionary significance. *Naturwissenschaften* 98: 855–62.

Schmidt, J O (1998) Mass action in honey bees: alarm, swarming and role of releaser pheromones. In Vander Meer, R K, Breed, M D, Espelie, K E & Winston, M L (eds.) *Pheromone Communication in Social Insects: Ants, Wasps, Bees, and Termites.* pp. 257–92. Boulder, CO: Westview Press.

Schmidt, M & Mellon, D (2011) Neuronal processing of chemical information in crustaceans. In Breithaupt, T & Thiel, M (eds.) *Chemical Communication in Crustaceans.* pp. 123–47. New York: Springer.

Schneider, D (1999) Insect pheromone research: some history and 45 years of personal recollections. *IOBC-WPRS Bull* 22. Available at http://phero.net/iobc/dachau/bulletin99/schneider.pdf [Accessed 21January 2013].

Schöne, H (1984) *Spatial Orientation: the Spatial Control of Behavior in Animals and Man.* Princeton: Princeton University Press.

Schönrogge, K, Gardner, M G, Elmes, G *et al.* (2006) Host propagation permits extreme local adaptation in a social parasite of ants. *Ecol Lett* 9: 1032–40.

Schoon, G A A (1997) Scent identifications by dogs (*Canis familiaris*): a new experimental design. *Behaviour* 134: 531–50.

Schulte, B A (1998) Scent marking and responses to male castor fluid by beavers. *J Mammal* 79: 191–203.

Schulz, S (2004) Semiochemistry of spiders. In Cardé, R & Millar, J G (eds.) *Advances in Insect Chemical Ecology.* pp. 110–50. Cambridge: Cambridge University Press.

Schulz, S (2009) Alkaloid-derived male courtship pheromones. In Conner, W E (ed.) *Tiger Moths and Woolly Bears: Behavior, Ecology, and Evolution of the Arctiidae.* pp. 145–54. Oxford: Oxford University Press.

Schulz, S, Estrada, C, Yildizhan, S, Boppré, M & Gilbert, L E (2008) An antiaphrodisiac in *Heliconius melpomene* butterflies. *J Chem Ecol* 34: 82–93.

Schwander, T, Lo, N, Beekman, M, Oldroyd, B P & Keller, L (2010) Nature versus nurture in social insect caste differentiation. *Trends Ecol Evol* 25: 275–82.

Schwarz, M P, Richards, M H & Danforth, B N (2007) Changing paradigms in insect social evolution: insights from halictine and allodapine bees. *Annu Rev Entomol* 52: 127–50.

Schwende, F J, Wiesler, D, Jorgenson, J W, Carmack, M & Novotny, M (1986) Urinary volatile constituents of the house mouse, *Mus musculus*, and their endocrine dependency. *J Chem Ecol* 12: 277–96.

Schwenk, K (1994) Why snakes have forked tongues. *Science* 263: 1573–7.

Scordato, E S & Drea, C M (2007) Scents and sensibility: information content of olfactory signals in the ring-tailed lemur, *Lemur catta. Anim Behav* 73: 301–14.

Searcy, W A & Nowicki, S (2005) *The Evolution of Animal Communication: Reliability and Deception in Signalling Systems.* Princeton: Princeton University Press.

Seeley, T D (1979) Queen substance dispersal by messenger workers in honey bee colonies. *Behav Ecol Sociobiol* 5: 391–415.

Seeley, T D (1985) *Honeybee Ecology: a Study of Adaptation in Social Life.* Princeton: Princeton University Press.

Seeley, T D (1995) *The Wisdom of the Hive. The Social Physiology of Honey Bee Colonies.* Cambridge, MA: Harvard University Press.

Seeley, T D (2002) When is self-organization used in biological systems? *Biol Bull* 202: 314–18.

Seeley, T D (2010) *Honeybee Democracy.* Princeton: Princeton University Press.

Seenivasagan, T & Vijayaraghavan, R (2010) Oviposition pheromones in haematophagous insects. In Gerald, L (ed.) *Pheromones.* pp. 597–630. London: Academic Press.

Serguera, C, Triaca, V, Kelly-Barrett, J, Al Banchaabouchi, M & Minichiello, L (2008) Increased dopamine after mating impairs olfaction and prevents odor interference with pregnancy. *Nat Neurosci* 11: 949–56.

Serrano, R, Barata, E, Birkett, M *et al.* (2008) Behavioral and olfactory responses of female *Salaria pavo* (Pisces: Blenniidae) to a putative multi-component male pheromone. *J Chem Ecol* 34: 647–58.

Serrão, E A & Havenhand, J (2009) Fertilization strategies. In Whal, M (ed.) *Marine Hard Bottom Communities: Patterns, Dynamics, Diversity, and Change.* pp. 149–64. Dordrecht: Springer.

Setchell, J M & Huchard, E (2010) The hidden benefits of sex: evidence for MHC-associated mate choice in primate societies. *Bioessays* 32: 940–8.

Setchell, J M, Kendal, J & Tyniec, P (2011) Do non-human primates synchronise their menstrual cycles? A test in mandrills. *Psychoneuroendocrinology* 36: 51–9.

Settle, R H, Sommerville, B A, McCormick, J & Broom, D M (1994) Human scent matching using specially trained dogs. *Anim Behav* 48: 1443–8.

Seybold, S J, Huber, D, Lee, J, Graves, A & Bohlmann, J (2006) Pine monoterpenes and pine bark beetles: a marriage of convenience for defense and chemical communication. *Phytochem Rev* 5: 143–78.

Shanbhag, S R, Müller, B & Steinbrecht, R A (1999) Atlas of olfactory organs of *Drosophila melanogaster*: 1. Types, external organization, innervation and distribution of olfactory sensilla. *Int J Insect Morphol Embryol* 28: 377–97.

Sharma, K, Vander Meer, R K & Fadamiro, H Y (2011) Phorid fly, *Pseudacteon tricuspis*, response to alkylpyrazine analogs of a fire ant, *Solenopsis invicta*, alarm pheromone. *J Insect Physiol* 57: 939–44.

Sheehan, M J & Tibbetts, E Λ (2011) Specialized face learning is associated with individual recognition in paper wasps. *Science* 334: 1272–5.

Shelley, W B, Hurley, H J & Nichols, A C (1953) Axillary odour: experimental study of the role of bacteria, apocrine sweat, and deodorants. *Arch Derm Syphilol* 68: 430–46.

Shelly, T E & Kennelly, S S (2007) Settlement patterns of Mediterranean fruit flies in the tree canopy: an experimental analysis. *J Insect Behav* 20: 453–72.

Shelly, T E, Edu, J & Pahio, E (2007) Condition-dependent mating success in male fruit flies: ingestion of a pheromone precursor compensates for a low-quality diet. *J Insect Behav* 20: 347–65.

Shelly, T W & Whittier, T S (1997) Lek behavior of insects. In Choe, J C & Crespi, B J (eds.) *The Evolution of Mating Systems in Insects and Arachnids.* pp. 273–93. Cambridge: Cambridge University Press.

Shepherd, G M (2004) The human sense of smell: are we better than we think? *PLoS Biol* 2: 572–5.

Shepherd, G M (2005) Outline of a theory of olfactory processing and its relevance to humans. *Chem Senses* 30: I3–I5.

Shepherd, G M (2006) Smell images and the flavour system in the human brain. *Nature* 444: 316–21.

Shepherd, G M (2010) New perspectives on olfactory processing and human smell. In Menini, A (ed.) *The Neurobiology of Olfaction.* Boca Raton, Fl.: CRC Press. Available online at www.ncbi.nlm.nih.gov/books/NBK55977.

Shepherd, G M (2012) *Neurogastronomy: How the Brain Creates Flavor and Why it Matters.* New York: Columbia University Press.

Sherborne, A L, Thom, M D, Paterson, S *et al.* (2007) The basis of inbreeding avoidance in house mice. *Curr Biol* 17: 2061–6.

Sherman, P W, Lacey, E A, Reeve, H K & Keller, L (1995) The eusociality continuum. *Behav Ecol* 6: 102–8.

Sherman, P W, Reeve, H K & Pfennig, D W (1997) Recognition systems. In Krebs, J R & Davies, N B (eds.) *Behavioural Ecology: an Evolutionary Approach*, 4th edn., pp. 69–96. Oxford: Blackwell Science.

Shine, R & Mason, R T (2012) An airborne sex pheromone in snakes. *Biol Lett* 8: 183–5.

Shirangi, T R, Dufour, H D, Williams, T M & Carroll, S B (2009) Rapid evolution of sex pheromone-producing enzyme expression in *Drosophila. PLoS Biol* 7: e1000168.

Shirasu, M & Touhara, K (2011) The scent of disease: volatile organic compounds of the human body related to disease and disorder. *J Biochem* 150: 257–66.

Shuster, S M (2009) Sexual selection and mating systems. *Proc Natl Acad Sci USA* 106: 10009–16.

Shuster, S M (2010) Alternative mating strategies. In Westneat, D F & Fox, C W (eds.) *Evolutionary Behavioral Ecology*. pp. 434–50. New York: Oxford University Press.

Siefkes, M J, Scott, A P, Zielinski, B, Yun, S S & Li, W (2003) Male sea lampreys, *Petromyzon marinus* L., excrete a sex pheromone from gill epithelia. *Biol Reprod* 69: 125–32.

Silbering, A F & Benton, R (2010) Ionotropic and metabotropic mechanisms in chemoreception: 'chance or design'? *EMBO Rep* 11: 173–9.

Silbering, A F, Rytz, R, Grosjean, Y *et al.* (2011) Complementary function and integrated wiring of the evolutionarily distinct *Drosophila* olfactory subsystems. *J Neurosci* 31: 13357–75.

Siljander, E, Penman, D, Harlan, H & Gries, G (2007) Evidence for male- and juvenile-specific contact pheromones of the common bed bug *Cimex lectularius. Entomol Exp Appl* 125: 215–19.

Siljander, E, Gries, R, Khaskin, G & Gries, G (2008) Identification of the airborne aggregation pheromone of the common bed bug, *Cimex lectularius. J Chem Ecol* 34: 708–18.

Sillén-Tullberg, B & Møller, A P (1993) The relationship between concealed ovulation and mating systems in anthropoid primates – a phylogenetic analysis. *Am Nat* 141: 1–25.

Sillero-Zubiri, C & Macdonald, D W (1998) Scent-marking and territorial behaviour of Ethiopian wolves *Canis simensis. J Zool* 245: 351–61.

Silverstein, R M (1990) Practical use of pheromones and other behavior-modifying compounds: overview. In Ridgeway, R L, Silverstein, R M & Inscoe, M N (eds.) *Behavior-modifying Chemicals for Insect Management.* pp. 1–8. New York: Marcel Dekker.

Simerly, R B (2002) Wired for reproduction: organization and development of sexually dimorphic circuits in the mammalian forebrain. *Annu Rev Neurosci* 25: 507–36.

Simpson, S J, Sword, G A & Lo, N (2011) Polyphenism in insects. *Curr Biol* 21: R738–R749.

Singer, A G, Beauchamp, G K & Yamazaki, K (1997) Volatile signals of the major histocompatibility complex in male mouse urine. *Proc Natl Acad Sci USA* 94: 2210–14.

Singh, D & Bronstad, P M (2001) Female body odour is a potential cue to ovulation. *Proc R Soc B* 268: 797–801.

Sirugue, D, Bonnard, O, LeQuere, J L, Farine, JP & Brossut, R (1992) 2-Methylthiazolidine and 4-ethylguaiacol, male sex-pheromone components of the cockroach *Nauphoeta cinerea* (Dictyoptera, Blaberidae) – a reinvestigation. *J Chem Ecol* 18: 2261–76.

Slatyer, R A, Mautz, B S, Backwell, P R Y & Jennions, M D (2012) Estimating genetic benefits of polyandry from experimental studies: a meta-analysis. *Biol Rev* 87: 1–33.

Slessor, K N, Winston, M L & Le Conte, Y (2005) Pheromone communication in the honeybee (*Apis mellifera* L.). *J Chem Ecol* 31: 2731–45.

Sliwa, A & Richardson, P R K (1998) Responses of aardwolves, *Proteles cristatus*, Sparrman 1783, to translocated scent marks. *Anim Behav* 56: 137–46.

Slonim, D K & Yanai, I (2009) Getting started in gene expression microarray analysis. *PLoS Comput Biol* 5: e1000543.

Smadja, C & Butlin, R K (2009) On the scent of speciation: the chemosensory system and its role in premating isolation. *Heredity* 102: 77–97.

Smadja, C & Ganem, G (2008) Divergence of odorant signals within and between the two European subspecies of the house mouse. *Behav Ecol* 19: 223–30.

Smallegange, R C, Verhulst, N O & Takken, W (2011) Sweaty skin: an invitation to bite? *Trends Parasitol* **27**: 143–8.

Smith, A A, Hölldobler, B & Liebig, J (2012) Queen-specific signals and worker punishment in the ant *Aphaenogaster cockerelli*: the role of the Dufour's gland. *Anim Behav* **83**: 587–93.

Smith, B H & Breed, M D (1995) The chemical basis for nest-mate recognition and mate discrimination in social insects. In Cardé, R T & Bell, W J (eds.) *Chemical Ecology of Insects 2*. pp. 287–317. London: Chapman and Hall.

Smith, C R, Tóth, A L, Suarez, A V & Robinson, G E (2008) Genetic and genomic analyses of the division of labour in insect societies. *Nat Rev Genet* **9**: 735–48.

Smith, J L, Cork, A, Hall, D R & Hodges, R J (1996) Investigation of the effect of female larger grain borer, *Prostephanus truncatus* (Horn) (Coleoptera: Bostrichidae), and their residues on the production of aggregation pheromone by males. *J Stored Prod Res* **32**: 171–81.

Snell, T W (2011a) A review of the molecular mechanisms of monogonont rotifer reproduction. *Hydrobiologia* **662**: 89–97.

Snell, T W (2011b) Contact chemoreception and its role in zooplankton mate recognition. In Breithaupt, T & Thiel, M (eds.) *Chemical Communication in Crustaceans*. pp. 451–66. New York: Springer.

Šobotník, J, Jirosová, A & Hanus, R (2010) Chemical warfare in termites. *J Insect Physiol* **56**: 1012–21.

Šobotník, J, Bourguignon, T, Hanus, R *et al.* (2012) Explosive backpacks in old termite workers. *Science* **337**: 436.

Soini, H A, Bruce, K E, Wiesler, D *et al.* (2005) Stir bar sorptive extraction: a new quantitative and comprehensive sampling technique for determination of chemical signal profiles from biological media. *J Chem Ecol* **31**: 377–92.

Solomon, N G & Getz, L L (1997) Examination of alternative hypotheses for cooperative breeding in rodents. In Solomon, N G & French, J A (eds.) *Cooperative Breeding in Mammals*. pp. 199–230. Cambridge: Cambridge University Press.

Solomon, N G & Keane, B (2007) Reproductive strategies in female rodents. In Wolff, J O & Shermann, P W (eds.) *Rodent Societies: an Ecological and Evolutionary Perspective*. pp. 42–56. Chicago: Chicago University Press.

Sonenshine, D E (2004) Pheromones and other semiochemicals of ticks and their use in tick control. *Parasitology* **129**: S405–S425.

Sonenshine, D E (2006) Tick pheromones and their use in tick control. *Annu Rev Entomol* **51**: 557–80.

Soo, M L M & Stevenson, R J (2007) The moralisation of body odor. *Mankind Q* **47**: 25–56.

Sorensen, P W & Hoye, T R (2007) A critical review of the discovery and application of a migratory pheromone in an invasive fish, the sea lamprey *Petromyzon matinus* L. *J Fish Biol* **71**: 100–14.

Sorensen, P W & Wisenden, B D (eds) (2014) *Fish Pheromones and Related Conspecific Chemical Cues*. Chichester: Wiley Blackwell.

Sorensen, P W, Christensen, T A & Stacey, N E (1998) Discrimination of pheromonal cues in fish: emerging parallels with insects. *Curr Opin Neurobiol* **8**: 458–67.

Sorensen, P W, Pinillos, M & Scott, A P (2005) Sexually mature male goldfish release large quantities of androstenedione into the water where it functions as a pheromone. *Gen Comp Endocrinol* **140**: 164–75.

Soucy, E R, Albeanu, D F, Fantana, A L, Murthy, V N & Meister, M (2009) Precision and diversity in an odor map on the olfactory bulb. *Nat Neurosci* **12**: 210–20.

Soussignan, R, Schaal, B, Marlier, L & Jiang, T (1997) Facial and autonomic responses to biological and artificial olfactory stimuli in human neonates: re-examining early hedonic discrimination of odors. *Physiol Behav* **62**: 745–58.

South, A & Lewis, S M (2011) The influence of male ejaculate quantity on female fitness: a meta-analysis. *Biol Rev* **86**: 299–309.

South, S H, House, C M, Moore, A J, Simpson, S J & Hunt, J (2011) Male cockroaches prefer a high carbohydrate diet that makes them more attractive to females: implications for the study of condition dependence. *Evolution* **65**: 1594–606.

Spannhoff, A, Kim, Y K, Noel, J *et al.* (2011) Histone deacetylase inhibitor activity in royal jelly might facilitate caste switching in bees. *EMBO Rep* **12**: 238–43.

Spehr, M & Munger, S D (2009) Olfactory receptors: G protein-coupled receptors and beyond. *J Neurochem* **109**: 1570–83.

Spehr, M, Spehr, J, Ukhanov, K *et al.* (2006a) Parallel processing of social signals by the mammalian main and accessory olfactory systems. *Cell Mol Life Sci* **63**: 1476–84.

Spehr, M, Kelliher, K R, Li, X H *et al.* (2006b) Essential role of the main olfactory system in social recognition of major histocompatibility complex peptide ligands. *J Neurosci* **26**: 1961–70.

Sreng, L (1990) Seducin, male sex-pheromone of the cockroach *Nauphoeta cinerea* – isolation, identification, and bioassay. *J Chem Ecol* **16**: 2899–912.

Srinivasan, J, Kaplan, F, Ajredini, R *et al.* (2008) A blend of small molecules regulates both mating and development in *Caenorhabditis elegans*. *Nature* **454**: 1115–18.

Srinivasan, J, von Reuss, S H, Bose, N *et al.* (2012) A modular library of small molecule signals regulates social behaviors in *Caenorhabditis elegans*. *PLoS Biol* **10**: e1001237.

Stacey, N E & Sorensen, P W (2006) Reproductive pheromones. In Sloman, KA, Wilson, RW & Balshine, S (eds.) *Fish Physiology, Volume 24: Behaviour and Physiology of Fish*. pp. 359–412. Amsterdam: Academic Press, Elsevier.

Stacey, N E & Sorensen, P W (2009) Fish hormonal pheromones. In Pfaff, D W, Arnold, A P, Fahrbach, S E, Etgen, A M & Rubin, R T (eds.) *Hormones, Brain and Behavior*, 2nd edn., pp. 639–81. San Diego, CA: Academic Press.

Stacey, N E & Sorensen, P W (2011) Hormonal pheromones. In Farrell, A P (ed.) *Encyclopedia of Fish Physiology: From Genome to Environment*. pp. 1553–62. San Diego, CA: Academic Press.

Stadler, B & Dixon, A F G (2005) Ecology and evolution of aphid–ant interactions. *Annu Rev Ecol Evol Syst* **36**: 345–72.

Stamps, J (1994) Territorial behavior – testing the assumptions. *Adv Study Behav* **23**: 173–232.

Stapley, J, Reger, J, Feulner, P G D *et al.* (2010) Adaptation genomics: the next generation. *Trends Ecol Evol* **25**: 705–12.

Steel, E & Keverne, E B (1985) Effect of female odor on male hamsters mediated by the vomeronasal organ. *Physiol Behav* **35**: 195–200.

Steiger, S, Franz, R, Eggert, A K & Muller, J K (2008a) The Coolidge effect, individual recognition and selection for distinctive cuticular signatures in a burying beetle. *Proc R Soc B* **275**: 1831–8.

Steiger, S, Schmitt, T & Schaefer, H M (2011) The origin and dynamic evolution of chemical information transfer. *Proc R Soc B* **278**: 970–9.

Steiger, S S, Fidler, A E, Valcu, M & Kempenaers, B (2008b) Avian olfactory receptor gene repertoires: evidence for a well-developed sense of smell in birds? *Proc R Soc B* **275**: 2309–17.

Stein, B E & Meredith, M A (1993) *The Merging of the Senses*. Cambridge, MA: MIT Press.

Stern, D L & Foster, W A (1996) The evolution of soldiers in aphids. *Biol Rev* **71**: 27–79.

Stern, K & McClintock, M K (1998) Regulation of ovulation by human pheromones. *Nature* **392**: 177–9.

Stevenson, R J (2010) An initial evaluation of the functions of human olfaction. *Chem Senses* **35**: 3–20.

Stoddard, P K & Salazar, V L (2011) Energetic cost of communication. *J Exp Biol* **214**: 200–5.

Stoddart, D M (1990) *The Scented Ape. The Biology and Culture of Human Odour*. Cambridge: Cambridge University Press.

Stoeffler, M, Tolasch, T & Steidle, J L M (2011) Three beetles – three concepts. Different defensive strategies of congeneric myrmecophilous beetles. *Behav Ecol Sociobiol* **65**: 1605–13.

Stökl, J, Brodmann, J, Dafni, A, Ayasse, M & Hansson, B S (2011) Smells like aphids: orchid flowers mimic aphid alarm pheromones to attract hoverflies for pollination. *Proc R Soc B* **278**: 1216–22.

Storer, A J, Wainhouse, D & Speight, M R (1997) The effect of larval aggregation behaviour on larval growth of the spruce bark beetle *Dendroctonus micans*. *Ecol Entomol* **22**: 109–15.

Störtkuhl, K F & Fiala, A (2011) The smell of blue light: a new approach towards understanding an olfactory neuronal network. *Front Neurosci* **5**: 72.

Stowers, L, Holy, T E, Meister, M, Dulac, C & Koentges, G (2002) Loss of sex discrimination and male–male aggression in mice deficient for TRP2. *Science* **295**: 1493–500.

Strausfeld, N J (2009) Brain organization and the origin of insects: an assessment. *Proc R Soc B* **276**: 1929–37.

Strausfeld, N J & Hildebrand, J G (1999) Olfactory systems: common design, uncommon origins? *Curr Opin Neurobiol* 9: 634–9.

Stuart, A M (1969) Social behavior and communication. In Krishna, K (ed.) *The Biology of Termites*. pp. 193–232. New York: Academic Press.

Sturgis, S J & Gordon, D M (2012) Nestmate recognition in ants (Hymenoptera: Formicidae): a review. *Myrmecol News* 16: 101–10.

Su, C Y, Menuz, K & Carlson, J R (2009) Olfactory perception: receptors, cells, and circuits. *Cell* 139: 45–59.

Su, C Y, Martelli, C, Emonet, T & Carlson, J R (2011) Temporal coding of odor mixtures in an olfactory receptor neuron. *Proc Natl Acad Sci USA* 108: 5075–80.

Suckling, D M, Peck, R W, Stringer, L D, Snook, K & Banko, P C (2010) Trail pheromone disruption of Argentine ant trail formation and foraging. *J Chem Ecol* 36: 122–8.

Suckling, D M, Woods, B, Mitchell, V J et al. (2011) Mobile mating disruption of light-brown apple moths using pheromone-treated sterile Mediterranean fruit flies. *Pest Manag Sci* 67: 1004–14.

Suckling, D M, Tobin, P C, McCullough, D G & Herms, D A (2012a) Combining tactics to exploit Allee effects for eradication of alien insect populations. *J Econ Entomol* 105: 1–13.

Suckling, D M, Stringer, L D, Corn, J E et al. (2012b) Aerosol delivery of trail pheromone disrupts the foraging of the red imported fire ant, *Solenopsis invicta*. *Pest Manag Sci* 68: 1572–8.

Sugahara, M & Sakamoto, F (2009) Heat and carbon dioxide generated by honeybees jointly act to kill hornets. *Naturwissenschaften* 96: 1133–6.

Sullivan, T P & Crump, D (1984) Influence of mustelid scent gland compounds on suppression of feeding by snowshoe hares (*Lepus americanus*). *J Chem Ecol* 10: 903–19.

Sumpter, D J T (2006) The principles of collective animal behaviour. *Phil Trans R Soc B* 361: 5–22.

Sumpter, D J T (2010) *Collective Animal Behavior*. Princeton: Princeton University Press.

Sumpter, D J T, Mann, R P & Perna, A (2012) The modelling cycle for collective animal behaviour. *Interface Focus* 2: 764–73.

Sun, L X & Müller-Schwarze, D (1997) Sibling recognition in the beaver: a field test for phenotype matching. *Anim Behav* 54: 493–502.

Sun, L X & Müller-Schwarze, D (1998) Beaver response to recurrent alien scents: scent fence or scent match? *Anim Behav* 55: 1529–36.

Sun, L X & Müller-Schwarze, D (1999) Chemical signals in the beaver: one species, two secretions, many functions? In Johnston, R E, Müller-Schwarze, D & Sorensen, P W (eds.) *Advances in Chemical Signals in Vertebrates*. pp. 281–8. New York: Kluwer Academic/Plenum Press.

Sunamura, E, Suzuki, S, Nishisue, K et al. (2011) Combined use of a synthetic trail pheromone and insecticidal bait provides effective control of an invasive ant. *Pest Manag Sci* 67: 1230–6.

Sundberg, H, Døving, K, Novikov, S & Ursin, H (1982) A method for studying responses and habituation to odors in rats. *Behav Neural Biol* 34: 113–19.

Süskind, P (1986) *Perfume. The Story of a Murderer*. London: Hamish Hamilton. Translated by J E Woods.

Swaisgood, R & Schulte, B (2010) Applying knowledge of mammalian social organization, mating systems, and communication to management. In Kleiman, D, Thompson, K & Baer, C (eds.) *Wild Mammals in Captivity: Principles and Techniques for Zoo Management*, 2nd edn., pp. 329–43. Chicago: University of Chicago Press.

Swaney, W T & Keverne, E B (2011) Genomic imprinting and sexual experience-dependent learning in the mouse. In Clelland, J D (ed.) *Genomics, Proteomics, and the Nervous System*. pp. 195–225. New York: Springer.

Swaney, W T, Curley, J P, Champagne, F A & Keverne, E B (2007) Genomic imprinting mediates sexual experience-dependent olfactory learning in male mice. *Proc Natl Acad Sci USA* 104: 6084–9.

Swann, J, Fabre-Nys, C & Barton, R (2009) Hormonal and pheromonal modulation of the extended amygdala: implications for social behavior. In Pfaff, D W, Arnold, A P, Fahrbach, S E, Etgen, A M & Rubin, R T (eds.) *Hormones, Brain and Behavior*. pp. 441–72. San Diego: Academic Press.

Symonds, M R E & Elgar, M A (2008) The evolution of pheromone diversity. *Trends Ecol Evol* 23: 220–8.

Symonds, M R E, Johnson, T L & Elgar, M A (2012) Pheromone production, male abundance, body size, and the evolution of elaborate antennae in moths. *Ecol Evol* 2: 227–46.

Számadó, S (1999) The validity of the handicap principle in discrete action–response games. *J Theor Biol* 198: 593–602.

Számadó, S (2003) Threat displays are not handicaps. *J Theor Biol* 221: 327–48.

Számadó, S (2008) How threat displays work: species-specific fighting techniques, weaponry and proximity risk. *Anim Behav* 76: 1455–63.

Számadó, S (2011a) The rise and fall of handicap principle: a commentary on the "Modelling and the fall and rise of the handicap principle". *Biol Philos* 27: 279–86.

Számadó, S (2011b) The cost of honesty and the fallacy of the handicap principle. *Anim Behav* 81: 3–10.

Székely, T, Moore, A & Komdeur, J (eds.) (2010) *Social Behaviour: Genes, Ecology and Evolution.* Cambridge: Cambridge University Press.

Taborsky, M, Oliveira, R F & Brockmann, H J (2008) The evolution of alternative reproductive tactics: concepts and questions. In Oliveira, R F, Taborsky, M & Brockmann, H J (eds.) *Alternative Reproductive Tactics: an Integrative Approach* pp. 1–22. Cambridge: Cambridge University Press.

Tallamy, D W (2005) Egg dumping in insects. *Annu Rev Entomol* 50: 347–70.

Tallamy, D W & Denno, R F (1982) Maternal care in *Gargaphia solani* (Hemiptera: Tingidae). *Anim Behav* 29: 771–8.

Tarver, M R, Zhou, X G & Scharf, M E (2010) Socio-environmental and endocrine influences on developmental and caste-regulatory gene expression in the eusocial termite *Reticulitermes flavipes*. *BMC Mol Biol* 11: 28.

Teal, P E A, Gomez-Simuta, Y & Proveaux, A T (2000) Mating experience and juvenile hormone enhance sexual signaling and mating in male caribbean fruit flies. *Proc Natl Acad Sci USA* 97: 3708–12.

Temeles, E J (1994) The role of neighbours in territorial systems: when are they 'dear enemies?'. *Anim Behav* 47: 339–50.

ten Cate, C, Verzijden, M N & Etman, E (2006) Sexual imprinting can induce sexual preferences for exaggerated parental traits. *Curr Biol* 16: 1128–32.

Theis, K R, Schmidt, T M & Holekamp, K E (2012) Evidence for a bacterial mechanism for group-specific social odors among hyenas. *Sci Rep* 2.

Theisen, B, Zeiske, E, Silver, W L, Marui, T & Caprio, J (1991) Morphological and physiological studies on the olfactory organ of the striped eel catfish, *Plotosus lineatus*. *Mar Biol* 110: 127–35.

Thesen, A, Steen, J B & Døving, K B (1993) Behavior of dogs during olfactory tracking. *J Exp Biol* 180: 247–51.

Thewissen, J, George, J, Rosa, C & Kishida, T (2011) Olfaction and brain size in the bowhead whale (*Balaena mysticetus*). *Mar Mamm Sci* 27: 282–94.

Thistle, R, Cameron, P, Ghorayshi, A, Dennison, L & Scott, K (2012) Contact chemoreceptors mediate male–male repulsion and male–female attraction during *Drosophila* courtship. *Cell* 149: 1140–51.

Thom, C, Gilley, D C, Hooper, J & Esch, H E (2007) The scent of the waggle dance. *PLoS Biol* 5: e228.

Thom, M D & Hurst, J L (2004) Individual recognition by scent. *Ann Zool Fenn* 41: 765–87.

Thom, M D, Stockley, P, Jury, F *et al.* (2008) The direct assessment of genetic heterozygosity through scent in the mouse. *Curr Biol* 18: 619–23.

Thomas, J A, Knapp, J J, Akino, T *et al.* (2002) Insect communication: parasitoid secretions provoke ant warfare. *Nature* 417: 505–6.

Thomas, J A, Schönrogge, K & Elmes, G W (2005) Specializations and host associations of social parasites of ants. In Fellowes, M D E, Holloway, G J & Rolff, J (eds.) *Insect Evolutionary Ecology.* pp. 479–518. Wallingford: CABI.

Thomas, J H (1993) Chemosensory regulation of development in *C. elegans*. *Bioessays* 15: 791–7.

Thomas, M L (2011) Detection of female mating status using chemical signals and cues. *Biol Rev* 86: 1–13.

Thomas, M L & Simmons, L W (2009) Male-derived cuticular hydrocarbons signal sperm competition intensity and affect ejaculate expenditure in crickets. *Proc R Soc B* 276: 383–8.

Thomas, M L & Simmons, L W (2011) Crickets detect the genetic similarity of mating partners via cuticular hydrocarbons. *J Evol Biol* 24: 1793–800.

Thompson, G J, Kucharski, R, Maleszka, R & Oldroyd, B P (2006) Towards a molecular definition of worker sterility: differential gene expression and reproductive plasticity in honey bees. *Insect Mol Biol* 15: 537–644.

Thompson, J N (2009) The coevolving web of life. *Am Nat* 173: 125–40.

Thornhill, R (1979) Male pair formation pheromones in *Panorpa* scorpionflies (Mecoptera: Panorpidae). *Environ Entomol* 8: 886–9.

Thornhill, R & Alcock, J (1983) *The Evolution of Insect Mating Systems*. Cambridge, MA: Harvard University Press.

Thoss, M, Ilmonen, P, Musolf, K & Penn, D J (2011) Major histocompatibility complex heterozygosity enhances reproductive success. *Mol Ecol* 20: 1546–57.

Thysen, B, Elliott, W H & Katzman, P A (1968) Identification of estra-1, 3, 5 (10), 16-tetraen-3-ol (estratetraenol) from the urine of pregnant women. *Steroids* 11: 73–87.

Tibbetts, E A (2004) Complex social behaviour can select for variability in visual features: a case study in *Polistes* wasps. *Proc R Soc B* 271: 1955–60.

Tibbetts, E A & Dale, J (2007) Individual recognition: it is good to be different. *Trends Ecol Evol* 22: 529–37.

Tinbergen, N (1952) "Derived" activities; their causation, biological significance, origin, and emancipation during evolution. *Q Rev Biol* 27: 1–32.

Tirindelli, R, Dibattista, M, Pifferi, S & Menini, A (2009) From pheromones to behavior. *Physiol Rev* 89: 921–56.

Toda, H, Zhao, X & Dickson, B J (2012) The *Drosophila* female aphrodisiac pheromone activates ppk23+ sensory neurons to elicit male courtship behavior. *Cell Reports* 1: 599–607.

Todrank, J, Heth, G & Restrepo, D (2011) Effects of in utero odorant exposure on neuroanatomical development of the olfactory bulb and odour preferences. *Proc R Soc B* 278: 1949–55.

Tompkins, L, McRobert, S P & Kaneshiro, K Y (1993) Chemical communication in Hawaiian *Drosophila*. *Evolution* 47: 1407–19.

Toonen, R & Pawlik, J (2001) Foundations of gregariousness: a dispersal polymorphism among the planktonic larvae of a marine invertebrate. *Evolution* 55: 2439–54.

Tóth, A L & Robinson, G E (2007) Evo-devo and the evolution of social behavior. *Trends Genet* 23: 334–41.

Touhara, K (ed.) (2013) *Pheromone Signaling: Methods and Protocols*. New York, NY: Humana Press (Springer).

Touhara, K & Vosshall, L B (2009) Sensing odorants and pheromones with chemosensory receptors. *Annu Rev Physiol* 71: 307–32.

Toyoda, F, Yamamoto, K, Iwata, T *et al.* (2004) Peptide pheromones in newts. *Peptides* 25: 1531–6.

Trabalon, M & Bagnères, A-G (2010) Contact recognition pheromones in spiders and scorpions. In Blomquist, G J & Bagnères, A-G (eds.) *Insect Hydrocarbons: Biology, Biochemistry, and Chemical Ecology*. pp. 344–74. Cambridge: Cambridge University Press.

Traniello, J F A & Robson, S K (1995) Trail and territorial communication in insects. In Cardé, R T & Bell, W J (eds.) *Chemical Ecology of Insects 2*. pp. 241–86. London: Chapman and Hall.

Trematerra, P (2012) Advances in the use of pheromones for stored-product protection. *J Pest Sci* 85: 285–99.

Troccaz, M, Borchard, G, Vuilleumier, C *et al.* (2009) Gender-specific differences between the concentrations of nonvolatile $(R)/(S)$-3-methyl-3-sulfanylhexan-1-ol and $(R)/(S)$-3-hydroxy-3-methyl-hexanoic acid odor precursors in axillary secretions. *Chem Senses* 34: 203–10.

Trumble, J T (1997) Integrating pheromones into vegetable crop production. In Cardé, R T & Minks, A K (eds.) *Pheromone Research: New Directions*. pp. 397–410. New York: Chapman and Hall.

Tsutsui, N D (2004) Scents of self: the expression component of self/non-self recognition systems. *Ann Zool Fenn* 41: 713–27.

Tumlinson, J H, Silverstein, R M, Moser, J C, Brownlee, R G & Ruth, J M (1971) Identification of the trail pheromone of a leaf-cutting ant, *Atta texana*. *Nature* 234: 348–9.

Ungerfeld, R (2007) Socio-sexual signalling and gonadal function: opportunities for reproductive management in domestic ruminants. *Soc Reprod Fertil Suppl* 64: 207–21.

Vahed, K (2007) All that glisters is not gold: sensory bias, sexual conflict and nuptial feeding in insects and spiders. *Ethology* 113: 105–27.

van der Pers, J N C & Minks, A K (1997) Measuring pheromone dispersion in the field with the single sensillum recording technique. In Cardé, R T & Minks, A K (eds.) *Pheromone Research: New Directions*. pp. 359–71. New York: Chapman and Hall.

van Djiken, M J, van Stratum, P & van Alphen, J J M (1992) Recognition of individual-specific marked parasitized hosts by the solitary parasitoid *Epidinocarsis lopezi*. *Behav Ecol Sociobiol* **30**: 77–82.

Van Dongen, S (2011) Associations between asymmetry and human attractiveness: possible direct effects of asymmetry and signatures of publication bias. *Ann Hum Biol* **38**: 317–23.

van Lenteren, J C (ed.) (2012) *IOBC Internet Book of Biological Control*, 6th edn. Wageningen, the Netherlands: Available from: www.iobc-global.org/publications_iobc_internet_book_of_biological_control.html.

van Wilgenburg, E, Symonds, M R E & Elgar, M A (2011) Evolution of cuticular hydrocarbon diversity in ants. *J Evol Biol* **24**: 1188–98.

van Wilgenburg, E, Felden, A, Choe, D H *et al.* (2012) Learning and discrimination of cuticular hydrocarbons in a social insect. *Biol Lett* **8**: 17–20.

van Zweden, J S & d'Ettorre, P (2010) Nestmate recognition in social insects and the role of hydrocarbons. In Blomquist, G J & Bagnères, A-G (eds.) *Insect Hydrocarbons: Biology, Biochemistry, and Chemical Ecology*. pp. 222–43. Cambridge: Cambridge University Press.

van Zweden, J S, Brask, J B, Christensen, J H *et al.* (2010) Blending of heritable recognition cues among ant nestmates creates distinct colony gestalt odours but prevents within-colony nepotism. *J Evol Biol* **23**: 1498–508.

van Zweden, J S, Gruter, C, Jones, S M & Ratnieks, F L (2011) Hovering guards of the stingless bee *Tetragonisca angustula* increase colony defensive perimeter as shown by intra- and inter-specific comparisons. *Behav Ecol Sociobiol* **65**: 1277–82.

Vander Meer, R K & Alonso, L E (1998) Pheromone directed behavior in ants. In Vander Meer, R K, Breed, M D, Espelie, K E & Winston, M L (eds.) *Pheromone Communication in Social Insects: Ants, Wasps, Bees, and Termites*. pp. 159–92. Boulder, CO: Westview Press.

Vander Meer, R K & Morel, L (1998) Nestmate recognition in ants. In Vander Meer, R K, Breed, M D, Espelie, K E & Winston, M L (eds.) *Pheromone Communication in Social Insects: Ants, Wasps,*

Bees, and Termites. pp. 79–103. Boulder, CO: Westview Press.

Vargo, E L (1992) Mutual pheromonal inhibition among queens in polygyne colonies of the fire ant *Solenopsis invicta*. *Behav Ecol Sociobiol* **31**: 205–10.

Vargo, E L (1998) Primer pheromones in ants. In Vander Meer, R K, Breed, M D, Espelie, K E & Winston, M L (eds.) *Pheromone Communication in Social Insects: Ants, Wasps, Bees, and Termites*. pp. 293–313. Boulder, CO: Westview Press.

Vargo, E L & Husseneder, C (2009) Biology of subterranean termites: insights from molecular studies of Reticulitermes and Coptotermes. *Annu Rev Entomol* **54**: 379–403.

Vaziri, A & Emiliani, V (2012) Reshaping the optical dimension in optogenetics. *Curr Opin Neurobiol* **22**: 128–37.

Venkataraman, A B, Swarnalatha, V B, Nair, P & Gadagkar, R (1988) The mechanism of nestmate discrimination in the tropical social wasp *Ropalidia marginata* and its implications for the evolution of sociality. *Behav Ecol Sociobiol* **23**: 271–9.

Vereecken, N J & McNeil, J N (2010) Cheaters and liars: chemical mimicry at its finest. *Can J Zool* **88**: 725–52.

Vereecken, N J & Schiestl, F P (2008) The evolution of imperfect floral mimicry. *Proc Natl Acad Sci USA* **105**: 7484–8.

Vergoz, V, McQuillan, H J, Geddes, L H *et al.* (2009) Peripheral modulation of worker bee responses to queen mandibular pheromone. *Proc Natl Acad Sci USA* **106**: 20930–5.

Verzijden, M N & Rosenthal, G G (2011) Effects of sensory modality on learned mate preferences in female swordtails. *Anim Behav* **82**: 557–62.

Verzijden, M N, ten Cate, C, Servedio, M R *et al.* (2012) The impact of learning on sexual selection and speciation. *Trends Ecol Evol* **27**: 511–19.

Via, S (2009) Natural selection in action during speciation. *Proc Natl Acad Sci USA* **106**: 9939–46.

Vickers, N J (2000) Mechanisms of animal navigation in odor plumes. *Biol Bull* **198**: 203–12.

Vickers, N J (2006) Winging it: moth flight behavior and responses of olfactory neurons are shaped by pheromone plume dynamics. *Chem Senses* **31**: 155–66.

Vickers, N J & Baker, T C (1991) The effects of unilateral antennectomy on the flight behavior of male

Heliothis virescens in a pheromone plume. *Physiol Entomol* 16: 497–506.

Vickers, N J, Christensen, T A & Hildebrand, J G (1998) Integrating behavior with neurobiology: odor-mediated moth flight and olfactory discrimination by glomerular arrays. *Integr Biol* 1: 224–30.

Viitala, J, Korpimaki, E, Palokangas, P & Koivula, M (1995) Attraction of kestrels to vole scent marks visible in ultraviolet-light. *Nature* 373: 425–7.

Vogel, S (1983) How much air flows through a silkmoth's antenna? *J Insect Physiol* 29: 597–602.

Vogel, S (1994) *Life in Moving Fluids: the Physical Biology of Flow*, 2nd edn. (1996 paperback printing) Princeton: Princeton University Press.

Voigt, C C, Caspers, B & Speck, S (2005) Bats, bacteria, and bat smell: sex-specific diversity of microbes in a sexually selected scent organ. *J Mammal* 86: 745–9.

Vosshall, L B & Hansson, B S (2011) A unified nomenclature system for the insect olfactory coreceptor. *Chem Senses* 36: 497–8.

Vosshall, L B & Stocker, R E (2007) Molecular architecture of smell and taste in *Drosophila*. *Annu Rev Neurosci* 30: 505–33.

Vosshall, L B, Amrein, H, Morozov, P S, Rzhetsky, A & Axel, R (1999) A spatial map of olfactory receptor expression in the *Drosophila* antenna. *Cell* 96: 725–36.

Vrieze, L A, Bergstedt, R A & Sorensen, P W (2011) Olfactory-mediated stream-finding behavior of migratory adult sea lamprey (*Petromyzon marinus*). *Can J Fish Aquat Sci* 68: 523–33.

Vyas, A (2013) Parasite-augmented mate choice and reduction in innate fear in rats infected by *Toxoplasma gondii*. *J Exp Biol* 216: 120–6.

Wabnitz, P A, Bowie, J H, Tyler, M J, Wallace, J C & Smith, B P (1999) Aquatic sex pheromone from a male tree frog. *Nature* 401: 444–5.

Wachowiak, M (2010) Active sensing in olfaction. In Menini, A, (ed.) *The Neurobiology of Olfaction*. Boca Raton, FL: CRC Press. Available online at www.ncbi.nlm.nih.gov/books/NBK55978.

Wagner, C M, Jones, M L, Twohey, M B & Sorensen, P W (2006) A field test verifies that pheromones can be useful for sea lamprey (*Petromyzon marinus*) control in the Great Lakes. *Can J Fish Aquat Sci* 63: 475–9.

Waldman, B, Frumhoff, P C & Sherman, P W (1988) Problems of kin recognition. *Trends Ecol Evol* 3: 8–13.

Walker, D B, Walker, J C, Cavnar, P J *et al.* (2006) Naturalistic quantification of canine olfactory sensitivity. *Appl Anim Behav Sci* 97: 241–54.

Walker, J C, Hall, S B, Walker, D B *et al.* (2003) Human odor detectability: new methodology used to determine threshold and variation. *Chem Senses* 28: 817–26.

Wang, J, Ross, K & Keller, L (2008a) Genome-wide expression patterns and the genetic architecture of a fundamental social trait. *PLoS Genet* 4: e1000127.

Wang, L M, Han, X Q, Mehren, J *et al.* (2011) Hierarchical chemosensory regulation of male-male social interactions in *Drosophila*. *Nat Neurosci* 14: 757–62.

Wang, S P, Sato, K, Giurfa, M & Zhang, S W (2008b) Processing of sting pheromone and its components in the antennal lobe of the worker honeybee. *J Insect Physiol* 54: 833–41.

Wang, Y, Kocher, S D, Linksvayer, T A *et al.* (2012) Regulation of behaviorally associated gene networks in worker honey bee ovaries. *J Exp Biol* 215: 124–34.

Wanner, K W, Nichols, A S, Walden, K K O *et al.* (2007) A honey bee odorant receptor for the queen substance 9-oxo-2-decenoic acid. *Proc Natl Acad Sci USA* 104: 14383.

Watelet, J B, Strolin-Benedetti, M & Whomsley, R (2009) Defence mechanisms of olfactory neuro-epithelium: mucosa regeneration, metabolising enzymes and transporters. *B-Ent* 5: Suppl. 13, 21–37.

Watson, P J (1986) Transmission of a female sex pheromone thwarted by males in the spider *Linyphia litogiosa* Keyserling (Linyphiidae). *Science* 233: 219–21.

Webb, B, Harrison, R R & Willis, M A (2004) Sensorimotor control of navigation in arthropod and artificial systems. *Arthropod Struct Dev* 33: 301–29.

Webster, D R & Weissburg, M J (2001) Chemosensory guidance cues in a turbulent chemical odor plume. *Limnol Oceanogr* 46: 1034–47.

Webster, D R & Weissburg, M J (2009) The hydrodynamics of chemical cues among aquatic organisms. *Annu Rev Fluid Mech* 41: 73–90.

Wedekind, C & Füri, S (1997) Body odour preferences in men and women: do they aim for specific MHC combinations or simply heterozygosity? *Proc R Soc B* **264**: 1471–9.

Wedekind, C, Seebeck, T, Bettens, F & Paepke, A J (1995) MHC-dependent mate preferences in humans. *Proc R Soc B* **260**: 245–9.

Wedell, N (2005) Female receptivity in butterflies and moths. *J Exp Biol* **208**: 3433–40.

Weeks, E N I, Birkett, M A, Cameron, M M, Pickett, J A & Logan, J G (2011) Semiochemicals of the common bed bug, *Cimex lectularius* L. (Hemiptera: Cimicidae), and their potential for use in monitoring and control. *Pest Manag Sci* **67**: 10–20.

Wegner, K M, Kalbe, M, Kurtz, J, Reusch, T B H & Milinski, M (2003) Parasite selection for immuno-genetic optimality. *Science* **301**: 1343.

Weil, T, Hoffmann, K, Kroiss, J, Strohm, E & Korb, J (2009) Scent of a queen: cuticular hydrocarbons specific for female reproductives in lower termites. *Naturwissenschaften* **96**: 315–19.

Weiner, S A & Toth, A L (2012) Epigenetics in social insects: a new direction for understanding the evolution of castes. *Genetics Research International* **2012**: 11 doi:10.1155/2012/609810.

Weissburg, M J (2000) The fluid dynamical context of chemosensory behavior. *Biol Bull* **198**: 188–202.

Weissburg, M J (2011) Waterborne chemical communication: stimulus dispersal dynamics and orientation strategies in crustaceans. In Breithaupt, T & Thiel, M (eds.) *Chemical Communication in Crustaceans*. pp. 63–83. New York: Springer.

Weissburg, M J, Doall, M H & Yen, J (1998) Following the invisible trail: kinematic analysis of mate-tracking in the copepod *Temora longicornis*. *Phil Trans R Soc B* **353**: 701–12.

Wells, M J & Buckley, S K L (1972) Snails and trails. *Anim Behav* **20**: 345–55.

Welsh, R G & Müller-Schwarze, D (1989) Experimental habitat scenting inhibits colonization by beaver, *Castor canadensis*. *J Chem Ecol* **15**: 887–93.

Wenhold, B A & Rasa, O A E (1994) Territorial marking in the yellow mongoose *Cynictis penicillata* – sexual advertisement for subordinates. *Z Saugetierkd* **59**: 129–38.

Wenseleers, T, Billen, J & Hefetz, A (2002) Territorial marking in the desert ant *Cataglyphis niger*: does it pay to play bourgeois? *J Insect Behav* **15**: 85–93.

Wertheim, B, Van Baalen, E J A, Dicke, M & Vet, L E M (2005) Pheromone-mediated aggregation in nonsocial arthropods: an evolutionary ecological perspective. *Annu Rev Entomol* **50**: 321–46.

West, S A & Gardner, A (2010) Altruism, spite, and greenbeards. *Science* **327**: 1341–4.

White, A M, Swaisgood, R R & Zhang, H (2002) The highs and lows of chemical communication in giant pandas (*Ailuropoda melanoleuca*): effect of scent deposition height on signal discrimination. *Behav Ecol Sociobiol* **51**: 519–29.

White, T L (2009) A second look at the structure of human olfactory memory. *Ann N Y Acad Sci* **1170**: 338–42.

Whitfield, C W, Cziko, A M & Robinson, G E (2003) Gene expression profiles in the brain predict behavior in individual honey bees. *Science* **302**: 296–9.

Whitman, M C & Greer, C A (2009) Adult neurogenesis and the olfactory system. *Prog Neurobiol* **89**: 162–75.

Whittaker, D J, Soini, H A, Atwell, J W (2010) Songbird chemosignals: volatile compounds in preen gland secretions vary among individuals, sexes, and populations. *Behav Ecol* **21**: 608–14.

Whittier, T S & Kaneshiro, K Y (1995) Intersexual selection in the Mediterranean fruit-fly – does female choice enhance fitness. *Evolution* **49**: 990–6.

Whittier, T S, Nam, F Y, Shelly, T E & Kaneshiro, K Y (1994) Male courtship success and female discrimination in the Mediterranean fruit-fly (Diptera, Tephritidae). *J Insect Behav* **7**: 159–70.

Wicker-Thomas, C (2007) Pheromonal communication involved in courtship behavior in *Diptera*. *J Insect Physiol* **53**: 1089–100.

Wilburn, D B, Bowen, K E, Gregg, R G et al. (2012) Proteomic and UTR analyses of a rapidly evolving hypervariable family of vertebrate pheromones. *Evolution* **66**: 2227–39.

Wiley, C, Ellison, C K & Shaw, K L (2011) Widespread genetic linkage of mating signals and preferences in the Hawaiian cricket *Laupala*. *Proc R Soc B* **279**: 1203–9.

Wiley, R H (2013) Specificity and multiplicity in the recognition of individuals: implications for the evolution of social behaviour. *Biol Rev* **88**: 179–95.

Wilke, K, Martin, A, Terstegen, L & Biel, S (2007) A short history of sweat gland biology. *Int J Cosmet Sci* **29**: 169–80.

Willis, M A (2008a) Chemical plume tracking behavior in animals and mobile robots. *Navigation* 55: 127–35.

Willis, M A (2008b) Odor plumes and animal orientation. In Firestein, S & Beauchamp, G (eds.) *Olfaction and Taste*. pp. 771–81. San Diego: Academic Press.

Willis, M A & Avondet, J L (2005) Odor-modulated orientation in walking male cockroaches *Periplaneta americana*, and the effects of odor plumes of different structure. *J Exp Biol* 208: 721–35.

Wilms, J & Eltz, T (2008) Foraging scent marks of bumblebees: footprint cues rather than pheromone signals. *Naturwissenschaften* 95: 149–53.

Wilson, A D & Baietto, M (2011) Advances in electronic-nose technologies developed for biomedical applications. *Sensors* 11: 1105–76.

Wilson, D A & Stevenson, R J (2006) *Learning to Smell: Olfactory Perception from Neurobiology to Behavior*. Baltimore, MD: Johns Hopkins University Press.

Wilson, E O (1962) Chemical communication among workers of the fire ant *Solenopsis saevissima* (Fr. Smith). 1: the organization of mass foraging. *Anim Behav* 10: 134–47.

Wilson, E O (1970) Chemical communication within animal species. In Sondheimer, E, (ed.) *Chemical Ecology*. pp. 133–55. New York: Academic Press.

Wilson, E O (1971) *The Insect Societies*. Cambridge, MA: Belknap Press.

Wilson, E O & Bossert, W H (1963) Chemical communication among animals. *Recent Prog Horm Res* 19: 673–716.

Wilson, M (2008) *Bacteriology of Humans: an Ecological Perspective*. Oxford: Blackwell.

Wilson, R I & Mainen, Z F (2006) Early events in olfactory processing. *Annu Rev Neurosci* 29: 163–201.

Winston, M L (1987) *The Biology of the Honey Bee*. Cambridge, MA: Harvard University Press.

Winston, M L (1997) *Nature Wars: People vs. Pests*. Cambridge, MA: Harvard University Press.

Winston, M L & Slessor, K N (1998) Honey bee primer pheromones and colony organization: gaps in our knowledge. *Apidologie* 29: 81–95.

Wisenden, B D (1999) Alloparental care in fishes. *Rev Fish Biol Fish* 9: 45–70.

Wisenden, B D (2014) Chemical cues that indicate risk of predation. In Sorensen, P W & Wisenden, B D (eds.) *Fish Pheromones and Related Conspecific Chemical Cues*. Chichester: Wiley-Blackwell.

Wittmann, D, Radtke, R, Zeil, J, Luebke, G & Francke, W (1990) Robber bees (*Lestrimelitta limao*) and their host: chemical and visual cues in nest defense by *Trigona angustula* (Apidae: Meliponinae). *J Chem Ecol* 16: 631–42.

Witzgall, P, Kirsch, P & Cork, A (2010) Sex pheromones and their impact on pest management. *J Chem Ecol* 36: 80–100.

Wolff, J O (2003) Laboratory studies with rodents: facts or artifacts? *Bioscience* 53: 421–7.

Wolff, J O & Sherman, P W (eds.) (2007) *Rodent Societies: an Ecological and Evolutionary Perspective*. Chicago: Chicago University Press.

Wolff, J O, Dunlap, A S & Ritchhart, E (2001) Adult female prairie voles and meadow voles do not suppress reproduction in their daughters. *Behav Processes* 55: 157–62.

Wolfner, M F (2009) Battle and ballet: molecular interactions between the sexes in *Drosophila*. *J Hered* 100: 399–410.

Wood, D L (1982) The role of pheromones, kairomones, and allomones in the host selection and colonization behavior of bark beetles. *Annu Rev Entomol* 27: 411–46.

Woodard, S H, Fischman, B J, Venkat, A *et al.* (2011) Genes involved in convergent evolution of eusociality in bees. *Proc Natl Acad Sci USA* 108: 7472–7.

Woodley, S K (2010) Pheromonal communication in amphibians. *J Comp Physiol A* 196: 713–27.

Workman, J & Weyer, L (2012) *Practical Guide and Spectral Atlas for Interpretive Near-infrared Spectroscopy*, 2nd edn. Boca Raton, FL: CRC Press.

Wu, M V & Shah, N M (2011) Control of masculinization of the brain and behavior. *Curr Opin Neurobiol* 21: 116–23.

Wurm, Y, Wang, J & Keller, L (2010) Changes in reproductive roles are associated with changes in gene expression in fire ant queens. *Mol Ecol* 19: 1200–11.

Wyatt, T D (1997) Putting pheromones to work: paths forward for direct control. In Cardé, R T & Minks, A K (eds.) *Pheromone Research: New Directions*. pp. 445–59. New York: Chapman and Hall.

Wyatt, T D (2003) *Pheromones and Animal Behaviour: Communication by Smell and Taste.* Cambridge: Cambridge University Press.

Wyatt, T D (2005) Pheromones: convergence and contrasts in insects and vertebrates. In Mason, R T, LeMaster, M P & Müller-Schwarze, D (eds.) *Chemical Signals in Vertebrates 10.* pp. 7–20. New York: Springer.

Wyatt, T D (2009) Fifty years of pheromones. *Nature* 457: 262–3.

Wyatt, T D (2010) Pheromones and signature mixtures: defining species-wide signals and variable cues for identity in both invertebrates and vertebrates. *J Comp Physiol A* 196: 685–700.

Wyatt, T D (2011) Pheromones and behavior. In Breithaupt, T & Thiel, M (eds.) *Chemical Communication in Crustaceans.* pp. 23–38. New York: Springer.

Wynn, E, Sanchez-Andrade, G, Carss, K & Logan, D (2012) Genomic variation in the vomeronasal receptor gene repertoires of inbred mice. *BMC Genomics* 13: 415.

Wysocki, C J & Beauchamp, G K (1984) Ability to smell androstenone is genetically determined. *Proc Natl Acad Sci USA* 81: 4899–902.

Wysocki, C J & Preti, G. (2009) Human pheromones: what's purported, what's supported. *A Sense of Smell Institute White Paper* [Online]. Available: www.senseofsmell.org/research/C.Wysocki-White-Paper-Human_Pheromones.pdf [Accessed 29 Oct 2012].

Wysocki, C J, Dorries, K M & Beauchamp, G K (1989) Ability to perceive androstenone can be acquired by ostensibly anosmic people. *Proc Natl Acad Sci USA* 86: 7976–8.

Xu, F Q, Schaefer, M, Kida, I *et al.* (2005) Simultaneous activation of mouse main and accessory olfactory bulbs by odors or pheromones. *J Comp Neurol* 489: 491–500.

Xu, S, Schlüter, P M & Schiestl, F P (2012) Pollinator-driven speciation in sexually deceptive orchids. *Int J Ecol* 2012: doi:10.1155/2012/285081.

Xu, Y, Gong, F, Dixon, SJ *et al.* (2007) Application of dissimilarity indices, principal coordinates analysis, and rank tests to peak tables in metabolomics of the gas chromatography/mass spectrometry of human sweat. *Anal Chem* 79: 5633–41.

Xue, B Y, Rooney, A P, Kajikawa, M, Okada, N & Roelofs, W L (2007) Novel sex pheromone desaturases in the genomes of corn borers generated through gene duplication and retroposon fusion. *Proc Natl Acad Sci USA* 104: 4467–72.

Yamagata, N, Fujiwara-Tsujii, N, Yamaoka, R & Mizunami, M (2005) Pheromone communication and the mushroom body of the ant, *Camponotus obscuripes* (Hymenoptera : Formicidae). *Naturwissenschaften* 92: 532–6.

Yamagata, N, Nishino, H & Mizunami, M (2006) Pheromone-sensitive glomeruli in the primary olfactory centre of ants. *Proc R Soc B* 273: 2219–25.

Yamagata, N, Nishino, H & Mizunami, M (2007) Neural pathways for the processing of alarm pheromone in the ant brain. *J Comp Neurol* 505: 424–42.

Yamamoto, M E, Araújo, A, de Sousa, M B C & Arruda, M d F (2010) Social organization in *Callithrix jacchus*: cooperation and competition. *Adv Study Behav* 42: 259–73.

Yamamoto, Y & Matsuura, K (2011) Queen pheromone regulates egg production in a termite. *Biol Lett* 7: 727–9.

Yamazaki, K & Beauchamp, G K (2007) Genetic basis for MHC-dependent mate choice. *Adv Genet* 59: 129–45.

Yamazaki, K, Beauchamp, G K, Wysocki, C J *et al.* (1983) Recognition of H-2 types in relation to the blocking of pregnancy in mice. *Science* 221: 186–8.

Yambe, H, Kitamura, S, Kamio, M *et al.* (2006) L-Kynurenine, an amino acid identified as a sex pheromone in the urine of ovulated female masu salmon. *Proc Natl Acad Sci USA* 103: 15370–4.

Yang, C H, Rumpf, S, Xiang, Y *et al.* (2009) Control of the postmating behavioral switch in *Drosophila* females by internal sensory neurons. *Neuron* 61: 519–26.

Yang, Z & Schank, J (2006) Women do not synchronize their menstrual cycles. *Hum Nat* 17: 434–47.

Yarmolinsky, D A, Zuker, C S & Ryba, N J P (2009) Common sense about taste: from mammals to insects. *Cell* 139: 234–44.

Yeargan, K V & Quate, L W (1997) Adult male bolas spiders retain juvenile hunting tactics. *Oecologia* 112: 572–6.

Yen, J & Lasley, R (2011) Chemical communication between copepods: finding the mate in a fluid environment. In

Breithaupt, T & Thiel, M (eds.) *Chemical Communication in Crustaceans*. pp. 177–97. New York: Springer.

Yen, J, Weissburg, M J & Doall, M H (1998) The fluid physics of signal perception by mate-tracking copepods. *Phil Trans R Soc B* 353: 787–804.

Yew, J Y, Dreisewerd, K, Luftmann, H *et al.* (2009) A new male sex pheromone and novel cuticular cues for chemical communication in *Drosophila. Curr Biol* 19: 1245–54.

Yew, J Y, Soltwisch, J, Pirkl, A & Dreisewerd, K (2011) Direct laser desorption ionization of endogenous and exogenous compounds from insect cuticles: practical and methodologic aspects. *J Am Soc Mass Spectrom* 22: 1273–84.

Yizhar, O, Fenno, L E, Davidson, T J, Mogri, M & Deisseroth, K (2011) Optogenetics in neural systems. *Neuron* 71: 9–34.

Yoder, J A & Grojean, N C (1997) Group influence on water conservation in the giant Madagascar hissing-cockroach, *Gromphadorhina portentosa* (Dictyoptera: Blaberidae). *Physiol Entomol* 22: 79–82.

Yoon, H Y, Enquist, L W & Dulac, C (2005) Olfactory inputs to hypothalamic neurons controlling reproduction and fertility. *Cell* 123: 669–82.

Yoshiura, K, Kinoshita, A, Ishida, T *et al.* (2006) A SNP in the ABCC11 gene is the determinant of human earwax type. *Nat Genet* 38: 324–30.

Young, A J (2009) The causes of physiological suppression in vertebrate societies: a synthesis. In Hager, R & Jones, C B (eds.) *Reproductive Skew in Vertebrates: Proximate and Ultimate Causes*. pp. 397–436. Cambridge: Cambridge University Press.

Zahavi, A (1975) Mate selection: a selection for a handicap. *J Theor Biol* 53: 205–14.

Zahavi, A (2008) The handicap principle and signalling in collaborative systems. In d'Ettorre, P & Hughes, D P (eds.) *Sociobiology of Communication: an Interdisciplinary Perspective*. pp. 1–9. Oxford: Oxford University Press.

Zahavi, A & Zahavi, A (1997) *The Handicap Principle. A Missing Piece of Darwin's Puzzle*. Oxford: Oxford University Press.

Zala, S M & Penn, DJ (2004) Abnormal behaviours induced by chemical pollution: a review of the evidence and new challenges. *Anim Behav* 68: 649–64.

Zala, S M, Potts, W K & Penn, D J (2004) Scent-marking displays provide honest signals of health and infection. *Behav Ecol* 15: 338–44.

Zampiga, E, Gaibani, G, Csermely, D, Frey, H & Hoi, H (2006) Innate and learned aspects of vole urine UV-reflectance use in the hunting behaviour of the common kestrel *Falco tinnunculus. J Avian Biol* 37: 318–22.

Zanen, P O, Sabelis, M W, Buonaccorsi, J P & Cardé, R T (1994) Search strategies of fruit-flies in steady and shifting winds in the absence of food odors. *Physiol Entomol* 19: 335–41.

Zavazava, N & Eggert, F (1997) MHC and behavior. *Immunol Today* 18: 8–10.

Zayed, A & Robinson, G E (2012) Understanding the relationship between brain gene expression and social behavior: lessons from the honey bee. *Annu Rev Genet* 46: 591–615.

Zeng, X N, Leyden, J J, Lawley, H J *et al* (1991) Analysis of characteristic odors from human male axillae. *J Chem Ecol* 17: 1469–92.

Zhang, J-X, Zuo, M X & Sun, L (2009) The volatile composition of uropygial glands contains information about sex, individual, and species in Bengalese finches, *Lonchura striata. Curr Zool* 55: 357–65.

Zhang, J-X, Wei, W, Zhang, J-H & Yang, W-H (2010) Uropygial gland-secreted alkanols contribute to olfactory sex signals in budgerigars. *Chem Senses* 35: 375–82.

Zhang, X M & Firestein, S (2002) The olfactory receptor gene superfamily of the mouse. *Nat Neurosci* 5: 124–33.

Zhang, X M & Firestein, S (2009) Genomics of olfactory receptors. *Results Probl Cell Differ* 47: 239–55.

Zhao, C H, Löfstedt, C & Wang, X Y (1990) Sex-pheromone biosynthesis in the Asian corn-borer *Ostrinia furnacalis* (II) – biosynthesis of (E)-12-tetradecenyl and (Z)-12-tetradecenyl acetate involves delta-14 desaturation. *Arch Insect Biochem Physiol* 15: 57–65.

Zhou, J-J (2010) Odorant-binding proteins in insects. In Gerald, L (ed.) *Pheromones*. pp. 241–72. London: Academic Press.

Zhou, W & Chen, D (2008) Encoding human sexual chemosensory cues in the orbitofrontal and fusiform cortices. *J Neurosci* 28: 14416.

Zhuang, J J & Hunter, C P (2011) RNA interference in *Caenorhabditis elegans*: uptake, mechanism, and regulation. *Parasitology* 139: 560–73.

Zimmer-Faust, R K, Finelli, C M, Pentcheff, N D & Wethey, D S (1995) Odor plumes and animal navigation in turbulent water-flow – a field-study. *Biol Bull* 188: 111–16.

Zimmermann, Y, Ramírez, S & Eltz, T (2009) Chemical niche differentiation among sympatric species of orchid bees. *Ecology* 90: 2994–3008.

Zou, D J, Chesler, A & Firestein, S (2009) How the olfactory bulb got its glomeruli: a just so story? *Nat Rev Neurosci* 10: 611–18.

Zube, C, Kleineidam, C J, Kirschner, S, Neef, J & Rossler, W (2008) Organization of the olfactory pathway and odor processing in the antennal lobe of the ant *Camponotus floridanus*. *J Comp Neurol* 506: 425–41.

Zuk, M & Kolluru, G R (1998) Exploitation of sexual signals by predators and parasitoids. *Q Rev Biol* 73: 415–38.

CREDITS

I am grateful to the following copyright holders for permission to reproduce their figures. For full details of the sources see the respective caption and references. The list does not include non-copyright material, or individual acknowledgements (which are made in captions where appropriate).

AAAS (American Association for the Advancement of Science): Reprinted with permission from the American Association for the Advancement of Science. Figures 3.11b, 9.10; American Chemical Society: Figure 13.2 GC trace; Annual Reviews: with permission from the respective Annual Review © year of publication by Annual Reviews www.Annual Reviews.org Figures 3.11, 9.1ai & aii, 10.6, Figure in Box 10.3; Begell House: Figure 3.9; CAB International (CABI Publishing): Figure 7.2 (right); Cambridge University Press: Figures 3.6a, 6.3, Figure in Box 6.1, 8.2a, 9.1bii, 10.1, 13.5; Chicago University Press: Figure 5.3; Company of Biologists Ltd: Figures 10.3, 10.5cde; Cooper Ornithological Society: Figure 10.9b; Dadent & Sons: Figure in Box 6.3; Elsevier: Figures 1.4, 1.8, 2.2a, 2.6, Figures in Box 3.1, 6.1, 7.5, 7.7, Figure in Box 7.3, 9.2c, 9.3 left hand side, 9.5, 9.7, 9.11, 9.14, 10.4ab, 10.5a, Figure in Box 10.1, 13.5, 13.8; Great Lakes Fishery Commission: Figure 12.3c; INRA: Table 1.2, part; Institute of Navigation: Figure 10.5; John Wiley & Sons Inc: Figures 1.9, 2.3, 5.1, 5.2, 5.4ab, 7.3, 7.6, 9.4, 9.13a, Figures (c) in Box 9.2, 11.3, 13.3; Marine Biological Laboratory, Woods Hole: Figure 10.5f; National Academy of Sciences, USA: Figures 3.7, Figure in Box 6.2; National Research Council of Canada Press: Figures 3.3, 12.3b; Nature Publishing Group: Figures in Box 1.5, 3.5a, 3.8a, 3.11a, 3.16abcd, 8.4c, 9.2a, 9.8a,b, 9.12, 9.15, 10.1, 10.2, 10.7a, 11.4, 13.1, 13.7; Ohio State University: Figure 8.1; Oxford University Press: Figures 1.2ab, 3.4, 3.5, 3.14, 5.2, 5.5, 5.6, 6.1, 11.6, Figure in Box 11.1, Table 13.1, 13.7, A.11, A.12; Perseus Press: Figure in Box 6.1; Royal Society of London: Figures 1.2d, 3.5b, 3.10, 7.1f, 9.9; Sinauer Associates: Figure 1.5a; Springer: Figures 1.1, 1.2c, 1.3, 1.7, Box 1.2, 2.1b, 2.2b, 2.4, 3.12, 3.13, 4.2, Figure in Box 5.1, 7.1, 7.6, Figure in Box 7.1, 9.2cii, 10.5b, 10.7, 10.8, 10.9a, 11.1, 12.1; Tables 1.1, 7.1, 11.2, A.1; Taylor & Francis: Figures 7.1ef; University of Kansas Press: Figure 7.6 part.

All reasonable efforts have been made by the author to trace the copyright owners of the images reproduced in this book. In the event that the author or publishers are notified of any mistakes or omissions by copyright owners after the publication of this book, the author and the publisher will endeavor to rectify the position accordingly for any subsequent printing.

INDEX

Printed in the United States
By Bookmasters